Developments in Geotechnical Engineering 10

ENGINEERING GEOLOGY

Czechoslovak Academy of Sciences

Further titles in this series:

1. *G. SANGLERAT*
THE PENETROMETER AND SOIL EXPLORATION

2. *Q. ZÁRUBA AND V. MENCL*
LANDSLIDES AND THEIR CONTROL

3. *E. E. WAHLSTROM*
TUNNELING IN ROCK

4A. *R. SILVESTER*
COASTAL ENGINEERING, I
Generation, Propagation and Influence of Waves

4B. *R. SILVESTER*
COASTAL ENGINEERING, II
Sedimentation, Estuaries, Tides, Effluents, Modelling

5. *R. N. YONG AND B. P. WARKENTIN*
SOIL PROPERTIES AND BEHAVIOUR

6. *E. E. WAHLSTROM*
DAMS, DAM FOUNDATIONS, AND RESERVOIR SITES

7. *W. F. CHEN*
LIMIT ANALYSIS AND SOIL PLASTICITY

8. *L. N. PERSEN*
ROCK DYNAMICS AND GEOPHYSICAL EXPLORATION
Introduction to Stress Waves in Rock

9. *M. D. GIDIGASU*
LATERITE SOIL ENGINEERING
Pedogenesis and Engineering Principles

11. *H. K. GUPTA AND B. K. RASTOGI*
DAMS AND EARTHQUAKES

12. *F. H. CHEN*
FOUNDATIONS ON EXPANSIVE SOILS

13. *L. HOBST AND J. ZAJÍC*
ANCHORING IN ROCK

14. *B. VOIGHT*, *Editor*
ROCKSLIDES AND AVALANCHES

Developments in Geotechnical Engineering 10

ENGINEERING GEOLOGY

by

QUIDO ZÁRUBA

Emeritus Professor of Geology at the Technical University, Prague, Czechoslovakia

and

VOJTĚCH MENCL

Emeritus Professor at the Technical University Brno, Czechoslovakia

ELSEVIER SCIENTIFIC PUBLISHING COMPANY

Amsterdam • Oxford • New York • 1976

Scientific Editor
Doc. Ing. Jiří Škopek, CSc.

Published in co-edition with
ACADEMIA, Publishing House of the Czechoslovak Academy of Sciences, Prague

Distribution of this book is being handled by the following publishers

for the U.S.A. and Canada
American Elsevier Publishing Company, Inc.,
52 Vanderbilt Avenue
New York, New York 10017

for the East European Countries, China, Northern Korea, Cuba, Vietnam and Mongolia
Academia, Publishing House of the Czechoslovak Academy of Sciences, Prague

for all remaining areas
Elsevier Scientific Publishing Company
335 Jan van Galenstraat
P.O. Box 211, Amsterdam, The Netherlands

ISBN 0-444-99877-2
Library of Congress Card Number 74-21871
With 380 Illustrations

Printed in Czechoslovakia

CONTENTS

PREFACE

The ever increasing importance of engineering geology has entailed the publication of a number of books dealing with this branch of applied geology. Some of them, however, are rather textbooks of general geology for engineers, others are directed mainly to geotechnique or methods of soil mechanics and very little space is left for engineering geology proper.

We have endeavoured to write a book containing sufficient geological information for the understanding of relations between the geology of a building site and the engineering structure. The examples, taken for the most part from our experience and practice, have been selected to provide evidence for the significance of engineering geology in planning, design, construction and maintenance of engineering structures.

In order to concentrate on this comprehensive subject, we have abstained from a systematic treatment of general geology and petrography, subjects which are dealt with in many excellent textbooks. In the field of soil and rock mechanics we explain the relationship between physical processes and their rheological effects which play an important role in the behaviour of the rock environment. In our opinion, it will provide civil engineering students with an insight into the physical basis of processes which they have learned to express mathematically. On the other hand, students of geological sciences may obtain a better idea of mathematical formulation of the problems, particularly with respect to modern computation methods.

The book is intended particularly for civil engineering students and students of engineering geology in the university faculties of natural sciences. In view of this dual purpose we have tried to describe geological features so as to be comprehensible to Technical College students and to explain construction problems intelligibly for geology students. We hope the book may also be of assistance to planners, civil engineers and graduate engineering geologists.

For the English translation the third Czech edition has been used, which was reworked and complemented during the years 1971 and 1972. The book, which is largely the joint effort of the two authors, comprises fourteen chapters, eight of which (1, 2, 4, 6, 7, 9, 13 and 14) were prepared by Q. Záruba and translated by Mrs. H. Zárubová and deal predominantly with the geological aspects of the subjects, while the other six chapters (3, 5, 8, 10, 11 and 12), which were written and translated by V. Mencl, discuss the subject from the point of view of soil and rock mechanics.

Acknowledgements: The writers wish to express their thanks to Doc. J. Škopek for his constructive comments and discussion, to Dr. R. F. Legget, who kindly revised the English translation of Chapter 13, and to Dr. A. Absolon for the meticulous drawing of most of the new figures.

Q. Z., V. M.

Introduction

CO-OPERATION OF GEOLOGISTS WITH ENGINEERS

Planned economy presumes that all undertakings are premeditated and prepared in detail. The main condition for good planning is therefore reliable preparation of materials and a comprehensive knowledge of the problem, with due consideration of every factor involved.

All engineering projects must be assessed in the light of the geological structure of the area, since each cubic metre of earth excavated, every loading of the land surface and each pile driven into the ground will interfere with the existing natural state of equilibrium. To design an engineering work without knowing the type of soil on which it is to be built, would be as unreasonable as to design a structure from material of unknown properties. Knowledge of natural conditions is therefore a prerequisite of all planning. Although modern methods make it possible to accomplish very bold engineering works under adverse conditions, good planning should aim at economic performance, without superfluous expense or unnecessary difficulties. To attain this objective the designers must take into consideration the geological setting of the area.

The first stage of a construction project is therefore a detailed investigation of the area to be built upon. This is where the field of engineering geology mediates between the engineer and nature, concerning the geological and tectonic structure of the area, the petrographic character of rocks and their mechanical properties, and contemporary geodynamical processes with respect to the designed work.

Although an engineering geologist should submit the results of his investigations in a form intelligible to the engineer, his cooperation with an engineer is more fruitful when the latter has at least a basic knowledge of geology. Only in this case can the engineer fully comprehend geological reports and interpret them correctly, and define the problems that should be solved with the help of geological investigation. From this it follows that a course of general and engineering geology should not be omitted from the curriculum of students of engineering and architecture. In some countries such lectures were introduced as early as the nineteenth century; at the Technical University in Prague, Dr. J. Krejčí was appointed professor of geology in 1864. It should be noted that lectures on geology in engineering faculties should emphasize the mutual relations between the geological features of an area and engineering structure rather than „cram" the students with a heap of geological data. This makes great demands on a competent lecturer of engineering geology: he must be acquainted equally with the geological and the engineering and economic viewpoints.

The importance of engineering geology for many industries, and various branches of engineering, is now universally acknowledged. Geological investigation is today indispensable in the search for and appraisal of building sites, in proposing construction methods and in the building process itself. The systematic application of geology has taken place only in the last few decades. Appreciation of geology in the building industry can be accounted for by two reasons: (1) engineers' requirements for a good knowledge of geological conditions have steadily increased as ever larger and bolder structures, such as dams, bridges, tunnels, are undertaken at less and less suitable sites, and (2) the ability of geologists to adapt their way of thinking and the presentation of their findings to the demands of practice. As a result, the earlier occasional cooperation of geologists in difficult cases and building failures has developed into continuous collaboration of engineering geologists in planning, design, construction, operation, and maintenance of engineering structures.

Today the responsibility for the reliability and completeness of engineering geological investigation rests with engineering geologists. They should be mainly competent geologists, who are able to prepare geological reports for a planned engineering work in such a way that they provide reliable and comprehensible information for the engineer-designer and other collaborators. To perform this task, the engineering geologist should possess, in addition to comprehensive education in the geological sciences, particularly in petrography, tectonics and regional geology, a working knowledge of engineering disciplines, mainly of foundation and civil engineering. Engineering-geological investigation also involves the use of soil and rock mechanics, geophysics, geochemistry, clay mineralogy and other disciplines. With regard to the wide scope of these branches of science, it is impossible for an engineering geologist to master all of them, but he should know when recourse to a specialist is necessary, and should include the results of these special analyses in the final report.

Geological conditions are often so complicated that they do not permit investigations to be conducted as routine work. Several boreholes can never replace a detailed study of the area by an expert engineering geologist. The construction site may be endangered by a geological process of a higher order of magnitude than can be recorded by local boring. When examining the area, the engineering geologist should endeavour to conceive its general geological structure and grasp the geological processes that led to its development and the geomorphological forms. He should visualize the effect of the engineering work on the natural environment in the sense of whether it might cause an unwelcome acceleration of geological processes, e.g. by pumping of water, undercutting of slopes and the like. It follows from this that the engineering geologist should possess knowledge and experience enough to be able to discern relevant problems connected with the geological conditions in relation to the proposed work. He should therefore be acquainted with the project in such measure that he can correctly assess the factors to be considered. This requires close contact between the designer and the engineering geologist, a mutual understanding without which no engineering geological investigation can be successfully accomplished.

The introduction of geological investigations into engineering and economic planning has contributed to the improvement of plans and to the safety and economy of many engineering works. The participation of geologists in regional planning and heavy construction is increasingly important, but it is necessary that engineering geologists should become aware of their responsibility and of the importance of their position, i.e. they should endeavour to associate their knowledge with the solution of practical problems, and engineers should endeavour to understand geology and to respect the results of geological investigations in their work.

Chapter 1

GEOLOGICAL INVESTIGATIONS

The task and character of geological investigations differ depending on the stage of the development of an engineering project, on the importance of the designed engineering work, on the geological structure and on the state of basic geological investigation of the project area. On this basis we distinguish:

(1) The *reconnaissance investigation* carried out for the initial project study.

(2) The *detailed (*or *feasibility) investigation,* which should produce a reliable basis for the development of the project, the plans, budget and working schedule for construction.

(3) The *investigation during construction,* particularly during the excavation of foundations, in order to check the conclusions from the reconnaissance and detailed investigations, and to provide the construction engineer with the desired information.

These geological investigations carried out for an engineering work derive from the basic geological investigation which is directed to the systematic geological and hydrogeological study and mapping of the whole country. Basic geological investigation involves the study of the geological and tectonic structure of large regions and the solution of problems of stratigraphy and the genesis and age of rocks from a wider point of view. The results of this investigation are compiled into geological maps which are published by Geological Services of individual countries. A detailed knowledge of the geology of the area, including the building site, is important to avoid building upon, for example, mineral deposits, catchment areas of major ground-water sources, or protection zones around mineral springs. These questions are dealt with in Chapter 14 on geological investigation for regional planning.

1.1. Reconnaissance investigation is generally a rather difficult but very responsible task because it is necessary to decide on the location and construction method for the structure on the basis of available, but very often incomplete, data. The reconnaissance investigation should therefore be conducted by an engineering geologist who has experience and preliminary knowledge of the geological conditions of the area in question.

The means used in reconnaissance study are similar to those applied in basic geological investigation. They involve the following:

(a) Results of basic geological investigations, i.e. geological maps (see Chapter 2), reports published or deposited in the archives, and engineering geological maps which are a very useful tool at this stage of the investigation.

(b) A thorough inspection of the area, involving a study of all outcrops and exposures, morphological forms, and all phenomena which can throw any light on the geological structure and the development of the surface relief.

(c) Information obtainable from man's activities: wells, quarries, mines, structures, or from local workers on groundwater conditions, experiences with building foundations, etc.

From all this information the engineering geologist tries to deduce an understanding of the geological structure of the project area. At this point in the study his reasoning must be both inductive and deductive. Should further investigation be economic and reasonable, it must be carefully planned and for this purpose a preliminary working hypothesis must be developed.

Since the information gathered from the above sources is usually not completely satisfactory, the first conception is more or less an assumption which may prove to be true but may also be found to differ from actual conditions.

(d) The correctness of this assumption should be verified by suitably located pits or boreholes. They provide, at the same time, quantitative information such as the thickness of surficial deposits and the depth of weathering. At this stage of the investigation, the extent of boring activity should be governed by economic points of view.

(e) Geophysical methods can be applied to a preliminary determination of the depth of bedrock, of the ground-water table and analogous features.

(f) Examination of the chemistry of ground and surface water which will come into contact with the constructions is also part of the reconnaissance investigation. If the water is suspect, samples should be taken and sent to a research institute for detailed chemical analysis. The necessary measures based on the result of the analysis can then be decided upon to protect the foundation structures.

The results of preliminary geological investigation are not achieved by precise working methods and some differences between the assumed and actual geological conditions cannot be excluded. The reconnaissance geological study should provide the designer with general information only about the feasibility of the work, assist him in selecting the building site, and serve to suggest a plan of detailed investigation. Preliminary geological reports are not a sufficiently reliable basis for the detailed elaboration of a project, building plans, or the assessment of cost and the construction schedule.

1.2. Detailed geological investigation. The programme for, and volume of, detailed geological investigations are prepared on the basis of the results of the reconnaissance studies and with regard to the layout of the projected engineering work. Detailed investigation usually comprises the survey and detailed engineering geological study of the construction site and its vicinity, and systematic subsurface exploration using adequate methods. The exact programme of subsurface exploration will depend on the type of the structure to be built, and is intended to determine the behaviour of rocks with respect to excavation, to estimate foundation conditions and to search for appropriate building materials. The bearing capacity of the foundation beds is examined either through the use of undisturbed samples in the laboratory or by loading tests directly on site (see Chapter 10). Engineering geological investigations must be complemented by a detailed study of the hydrogeological conditions of the project area.

The results of these detailed investigations are integrated and assessed by the engineering geologist, the designer and other group-study members. Their report should comprise a detailed description of geological conditions, analysis and evaluation of subsurface exploration, determination of foundation depth and method, and of special measures, if necessary, e.g. directions for the stabilization of hazardous slopes or of foundations for high fills.

The detailed investigation must be made in good time: before the preparation of the definitive project and before the start of construction work. One must not forget that the detailed geological investigation requires sufficient time to undertake all the individual tasks which are often rather laborious.

If the geological investigation is to be conducted economically, it must proceed in stages, from the general to more detailed studies; one stage should be completed before another is commenced. It would not be advisable to omit the preliminary stage in an attempt to shorten the duration of investigations by using a large amount of technical equipment, drilling rigs and personnel. One would run the risk that the expensive detailed investigations might be useless if the site or the layout of the structure proved unsuitable from the geological, or any other, standpoint.

It is an imprudent policy to save on detailed investigation, particularly on subsurface exploration. A thorough and reliable investigation in the initial stage of planning and before construction begins is the best protection for the designer from wrong assumptions. If the working schedule is to be accomplished as planned, it is necessary that all the information, on the basis of which the engineer makes his calculations, construction designs and budgets, is as correct and reliable as possible.

1.3. Geological investigation during construction. At all major engineering structures, such as dams, tunnels and bridges, it is also advisable to carry out engineering geological examinations during construction. The engineering geologist should inspect all excavations, document all temporary exposures in the foundations, and correlate the assumptions and conclusions of both the reconnaissance and detailed investigations with the actual conditions.

The engineering geologist cooperates in determining the depth of foundation, in the selection of natural building materials, in inspecting blasting and grouting procedures, etc. He procures samples of rocks for technological tests and is responsible for systematic geological documentation, so that all geological features are thoroughly and comprehensively recorded. He assists in all tests performed at the site, and in geophysical measurements. Grouting should be carried out under his supervision. One of his most important tasks is to anticipate possible emergencies caused by geological factors (formation of fissures, water seepage, subsidence) and to advise where to locate measuring instruments.

The engineering geologist should collaborate in measuring subsidence and other deformations and in interpreting the measurement results. At major engineering undertakings, the engineering geologist should, if possible, observe the effectiveness of measures taken for the protection of foundations, for ensuring the impermeability of ground, etc.

The cooperation of engineering geologists with engineers in this way will contribute greatly to the appreciation of geology by engineers, and conversely, will increase the engineering knowledge of the geologist.

During construction work, the engineering geologist should submit progress reports, accompanied by detailed graphical records to the engineer responsible for construction. Copies of these reports should be deposited in the appropriate archives so that the experience gained can be utilized for other works and the information can be readily available for regional geological studies.

Chapter 2

GEOLOGICAL MAPS AND SECTIONS

2.1. Types of geological maps. Geological maps represent the geological conditions and structure of the area as they appear at the earth's surface. They are constructed so that the data obtained by systematic investigation are plotted in a suitable topographic map. Geological mapping and field investigation form the true basis of all geological work. A good geological map does not only show which rocks crop out at the land surface but also enables the stratigraphy and tectonic structure of the area to be visualized. In showing the occurrences of economically important rocks and minerals it helps to appraise the mineral wealth of the area. A geological map is indispensable when prospecting for ore, coal and oil deposits, and construction materials such as building stone, sand and gravel.

Geological maps are prepared for various purposes and on various scales. Maps on a scale of 1 : 100,000 are designated as synoptic maps, those on scales of 1 : 20,000 to 1 : 50,000 as basic maps, and those on scales 1 : 2,000 to 1 : 10,000 as detailed geological maps. Geological maps representing the pre-Quaternary rocks without cover deposits are called solid or bedrock maps. All superficial deposits are plotted on drift maps.

The earlier geological maps were constructed only on the basis of field investigation. When the number and quality of rock outcrops were sufficient and the geologist was well instructed, the maps were satisfactory. If, however, the area was covered largely by younger deposits and weathering material, the map did not always correspond to actual natural conditions.

The accuracy and reliability requirements on the geological maps increase concomitantly with their wider usage for technical and economic studies. For this reason, the modern geological maps, and in particular engineering geological maps, are based not only on detailed field studies, but also on extensive boring and other subsurface explorations, complemented occasionally by geophysical measurements. These exploratory works are necessary especially in the areas covered by surficial deposits of great thickness.

The first maps used for engineering practice were so-called maps of soil conditions, which represented both the soil types and the pre-Quaternary rocks beneath. The depth of the bedrock surface was shown in the maps by the banding method (Fig. 2-1). Pale colours expressed the soil types and deep colours the outcrops of the bedrock at the surface. Where the bedrock was less than 4 m below the ground surface, the depth of it was shown by the width of the band in the colour of the pertinent rock. A broad band denoted the bedrock surface

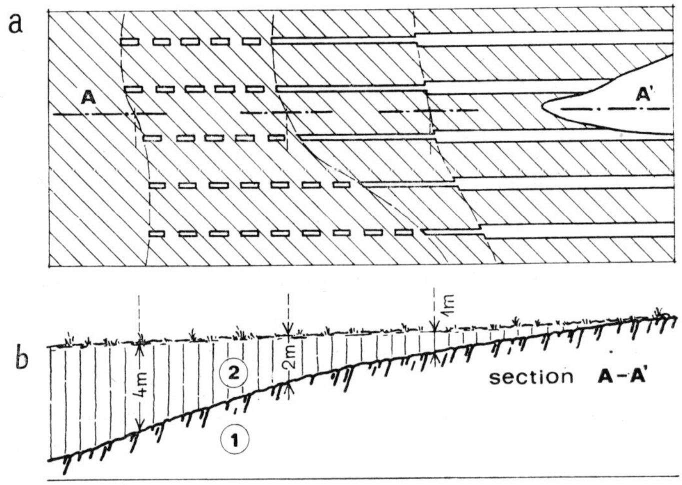

2-1. Depth of bedrock below the surface shown by the banding method, used in the first engineering geological maps. a — Plan view, b — A-A' section, 1 — bedrock, 2 — overburden.

to 1 m depth and a narrow band its depth between 1—2 m. An interrupted band implied that the bedrock surface is 2—4 m below the ground. For the construction of these first maps sounding to 2—4 m depth was used, and wide-spaced hand augering when the surficial deposits were of a greater thickness. In the 1950s, maps of several towns in Bohemia were prepared by this method on scales of 1 : 5,000 and 1 : 10,000. Some of them were published, but most of them appeared only in several copies to be used for the new city and building plans. At the time of their appearance these maps were quite progressive but soon afterwards they were superseded by maps based on a more comprehensive field investigation and more extensive subsurface exploration to fulfil the increasing requirements of engineers and planners. The geological maps became specialized; among others, engineering geological maps have evolved, which are indispensable aids to economic urban and regional planning and to the design of large hydraulic, transport and industrial structures.

2.2. Engineering geological maps. The objective of an engineering geological map is to provide a comprehensive and clear picture of geological, morphological and hydrogeological conditions of the area, from which the information needed for the proposed development can be drawn. An engineering geological map should, therefore, be based on a good geological map, represent both superficial deposits and the pre-Quaternary bedrock, and intelligibly show various dynamic phenomena (landslides, erosion and karst features, etc.) and the principal hydrogeological data.

Since the engineering geological map has to carry a large amount of information, great care must be devoted to graphical representation. The compilation of an engineering geological map is a very responsible task, because those engaged in planning and construction can observe the results of geological investigations only if they comprehend them. In order to ensure that the map is clearly understood, it must not be „overloaded" with information; this implies that all findings cannot be plotted on one map sheet.

Experience has shown that an adequate engineering geological map should consist of three sheets. The first sheet presents the *engineering geological conditions*: the pre-Quaternary bedrock is shown by hatching, the superficial deposits are shown in colours, the beds outcropping at the surface by coloured areas and the lower layer by coloured stripes. The thickness of superficial deposits is expressed by the intensity (saturation) of colour or by Roman numerals (Fig. 2-2). Important geodynamic phenomena (landslides, erosion gullies, karst sinkholes) are shown in red and by appropriate symbols. Every engineering geological map should be supplemented by several geological sections across important parts of the mapped area.

The second sheet is a *hydrogeological map*; it presents data on the surface and

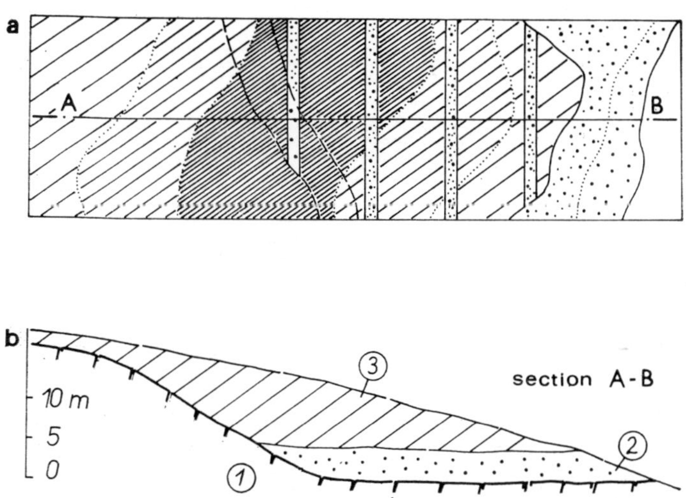

2-2. The thickness of overburden is shown in new engineering geological maps by the intensity of colour and the lower bed (gravel) by coloured stripes. The colours are here replaced by hatching. a — Plan view, b — A-B section, 1 — bedrock, 2 — sandy gravel, 3 — loess.

ground waters, on the permeability of rocks, areas of inundation, springs, wet ground, etc. Chemical properties of water, especially the grade of corrosivity, are designated by point markings at the localities of sampling.

The third sheet is the *documentation map*, in which all outcrops, test pits, boreholes, wells and other reference points used for the compilation of the map are plotted.

From these three sheets it is possible to compile maps for various purposes, e.g. maps of foundation grounds, in which areas suitable and unsuitable for building-up are differentiated. Deposits of mineral resources, natural reserves and protective zones of water sources, which should not be built up are also indicated. In „*zoned maps*" the mapped area is divided into zones of approximately the same engineering properties according to their suitability for certain kinds of construction.

Engineering geological maps are constructed on various scales depending on the complexity of geological conditions and on the purpose they have to serve. In Czechoslovakia, where the geological conditions are rather complicated, engineering geological maps on a scale of 1 : 25,000 are used for regional planning. For urban planning and design of industrial plants and large civil engineering constructions, engineering geological maps to a scale of 1 : 5,000 are usually prepared.

Before commencing field mapping, the geological literature and archive material concerning the area to be mapped should be thoroughly studied. It is generally necessary to conduct new subsurface exploration for engineering geological maps, which would show even the vertical distribution of rocks as accurately as possible. However, we cannot stress enough that the engineering geological maps should be compiled using geological methods, i.e. by detailed investigation of the area, and should not to be constructed from a network of boreholes only, since the boundaries between rocks are controlled by geological and geomorphological conditions and can only be defined reliably by field investigation. In field

Table 2.1

Map scale	Engineering geological conditions	
	simple	complicated
	Approx. number of reference points/km^2	
1 : 50,000	4	6
1 : 25,000	8	14
1 : 10,000	15	25
1 : 5,000	25	50

mapping the use of a base topographic map with a scale twice as large as that of the map to be prepared is recommended.

Every engineering geological map must be properly documented. All observations and measurements of exposures, pits and boreholes must be recorded, numbered and plotted on the documentation map. Geological Surveys of some countries publish instructions concerning the number of reference points (natural outcrops and boreholes, etc.) required for an area of 1 km^2. The number of points depends on the map scale, the complexity of geological structure and also on the accessibility of the terrain. Table 2.1 (p. 22) gives the number of reference points required for engineering geological maps following instructions of the Czech Geological Office.

The descriptions of outcrops and boreholes should be complemented by samples of all important rocks. The results of petrographical analysis and of laboratory tests performed on the rock samples are summarized in a table, in which the number of tests and scatter of numerical values are recorded.

Photographs should be taken of important exposures; the topographical character of the area is also best demonstrated by photographs.

A good engineering geological map provides a reliable basis for the optimum land-use in regional planning, for the correct location of industry and settlements, as well as for hydraulic and communication studies, especially in the first stage of planning. This instant, accurate information of, for example, the extent of mineral resources, prevents the ore-bearing areas from being urbanized.

If the planner is acquainted with the geological conditions of the building site in time, he can base the very first project studies on factual data, thus ensuring that no basic changes are necessary in the later stage of planning. The subsequent detailed investigation and subsurface exploration are necessary for refining the first assumption.

It is of primary importance to have an engineering geological map available before the beginning of work on large structures. A carefully prepared and well documented engineering geological map reduces the cost of construction, as it ensures a reliable basis for the project.

The preparation of detailed engineering geological maps is relatively laborious and expensive; therefore they are made mainly for the areas of economic importance, where a large urban and industrial development or intensive working of mineral resources are planned, or the reclamation of devastated areas is considered.

2.3. Maps of geologically hazardous zones. The endeavour to provide speedily material necessary for the economic planning of new settlements in some mountain regions led the French Geological Service to construct maps showing zones where geological hazards are to be expected (Cartes de risques géologiques). Slopes threatened by rockfalls, landslides, erosion, debris and earth flows are plotted on the topographic map on a scale of 1 : 20,000. The hazardous places (head scarps) are shown by red shading and symbols, and green-coloured areas represent places of possible accumulation of moving masses.

On the map of the town of Peisey-Nancroix in the Savoy Alps, areas of gypsum deposits, which are threatened by subsidence, undermined areas, and dead ice covered by detritus are also represented. The maps are compiled from the results of field mapping and from the study of air photos without new subsurface exploration works. In comparison with the engineering geological maps these maps can be prepared at much less cost and within a shorter time. Their quality of course depends on the experience and expertise of the mapping geologist. They cannot fully replace engineering geological maps, but they are a suitable aid for avoiding the building-up of geologically hazardous areas, chiefly in the mountain regions (J. Goguel, M. Humbert,1972).

For a similar purpose, maps of the existing and potential slide areas on a scale of 1 : 25,000 have been constructed for the entire territory of Czechoslovakia, on the basis of systematic registration of sliding-prone areas. This action carried out during 1962-1963 was stimulated by the disastrous Handlová landslide in 1961.

2.4. Topographic base maps for geological mapping. The reliability of geological maps depends chiefly on the accuracy of topographic base maps. For geological mapping, therefore, use is made of the most detailed maps that are available for the given area. In Czechoslovakia, maps on a scale of 1 : 5,000 are used for detailed geological investigations, in urban planning, prospecting for mineral resources, etc.

Geological conditions of damsites, areas to be flooded by reservoirs, navigation canals, tunnels and similar large structures are plotted in plans prepared for this purpose to a scale of 1 : 2,000 to 1 : 500, with contour intervals of 1 m. The plans are constructed on the basis of photogrammetric or tachymetric measurement. Good photogrammetric plans are particularly suitable for geological investigations, since they clearly show all prominent edges and isolated rocks, so that the outcrops can be plotted accurately in them. On the other hand, some tachymetric plans constructed only schematically from a thin network of surveyed points, are little suitable — if not unusable — for geological investigations.

The usual drawback to plans prepared for a given structure is their small areal extent. The cost of surveying is relatively high and the survey is therefore generally limited to the building site and its vicinity. This is of course insufficient for geological mapping and recognition of the geological structure of the area. The geologist who is charged with the research of the building site should act as an adviser in determining the size of the area to be surveyed because the additional measurement is expensive and often impossible due to the shortage of time.

2.5. Use of aerial photographs for geological mapping. Geological investigation with the aid of aerial photographs has increased considerably during the last decades. It is applied mainly in little investigated regions with difficult access, where ample geological information can thus be obtained without cumbersome and time-consuming field work. Aerial photography is used especially in arid and semiarid regions, where the vegetation does not conceal the geological structure.

Aerial photos can be used universally to supplement geological research, for example, when mapping slope movements, studying erosion features, mapping areas to be flooded by impounded streams, and observing erosion and abrasion processes on the banks of lakes and reservoirs. In Czechoslovakia, aerial photography was first used in 1925 for the survey of the Slapy reservoir and investigation of its banks.

For geological investigations black and white, coloured and multispectral aerial photos are used. *Black and white* vertical photos are applied most frequently. They are taken so that, when observed through a stereoscope, they provide an exaggerated spatial model of the area showing a clear picture of the geomorphology. Geological features are deciphered from the aerial photos using the diagnostic characteristics, such as the photographic tone (i.e. different hues of grey colour), the structure of the picture, the shape of the features observed, and their dimensions and interrelationships. A systematic study of stereo-pairs based on diagnostic characteristics makes it possible to analyse topographic forms, the configuration of the drainage network, erosion processes, types of vegetation cover and land-use, and infer from this information the basic knowledge of geological conditions of the area studied.

The density and arrangement of the stream pattern, for example, enable the identification of the rock types which make up the area and its geological structure; the transverse and longitudinal profiles of erosion gullies indicate the type of surficial deposits. Slope movements, debris fans and alluvial cones, moraines, lava flows, etc. are also discernible from the topographic forms. From the photographic tone the depth of the ground-water table, the presence of abandoned channels in the flood plains, the location and extent of back-filled disused quarries and sandpits, can often be inferred.

Aerial photos are also used to advantage in tectonic studies, because they often reveal tectonic lines and structures, which are difficult to determine in the field.

For the reconnaissance regional studies aerial photos on scales of 1 : 30,000 to 1 : 60,000 are generally used. Photos to a scale 1 : 8,000 or 1 : 12,000 are needed for detailed geological research.

The main advantage of geological aerial investigation compared with conventional research methods is its lower cost and time saving. The methods of detailed geological interpretation of black and white aerial photos have been described in detail in many textbooks, to which reference is made for further information.

The important diagnostic characteristic for the interpretation of coloured aerial photographs is the colour of individual complexes. Coloured photos have proved suitable mainly when the soil or rock types differ in colour. In engineering geological investigation aerial photographs are often used in the study of landslides, the assessment of potential slope failures, and in the prospecting for deposits of building materials.

Multispectral false-colour photographs are taken on films with three emulsion layers sensitive to green, red and infrared rays, where the positive does not show the area in natural colour hues. In engineering geological practice multispectral false-colour photographs are used, e.g. for delimiting deposits of organic materials (in abandoned channels, etc.), for wet grounds and glaciofluvial sediments, for discriminating various rock types, or tracing major fault lines. The diagnostic features are the colours of the objects which, however, are distorted. False-colour photos may also be used for special purposes; as an example we may cite the delimitation of waste tips, which are the source of stream pollution owing to the decomposition of pyrite they contain. Characteristic colours of the studied phenomena must naturally be checked by field investigation.

The results of aerial mapping are unquestionably a very useful guide for more detailed investigation, but it should be emphasized that they cannot substitute for field study, even if the photos are as perfect as possible and their interpretation is most careful.

2.6. Geological sections. A geological section is a graphic representation of the sequence of rocks or of the geological structure of the study area. It generally shows the course of beds and tectonic lines in a vertical plane.

For drawing a geological section we need to know the form of the land surface in the section line and pertinent geological data. The form of the land surface can be obtained from a topographical or geological map, provided that it is precise enough, or by surveying the ground surface along the section. Accordingly, measured and constructed sections are distinguished.

Measured geological sections. All vertical or steeply inclined outcrops or artificial exposures, e.g. test pits and trenches, furnish sections that can be surveyed and drawn. Deep-cut valleys of rivers and brooks with steep slopes free of vegetation are suitable for the compilation of large-scale measured sections. Natural outcrops are not always perpendicular to the strike of beds; in European geological literature they are often termed as "*defilés*".

In many civil engineering structures for which long and deep cuttings are made, important geological sections can be obtained from the exposures on excavation walls. These sections are of great documentary value because the excavations are often only temporary.

In surveying the geological section of an excavation wall or quarry face the base line at the bottom of the wall is chosen first, along which stakes are set 2—4 m apart, depending on the complexity of the section. The line preferably should be horizontal. The stakes divide the wall into small compartments, which are successively measured and plotted. When the wall is inclined, the lengths are measured on an oblique area, so that the drawing prepared in the field is somewhat exaggerated. It is therefore necessary to measure and note the angle of slope, so that the heights can be recalculated to a vertical section. The section should be supplemented by detailed explanations, description of beds, joints and by other data, in particular by the disposition of the section with respect to the cardinal points. The elevation above sea level of the base line should be defined as accurately as possible. The section line should always be plotted on an accompanying map.

When the geological conditions are complicated, it is advisable to take photographs of the outcrops and draw the detailed features with coloured pencils on an enlarged print on mat paper. Prominent objects, such as buildings, trees or posts, should be shown, if possible, on the photograph and in the section for easier orientation.

Precisely measured sections can be obtained with the use of *photogrammetry*. This method is especially suitable for the survey of geological sections on inaccessible mountain slopes, which can be readily photographed from the opposite valley slopes or from the air (Fig. 2-3a,b; Schweizerische Bauzeitung 1939).

Constructed sections are not measured in the field but are prepared from geological maps and/or information gathered from boreholes, wells, mining works, etc.* For the construction of sections we need to know the form of the land surface in the line of section, the boundaries of individual beds and their strike and dip. All these data are plotted on the geological maps so that we can construct a section from a sufficiently detailed geological map provided that we are well acquainted with the stratigraphy and regional geology of the area.

The question may arise as to why geological sections are constructed when all the data are given in the geological map. The following figures show the significance of the sections. On the map shown in Fig. 2-4 the mode of deposition of the beds seems complicated but the section shows that they have a uniform moderate dip of the same sense. The geological map presents the boundaries of beds as they intersect the land surface. When the surface is dissected and the bedding planes are contorted, as is often the case in folded complexes, the mode of deposition of the rocks is not always easy to visualize from the geological map. In some cases, e.g. in engineering geological research for mountain tunnels, the construction of geological sections is prerequisite for a true assessment of the geological structure of the area.

According to the direction of a section to the strike of beds, cross and longitudinal

* For the construction of geological sections from boreholes see Chapter 4, p. 132.

a

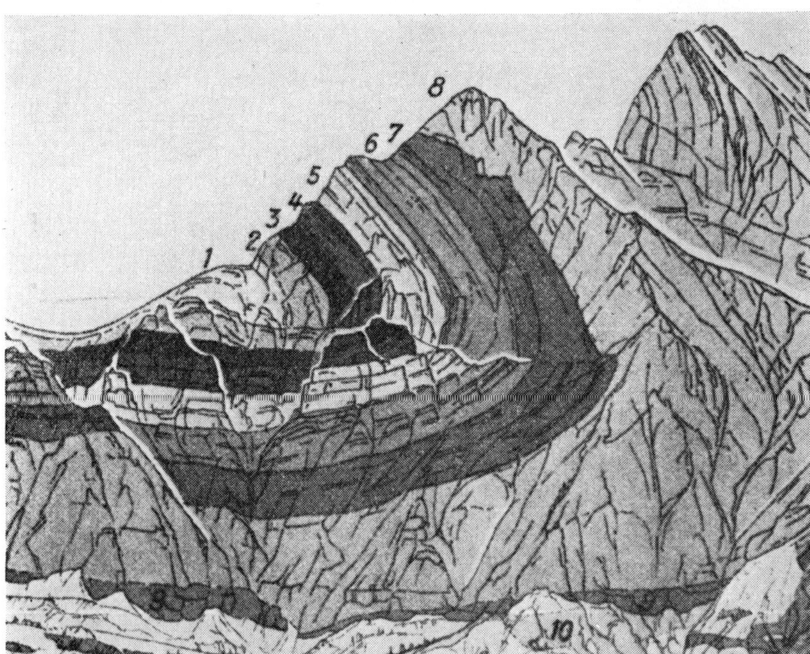

b

2-3. Photogrammetric measurement of geological sections. a — Photograph, b — photogrammetric interpretation. (After Schweizerische Bauzeitung 1939.)

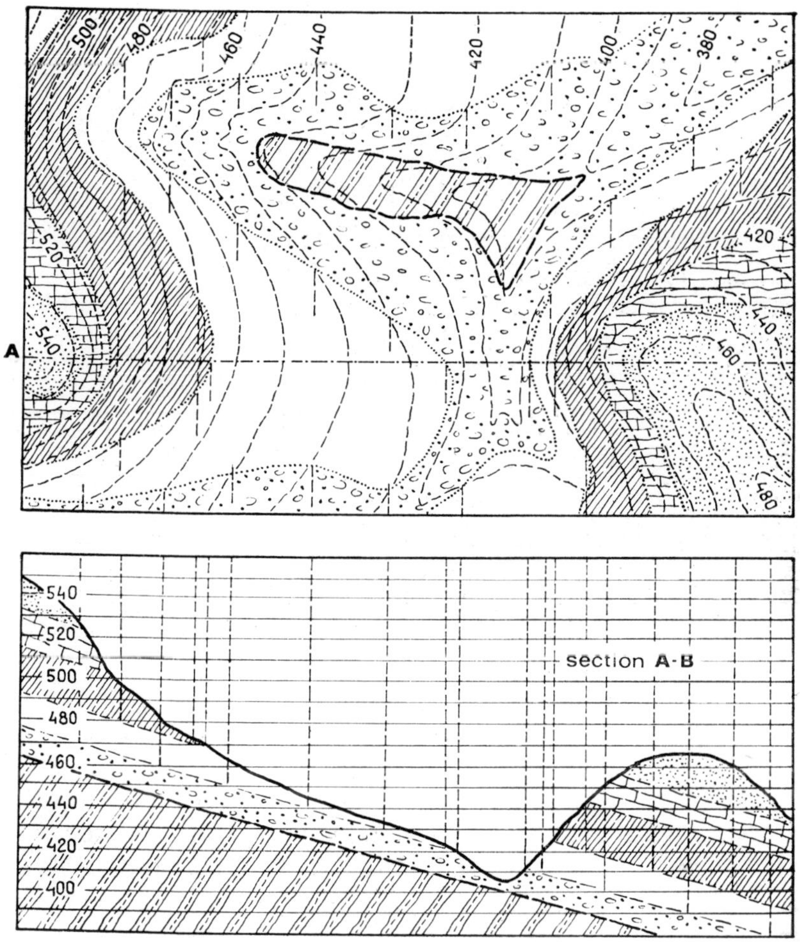

2-4. The geological map of an area with an uneven surface appears complicated even with a uniform dip of beds; this is shown in geological section A-B.

sections are distinguished. The former are drawn at right angles to the strike of beds and the latter parallel to it. Oblique sections are also sometimes constructed. Cross-sections are the most suitable means by which to recognize the geological structure of the area. The correctness of the geological map can be checked by a set of parallel cross-sections.

Longitudinal sections are constructed when the lateral extent and dip of beds need to be established. They are used in the study of hydrogeological problems and as a basis for the appraisal of mineral resources. In drawing a section across strongly folded beds, a part of it must sometimes be drawn obliquely to the strike of beds. In this case the dip of beds towards the section line must be correctly determined. In some cases the section line is somewhat deflected so that the section can be drawn perpendicular to the beds. The points of deflection must be denoted on both the map and section.

Geological sections are generally constructed on the same scale for lengths and heights. Exaggerated sections are used for complexes with a very small dip of beds, in hydrogeological investigation, and in the study of Quaternary sediments (e.g. for river terraces, peat bogs, etc.). The exaggeration of sections should be distinctly expressed. It is not recommended to exaggerate geological sections through folded, bedded or schistose rocks. Such a section greatly distorts the picture of the geological structure of the area and the dips of beds must be recalculated with respect to exaggeration. Table 2.2 provides a clue how to adjust the dip of beds in exaggerated sections. Figure 2-5a shows a geological section of

Table 2.2

Angle in degrees	Exaggerated section		
	2×	3×	4×
5	9°55′	14°45′	19°15′
10	19°25′	27°50′	35°10′
20	36°	47°30′	55°30′
30	49°05′	60°	66°30′
45	63°30′	71°40′	76°
60	73°55′	79°05′	81°45′
75	82°30′	84°50′	86°10′
85	87°30′	88°20′	88°45′

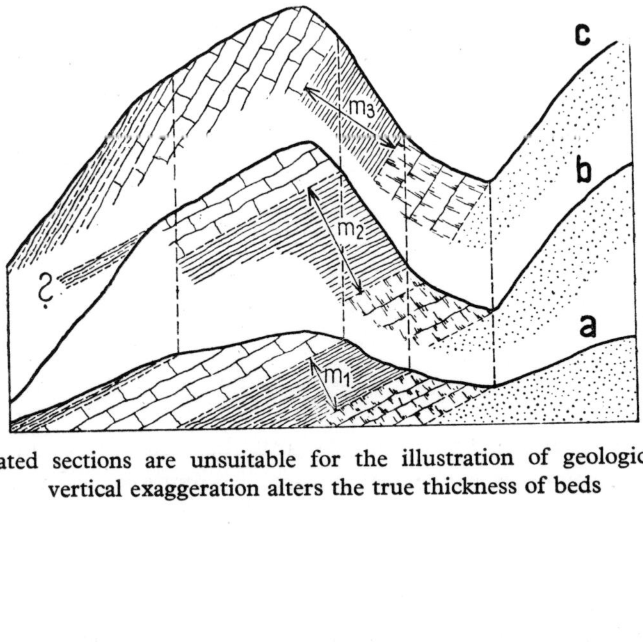

2-5. Exaggerated sections are unsuitable for the illustration of geological conditions; vertical exaggeration alters the true thickness of beds

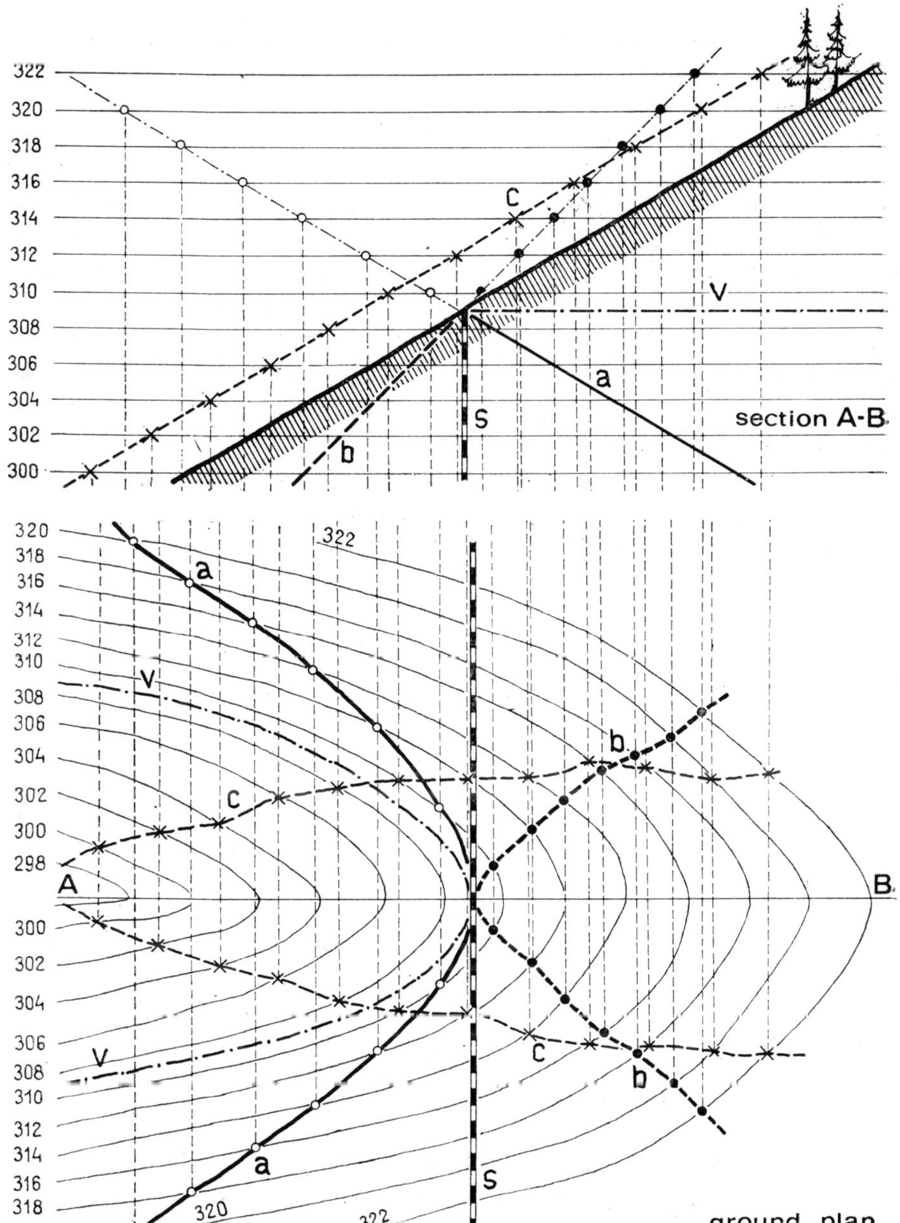

2-6. Outcrops of beds with different dips in a valley head, as seen on a map. (After Bonte, 1969.)

moderately inclined beds drawn to a true scale, and above it is the same section, the vertical scale of which is three times exaggerated. In Fig. 2-5b the dip of beds has not been exaggerated and therefore beds whose dip is smaller than the exaggerated inclination of the slope could not be plotted in the figure (on the left). Even if the dip of beds is exaggerated to the same ratio (Fig. 2-5c), the thickness of beds is not true. The use of geological sections of exaggerated vertical scale are usually avoided unless absolutely necessary.

In constructing sections from a geological map it is necessary to realize how the outcrops of beds of different dip appear in the map. Figure 2-6 shows a valley head with several outcrops of beds dipping at various angles. Vertical beds, dykes and faults are rectilinear in the map and horizontal beds follow the contours. When the beds dip into the slope, the outcrop is concave against the gradient of the valley; when the dip of beds is downslope, the outcrop is concave along the gradient of the valley floor. The beds striking parallel to the valley floor crop out in the slope and run roughly parallel to its direction in the plan.

For a correct construction of geological sections it is necessary to determine the true thickness of beds. It can be readily computed from the width of the bed shown in the map, provided that the dip of the bed and the angle of slope are known.

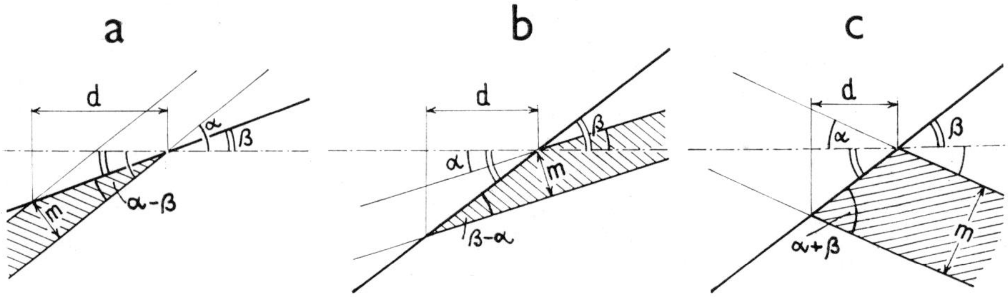

2-7. Determination of the thickness of beds with different dips. m — Thickness of the bed, α — dip of the bed, β — angle of slope.

If the bed dips downslope and the dip exceeds the angle of slope (Fig. 2-7a), then the thickness of the bed is:

$$m = \frac{d \sin(\alpha - \beta)}{\cos \beta};$$

if the bed dips downslope and the dip is smaller than the angle of slope, the thickness of the bed (Fig. 2-7b) is:

$$m = \frac{d \sin(\beta - \alpha)}{\cos \beta};$$

in case the bed dips into the slope (Fig. 2-7c), its thickness is:

$$m = \frac{d \sin(\alpha + \beta)}{\cos \beta}.$$

For engineering structures it is often necessary to construct detailed sections even across the superficial deposits. Great mistakes are frequently made in drawing sections through Quaternary sediments, which may adversely influence the design of the structure. It is very important to understand the interrelations, age and

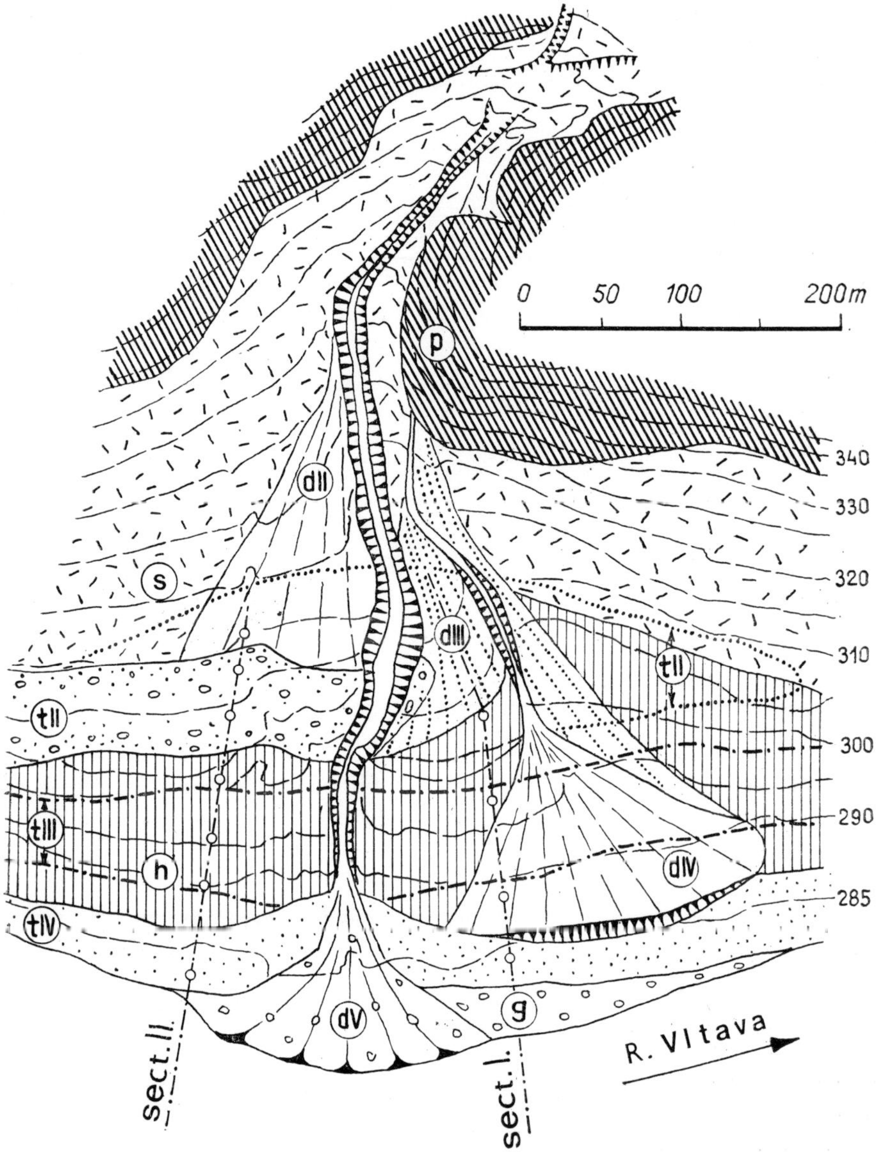

2-8. Part of a geological map showing superficial deposits on the slopes of the Vltava river valley: p — porphyry, t II, t III, t IV — Pleistocene terraces, h — loess loam, d II — d V — alluvial cones, g — gravel washed-out from the youngest cone, s — slope debris.

development of these youngest deposits. Part of the geological map and two sections from the middle reach of the Vltava valley are given here as an example. In the widened river valley three terraces (designated t II, t III, t IV in Fig. 2-8) have been preserved. They are covered in part by slope loam, detritus, loess loam and alluvial fans (d II, d III, d IV, d V). In order to draw correctly the section through the slope, the age succession of individual Quaternary sediments must be determined (Fig. 2-9). The very thick middle terrace is the oldest and was overlaid

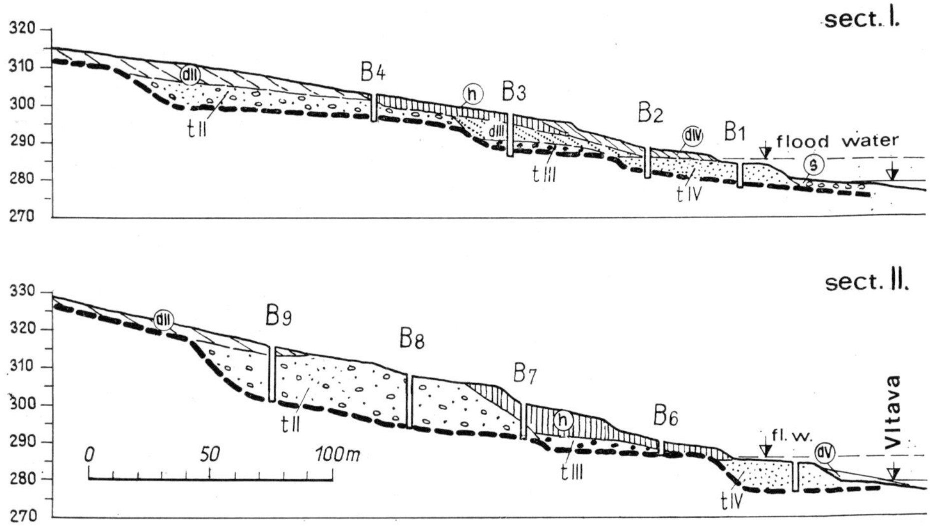

2-9. Cross-sections constructed from the map shown in Fig. 2-8. For explanation see Fig. 2-8.

by the oldest part of the alluvial fan (d II). Terrace deposits (t III) laid down after the downcutting of the valley are not seen in the map; they have been found in boreholes below loess loams and the younger alluvial fan (d III), which in this stage was shifted somewhat to the north. Subsequent to the following period of erosion, sand and gravel of the valley terrace were deposited (t IV) and then overlaid by the youngest sandy flood sediments. The frontal part of the alluvial cone (d IV) lying on the valley terrace has been eroded by recent floods. The youngest element is an erosion gorge, which in historical times cut through all Quaternary sediments and deposited the material in the youngest (Holocene) alluvial cone (d V), which encroaches upon the present channel of the Vltava.

Without understanding the geological-historical development of Quaternary deposits, it is very difficult to compile a section that would correspond to reality, because subsurface exploration is rarely so detailed as to record the boundaries of all strata.

Chapter 3

MECHANICAL PROPERTIES OF ROCKS

When co-operating in the design of engineering structures, the engineering geologist is concerned with changes which occur in the rock due to the construction. These changes may be of a physical or chemical nature and their development depends on the properties of the rocks at the particular site. Several of the properties have been investigated by geological sciences and we shall meet them in the other chapters of this book (for instance the questions of rock weathering, the gas content of rocks, etc.). However, a considerable portion of the activity of engineering geologists is directed towards the study of the mechanical properties of rocks as well as of the modes in which the rock will respond to the changes of loading.

The engineering geologist is expected to be aware of the mechanical behaviour of rocks for several reasons:

(1) In particular cases he has to provide himself with static analyses, e.g. when investigating the stability of natural slopes and sometimes even when designing artificial slopes.

(2) In other cases he has to provide the designer with the numerical values of mechanical properties of rocks. These values are often obtained by tests which cannot be conducted and evaluated whithout knowledge of the principles of mechanics.

(3) However, the task of the engineering geologist is greater than that of the designer. The latter will analyse the problem from a rather phenomenological point of view, as he will only consider the mechanical relevance of the parameters. In contrast, the engineering geologist also has to deal with the physical background of the mechanical changes, namely the changes in the rock structure. In the latter activity he is entirely dependent on his own observations.

(4) The engineering geologist is expected to submit his own opinion of the project, independent of the results of the static analysis of the designer. He will not base his opinion on the results of subtle static calculations, but on his experience, his capability of judgement and on simple "engineering" static analyses. His report will have the value of being a counterbalance to the results based on more exact static analyses. The discrepancy between the two conclusions may draw attention to several important facts which might remain unnoticed in the static analysis or could not be built into the solution in an appropriate manner.

3.1. Physical background of the mechanical behaviour of rocks. In natural conditions most rocks do not meet the requirements of a continuum model in terms of mechanics. On the contrary, they consist of larger or smaller blocks or particles and therefore can be considered as discontinuous media. However, practical solutions, based on the mechanical relations between blocks or particles, have not been employed due to their complexity. At best, partial solutions appear which idealize the rock blocks as parallelepipeds (Trollope, 1968, Goodman et al., 1968), or the sand grains as spherical elements. Unless a sufficient number of such solutions are available, the methods of continuum mechanics will prevail and only significant surfaces are assumed to require handling as discontinuity planes. The study of the rock structure then serves to explain the inferior mechanical properties compared to solid rock.

On account of the existence of planes of discontinuity, rocks are capable of transmitting tensile forces only to a limited degree. On the other hand, they can resist compressive and shear forces, or in other words, they exhibit both compression and *shear strength*. As will be outlined in Sections 3.5 and 3.6, large compressive loading frequently causes a shear failure. Why is it, then, that rocks are capable of resisting shear strain?

With materials composed of particles, such as sand, internal friction between particles naturally comes into consideration. It is governed by laws similar to the shear friction between two bodies, i.e.:

$$T = f \, . \, N,$$

where T is the peak *shear resistance*, f is the *friction coefficient*, and N is the normal force, which thrusts the bodies against one another. A practice prevails in soil and rock mechanics of replacing f by tan Φ, where Φ is the angle between the resultant R and its normal component N (Fig. 3-1). Because of the surface roughness

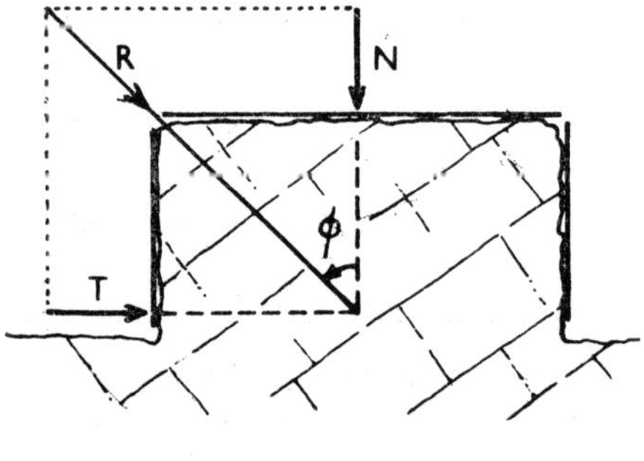

3-1. Significance of the angle Φ in the analysis of the resistance of rocks against shearing forces.

and unevenness of the contact areas, this angle is quite large, e.g. 25—30° with sand.

With well-interlocked and resistant particles, like those present in dense rocks and soils, the portions of the body on both sides of the shear surface move from interlocking during shear straining and in this way the slip surface can develop. With coarse dense sand this separation can amount to 5 mm and therefore takes on a macroscopic character. For this reason, and also because it occurs only with dense material, this phenomenon and its mechanical effects are separated from the concept of friction and are termed *dilatancy*. During the development of dilatancy the point of application of the normal force elevates in the direction opposite to that of the force. Therefore the shear force has not only to exceed the resistance arising from friction, but also to produce the work necessary to elevate the point of application of the normal force. As a result, the portion of resistance related to dilatancy increases with increasing normal stress. The angle of shear resistance is greater in materials exhibiting dilatancy than in those where dilatancy is absent. For instance, this difference is of the order of 10° when comparing the resistance of dense sand with that of loose. Apparently the degree of resistance due to particle interlocking is contingent upon the particle strength. Therefore dilatancy, as well as the resulting resistance, is dependent on the strength of grains in sand as well as of pebbles in gravel. In this way the importance of the study of the strength of grains and fragments becomes apparent.

The mechanism of dilatancy which occurs during the small twisting and other rearrangement of blocks in a rock mass due to the shear straining is basically identical. The phenomenon of dilatancy can therefore be observed in rocks as well as in sands and gravels (see Sections 3.4 and 3.5). Its occurrence is governed by the *strength of blocks* (or in other words by the strength of the *rock material*) which has to be sufficient to prevent the crushing or breaking of block edges. Moreover, the interlocking of blocks contributes to the rock resistance in another manner. In many cases the pattern of interlocking is complex, the slip surface cannot develop by block-to-block movement and shearing of blocks is necessary. This produces a resistance which is caused by the integrity of the material, and with rocks reflects the *cohesion* of the rock material. It is to be expected that its magnitude does not change with the increasing normal force when solid rocks are considered. It therefore represents the constant component of the shear strength. Its magnitude results from the nature of the structural *crystalline bonds* within mineral grains as well as from the linkage between grains (crystalline bonds between the grains of igneous rocks, cementation bonds between the grains of sedimentary rocks, etc).

These bonds have different stability so that their failure also varies. For instance, failure occurs at 1 per cent contraction in crystalline limestone under uniaxial compression. The contraction of granite hardly attains two-thirds of the same value. Less rigid are the cementation bonds due to various carbonates, silicates, etc. (microcrystalline or cryptocrystalline nature of the binder). For instance, the strength of a dry argillite is reached with a contraction of 1.5 per cent in uniaxial compression.

Maximum ductility is encountered with soils containing secondary molecular bonds such as prevail in wet clay. Failure occurs with uniaxial contraction of 4 per cent and even more. *Clay* possesses a small cohesion, but its sphere of action extends to a greater distance than that of primary bonds. In principle the molecular bonds in clay originate from the residual and not fully counterbalanced charges associated with the irregular arrangement of molecules on the surface of the crystalline particles of clay minerals. Edges of particles are charged positively whereas the flat faces exhibit negative charges which give rise to interparticle bonds. Moreover, the cations adsorbed on the particle surface serve as a linkage between the particles. On the other hand, water molecules drawn into the areas between the cations by osmotic forces separate the particles and decrease the soil strength. Consequently, the cohesion of clayey soils decreases with increasing water content. As water can be pressed out from within the pores, clay particles come close to one another (the soil *consolidates* under pressure) and cohesion increases. Clayey soils are therefore characteristic in that their cohesion increases with increasing pressure. Hence the increase in their shear resistance with the growing normal stress is produced not only by friction, but also by an increase in cohesion, and up to a certain limit by the dilatancy component.

A peculiar source of strength is characteristic of the so-called untrue cohesion. This is encountered in wet, fine grained soils and is a result of the surface tension in menisci of capillary water present in the corners between the grains. Untrue cohesion yields a passing component of strength because it vanishes with both soil saturation and desiccation. Its consequences are therefore regarded as disadvantageous, particularly with regard to excavation works.

3.2. Character of the planes of discontinuity and methods of representation. As pointed out at the beginning of Section 3.1, minor and frequent discontinuity surfaces are not considered as separate units when dealing with several statical problems of a rock mass, for instance, when investigating the state of stress or deformation. The rock mass is then treated as a continuum. The presence of the surfaces is reflected in the inferiority of the mechanical properties. On the other hand, important faults and fault zones are considered as separate units, but even unimportant planes of weakness cannot be neglected if they might endanger the stability of a rock mass; for instance, a fissure may promote the collapse of a steep rock slope. This is why so much space is devoted to the planes of weakness or discontinuity existing in rocks, as well as to the methods of their representation in engineering geological studies.

The presence of the planes and their orientation can be recognized over a limited area or over a short distance only, for instance on the walls of an exploration gallery. Therefore questions arise as to the size and continuity of a plane, whether it is limited to the ground surface or if it extends to great depth, etc. For this purpose the origin of the planes of weakness and discontinuity should be studied.

3-2. Parallel orientation of granodiorite, along which cleavage occurs under seismic loading, thus threatening the transformer platform of the hydro-electric power plant at Huallanca, Peru. (Photograph by Mencl.)

These planes may either be of *tectonic* origin (such as faults and fractures developed during mountain-building) or of *atectonic* origin, such as:

Separation planes originating from internal stresses and due to cooling and contraction (for instance sheet jointing in granite, columnar jointing in basalt, or slabbing of a phonolite).

Sheet joints formed roughly parallel to the ground surface as a result of stress relief caused by removal of the overburden.

Sets of fractures and slip surfaces accompanying gravity movements or vertical subsidence caused by nonuniform compaction of the underlying rocks.

Bedding planes and sets of cracks accompanying the diagenetic lithification of deposits.

Crystalline rocks exhibit the first indications of mechanical weakness in the direction parallel to the elongation of mineral grains. The rock possesses a decreased strength along these planes, which may be transformed into planes of mechanical discontinuity if exposed to the action of surface agents (Fig. 3-2). *Metamorphic rocks* are weakened along the foliation planes, although a large amount of crystalline intergrowth cohesion may still be present. Thus, for example, the shear strength measured along the foliation planes of the amphibolite on the site of the Dalešice dam in Moravia amounts to about half of that measured across the planes. Tectonic deformation produced faults and fault zones, both longitudinal and transversal. The former originated in layers of low strength derived from the primary weak rocks that were kneaded between the harder beds during the development of foliation. Polished and striated surfaces often accompany the faults, and cataclastic rock (fault breccia, fault gouge) and mylonite give characteristic features to the fault zones.

Igneous and metamorphic rocks are usually traversed by more than one set of joints which run in several directions and contain different infilling materials. Several fissures are often healed a posteriori, by material originated from the magma or by silica or ore veins, etc. Some healed fissures possess a large strength so that they do not represent planes of weakness at all. On the other hand the rock may be decomposed in the vicinity of joints, e.g. by hydrothermal activity, and such joints represent important discontinuity planes. Near the ground surface the open joints are usually filled with products of weathering or by loamy infiltrations.

Joint systems open if the rock is exposed to the action of surface agents. As a result of temperature changes and horizontal stress release joints develop parallel to the ground surface. Old joints may open on mountain flanks and valley slopes due to the stress release as well as to the gravity forces. As a valley deepens, new surfaces of discontinuity arise, which may develop into large open cracks on the upper parts of the slopes and cause gravitational slides (Fig. 3-3).

Folding also produced surfaces of weakness in rocks. In addition to the remnants of older surfaces, sets of joints are encountered in folded rocks, e.g. sets of joints which accompany the fold axes. The relationship of the joint sets to the fold and fault structures will not be dealt with here because the problems involved fall in the field of structural geology. The reader is referred to the relevant textbooks.

The relationship between the orientation of the faults and the direction of the acting forces will be treated in Sections 3.4 and 3.5.

The most important surfaces of discontinuity in *sedimentary rocks* are the bedding planes and tectonic dislocations. Because of the relatively weak character of these rocks broad zones of tectonic disturbance occur. Compared with igneous and metamorphic rocks they have an advantage, namely, that most joints are not continuous but end when they reach a relatively stronger layer. The bedding conditions often change in the vicinity of transverse faults, because the beds are often bent and sheared along the fault. The bending may extend for a distance of several tens of metres and the blocks of stronger rocks are slightly turned. Displacements of blocks relative to each other are perceivable and are accompanied by slickensides. These polished and striated surfaces stand for the set of shears conjugate to the major system which is parallel with the fault.

The major system of joints in *clayey soils* is usually parallel to the bedding. In addition, frequent short fissures occur, most of them being perpendicular to the bedding. The two sets become evident after opening due to desiccation near the ground surface. In the zones of tectonic deformation and positions of differential subsidence of the underlying beds, several sets of numerous slicken-sided fissures, as well as minor faults and bend-glidings, are present.

Of special interest and importance are the features formed by the *stress release* resulting from the denudation of the overburden. Undulation of the horizontal ground surface, accompanied by a set of curved slip surfaces, which disappear at some depth, is not rare. The curved slip surfaces occur in two conjugate sets exhibiting opposite dip orientations.

The two conjugate sets of curved slip surfaces are also observed on valley slopes. They arise as a result of the removal of lateral forces during valley erosion. Their occurrence is comparable to the existence of the so-called Luder's surfaces which develop around a hole in a stressed steel body. In addition, slip surfaces of ancient landslides are present in the slopes.

3-3. Open cracks and gravitational slip surfaces in the walls of the Paria valley, one of the numerous transversal glacial valleys in the Cordillera Blanca in Peru. Stress relief in the massif is caused by a slip displacement of the neighbouring mass along an important fault which borders the mountains at the distance of 5 km to the right-hand side of the figure (Photograph by Mencl).

Most engineering-geological investigations (e.g. when dealing with the stability of slopes, tunnelling, or foundation of dams) require that the presence of the discontinuity surfaces is recognized and appropriately represented. In order to discern the various sets of surfaces, a number of measurements of their orientation and characteristics are necessary. These can be made conveniently on the walls of natural outcrops or artificial exposures, and in the exploratory drifts and galleries. Additional measurements may be made on drilling cores or borehole walls. The strikes and dips of joint surfaces are recorded, and a concise description of the studied joint or fault, e.g. uneven, rough, striated or polished surface, open or closed, with a clayey infilling, etc. should be given. The areal extension of the surface is investigated as far as possible and a description of its boundaries is compiled. On the basis of these observations a graphical representation is drawn, usually in the form of either radial or circular diagrams.

Radial diagrams represent the surface frequency and may be plotted in two ways. One method is to represent the strike of the surface by a radius vector of the same direction, the length of which illustrates the number of measured surfaces of the given strike (i.e. contained within the boundaries of deviation of 5 or 10 grades). The extreme points of the vectors are connected by a broken line. As the diagram is symmetrical, only half need be drawn. As this kind of diagram does not present a picture of the dips of a studied surface, a modification may be made. The number of the measured surfaces is represented by the length of a vector, having the same direction as the dips. The average value of the dip is indicated by a figure. This diagram is not symmetrical and must be drawn in full.

Circular diagrams are widespread nowadays, and are mostly of two types,

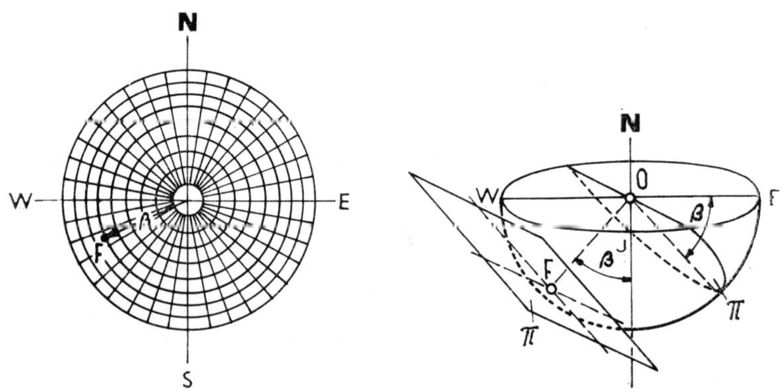

3-4. Representation of planes of weakness in Lambert equal-area projection. Any plane is represented by pole F, which is plotted into the equal-area net according to the strike and dip of the plane (Müller, 1953). When plotting (on transparent paper) a small circle of a diameter equal to one tenth of the diameter of the net circumference, the area of the small circle equals one hundredth of the total area of the net. Using the transparent net, the number of poles falling within a limit of one per cent scattering can be determined.

the simpler one being shown in Fig. 12-29. A plane having the same strike and dip as the observed surfacc, is sct through the centre of a sphere. The intersection of the plane with the outer edge of the lower hemisphere is plotted as it appears in the orthographic projection. A drawback to this system of representation is that the diagram often presents a maze of lines so that statistical analysis is not possible.

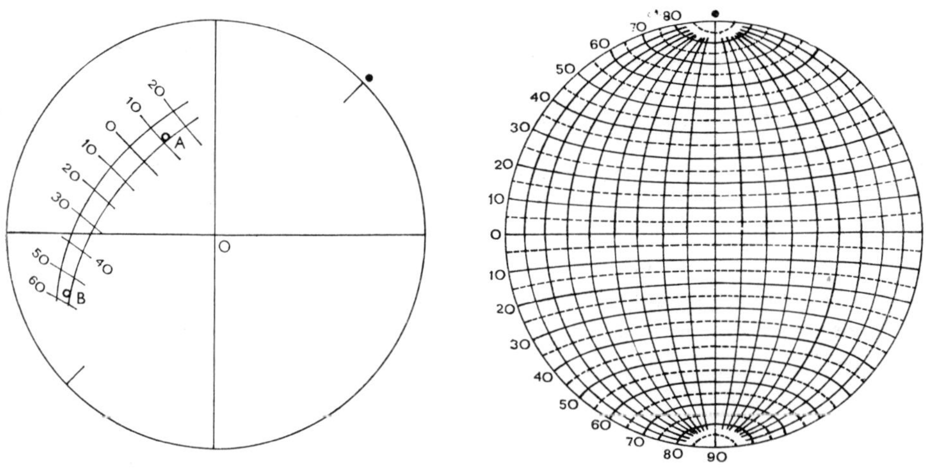

3-5. Determination of the angle made by directions OA and OB (Fig. a) which are plotted in Lambert equal-area projection. Schmidt's transparent net (b) is turned around the point 0 as long as points A and B are covered by a single meridian. The distance of the angle between points A and B is given on the meridian by the difference in the angle values of the parallels.

From the point of view of *statistical analysis* the Lambert equal-area projection of the hemisphere is convenient. As shown in Fig. 3-4, the orientation of a studied plane is represented by the point of intersection of the lower hemisphere with the line drawn from the centre normal to the given joint or fault. The point of intersection (pole) is plotted by using the Schmidt or equal-area net. The application of this net makes it possible to analyse directly the population of poles.

In many problems of engineering geology determination of the angle made by two rays represented by poles A and B (Fig. 3-5) is necessary. So, for example, with the foundation blocks of a concrete dam a need arises to determine the angle made by the resultant of acting forces (represented, e.g. by the pole A) with the perpendicular to a fault (pole B). A transparent Schmidt net (Fig. 3-5b) is used: it contains meridians and parallels plotted in equal-area projection, and can also be employed for the tracing of poles.

3.3. Physical and index properties of rocks. The analysis of mechanical properties of rocks cannot be segregated from the examination of several of their simple physical properties, which can be determined easily and often have such a direct relationship with the mechanical properties that the values of the latter can be established indirectly. Therefore some physical properties may serve as a basis for special classifications of rock and soil mechanics.

Moreover, several physical properties can be determined in an indirect way by using tests of conventional character. Thus, e.g. the porosity of sand, which is a typical physical property, can be determined by means of a penetration test and therefore the result of this test can be established as an index property. Nevertheless, this approach brings about the danger that the wealth of experience could become worthless if the test technique had to be changed. Therefore a tendency exists to unify the test methods on an international basis and not to vary the test techniques without conclusive reasons. Moreover, it is recommended that the basic physical property is considered as an index property, while the corresponding readings from the test serve as auxiliary values.

Due to different characteristics of bonds between particles or blocks in rocks, the importance of individual physical properties will differ from rock to rock. From this point of view, in the practice of engineering geology, rocks are grouped into four categories: *uncemented rocks* (cohesionless soils or fragmentary rock masses), *cohesive soils*, *weak rocks*, and *hard rocks*. Let us observe the physical properties of the four groups in the sequence of their practical importance, because the engineering geological investigation of the site should proceed in a similar manner.

3.3.1. Physical properties of cohesionless soils and fragmentary rock masses. In the practice of engineering geology these are represented by sand and gravel, but also by more fine-grained soils (silt) on the one hand, and by fragmentary rock masses (rock debris) on the other. From the point of view of their relationship with important mechanical properties, the physical properties may be ranked as follows:

(a) *Density*. This is of primary importance, as any change in density promotes a qualitative change in the behaviour of the mass. Its magnitude is given by the volume of voids in the mass. In order to characterize numerically the content of voids, two physical properties have been established: the *porosity*, n, which stands for the percentage of the volume of voids contained in a given volume of the mass, and the *void ratio*, e, which is defined as the relationship between the volume of voids and the volume of grains (or fragments, blocks, etc.). Both quantities are related by the formula: $n = 100\,e/(1 + e)$. As a guide to the understanding of the practical values of porosity the conception of a mass containing spheres of equal diameter is used: in the most dense array they exhibit $n = 26\,\%$, in the loosest, $n = 48\,\%$. Any structure of the mass of spheres packed in a loose state will collapse easily, especially when subjected to shocks, as will a structure of loose sand. Therefore large settlements (of the order of decimetres) can be generated, particularly by shock action, even if the thickness of the sand layer does not exceed a few metres. When the soil layer lies below a water surface, the volume of water filling the pores is pressed out as the porosity decreases. As this process does not proceed immediately, pore-water pressure develops. In the course of this process the sand grains are separated by water films; these generate the spontaneous liquefaction of the sand layer, and the phenomenon of *quicksand* occurs. Low permeability is another factor which increases pore-water pressure. The stability of loose sand is therefore impaired by the presence of silty and/or clayey particles, so that the pore-water pressure is active for a longer period of time.

The boundary between porosities of *dense* and *loose sand* and silt is of the order of 38—40 %. More precise values may be determined by tests. In these the densities and pressures are established, with which the phenomenon of dilatancy disappears (Sections 3.1 and 3.4). When dealing with natural soil or with rock debris, we classify them as dense and loose, whereas the terms *compacted* or *uncompacted* (badly compacted) are employed for fills. The majority of standards and manuals apply a more simplified criterion than that based on dilatancy. The so-called *relative density* D is determined, and is defined by the expression: $(e_{max} - e_n)/(e_{max} - e_{min})$, e_{max} and e_{min} being void ratios corresponding to the loosest and densest laboratory states of the given soil, and e_n being the void ratio of the soil in the field. Soil exhibiting D smaller than 0.33 is considered as loose, with a value of 0.33—0.67 it is medium dense, and with values above this it is a dense soil. In order to secure the soil mass against the danger of the collapse of the grain skeleton, a value of D greater than 0.7 is required.

Because the specific gravity of the grain material, γ_s, proves to be constant for a particular soil, the unit weight of the dry soil, $\gamma_d = (1 - n/100)\gamma_s$, may be used as a criterion of density. This method is frequently used with fills.

When the density of a studied soil is of great practical significance, the engineering-geological investigation is faced with the problem of either determining the values of porosity of the particular soil, or at least of reporting whether the soil is in a dense state or not. Should this information be lacking, the result of an investigation of cohesionless material would be almost worthless. As undisturbed sampling of these soils is not possible, other methods of investigation are employed. When the layer is accessible, the volume of grains can be deduced from the dry weight of an amount of soil taken from a carefully excavated small hole. The volume of the hole is measured by refilling it with either a calibrated sand (from a special cone device) or with water (in the latter case the hole is bedded with a thin plastic foil). This technique is cumbersome and therefore it is usually replaced by a series of loading, penetration, or geophysical tests. Penetration as well as geophysical testing methods were first developed for use in boreholes. Several methods exist, but particular attention must be given to the *standard penetration test*. This represents a typical dynamic penetration test, as the measuring consists of counting the number of blows necessary to drive the 5-cm diameter sampling tube into the ground. The tube is driven 5 cm into the bottom of the borehole and then the number of blows necessary to drive the tube for the next 30 cm are counted. The blow device consists of a drop ram weighing 70 kg which falls from a height of 75 cm. The soil is considered to be loose if the required number of blows (denoted by N) is less than 10, while with 30 blows the soil can be regarded as perfectly dense.

Geophysical tests usually measure the dissipation of the gamma radiation.

(b) *Grain-size composition* is second in importance as was demonstrated in Section 3.1, where the contribution of dilatancy to the shear resistance of soils was noted: the amount of dilatancy is greater with a coarse sand than with a fine silt, etc., or the density of a soil containing gravel and sand is greater than that of a gravel only. The plotting of a grain-size curve (e.g. Fig. 13-51), determined by sieving methods, serves as a good basis for analysis of the gradation of a studied soil. However, for a better understanding, the utilisation of names is also necessary. The various sizes of grains are therefore divided as follows: silt, with grains smaller than 0.05 mm; sand containing grains of up to 5 mm; gravel, which changes into cobbles if greater than 7.5 cm. These boundaries (as well as further delimitations within fractions on fine, medium, and coarse) are often modified for practical reasons. For example, a sieve used for materials on the borderline between silt and sand is the No. 200 sieve, giving a boundary of 0.074 mm. Similarly, the No. 4 sieve defines the boundary between sand and gravel as 4.76 mm.

From the points of view of engineering geology as well as of practice, names of coarse soils based only on grain size are not always appropriate. The term gravel, for example, is used also for fragments of rock debris, which leads to errors. Therefore several authors recommended the use of the petrographic names pelite, aleurite, psammite and psephite, but these terms are not part of the language of civil engineers.

Among other features which are studied with the aid of the grain-size curve, the "coefficient of *uniformity*" is often examined to illustrate how the various grain sizes are represented in the soil. It is defined as $U = d_{60}/d_{10}$, i.e. by the relation of the sieve sizes through which 60 and 10 % (by weight) of the grains can pass. When all grain sizes are present in a suitable percentage, U should be more than 4 with gravel and more than 5—6 with sand. Another characteristic, which indicates whether the medium-sized grains are not lacking, is the coefficient of *curvature*. It is defined by the relationship: $d_{30}^2/(d_{60} \cdot d_{10})$. With a good gradation its magnitude should range between 1 and 3.

(c) *Strength of the grain material.* This physical property follows in the sequence of importance. Partial weathering of grains or weak grains in general, reduce the strength of soils, turning them into a rather contractant mass, and as the breakage of grains proceeds, large settlement is generated. The products of weathering are removed easily by water and the subsidence increases. The engineering geologist should not omit a study of the soil origin and should divide the particular profile into zones of different degrees of decomposition; accounts of the petrographic characteristics and the nature of weathering of the grains are also important.

(d) *Particle shape.* This may be studied simultaneously with (c) above. Rounded, bulky grains are less favourable for the development of the shear resistance of the soil than the angular and rather elongated ones. The presence of flaky mica grains usually increases the porosity of sands but reduces the permeability.

(e) *Macrostructure.* The large-scale structural nature of soil layer is of great practical significance and should be studied from the point of view of both the soil genesis and its technical consequences. Thus, for example, thin, loamy layers in a sand or gravel deposit diminish the vertical permeability. When the deposit is excavated as a fill material, the technique of excavation should ensure either the separation of the loamy material, which is difficult, or the satisfactory mixing with the material of the other layers.

(f) *Water content* and *unit weight.* The study of water content in pure cohesionless soils is generally unnecessary. The soil layer is situated either below the water table and consequently is almost saturated, or it lies above the water table where its moisture content changes with the overall conditions. Care is necessary with cohesionless soils containing either wet loamy ingredients or thin layers of wet loam. When constructing an earth dam, they can make the compaction of sandy gravel very difficult.

When the soil is saturated by a rising water table, water occupies about 85 % of the pore volume in sand and about 96 % in gravel. After the level of the water table has fallen, the degree of saturation S_r drops to 20 and 10 %, respectively.

For statical purposes the magnitude of the *unit weight* of the soil γ is needed. It is equal to $\gamma = (1 - n/100)\,\gamma_s + S_r\gamma_w n/100$, where γ_s is the unit weight of grains, and γ_w the unit weight of water. However, the soil is often situated below the water table and then the so-called *submerged unit weight* is considered. In technical practice this term indicates that portion of the unit weight which presses the grains together. This is obtained by subtracting the uplift from the weight of the grains. Therefore it represents that portion of the weight which is effective when determining the shear strength of soils. As outlined in Section 3.5, any notation associated with the effective portion of the acting forces is identified by a stroke. In our case: $\gamma' = (1 - n/100)(\gamma_s - \gamma_w)$, which follows from Archimedes law.

3.3.2. Physical properties of cohesive soils. (a) In principle, a clayey soil is the stronger the less water it contains. It is also denser and therefore its compressibility is smaller, as water molecules separate the clay particles. Therefore, the *water content* is considered to be the most important physical property of cohesive soils. Nevertheless, the effect of water is not the same in soils containing a smaller amount of clay particles (a "lean" clay) than in those exhibiting a large one (a "fat" clay). Therefore another criterion has been introduced, namely the *consistency*.

In relation to its water content, a soil may exhibit different states of consistency:

(1) The *solid* or *hard state*, when the soil is so dry that its colour has changed (generally to a lighter shade), and it cannot be broken into pieces. .

(2) The *semisolid*, or *very stiff* state, when the soil is evidently wet, but not wet enough to prevent crumbling when kneaded.

(3) The *plastic* state, when the soil can be kneaded. This state covers a broad range, and is further subdivided:

the *stiff-plastic* state, when kneading requires considerable exertion;

the weak or *soft-plastic* state, when kneading is easy;

the very soft or *pasty-plastic* state, when the soil can be squeezed out through the fingers of the clenched hand.

(4) The *liquid* state.

As previously stated, knowledge of the water content in a clayey soil does not provide sufficient information concerning its consistency. Therefore three numerical values are employed:

The *water content*, w, indicates the ratio (in %) of the weight of water present in the soil (determined by drying the soil to 105° C) to the dry weight of the soil.

The *plastic limit*, w_p, indicates the water content of the soil at the transition state when it passes from the plastic into the semisolid state. The definition of this boundary is arbitrary: the water content is determined when the soil begins to crumble when rolled into threads 3 mm thick.

The *liquid limit*, w_L, is a number which indicates the water content of the soil when it passes from the liquid into the plastic state. Again an arbitrary technique is used: a soil pat, placed in a dish, is divided by means of a grooving tool in two parts, which should make contact by the time the dish has been dropped twenty-five times. (Of course, an exact adherence to the specified techniques of this test is necessary.)

By a comparison of the water content of the studied soil with the two limits, its consistency can be established. Thus, for example, the London Clay (Eocene) has usually $w = 24$ to 34, $w_p = 24$ to 29 and $w_L = 43$ to 79 in field conditions, and the Neogene clay of the Carpathian Foredeep at Brno $w = 25$ to 28, $w_p = 25$, $w_L = 69$ (all quantities in %).

(b) However, the significance of the two *limits of consistency* (or Atterberg limits) is broader. The amount of water (the water content) necessary to transform the soil from the semisolid state (at w_p) to a liquid one (at w_L) can serve as an index of how much water the soil is capable of adsorbing. As this capacity is governed by the amount and petrographic character of the clay particles, as well as by the content of cations, the difference $I_p = w_L - w_p$ constitutes an important characteristic of the soil, called the *plasticity index*; it has become indispensable in geotechnique. A soil exhibiting a small value of I_p (less than about 15) will consolidate below the foundation within a relatively short time (Sections 3.2, 3.5, and 10.2). On the other hand, it tends to become slushy in the bottom of construction pits, for example. When used as a fill material, its compaction will be easier than that of a "fat" clay (which possesses a grater value of I_p). This is why the plasticity index serves as an index for classification of fine-grained soils (Section 3.9).

In order to take full advantage of this number from the geological point of view also, it is necessary to consider the two reasons for which the value of I_p may be low: the soil either contains a small amount of clay particles, or it contains a larger amount but with a lower capacity to bind water molecules.

(c) In order to answer the question why the value of I_p may be low, Skempton (1953) introduced the so-called *activity ratio*, which is equal to the ratio of I_p to the amount of particles smaller than 0.002 mm (the latter in per cent of the dry weight of the soil). Skempton established the activity ratio for several minerals, e.g.: quartz 0, calcite 0.18, muscovite 0.23, kaolinite 0.33, illite 0.92, Ca-montmorillonite 1.5, and Na-montmorillonite 7.2.

The quantity of particles smaller than 0.002 mm is therefore another important factor in the study of clayey-soil properties. Their relevance is far reaching: e.g. the magnitude of the so-called residual resistance of clays (Section 3.5) is governed by the amount of clay particles. However, one should not be misled into thinking that a particle size of 0.002 mm represents an absolute boundary line between clay and silt particles. The boundary is vague, but the value of 0.002 mm seems to be the most appropriate arbitrary limit and therefore should not be modified so that the wealth of experience is not lost. As in several cases of geotechnical practice the value of the activity ratio, determined by laboratory tests, seems to be unlikely, it is necessary to pay particular attention to the laboratory technique of the wet mechanical analysis.

Apparently the activity ratio has become a good link between genetic study and soil mechanics investigations. In several Neogene areas, for example, where sedimentation occurred under varying environmental conditions, as a rule influenced by a young volcanism, the activity ratio responds to the presence of montmorillonite. Weathering along joints, or remoulding along slip zones, etc. also influence the activity ratio. The study of the clay may be extended by chemical analyses, by differential thermal analysis and by X-ray analysis.

(d) Everyday practice, based on the experience of several generations, requires simple, visual arbitrary tests. The *crushing test* belongs in this category. A piece of a completely dry soil is broken and crumbled between the fingers. The observed strength is a measure of the character and quantity of the colloidal fraction contained in the soil. The dry strength increases with the plasticity index. Here and there this simple visual routine may occasion a surprise, if its results appear to be in contradiction with the magnitude of I_p. Thus, Mencl (1955) referred to a soil of Modrý Kámen (South Slovakia), which behaved as silty during the crushing test, although it exhibited a liquid limit of 58—72 and a plasticity index of 30—47. Similarly, Kazda (1963) studied clayey silty soils of an East-Slovakian reservoir site and found that the silty grains were in fact small clayey clusters, coated with amorphous silica. In the course of repeated kneading, which forms part of the consistency tests, both plasticity and liquid limits gradually increased.

(e) *Macrostructure* is another feature which should be studied with regard to mechanical properties. Several discontinuity planes were referred to in Section 3.2. But besides mechanical discontinuity, other factors may affect the mechanical behaviour of soils; e.g., water can penetrate along the joints deep into the soil mass and exert an uplift pressure which diminishes the stability. On the other hand a slope containing sets of dense joints can be drained in several ways.

Even when practical problems are concerned, the following questions arise: which macrostructural features were promoted by the original sedimentation (the primary structure), which belong to the secondary structure that develops in the process of diagenesis, and which are associated with the changes proceeding on the ground surface? Stratification is undoubtedly part of the primary structure. However, the differencies in the mineralogical character of clay particles may be of diagenetic origin, especially illitic clay is found to predominate, as this may be a final product of compression and dehydration processes.

Of large practical importance is the macrostructure of the *fissured clays* which are penetrated by a network of hair cracks usually oriented parallel and perpendicular to the bedding, although oblique directions are not infrequent. The compactness of this soil mass is thus seriously impaired. The origin of the cracks has not been explained in a satisfactory manner. The stiffer the clay, the more conspicuous is this phenomenon. The question arises, whether this structure is restricted to the surface layers or whether it continues into the depth. The observations made in the Neogene coal mines show that the cracks are also present at depth, the only difference being that they are closed. Near the ground surface they open which might be a result either of desiccation during Quaternary history or of shrinkage promoted by the increasing content of K-ions due to weathering.

(f) *Unit weight* and *water content*. In the statical analysis of large hydraulic structures founded on Tertiary clays the principal question is: Should it be assumed that the effective stress in clay is governed by the *uplift* deep below the valley bottom? First, it should be noted that a correct answer can be obtained only by measuring the *pore-water pressure* (Section 3.5). This has often yielded surprising results: e. g. the soil had not yet established its pore-water pressure in accordance with the actual height of overburden below a broad erosive valley with the result that the pore-water pressure exhibited negative values. Sometimes it was found that the soil was not fully saturated by water and that a large gas content (methane) was present, whose pressure was also to be regarded as detrimental to the strength. Investigation of this question requires, besides careful measurements, an account of the geological conditions. For example, in several cases clay is waterlogged by uplift arising from the underlying rock. This situation often occurs in the Neogene clay in the western section of the Carpathian Foredeep. The slip surfaces of the landslides extend to an unexpected depth (failure of the highway fill near Brno, Fig. 11-13). When the uplift is great and the thickness of the clay layer small, the clay may be much weakened, as appears in the denudation remnants of the Neogene clay seated in the underlying Cretaceous sandstone in East Bohemia.

When comparing the porosity with the unit weight it should be taken into account that the *specific weight* of clay particles is larger than that of sand grains. It is usually greater than 2.7 Mp/m^3, e. g. it equals 2.72 Mp/m^3 in the Neogene clay of the Carpathian Foredeep, where the porosity is about 43%. As the water content is between 25 and 28%, the degree of saturation is of the order of 98%.

(g) The presence of finely dispersed *organic matter* in the clay changes its colour to dark grey or black, and usually generates a conspicuous odour. The decrease in weight during annealing in the laboratory, as well as high Atterberg limits, are other characteristics. Clayey soils containing organic material are more compressible, but their shear strength is usually rather high. In several locations, e. g. in the flood-plain of the Bečva river in Moravia, the organic loam at the base of the Holocene deposits is very stiff due to the chemical compounds of the humic acid and calcium carbonate.

(h) *Calcium carbonate* content in a clayey soil decreases its plastic properties (Atterberg limits). The plasticity index is decreased to half with a carbonate content of 30 to 40% as compared with a corresponding clay free of carbonates. Soil slakes when wetted and disintegrates into small columns when desiccated, which helps the surface water to penetrate deep into the soil.

3.3.3. Rocks. As outlined in Section 3.2, mechanical properties of rocks are governed in general both by the strength of the rock material and by the geometry and other features of the discontinuity planes.

(a) *Strength of rock material.* Many testing methods have been developed in order to make a direct determination or correlative evaluation of the strength of rock samples. Several of the methods are even of arbitrary character. Indentation tests, rebound tests, bend tests and small-scale shear tests belong to this category. Several of them provide a good idea of the behaviour of the material under different conditions. Methods which make possible the study of the rock strength characteristics from point to point along the edges and on the corners of blocks are of particular value. Decrease in strength, produced by weathering or cataclastic weakening of rock at these particular spots has a large bearing on the resistance of the rock mass.

These characteristics can be satisfactorily studied by the application of the *Schmidt rebound hammer* (Fig. 3-6). The magnitude of the recoil of the hammer weight is dependent on the elastic properties of the rock material, provided that the impacts do not generate deformations along the joints. Therefore a light type of hammer is preferred (type *N*), with which the zone of influence is relatively small. The impact energy of this hammer is 2.2 kpcm. Because the elastic properties of the rock material correspond to the strength, the recoil value presents a good characteristic of the latter as well. A drawback to this method is that the sensitivity of recoil value is relatively small with a greater rock strength. Fortunately small magnitudes of strength are of first-rate interest. Up to a recoil value of 60 %, which corresponds to a compression strength of about 400 kp/cm^2, the relationship between the two parameters is usually rather linear. For the investigation of hard rocks a simultaneous application of both *N* and *L* hammer types (the impact energy of the latter is 7.5 kpcm) is expedient. Each new kind of rock requires

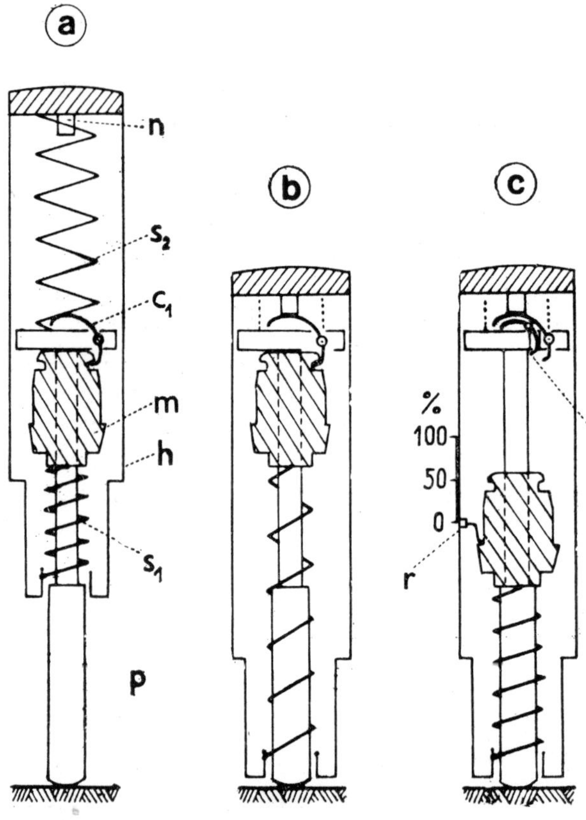

3-6. Principle of the Schmidt refraction hammer. When applying pressure to the hammer, the spring s_2 is contracted while s_1 is extended. As soon as the pin n bears on the pawl c_1, the mandrel m is released and, due to the tension of s_1, strikes the ending of piston p. The rebound displacement r of the mandrel is recorded on a sliding indicator. In order to repeat the test, the hammer is pressed until the pawl c_2, as well as the spring s_2, become free and the starting position of the hammer is obtained.

a new calibration of the hammer; this may be performed on rock fragments, which should be well fastened, and results are compared with those of the compression tests.

(b) *Water content* of the rock material and its influence on the strength. Usually the strength of wet rock is less than that of dry rock, a fact that can be explained by the action of pore-water pressure as well as by the effect of the water molecules on the bonds between the grains. Therefore a pronounced decrease in strength after saturation is observed with limestones and sandstones (often more than 25 %). As the absorption capacity increases due to weathering or other weakening, a tendency appears for a relationship to be established between the absorption capacity and the strength and compressibility of rock materials (Rocha et al., 1967). For this purpose, the *quality index*, $i = (P - p)/p$, has been introduced, P being the weight of the saturated sample and p that of the dried sample at 105° C.

(c) *Macrostructure*. In Section 3.2 several kinds of surfaces of weakness or discontinuity were referred to and their significance for the mechanical behaviour of rock emphasized. Representation of the geometrical pattern of the surfaces was discussed in Section 3.2. However, more information is needed for the purpose of statical analysis: The size and shape of blocks, the description of the rock material together with the characterization of the decrease of strength towards the edges and corners, geological evaluation of the surfaces, etc.

Following the tendency towards the establishing of numerical (index) values of rock properties as well as of their correlation with the mechanical properties, Stini (Müller, 1958) introduced the *density number k*, which indicates the number of joints corresponding to a length measured perpendicularly to the orientation of the joints. The inverse of k equals the average distance of the joints. As indicator of the continuity of joints Müller (1963) established the quantity $\varkappa$, denoting the ratio of the joint length to the total length of the rock exposure studied.

Moreover the width of the joints as well as the properties of their infilling material are to be registered according to the methods of petrography and soil mechanics. For this purpose samples are taken. The walls of joints should be carefully characterized as regards their coating, weakness, smoothness, etc.

(d) For the purpose of statical analysis knowledge of the elevation of the ground-water table is necessary.

3.3.4. Weak rocks. At the boundary between cohesive soils and hard rocks a group of weak rocks exists which gives rise to many difficulties at the sites. Definition of the term "weak rocks" is rather vague, the simplest being a rock with a compression strength of 500 kp/cm² or less. This definition does not illustrate the poor properties of these rocks. Various kinds of *claystones* and *marlstones*, *shales* and *argillites*, belong to this category. When wetted, they become soft, disintegrate, and exhibit large volume changes. Therefore any investigation should be adapted to this characteristic. Unfortunately, no uniform methods of investigation have been proposed. The usual measure of the intensity of disintegration in water, however reasonable it may be in principle, has the drawback that gravity also contributes to the destructive forces.

Petrographic analysis concentrates on the character of the bonds in the rock. When soluble salts are present, the problem is obvious. Much more intricate problems are presented by the following two cases: firstly, ionic carbonate or silicate bonds may interfere with the molecular clay bonds, as is shown, for example, in Cretaceous and Tertiary marls. Secondly, rocks containing clay have been compacted by mechanical forces and other diagenetic factors, so that clay minerals were transformed into chlorite and mica, as is shown with argillites of Carboniferous and Permian age.

Particle arrangement is another important feature. Often the rock contains very thin silty layers possessing a poor cementation, so that the disintegration is associated with suction of capillary water. The alignment of platy minerals results in the flat faces of particles being linked only by molecular bonds; this promotes disintegration after wetting. Even a higher carbonate content over 50 % will not guarantee permanence, as the distribution of carbonates and not only the quantity is important.

Two kinds of tests appear to be satisfactory for the practice of geotechnique:

Loss of strength in water. As relative loss is what matters, several arbitrary tests can be employed. With weak rocks, which often split into platy slabs, bending tests on plates are applicable.

The measurement of *swelling* and of the swelling pressure. Large swelling reveals predominance of clay particles.

3.4. Deformation properties of rocks. Recent methods of numerical analysis in the field of mechanics make it possible to follow the behaviour of structures and rock bodies continuously from the onset of loading until failure with a sufficient accuracy. In order to make these analyses possible, knowledge of the relation between loading and deformations is necessary, and its mathematical formulation must be established. This problem is the subject of this section. As the loading varies at different points on the body, it is first necessary to analyse the deformation behaviour of a small body.

3.4.1. Stress. In order to define the magnitude of stress at a studied point we consider it as an elementary force, which acts on a very small area with an equal intensity. This can be written: $\sigma = \lim F/A$ for $A \to 0$, where F is the force and A the area. Therefore stress has the dimension of force divided by area, e. g. kp/cm^2 or Mp/m^2 ($1\ kp/cm^2$ corresponds to $10\ Mp/m^2$). Sight should not be lost of the fact that in the field of engineering geology loadings are of a considerable size. A relatively small pressure of $2\ kp/cm^2$, as is used in the foundations resting on clay, is about six times greater than the pressure exerted by the sole of the human foot and corresponds to the pressure of a water column 20 m in height. Another question appearing in rock mechanics is whether it is possible to define the stress acting on any area at all, in view of the presence of smaller or larger voids. Engineering mechanics has settled this question in an arbitrary way, with regard to the requirements of practice, by defining the limit of magnitude up to which the area of a void should not be subtracted from the total area. With smaller voids the material is considered as quasi-continuous, while the existence of voids is reflected in the decreased values of the mechanical parameters of the material. The latter approach is therefore of rather phenomenological character, as it does not analyse stresses down to the minute structural details. This suggests that statical analysis of structural character should be advanced, and geological sciences above all are expected to encourage this.

In general, the direction of stress is oblique to the studied plane (Fig. 3-7a).

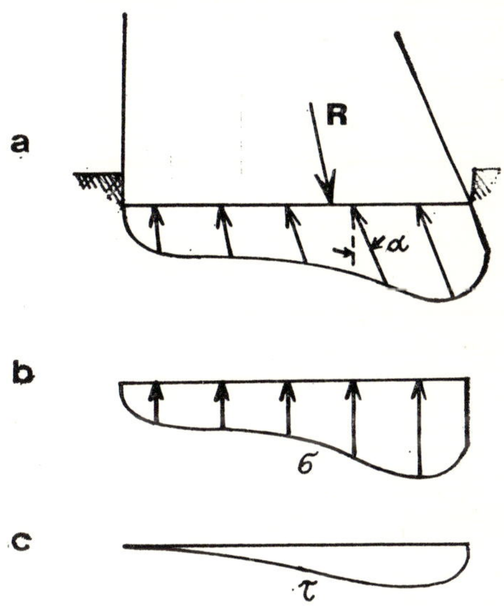

3-7. Contact stress in the foundation rock produced by the load R of a gravity dam. a — Total stress, b — normal component σ, c — tangential, shear component τ.

As with forces, we resolve the stress in its normal component (*normal stress*, σ) and tangential component (*tangential* or *shear stress*, τ). An example of a diagram of normal stress is plotted in Fig. 3-7b, and one of tangential stress in Fig. 3-7c. The magnitude of τ in the latter is plotted perpendicular to the studied plane. In contrast to the rules of classical mechanics, normal stress is denoted as positive if it is *compression*, and negative if *tension*, in the practice of soil- and rock mechanics. Nevertheless, in theoretical publications the rules of classic mechanics have been kept and with increasing application of recent computing methods they can be expected to prevail. This has the advantage that the sign of an increase in length or volume is positive.

With the development of computing methods the need for indicating the sign of the tangential stress also appears. A simple rule can be obtained from the conditions of equilibrium: Let us look normal to the studied plane, in the direction out from the studied part of the body. If this direction conforms with the positive direction of a co-ordinate axis, τ is positive, providing it has a direction identical with the positive direction of the other co-ordinate axis. If the normal to the studied plane is not parallel to any co-ordinate axis, the system of co-ordinate axes should be oriented by turning and the same rule as laid down above should be applied. As far as we shall adhere to the positive sign of the compressive stress, the best way of defining the sign of the tangential stress is by analyzing the orientation of the studied cross-section from its representation on the Mohr circle (see following).

Stresses acting on several planes of different orientations are usually studied at the same time and therefore an adequate notation is necessary. For example, a normal stress acting on the plane XY (Fig. 3-8) will be denoted σ_{zz}, as it acts

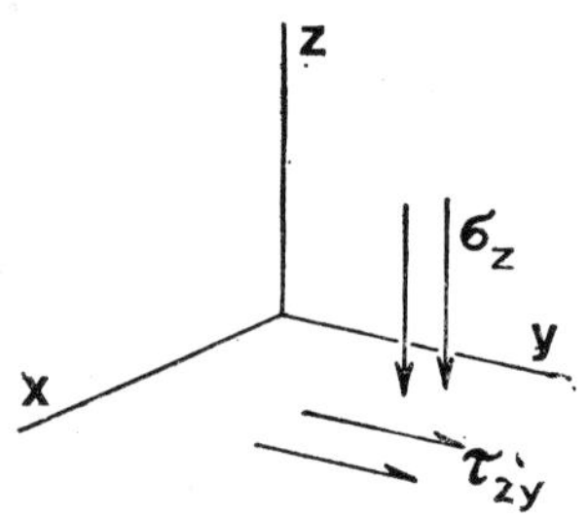

3-8. Systems of denotation of stress components; e. g. τ_{zy} denotes the tangential stress acting along a plane normal to *Z* axis in the direction of *Y* axis.

on a plane perpendicular to the *Z* axis, and has a direction parallel to the Z axis. Usually the more simple denotation σ_z is used. Similar simplification is not possible with τ_{zy} (it acts similarly on the plane normal to Z, but has the direction of Y).

Although the direction of stress acting on a plane (cross-section) in the body is generally oblique, three perpendicular planes exist at any point of the body along which the tangential stresses are absent. These are termed *principal planes*; normal stresses acting on them are called *principal stresses* and the directions of the latter are *principal axes*. In the sequence of their magnitude the principal stresses are called major principal stress (σ_1), medium principal stress (σ_2), and minor principal stress (σ_3). As σ_2 is only of secondary influence on the deformation and strength of the rock mass, many problems of geotechnique may be treated as two-dimensional, with σ_2 acting normal to the studied plane. Only a cross-section of the given body is then analyzed and the thickness normal to the studied plane should be 1 m.

One of the current analyses is to determine σ and τ acting on a plane if the stresses acting on other planes at the given point are known. Let us begin with the study of this analysis considering it as a two-dimensional problem. In a relatively small body a uniform stress field is assumed to exist. For example, when analyzing the results of an uni-axial compression test *(simple compression test)*, we assume, even if not quite correctly, that the stress conditions are identical over entire sample. Since it is known that the sample will be destroyed by slip fracture let us study the magnitude of σ and τ acting on a cross-section Π dipping at an angle β, e. g., for the sake of simplicity, with $\beta = 45°$.

If the forces acting on this plane must be determined (Fig. 3-9b), we resolve

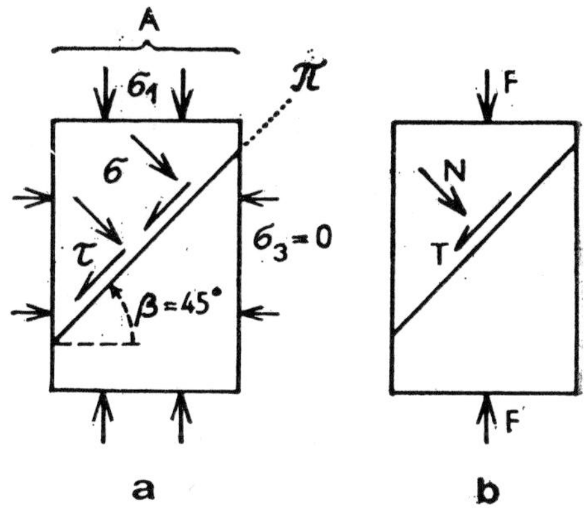

3-9. Resolution of a stress (Fig. a) or force (Fig. b) into the components which act on an oblique plane.

the force F into the components $N = F/\sqrt{2}$ and $T = F/\sqrt{2}$. However, the stress cannot be resolved in the same way. In the given case the area of the horizontal cross-section is A, but that along the plane Π is $A\sqrt{2}$. Yet $\sigma_1 = F/A$, and therefore $\sigma_{45} = N/(A\sqrt{2}) = \sigma_1/2$. Similarly $\tau_{45} = T/(A\sqrt{2}) = \sigma_1/2$. A similar analysis can be performed with other values of the angle β, only the relations become less simple, as sin β and cos β appear.

This is why Mohr (1914) transformed the mathematical solution into a simple graphical method, known as *Mohr's circle*. Its rules are simple: (a) Mohr's representation of stress is used (Fig. 3-10), i.e. all σ appear as abscissae and all τ as ordinates. (b) Mohr's circle K is drawn, delimited by the end points of the abscissae σ_1 and σ_3. (c) Co-ordinates σ and τ of each point of Mohr's circle represent the normal and tangential stress as acting on one of various planes Π. (d) In order to identify which point of the Mohr circle corresponds to a plane Π,

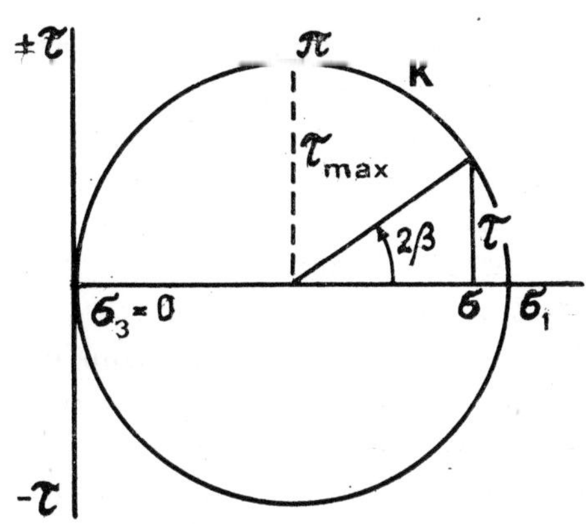

3-10. Representation of a plane stress state by means of a Mohr circle. The stress states correspond to the loading as shown in Fig. 3-9 (a).

it should be kept in mind that the angles on Mohr's circle are drawn with the same sense of rotation as in reality but with twice their magnitude. E.g., in order to find σ_{45} and τ_{45} acting on a plane inclined at $\beta = 45°$ (Fig. 3-9) we start from the point σ_1 (Fig. 3-10), which represents σ and τ (= 0) on the horizontal plane. Then we turn $2\beta = 90°$ anti-clockwise on Mohr's circle and obtain the point π. Its co-ordinates are $\sigma_{45} = \tau_{45} = \sigma_1/2$.

When plotting the Mohr circle it is not necessary to start with only the values σ_1 and σ_3; Mohr's circle is often defined by other points. However, the possibility of plotting the circle on the basis of the given points is an indicator of whether the state of stress has been defined by a sufficient number of quantities.

An analysis has been made for a small body exhibiting a uniform state of stress. A similar analysis can be made when analyzing the state of stress at a point of a large body in which the state of stress varies from point to point. In such a case the wedge shown in Fig. 3-9 is assumed to be cut out from the body at the studied point, so that the wedge is only imaginary and is assumed to be infinitely small. Nevertheless, the procedure for using Mohr's circle remains identical with that given above and provides the analysis of the state of stress at the given point. When the states of stress at other points of the body are known, the *net of trajectories of principal stresses* can be plotted (Fig. 3-11).

In the field of geotechnique the Mohr circle provides a good picture of the stress situation in the rock. It is evident that the further centre of the circle lies to the right in the σ-axis, the higher is the stress level of the body (in compression). The larger the diameter of Mohr's circle, the more dangerous is the shear stressing, as $\max \tau = (\sigma_1 - \sigma_3)/2$. In order to study the development of the stress situation in the course of a construction (e.g. of excavation), as well as during a laboratory test, it is useful to draw the Mohr circles for each step of loading, at least in an approximate fashion.

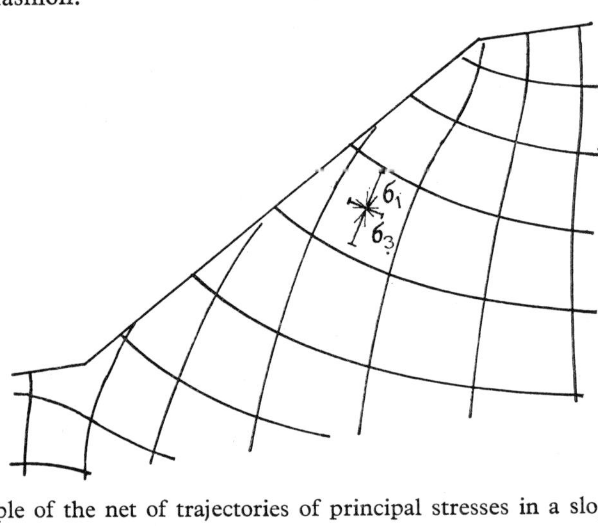

3-11. Example of the net of trajectories of principal stresses in a slope of a cutting.

3.4.2. Elastic deformation of rocks. Let us first observe the deformation-stress diagram of a uniformly stressed body. The diagram of a uni-axial compression test (simple compression test, unconfined compression test) on gneiss is presented in Fig. 3-12. ε_1 denotes the ratio $\Delta l_1/l_1$ and is called *normal*

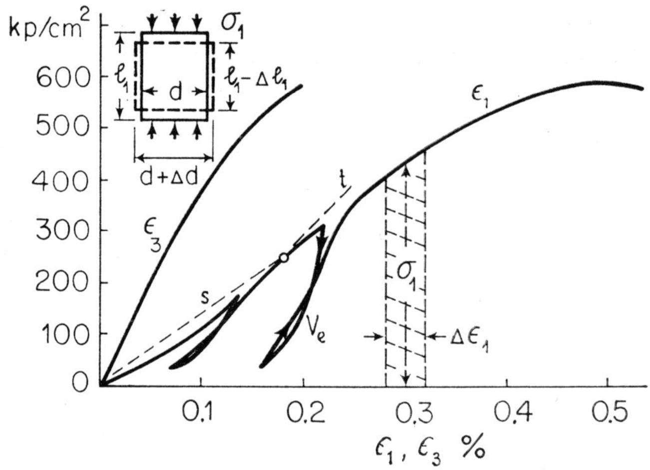

3-12. Stress versus deformation curve of an axially symmetrical compression test on gneiss. The stress-deformation ratio which defines the magnitude of the modulus of deformation, may be indicated either by the average, "secant" value (inclination of secant s) for a studied range of the curve, or by the tangential value (inclination of tangent t), applicable only for the particular stress state in question. The hatched area indicates the work introduced into the body when compressed by $\Delta\varepsilon_1$. The branch V_e indicates the elastic component of total deformation.

strain (here it is *contraction* as the length l_1 decreases; when the length increases it is *extension*). The behaviour of the material in this test is not quite *elastic*. The principal characteristic of an elastic deformation is that it is *reversible*, demonstrating that an amount of free, potential energy has been stored in the body, capable of producing work on unloading. That part of deformation which is *irreversible*, is called *plastic*. The ratio of elastic and plastic components of total deformation varies from rock to rock and also with varying stress conditions. E.g. with plate loading tests on a compact lydite the elastic deformation exceeded 80 % of the total deformation, even at a contact pressure as large as 200 kp/cm². Rock is compressed all around with this test, plastic deformations concentrating along the periphery of the loaded area. Conversely, with a dense soil, confined by all-around pressure, the elastic and plastic parts of the deformation exhibit the ratio of one to two even with small shear stressing.

Since plastic deformations are governed by principally different laws in their dependence upon the stress increments than the elastic ones are, it is necessary to separate the treating of elastic and plastic deformations whenever a high accuracy of results is required. As this procedure is intricate from the point of view of mathematical analysis, another approach is often used which integrates plastic deformations with the elastic ones, as long as the former are not too large. The so-called *pseudo-elastic analysis* is made in this way and will be treated in Section 3.4.3.

Fig. 3-12 shows that the elastic branches, e.g. denoted V_e, of the stress strain diagram, can be considered as straight lines parallel with one another. Then the laws of linear elasticity can be employed. This simplest and basic characterictic of elasticity makes the computations as easy as possible. For example, uniform parameters of a stress-strain relation may be applied with loading and unloading. However, recent statical analyses are not entirely dependent on the linear elasticity only, as the loading can be raised step by step. With each increment, new elastic parameters can be adopted. Nevertheless, as far as it is not quite incorrect, the possibility of application of the relations of linear elasticity will always be attractive.

The stress strain relation has the form $\varepsilon_1 = \sigma_1/E$, which is known as *Hooke's law*. The parameter E is called the *modulus of elasticity*. Its magnitude is, for example, about 2000 kp/cm^2 with well-compacted gravel, and even with rocks which are not extremely hard or compact it usually amounts to several tens of thousand kp/cm^2. For example E of a flysch sandstone with several joints is about 80,000 kp/cm^2. With hard rocks, as a gneiss, it attains values of several hundred thousand kp/cm^2.

When applying the relations of linear elasticity, the modulus of elasticity is treated as uniform throughout the entire range of the given loading. Its value is therefore a *secant value*, as the observance to the stress-strain diagram implies.

The compressed body in Fig. 3-12 also exhibits deformations in the orthogonal direction; it extends by $\Delta l_3 = \Delta d$. According to the foregoing statements, $\varepsilon_3 = \Delta l_3/l_3$ (where $l_3 = d$), and is negative with our system of signs. The relation $\varepsilon_3 = -\nu\varepsilon_1$ can be written and in this way another important parameter, *Poisson's ratio* ν, enters into the sphere of interest. Its magnitude stands only for the relationship between the magnitude of strain in the perpendicular direction and that in the direction of the applied stress. Therefore it is not produced by a result of the fact that the material simply deforms less or more. We can start with the evaluation of ν by considering that if the material was entirely uncompressible, the increase in volume of a cube would be nil so that $\varepsilon_1 = -(\varepsilon_2 + \varepsilon_3)$, as with water. As $\varepsilon_2 = \varepsilon_3$, with axially-symmetrical compression test, we obtain $\nu = 0.5$. Conversely with rocks which deform due to the decrease in volume of joints or pores, the magnitudes of ν are relatively low. They are about 0.13 with small loading and 0.2 after the voids have been partly closed. With repeated loading the values of ν increase. The presence of water in the voids also increases ν of rocks. When indicating the values of ν, it is necessary to note whether the changes in volume due to dilatancy have been included or not. As will be explained in the following subsections, the latter of the two alternatives is more realistic.

On the basis of the given relations and with the application of the superposition, Hooke's law for the loading in three directions can be written as:

$$\varepsilon_1 = (\sigma_1 - \nu\sigma_2 - \nu\sigma_3)/E$$

and by cyclic replacement the relations for ε_2 and ε_3 are found.

With the analyses of tests on rocks as well as with other analyses the *change in volume* due to loading should often be considered. The elastic relations give — as follows from the simple conception of the deformation of a cube:

$$\Delta V/V = \varepsilon_1 + \varepsilon_2 + \varepsilon_3 = (\sigma_1 + \sigma_2 + \sigma_3)(1 - 2\nu)/E$$

From the formulation of the foregoing relations it may be seen that the existence of σ_2 cannot be neglected in the analysis of deformations. Even if we do not work with σ_2 in any other way, a magnitude must be given to it. The majority of two-dimensional solutions in engineering geology have the character of *plane deformation*, i.e. no deformation can occur perpendicular to the studied profile. This assumption results in:

$$\varepsilon_2 = (\sigma_2 - \nu\sigma_1 - \nu\sigma_3)/E = 0$$

which yields:

$$\sigma_2 = \nu(\sigma_1 + \sigma_3)$$

How can E of a particular rock be established? In order to make a careful analysis of elastic and plastic deformations a large volume should be cut out of the rock and loaded. The volume should be of sufficient size to represent adequately the rock characteristics, including its macrostructure. Only with cohesive soils can the laboratory tests on samples be considered as representative, so that with rocks, and often with cohesive soils also, complex *field tests* are preferable. The term complex is used here to indicate that the straining of the considered body is not uniform. Two kinds of tests of this type have been employed in practice:

(a) Most widespread are the *loading tests*, which are executed on the bottom of a construction pit, in exploration shafts, in galleries etc. The tests in galleries are advantageous as the loading jacks can be propped against the roof or walls. With this simple arrangement the rock is loaded by means of a rigid plate and the final settlement (normal displacement) awaited after each load increment. In order to establish the elastic and plastic components of displacements, unloading and reloading follow after each load increment. An example of a result of a loading test is shown in Fig. 3-13.

To compute the modulus of elasticity, *Schleicher's formula* is used which indicates the compression value obtained when a plate is pressed into the surface of an elastic body:

$$w = mP(1 - \nu^2)/(E\sqrt{A})$$

where w is the displacement (m), P the total force of the jacks (Mp), E the modulus of deformation (Mp/m^2) and A the area of the plate (m^2). The factor m for a rigid circular plate equals 0.89.

As with the behaviour of rigid structures laid on compressible soil (Section 10.5) the *contact pressure* is larger along the circumference of the loaded area than at the centre under rigid testing plates as well. Therefore uniform contact pressure does not exist under the plate. Due to the great pressures along the circumference of the loaded area the plate is thrust into the surface of the rock and smaller values of E are obtained. Therefore testing technique bringing about a uniform

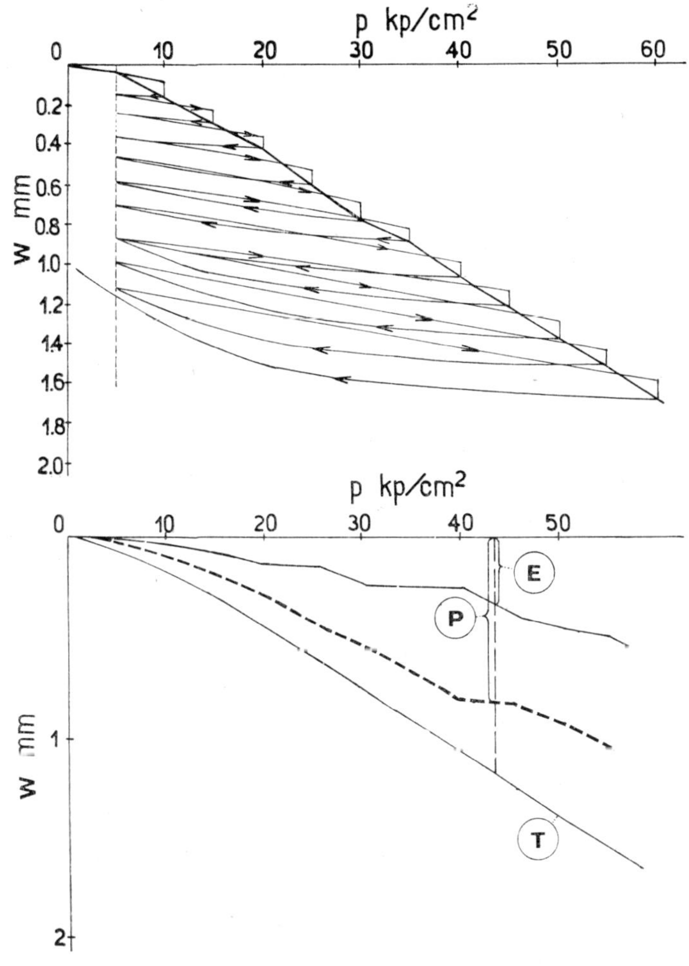

3-13. Field plate loading test diagram. The plate was unloaded after each loading stage. The values of the elastic parts (curve E), and of the plastic, irreversible parts (curve P) of the total displacement (T) can be obtained in this way. w — Vertical displacement of the loading plate, p — specific pressure on the rock.

distribution of normal pressure is preferred. This can be achieved by inserting a yielding flat jack (Freyssinet's flat jack) between the plate and the rock. The normal pressure is uniform, the middle point of the loaded area exhibits the settlement with $m = 1.13$, whereas that of the borderline is smaller, with $m =$

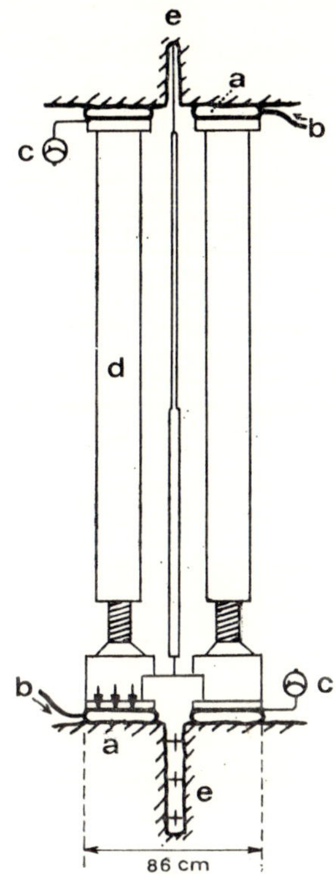

3-14. Plate loading test using an annulus-shaped plate (Shannon and Wilson, 1966). a — Freyssinet's cushion jack, b — supply of oil under pressure, c — pressure gauge, d — struts, e — recorders of the vertical rock displacements installed in the drill holes in the floor and in the roof.

$= 0.72$. Another method for eliminating the influence of local deformations on the contact between the loading plate and rock, as well for eliminating weakening of the rock near the surface, was introduced by Shannon and Wilson (in Deere et al., 1966). The loaded area has the shape of annulus and the vertical displacement is measured at several depths z under the centre (Fig. 3-14).

From the results the magnitude of E can be established, using a formula based on elastic solution.

The testing area should be investigated carefully from the point of view of geological structure and geotechnical index properties before the test is carried out. After the test, all changes in structure should be recorded.

(b) A relatively large volume of rock can be involved by using tests in which the periphery of a circular gallery is loaded. Since the deformation properties of rocks are of importance for the design of water-pressure tunnels (Section 12.8), a method was developed which simulates the action of water pressure on the rock. The theory of elasticity presents a simple formula:

$$\Delta a = (1 + \nu)\, ap/E,$$

where Δa is the elongation of radius a of the gallery and p is the pressure acting on the circumference. This formula is valid only when plane deformation can be considered. This can be satisfied when the tested segment is long enough; it should be at least five diameters of the gallery.

When the elastic components of deformation are considered, this formula can also be employed in the analysis of the so-called *pressuremeter tests* in boreholes. The Menard pressuremeter exerts pressure on the borehole walls by means of water pumped into a rubber cell inserted into the borehole. The change in borehole diameter is measured by means of the volume of the pumped water. This method is inexpensive. Eisenstein and Morrison (1973) presented an example of a successful application of this technique for the determination of the so-called *modulus of instantaneous deformation* of clay (Section 3.4.3).

Since the plate loading tests can be performed only on the surface of the rock, and since they are expensive, a tendency prevails to limit their number as much as possible. Therefore only several point values of E are obtained. These are completed by indirect tests. Thus, for example, the dependence of the recoil values of the large M type Schmidt hammer on the modulus of elasticity can be studied. Nevertheless, *seismic tests* are most widespread. They are based on the phenomenon that the velocity of seismic tremors in a rock varies in accordance with its physical properties. The elastic waves propagate from the point of impulse into the environment. If the studied zone lies deep enough below the ground surface, the surface waves are of less importance and two types of waves come into consideration: The longitudinal or compression waves (P waves) and the transverse or shear waves (S waves). The velocity of the P waves is 1.5 to 2 times greater than that of the S waves. Therefore both waves do not attain the registration point at the same time. The analysis of the equilibrium between the dynamic forces and the deformation forces acting on a volume unit yields the formula for the *velocity of waves*:

$$v_p = \sqrt{(1 - \nu)\, E_d/\varrho(1 + \nu)(1 - 2\nu)} \qquad \text{and} \qquad v_S = \sqrt{G_d/\varrho},$$

ϱ being the density of the rock (the unit weight divided by the gravity acceleration, therefore the dimension of ϱ is ps²/cm⁴ or rather Mp s²/m⁴). G_d is the *dynamical modulus of rigidity* the relation of which to E_d will be explained later. When measuring v_P and v_S both E_d and ν can be found. Small charges of explosive are therefore fired at different depths and the times of arrival of the waves are registered in the receiving holes. To measure both v_P and v_S, it is necessary that the receiver (geophone) is sensitive to the waves producing vibration in the direction of its axis only. Therefore several geophones are placed with different axial directions. The instant of the shot is marked on the graphs of the register by means of a circuit cable fastened to the cartridges.

The v_P values of dense gravel are of the order of 500 m/s, of dense flysch rocks 1500 m/s, and those of compact granite 5000 m/s.

The values of the *dynamic modulus of elasticity* E_d are not equal to the magnitude of E found by statical tests. They are greater, as fissures do not influence the seismic behaviour, in particular when water is present in the rock. Therefore the largest difference between E_d and E exists in rock containing joints, where E_d/E amounts to 6. Conversely, E_d/E is generally less than 2 with compact rocks.

Seismic investigation presents a good method to check the results of loading tests. A small weakening of rock during blasting is sufficient to reduce the results of loading tests to small values. These results then may be mistaken for the moduli of the entire rock mass. An example is shown in Fig. 3-15 of how an excessive

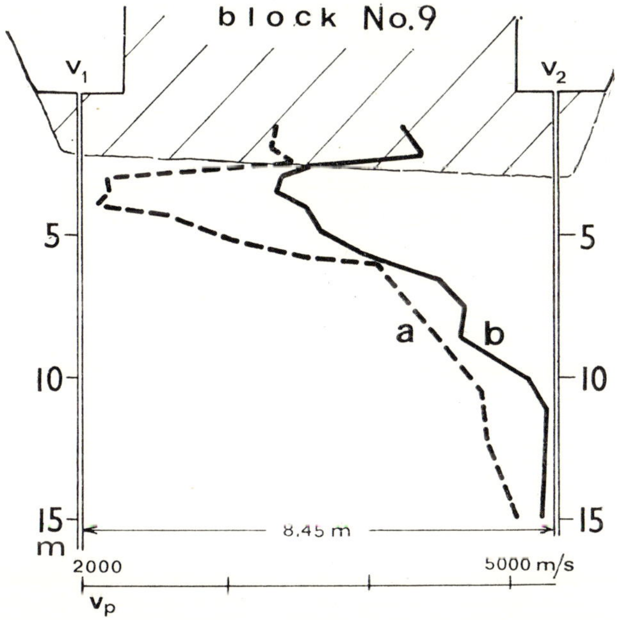

3-15. Results of seismic measurements in the gneiss below block 9 of the Vrchlice arch dam; V_1 and V_2 — drill holes between which the measurements were performed.

usage of explosives affected the v_P values of gneiss at the damsite of Vrchlice (Bohemian Massif) and how v_P increased with depth. Similarly weakened zones of 2—3 m in thickness develop at the periphery of investigation galleries as a result partly of blasting and partly of plastic yielding of the rock.

Until now we have attached the deformation of a loaded body to the directions of principal stresses. But these directions vary from point to point of the body and therefore they cannot serve as a basis of a co-ordinate system. Particularly for the purpose of the new computing methods the knowledge of stress-strain relations bearing on the conventional orthogonal co-ordinate system is needed. As its axes differ from the principal axes, a *shear stress* exists along

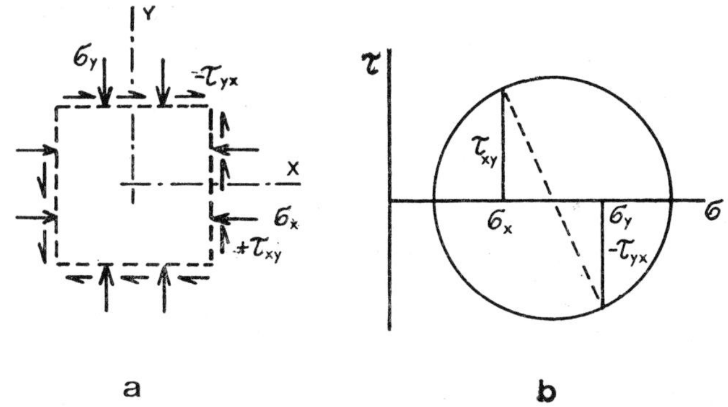

3-16. (a) Not only normal stresses σ_x and σ_y but also a shear stress $\tau_{xy} = \tau_{yx}$ exist in the planes parallel with axes X and Y of a loaded body. (b) Mohr's circle for the stress state shown in (a). Additional sign convention is necessary when employing Mohr's representation. Shear stresses as plotted in Fig. (a) should be taken as positive according to the general rule stated in Sect. 3.4.1. Nevertheless, shear stresses acting on the horizontal faces of the square are taken as negative, when plotting Mohr's circle, as they yield a clockwise moment (provided that compression stresses are taken as positive).

them (Fig. 3-16). The components of deformation produced by σ_x and σ_y can be computed in the same way as with principal stresses. But in addition the deformation caused by τ_{xy} is to be considered. How does the body deform when subjected to a *pure shear* $\tau = \tau_{xy} = \tau_{yx}$? The right angle between the axes X and Y changes by a shear strain γ (Fig. 3-17). Its value can be determined by transforming the analysis to deal with directions of the principal axes which are given by the diagonals of an imaginary square cut out of the body. Mohr's circle indicates that $\sigma_1 = \tau$ and $\sigma_3 = -\tau$. The length of the diagonals is $a\sqrt{2}$, the difference in the lengths produced by the action of both σ_1 and σ_3 stresses is $\varepsilon_1 a\sqrt{2} = -\varepsilon_3 a\sqrt{2} = a\sqrt{2}(1 + \nu)\,\tau/E$. The components of displacements of the corners of the square measured in the directions of the sides are $\pm a\sqrt{2}\,(1 + \nu)\,\tau/(E\sqrt{2}) = a(1 + \nu)\,\tau/E$. Since this quantity belongs to one half of the length of the side $a/2$, the shear strain corresponding to a unit length of the side is $\gamma = 2(1 + \nu)\,\tau/E$.

The term $E/2(1 + \nu)$ is called the *modulus of rigidity* and is denoted by G. It has the dimension of kp/cm^2 or Mp/m^2, as does E. By the outlined expression the dependence of the magnitudes of E, G and ν is established.

The largest shear strains occur along the planes of maximal values of τ, which make the angle of 45° with the directions of principal axes. It is easy to deduce that $\gamma_{max} = \varepsilon_1 - \varepsilon_3$.

As the plane deformation is considered, the studied body is stressed also by $\sigma_z = \nu(\sigma_x + \sigma_y)$ in the direction perpendicular to the studied plane. *Hooke's law* then yields:

$$\varepsilon_x = (\sigma_x - \nu^2\sigma_x - \nu^2\sigma_y - \nu\sigma_y)/E$$
$$\varepsilon_y = (\sigma_y - \nu^2\sigma_y - \nu^2\sigma_x - \nu\sigma_x)/E$$
$$\gamma_{xy} = \tau_{xy}/G = 2(1 + \nu)\,\tau_{xy}/E.$$

The values of stresses can be computed from the foregoing:

$$\sigma_x = [(1 - \nu)\,\varepsilon_x + \nu\varepsilon_y]\,E/(1 + \nu)(1 - 2\nu)$$
$$\sigma_y = [\nu\varepsilon_x + (1 - \nu)\,\varepsilon_y]\,E/(1 + \nu)(1 - 2\nu)$$
$$\tau_{xy} = (1/2 - \nu)\,\gamma_{xy}E/(1 + \nu)(1 - 2\nu).$$

For the purpose of recent computing methods the formulation of the relations in the *matrix form* is used:

$$\begin{pmatrix}\sigma_x\\ \sigma_y\\ \tau_{xy}\end{pmatrix} = E/(1+\nu)(1-2\nu)\begin{bmatrix}1-\nu & \nu & 0\\ \nu & 1-\nu & 0\\ 0 & 0 & \frac{1}{2}-\nu\end{bmatrix}\begin{Bmatrix}\varepsilon_x\\ \varepsilon_y\\ \gamma_{xy}\end{Bmatrix}.$$

The square matrix on the right side including the foregoing coefficients is denoted as the D matrix and contains parameters which are supplied by geotechnical investigations. It is the matrix of physical relations between stress and strain based on the laws of elasticity.

The expression for the *change of volume* as written earlier indicates that the

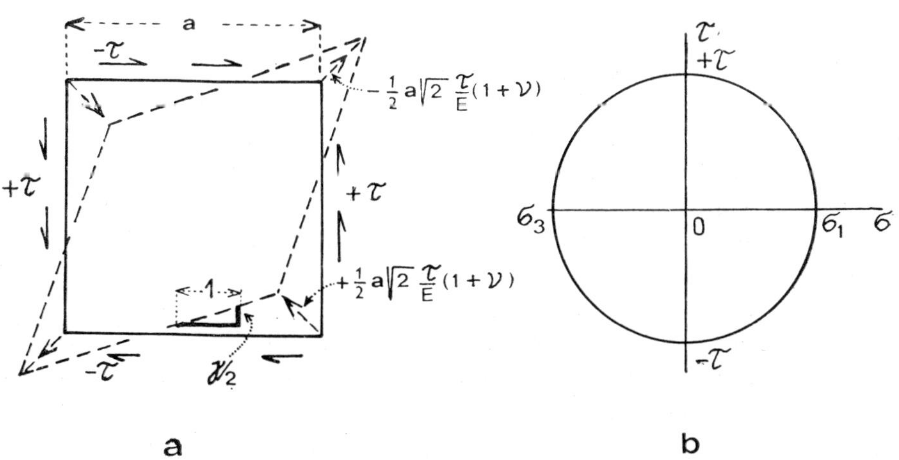

3-17. Determination of shear strain γ caused by a pure shear loading τ. a — Deformation of a small imaginary body cut from the loaded body, b — Mohr's circle.

volume should not change due to the action of the shear stress. Despite this a small decrease can be observed at the beginning of the shear tests (Fig. 3-19). This is partly due to the fact that the moduli of deformation with loading and unloading are not equal, in contrast to what is assumed when plotting Fig. 3-17. Another reason is that the postulate of objectivity is not respected with the block shear test, i.e. that the pure shear loading was not actually realized.

As mentioned, elastic deformations of rocks represent only a part of the total deformations, even with small loading. As the shear stressing increases, the *elastic part* decreases but does not disappear. The ratio of the *elastic* to plastic deformations is still of the order of 1 : 3.5 in dense sand with a factor of safety of 1.5, which is frequent when designing earth slopes. However, in clay this ratio is much more reduced, and is of the order of 1 : 6 under similar conditions. As will be outlined in Section 3.4.4, dilatancy, a phenomenon of great importance, is an elastic process to a considerable extent.

3.4.3. Pseudo-elastic solutions. Plastic deformations differ from elastic ones in that *slips* develop along planes more or less conform to the planes of maximum shear stresses. The shear strain, as studied in the preceding section, changes into slip along slip surfaces due to the increasing shear displacements. As will be seen in the following section, the rock volume may vary in the planes of slip movements but outside them it varies much less. For the purpose of design in the practice of geotechnique analyses are frequently applied with which the plastic components of deformation are associated with the elastic ones. The total deformations are therefore treated as if they were elastic, which makes the analysis less difficult. Attention must be given to the choice of magnitude of ν. This will be greater than with elastic deformations and will approach the limit of 0.5 since volume changes accompanying plastic deformations (outside the slip planes) are relatively small. By increasing ν the magnitude of G decreases. The consistency between E, G and ν must continue, as it is defined by the relations resulting from Fig. 3-17. Therefore, when slip deformations are simulated in pseudo-elastic analyses, decreased values of E should be applied in the slip zone.

The application of *pseudo-elastic analyses* will be the more justified the smaller are the plastic deformations (even if they are greater than the elastic ones). In addition it is desirable that the principal stresses increase continually without decreasing (the so-called monotonous loading in the free application of this term).

The first condition can be satisfied by limiting the degree of shear stressing up to which the analysis is appropriate (see Section 3.4.4). For the sake of clarity only, let us note that with rather dense rock loading should not exceed the state at which the shear stressing of rock amounts to about half of the shear strength. This stressing is not usually reached under the foundations of structures, since it is there exceeded under the borders only, but it is exceeded when designing earth slopes.

The second condition, that the principal stress increases continually without decreasing, is affected by the type of the problem studied. Again, it can be satisfied when computing the settlement of the foundation soil under a structure or a fill, but it can scarcely be considered as satisfied in a rock near a tunnel. The normal stresses increase in the directions tangential to the tunnel opening but decrease considerably in the normal directions (see Fig. 12-17). If elastic analyses are to be employed despite this, it is necessary to adjust the values of E, G and ν. As the stresses will change during the excavation of the opening, so the plastic deformations will vary and with them the *pseudo-elastic parameters* E, G and ν. Therefore the analysis should be performed in steps, for example by simulating the progress of excavation, and at each step attributing new magnitudes of E, G and ν to the actual stress state. The values of E, G and ν are, therefore, the tangent values, as they are defined by the local inclination of the stress-strain curve (Fig. 3-12). The increments in stress and strain against the preceding step appear in the stress-strain relations instead of total values.

As will be analyzed in the following sections, the difference between the elastic and plastic deformations is of fundamental importance and cannot be encompassed only by introducing varying magnitudes of E, G and ν, i.e. by applying the relations of *nonlinear elasticity*. Despite this, the incremental pseudo-elastic computations yield results approaching reality and have a widespread application in geotechnique. How the volume changes due to dilatancy are introduced will also be discussed in the following sections.

The parameters of rock deformation in pseudo-elastic analyses are usually determined by using identical testing techniques such as are described in Section 3.4.2, with the only difference that total deformations are studied instead of elastic components only. The *modulus of deformation* E_0 appears instead of the modulus of elasticity E; of course, the magnitudes of E_0 are smaller than those of E. For example, it is approximately 750 kp/cm^2 with a well-compacted gravel, approximately 2000 kp/cm^2 with a dense terrace gravel, or approximately 50,000 kp/cm^2 with a less jointed sandstone. E_0 increases to over 100,000 kp/cm^2 with sound crystalline rocks containing only closed fissures. Nevertheless, it would be wrong to expect that the values of E_0 of different rocks could be compiled in tables. An example of results of plate loading test is instructive: A jointed amphibolite loaded by a rigid plate 70 cm in diameter exhibits a settlement of $w = 0.15$ mm under a pressure $p = 13$ kp/cm^2. The latter corresponds to a compression under the downstream foot of a gravity dam about 65 m in height. The magnitude of $\nu = {}^1/_5$ can be estimated by experience since it is not of great significance for the result. Schleicher's formula (Section 3.4.2) gives $E_0 = 46\,000$ kp/cm^2. But if only one joint present in the rock should close by 1 mm after loading, w would increase to 1.15 mm and a value of $E_0 = 7\,700$ kp/cm^2 would be obtained.

Therefore the magnitude of E_0 depends not only on the kind of rock but also on the state of its jointing, the degree of weathering of its edges, etc. The applicability of results of the expensive field loading tests is therefore limited and the importance of experience increases. Often the sound judgement of an experienced engineering geologist may be nearer reality than the results of tests examined uncritically. The most valuable are the magnitudes of E_0 found from the recalculation of measured settlements of structures.

Experience has also shown that the moduli of deformation of bedded or foliated rocks are not isotropic. They are frequently smaller in the direction perpendicular to the foliation planes, etc., than in the parallel direction.

The results of *geophysical exploration* (Section 3.4.2) are also applied for the estimation of the magnitudes of E_0. Seismic tests make the generalization of the results of point tests possible. E.g. when studying a valley profile estimations may be made of the differences between the values of E_0 of the upper zones of slopes, those of subsequently denudated and therefore much less weathered rock of the lower zones, those of the rock confined below the valley bottom, and those of the fault zones. However not even the results of seismic tests can be applied without adjustment. The increased value ν is introduced when great seismic velocities are measured. The magnitude of the dynamic modulus E_d can be established, but its relation to E_0, i.e. E_d/E and E/E_0, is not constant. As noted in Section 3.4.2, the ratio E_d/E may amount to 6 with jointed rocks. Since E/E_0 is about 1.8, E_d/E_0 may achieve a value of 10. On the other hand, compact rocks exhibit E_d/E of about 2 and E/E_0 of about 1.5, which yields a value for E_d/E_0 of about 3 to 3.5.

The tendency for these relations to worsen where weak rocks are concerned is apparent from the results measured by Kujundžič et al. (1970, Fig. 3-18).

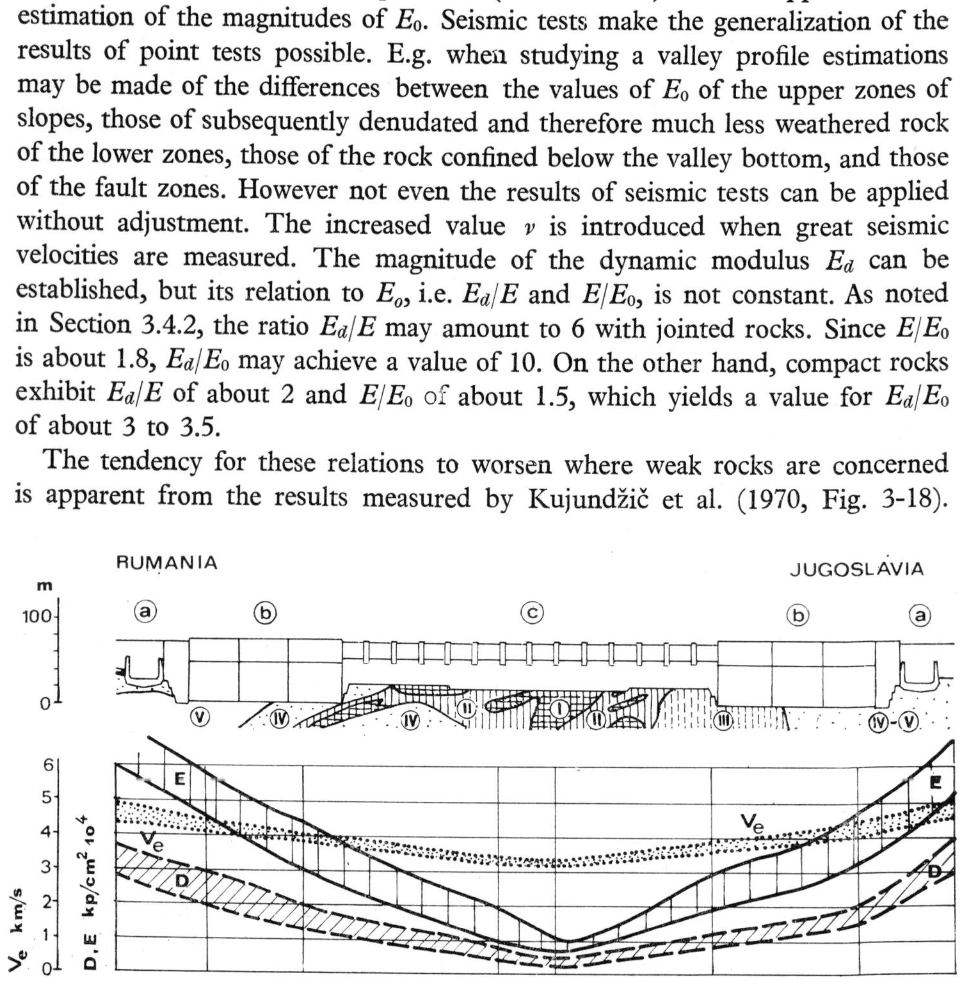

3-18. Geotechnical longitudinal profile of waterwork Derdap on the Danube. a — Lock chambers, b — power stations, c — weir, I—V — metamorphic schists divided according to the velocities of elastic waves (V_e), moduli of elasticity (E) and moduli of deformation (D). (After Kujundžič et al., 1970.)

When calculating E_d from the presented values of velocity, we obtain for a rather tectonically weakened gneiss (velocity 4.2 km/s) $E_d = 400\,000$ kp/cm². The modulus of elasticity established by direct measurement was $E = 53{,}000$ kp/cm² and the modulus of deformation $E_0 = D = 25{,}000$ kp/cm². Therefore E_d/E was 7.5 and E/E_0 was 2. As seen from Fig. 3—18, the ratios with more tectonically weakened rocks are so great that the analysis becomes meaningless. Relatively large velocities and the resulting values of dynamic moduli should probably be ascribed to the fact that the properties of rocks improve rapidly with depth. It is apparent that the values of E_d/E will exceed unity the less, the shorter is the spacing between the source and the recorder.

Deere et al. (1966) interpreted the values of E_d of unweathered rocks by applying the *modified core recovery ratio* as an index of the degree of jointing. The percentage of the total sum of the lengths of cores to the corresponding length of the borehole is established. A careful diamond drilling is necessary and only core pieces longer than 10 cm are considered (or shorter than 10 cm only when broken during drilling). The results can serve as an index number of the quality of the rock mass as well as an indicator of the ratio E_0/E_d. It is evident that E_0/E_d increases very slowly as the index value increases to 20 per cent, and only after attaining magnitudes of 50 % is the increase in E_0/E_d considerable.

With regard to the above discussion, two kinds of variations in the magnitudes of deformation parameters can be incorporated into pseudo-elastic solutions. Firstly, they vary from one location to another, and secondly they change with the state of stress. For example E_0 increases with $\sigma_1 + \sigma_3$ and decreases with $\sigma_1 - \sigma_3$ or with σ_1/σ_3. With soils these relations can be established by means of laboratory tests. Since the classical *triaxial tests* indicate exaggerated deformations, a more exact determination of deformation parameters is preferred by testing cubical samples in special triaxial cells (Ko and Scott, 1967).

Clayey soils exhibit a typical increase in E_0 with depth. The Neogene marl of the Brno basin at the site of the thermal power station can be characterized as follows: Under the cover of 9.5 m of alluvial sediments the first layer of 50 cm is partially loosened, probably due to the relief of horizontal stress, and its $E_0 = 50$ kp/cm². Yet at a depth of 11 m $E_0 = 70$ kp/cm², at a depth of 15 m $E_0 = 120$ kp/cm², at 20 m $E_0 = 160$ kp/cm², and at 60 m $E_0 = 250$ kp/cm². These magnitudes of the modulus of deformation are valid under the conditions of full *consolidation*, i.e. when the pore water is pressed out from the soil during contraction. When consolidation is not possible, e.g. with a short-term loading, the values of E_0 are about three times greater (the so-called *instantaneous modulus of deformation*).

The magnitudes of ν for the purpose of pseudo-elastic analyses are often obtained by estimation. With rocks magnitudes of 1/7 to 1/5 are frequent; with soils the rule is employed that ν is the greater the more the contraction of soil is associated with a lateral expansion. Thus ν of loose sand is about 0.3, that of dense sand 0.4, and with clay about 0.35, when the process of consolidation prevails, but almost 0.5 when slip deformation is to be considered. The magnitude of 0.5 must not be introduced into the *D*-matrix, since this would result in mathematical disorder. Therefore only magnitudes smaller than 0.48 are used.

For the same reason the increase in volume due to dilatancy cannot be simulated by increasing ν over 0.5. It is better to consider the value of ν as unaffected by dilatancy and to introduce the latter in a separate way.

3.4.4. Changes in structure and in mechanical properties due to increasing shear stressing. Rigid shear deformation of rocks. In examining diagrams of tests, e.g. as in Fig. 3-12, we find that the stress-deformation curve bends as the stressing increases and the rock begins to behave as if it were more deformable. This contrasts with the fact that the rock is still capable of resisting increasing loads in this domain until maximal resistance (the strength) is reached. This section of the stress-strain curve is called the *strain-hardening range* and material exhibiting this property is called strain-hardening or tough material. Rocks are strain-hardening materials. But irreversible plastic deformations predominate over elastic ones and changes in the rock structure appear. In order to follow the tendency emphasized in the introduction to the present chapter, the engineering geologist should devote his attention to structural changes before analyzing the mechanical consequences.

Nevertheless, before dealing with these questions let us first define the different *mode of failure* occurring in rocks:

(a) *Extension fracture*, i.e. fracture resulting from increasing extension.

A most distinct extension fracture occurs under the action of tensile stress and can be called *tensile fracture*. In the field of engineering geology this type of fracture is rather rare. We meet it, for example, in the roofs of tunnels or in areas near deep excavations in clays or rocks. Another example is tensile cracks which appear in samples subjected to laboratory compression tests when lubricant used for the purpose of decreasing friction on the faces of cylindrical samples intrudes into the pores of the rock.

Less evident is the extension fracture when the external stresses are compressions only. Two cases can occur in geotechnique: First, when extension fracture is associated with the development of internal tensile stresses promoted by the shape *(boundary conditions)* of the compressed body. Vertical cracks appearing on samples compressed uniaxially ($\sigma_1 > \emptyset$, $\sigma_2 = \sigma_3 = \emptyset$) or triaxially ($\sigma_1 > \sigma_2 = = \sigma_3 > 0$) when the confining stress σ_3 is small, belong into this category. The barrel-shaped deformation of cylindrical samples (if the faces have been clasped by friction) generates vertical extension cracks. Even when a shear failure follows, it is combined with the extension fracture.

The second case is connected with the *non-homogeneity of rocks*, since they contain pores, fissures and weak grains. The development of stress conditions in the vicinity of these irregularities can be illustrated by considering the stress state generating in the roof of an underground opening (see Section 12.6). Tensile stress develops in the roof as well as in the floor during the excavation of an opening in a rock stressed by the weight of overburden. When testing a rock sample, small pores, fissures or grains result in a similar stress state. New cracks generate in the areas of tension and they conjoin when compression of the sample increases. The sample is subjected to a compression, but fails due to extension cracks which are parallel to the direction of σ_1. Therefore this type of failure occurs even when the sample does not deform in a barrel-shaped manner and is therefore common when the compressed faces of the sample are arranged so as to be free from friction. In the practice of geotechnique tunnel walls are endangered by this type of failure which manifests itself by spalling. A special case of this failure was analyzed by Griffith (1920) after whom it also obtained its name.

(b) *Shear failure* or *slip failure* contrasts from the preceding types in that the deformation develops from a considerable shear strain into a slip failure along an inclined plane. Two mechanisms of shear failure may be distinguished:

Brittle shear failure or *shear fracture*, with which only one or a small number of thin outstanding *slip surfaces* develop (Fig. 3-19a).

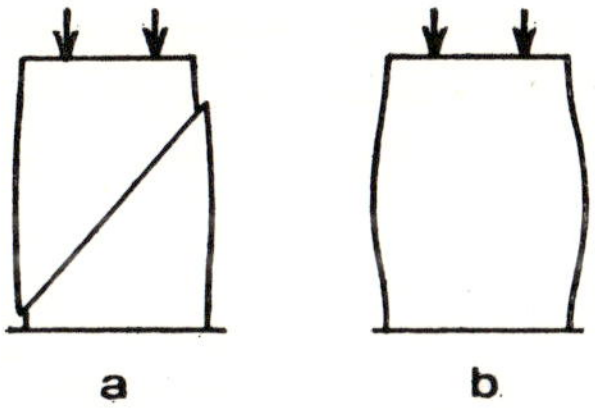

3-19. Shear fracture (a) and ductile shear failure (b).

Ductile shear failure (Fig. 3-19b) with which a system of a very large number of rather invisible slip surfaces develops, forming a *slip zone*.

The phenomenological comparison of the latter mechanism with the deformation of metals sometimes led to the idea that rocks subjected to large confining pressures can behave as plastic materials, for example mild steel. Yet the application of term plastic is wrong in the given example as the term plastic deformation stands for an irreversible deformation apart from whether the substance is ductile or brittle. Moreover the term plastic deformation as used in the past assumed a large deformation free from any weakening of the structure of the material. But rocks (with a partial exception of soils) undergo a weakening of the internal structure even if deformed in a ductile manner; this requires a more detailed discussion.

A large difference exists between small plastic deformation of rocks and metals. The physical nature of plastic yield seen in mild steel is understood. The cubic lattice of its single crystals as well as the configuration of free electrons allow large deformations owing partly to twinning and partly to gliding. During gliding a layer of atoms breaks its bonds with the adjacent atoms and glide into a new position where bonds form with the next atoms. This is the reason why a mild steel bar can be bent and re-straightened.

Nevertheless, the explanation of the phenomenon of strain-hardening is rather complicated; the force capable of moving an atom across the atomic distance would be extremely large. Therefore shorter atomic distances are suggested to exist and this is substantiated by the existence of inherent imperfections called *dislocations* of the crystal lattice. In dislocations the distances between the atoms are smaller, so that relatively small forces are sufficient to produce gliding and the deformation occurs without the internal bonds being disturbed. But as deformation increases due to increasing stresses, this type of gliding is hindered, because dislocations developed during the crystallization of the crystal mass. Dislocations therefore have limited areal dimensions. Moreover the slip runs out of the single crystal and meets with a more rigid intercrystalline material. As a result, larger forces are necessary to produce further deformation which manifests itself in strain-hardening. Near the strength limit the bonds begin to suffer, first in the intercrystalline material, and the resistance of the material begins to achieve its maximum. After this a decrease in resistance follows and a continuous slip surface develops.

The *ductility of wet clay* has its origin in the secondary, molecular bonds between the particles, if no cementation exists.

However, neither of these two deformable bonds exists in rock material where rigid ionic bonds prevail. A short time after the first bend of the stress-strain curve is reached a deformation without damage of structure probably occurs due to twinning. But only rock salt, mica, and perhaps also calcite, exhibit somewhat greater deformability without suffering damage. Therefore the toughness of a rock is limited although differencies between particular rocks are of significance. Tunnel builders may witness that the toughness of limestone is greater than that of granite, or in other words that the limestone has a larger "plastic reserve". But in general, plastic deformations of rocks in the domain of strain-hardening are associated with *cataclastic* damage of structure and the terms pseudoductile or cataclastic ductile are probably suitable.

The initiation of crack formation can be shown by several direct or indirect laboratory tests:

Direct observation of *microfissures*. Schlosser (1967) inspected polished surfaces of cubes of Carrara marble which were subjected to uniaxial compression. The first phenomenon of gliding within the crystal lattice, and therefore of rearrangement of the atoms, appeared at a loading equal to 30—40 % of the rock strength. True microfissures appeared after a loading of the order of 50—60 % of the rock strength and were of intergranular character. Contrary to the former they could not be effaced by polishing.

X-ray observations. Gramberg (1963) found that with a compressed marble the first cataclastic features had the character of extension fractures combined with short cataclastic slip surfaces. On the other hand only extension fractures appeared with the primary disturbance of granite.

Exhibition of a mechanical phenomenon which may be called the *Bauschinger effect* in a very free application of this term: Samples of dry claystone were exposed to different degrees of loading in the tri-axial cell, not all to the limit of failure. After this they were removed from the cell and additionally exposed to a different type of loading, e.g. to a simple compression. It was observed that the compression strength decreased due to preliminary tri-axial loading (Mencl, 1965).

The increase in absorption of fluids or in *permeability to air* (Habib, 1968).

By the changes in the propagation of *microseismic* and *ultra-sonic waves.*

Until now only the behaviour of the rock material was discussed. But similar characteristics are also manifested by the rock mass. Under small loading the deformation occurs mainly as a result of slipping along the existing surfaces (Fig. 3-20). As loads increase, slipping leads to a subtle interlocking, which

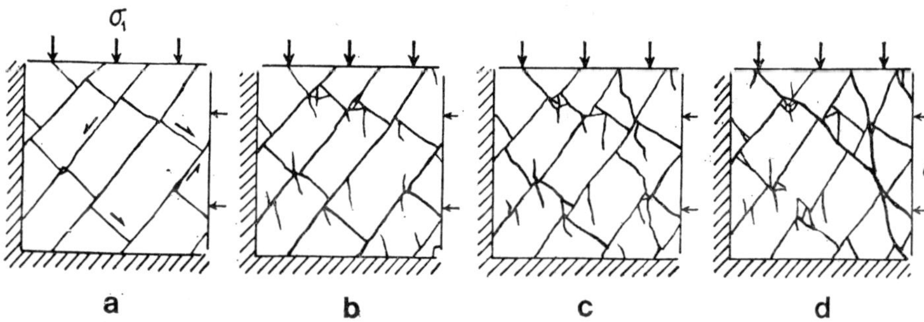

3-20. Progressive failure of a rock mass in the course of loading and the development of a slip surface. a — Up to a loading of 30 to 40 % of the total strength the blocks slip along the existing surfaces of weakness, b — up to 50 to 60 % of the total strength extension cracks develop, c — in the range of strain hardening partial slip surfaces begin to develop, which do not join, even when the strength of the mass has been achieved, d — only at the stage of "residual resistance" does a continuous slip surface develop.

together with natural interlocking contributes to strain-hardening. Cracks gradually appear, namely at the points of interlocking and in asperities; extension fractures also develop in blocks in the direction parallel to the direction of σ_1 (Fig. 3-20b). A spreading of the cracks follows and they join by means of oblique planes until the resistance of the rock is overcome (Fig. 3-20c). Near the stage of the so-called residual resistance a continuous slip surface develops (Fig. 3-20d).

Cataclastic deformation corresponding to stages (c) and (d) therefore induces a non-homogeneous process which results in a distinct slip surface (Fig. 19-a). The inclination of the latter is usually steeper than 45°.

The remainder of this subsection is restricted to the problems of *brittle shear fracture* only. Another phenomenon of practical importance appears which can also be associated with the progress of disturbance of the rock structure. This is the phenomenon of *dilatancy*. This term denotes an increase in volume associated with slip deformations, and was found in several internally structured materials. Therefore an increase in volume resulting from impending extension fracture cannot be regarded as dilatancy, a fact often neglected when analyzing test results. The best manifestation of dilatancy appears with shear tests made both on samples of the rock material and on large blocks of the rock (Fig. 3-21). The volume decreases during the first loading with these tests (curve *V*), a result of the fact

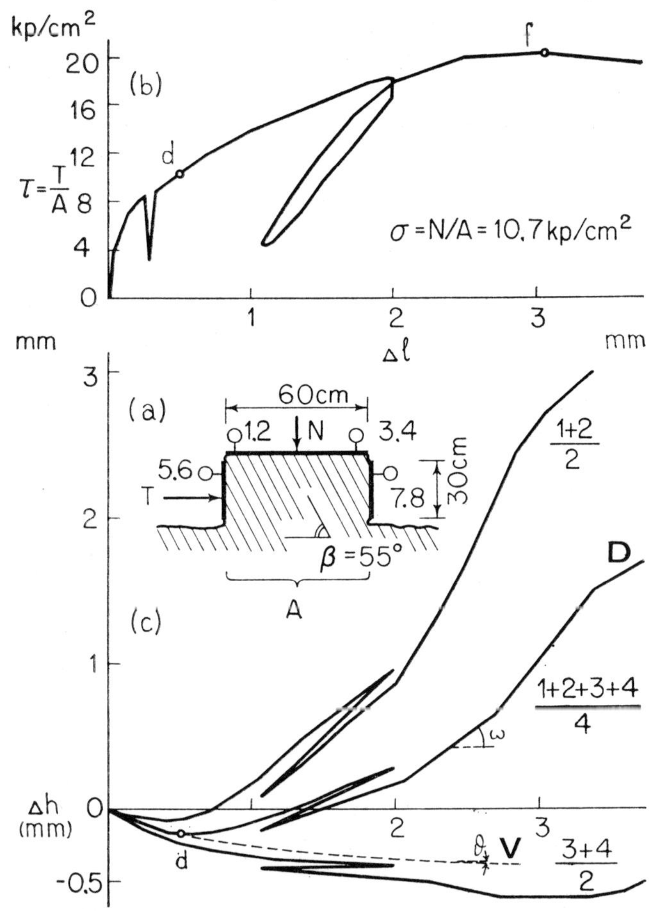

3-21. Diagram of a field block shear test on gneiss. a — Arrangement of the test, b — loading versus displacement diagram, c — increase in the height of the tested rock block begins to appear since the loading corresponding to point d has been achieved. 1, 2, 3, 4 — Dial gauges measuring the increase in the height, 5, 6, 7, 8 — dial gauges measuring the displacements of block.

that σ_1 increases while σ_3 decreases after applying tangential loading. The increase in σ_1 is accompanied by a greater reduction in volume than is the increase in volume owing to the decrease in σ_3. But starting from the point *d* (Fig. 3-21b) which is not far from the beginning of the strain-hardening section of the stress-deformation curve, the volume begins to increase. The curve *D* begins to veer off from the *V*-curve, indicating dilatancy.

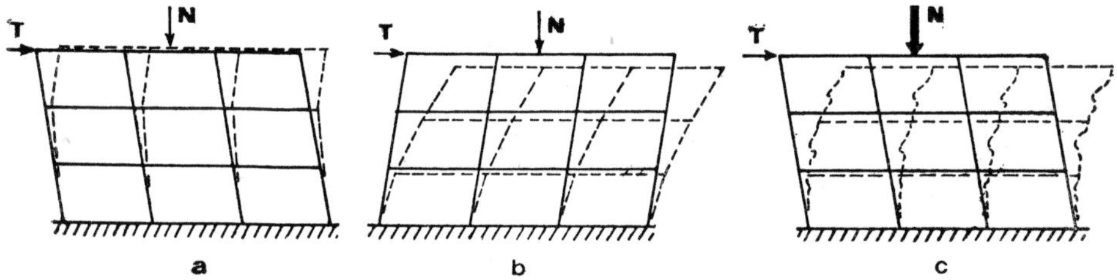

3-22. Symbolic synoptical model (a) of a rock block which increases in volume when sheared, (b) of a rock block which decreases in volume when sheared and (c) of a rock block which increases in volume when sheared under a small normal stress, but decreases in volume when sheared under a large normal stress.

The physical background of dilatancy is to be seen in the shifting of the asperities and interlockings, and in tilting of particles or blocks during the slip deformations along the slip surface. The phenomenon was first studied with dense sand.

Which is the practical importance of dilatancy? First, its development can be regarded as the onset of disturbance of the rock structure. Therefore the point *d* is of more physical importance than the *yield point*, which is defined by the first, often rather indistinct, bend of the stress-deformation curve (Fig. 3-21b). As far as the loading does not exceed the point *d*, preservation of the intact structure of the rock may be assumed. Therefore the stress at the *dilatancy limit* τ_d can be considered as an important parameter in the geotechnical analyses.

Work must be done during the separation due to dilatancy of the two parts of the rock mass, as the point of application of the *N*-force is displaced in a direction opposite to the direction of this force. This work is done by the *T*-force during the shear deformation. Therefore with dilatant materials the shear resistance is increased by this component, a fact which is reflected in the magnitude of the shear strength angle. The potential, free energy of the system is increased by this work. It is the potential energy in the gravity field which is concerned in the majority of geotechnical problems. These facts are of great practical relevance, as can be shown by observing the synoptical symbolic model plotted in Fig. 3-22(a). Due to the application of the *T*-force to the given system columns of all stories deform elastically, but only in one story they deform in a plastic way, as the work necessary to elevate the point of application of the *N*-force should be as small as possible. This yields the minimal increment in free, potential energy. An important conclusion follows: When the rock behaves in a dilatant manner, the *slip surface* develops as thinly as is possible, and the thickness is of a fraction of millimetre with clay, of several grains with sand and of several blocks with a rock mass.

Due to the onset of structural disturbance during the development of dilatancy the slip surface begins to form. Therefore the orientation of the slip surface is partially predestinated before the strength of the rock is reached. Each increment in loading takes advantage of this internal structural weakening and in this way an *induced anisotropy* develops. The geometry of large slip deformations may be analysed within the *theory of plastic flow*, one of the fundamentals of the theory of plasticity. There is no doubt that the basis of these analyses is to be found in the change in structure of the material. It is evident that dilatancy is an important factor in the relationship of increments of principal strains ε_1, ε_2, and ε_3, or of strains ε_x, ε_y, γ_{xy}, if general co-ordinates are considered. On the basis of strain increments the mathematical relations of the plastic flow can be formulated. How intensive the contribution of dilatancy is can be seen from Fig. 3-21c. The increase in the block height is equal almost to the displacement in the horizontal direction. Nevertheless, as the principles of plastic flow relate only to the plastic part of deformation, it is necessary to recognize which portion of *dilatancy* is *recoverable* and which is not.

While the decrease in volume according to the *V* line relates to the entire volume of the loaded body, the change represented by the *D* line corresponds to the thin slip zone (slip surface) only. This condition is one of the most neglected when evaluating the results of laboratory shear tests on soils and rocks.

It is to be recalled that in the synoptical model (Fig. 3-22) the columns of the symbolical frame structure are assumed to be strong enough to resist the resultant of the *T* and *N* forces. In an analogous way the properties of blocks, i.e. of the *rock material*, enter the field of interest. The quality of rock material on the edges of blocks especially affects the magnitude of dilatancy and hence that of the shear resistance in the region of strain-hardening.

The model also indicates that the plastic strains along the thin slip surface are relatively small in the region of strain-hardening as they are analogous to the displacement of the columns of one tilting story of the model. The remaining part of the model exhibits small and rather elastic displacements.

Nevertheless, there is a logical deficiency in the foregoing consideration: The lower columns in Fig. 3-22(a) are almost as loaded as those of the particular story which deforms in a plastic way. Why do they not deform also? This may be due to the fact that the structure begins to deform until one story has been deformed in a manner most favourable for the development of plastic yielding. Yet another question arises: let us suppose there are many stories. What is the limiting number of stories after which the aforesaid favourable conditions and the development of plastic yielding will be repeated? In a large rock mass slip surfaces may be assumed to develop at a certain frequency. These questions have not been studied until now. A possible method of investigation by the study of the phenomenon known as *kinking* in tectonics, with which the increase in volume is registered, is suggested.

Another important conclusion may be outlined, especially concerning underground works. It was supposed with Fig. 3-22 that one story was tilting and that the volume increased as a result. What occurs when we fix the upper horizontal beam of the frame and permit the T-force to act? By *suppression of dilatancy* the N-force will increase spontaneously, i.e. an induced compression will develop in the rock. According to several tests (Mencl, 1967) this stress amounts to many tens of kp/cm^2. The increased N augments the shear strength and therefore it provides an important contribution to the *arching of rocks.*

The engineering geologist should not only understand the phenomena associated with the dilatant behaviour of rocks but also be capable of measuring the magnitudes of the particular parameters in order that they may be incorporated into the physical equations. As a result of the foregoing the instructions may be set as follows: Because only a thin slip surface develops in rocks during the dilatant shear straining, tests may be recommended with which such a surface can develop and with which the magnitude of dilatancy can be measured. This may be e.g. a *box shear test* with soils or a *field block shear test* with rocks as shown in Fig. 3-21. The measured results should be interpreted in a separate way for plastic and elastic deformations. Therefore loading, unloading and reloading should be applied in turn.

The magnitudes of both elastic and plastic displacements along the developing slip surface should be determined, as well as the elastic and plastic increase in volume in the direction normal to the slip surface under different magnitudes of σ and τ. When interpreting the results for further treatment it should be remembered that dilatancy develops in the thin slip zone (slip surface) only and not in the entire volume of the rock mass. The width of the slip zone should also be registered.

One of the drawbacks of the field block shear tests is that the angle made between the directions of principal stresses and the direction of the slip surface cannot be measured directly. However, even with laboratory tests correct results are not obtained. The problem of the inclination of the slip surface will be discussed further in Section 3.5.

On the basis of the knowledge of the orientation and width of the developing slip surface, of the magnitude of shear displacement, as well as of the magnitude of dilatancy, the specialist in mechanics is capable of formulating the relations for plastic flow (Mejzlík and Mencl, 1973). As a result, the D-matrix (Section 3.4.2) will be enriched by new elements. As the static computation proceeds by increments representing the steps of the actual process in the field, it is necessary to introduce the changed shape of the studied body as a starting state into the calculation of the next step. In this way a correct location of appearance of dilatancy can be obtained. Such a procedure is named "the solution of second degree" in mechanics.

3.4.5. Ductile behaviour of rocks. What is the physical nature of ductile behaviour? In contrast to brittle behaviour it is linked to the decrease in volume during the shear straining (negative dilatancy or *contractancy*). An attempt to symbolize it in the form of a synoptical model is presented in Fig 3-22(b). The volume of the body decreases during plastic deformation and therefore the point of application of the *N*-force is displaced downwards. Maximal loss of potential energy will occur when all the stories tilt. Therefore it is not thin slip surface which is found in the rock, as in dilatant behaviour, but a *slip zone* which develops as thickly as possible. Sometimes such a behaviour is denoted as pseudo-plastic in geological publications; nevertheless the correct term is *ductile*. When encountered in the field, e.g. in the excavation of a sliding mass in a cutting, it is striking that instead of a distinct slip surface only a tangle of slickensides can be found in spite of the fact that clean-cut, thin slip surface outcrops in the root area of the slide.

What are the physical conditions necessary for the development of contractant behaviour? Two cases should be distinguished: At points near the surface, where *N*-forces are small, the contractant behaviour can be observed in loose rock masses. Such a behaviour is typical and well known with loose sand as well as with loose rock debris of the rock slopes. A ductile behaviour was observed in the tectonically loosened Permian arkose near Ulmanka in Slovakia; when crossing the area of an old rockslide by a tunnel a slip zone was found to extend for a length of 20 m.

With the exception of loose rocks, rock masses are dilatant under small normal loads but *contractant* when subjected to a large normal stress. An attempt to illustrate this process is shown in Fig. 3-22c. When the normal force is large, the columns collapse under the cooperation of both the forces and a reduction in volume results. A thick failure zone is obtained. It is apparent that not only the void content of the rock mass but also the strength of blocks are important factors.

A question arises as to what magnitude of *compression* is necessary to provoke the ductile, contractant behaviour of a rock. With solid clay or claystone this compression is of the order of 4 kp/cm^2, and with the clayey shale of Ordovician age it achieves about 6 to 7 kp/cm^2 when shearing occurs across the foliation planes. Phyllite and mica-schist exhibit contractant behaviour from about 50 kp/cm^2, again when sheared across the foliation planes.

What are the features of practical importance connected with the contractant behaviour of a rock mass? First, large shear displacements occur contrasting with relatively small displacements of a dilatant rock mass. After the rock achieves contractant behaviour under increasing loads, deformations increase considerably as the entire rock mass (entire thickness of the slip zone) contributes to the shear displacement. The difference can be studied from the results of *block shear tests* shown in Fig. 3-23. Under the normal stress of 1.07 kp/cm² the rock (Ordovician shale, slightly weathered) behaves in a dilatant way, as can be seen in the lower

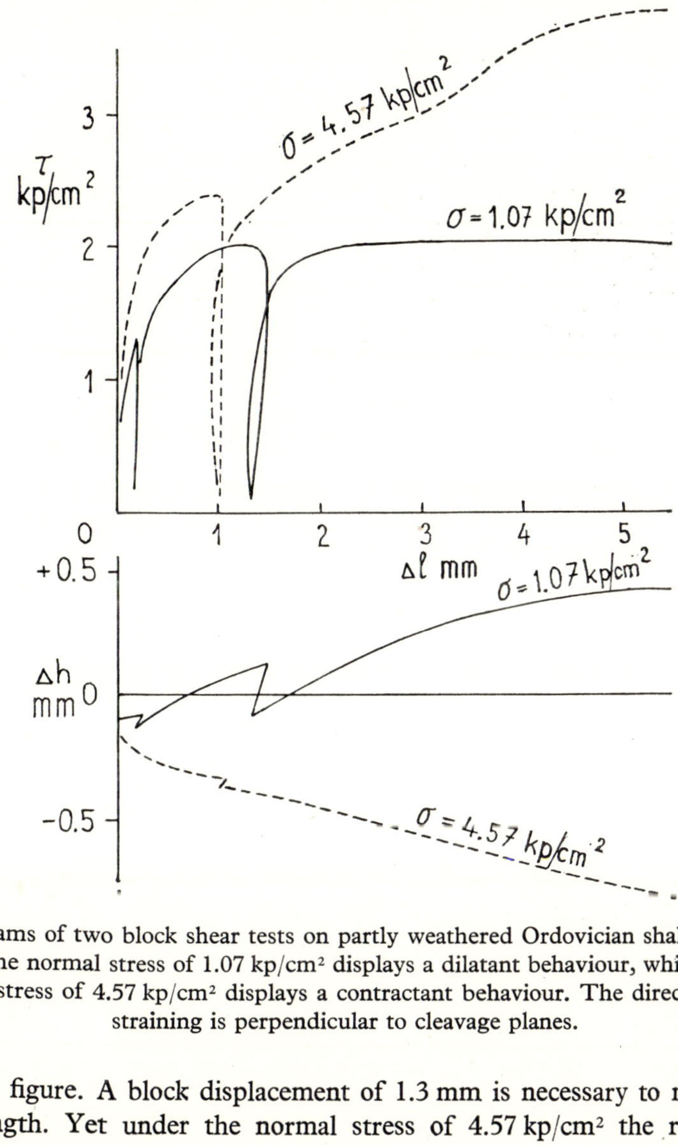

3-23. Diagrams of two block shear tests on partly weathered Ordovician shale. The shear test under the normal stress of 1.07 kp/cm² displays a dilatant behaviour, while that under the normal stress of 4.57 kp/cm² displays a contractant behaviour. The direction of shear straining is perpendicular to cleavage planes.

part of the figure. A block displacement of 1.3 mm is necessary to mobilize the shear strength. Yet under the normal stress of 4.57 kp/cm² the rock already behaves in a contractant way and the displacement at the strength limit is about 5.4 mm.

Owing to large displacements, rocks exhibiting contractant behaviour are dangerous for engineering structures:

As displacements are increased due to the thick slip zone, the *long-term deformation* also becomes expressive. This is dangerous with foundations of dams, in the abutments of arched bridges, etc. In the foundations of masonry dams on rather weak rocks resistance against the water pressure can be obtained by increasing the weight of the dam. However, this should be avoided since contractant behaviour could result associated with large displacements of blocks and this could be dangerous, e.g. to the behaviour of the grout curtain.

In underground works contractancy results in the "plastic" behaviour of rock, a situation known in mining as "*working at critical depth*".

Large displacements associated with ductile plastic deformation can also become dangerous when only a part of the body is subjected to a large normal stress and deforms in the ductile way, while another part exhibits dilatant, brittle shear failure under the action of smaller stress. Due to the difference in displacements a typical *progressive failure* develops during the stage of strain hardening. Deep sliding phenomena of slopes present typical examples (Section 11.4).

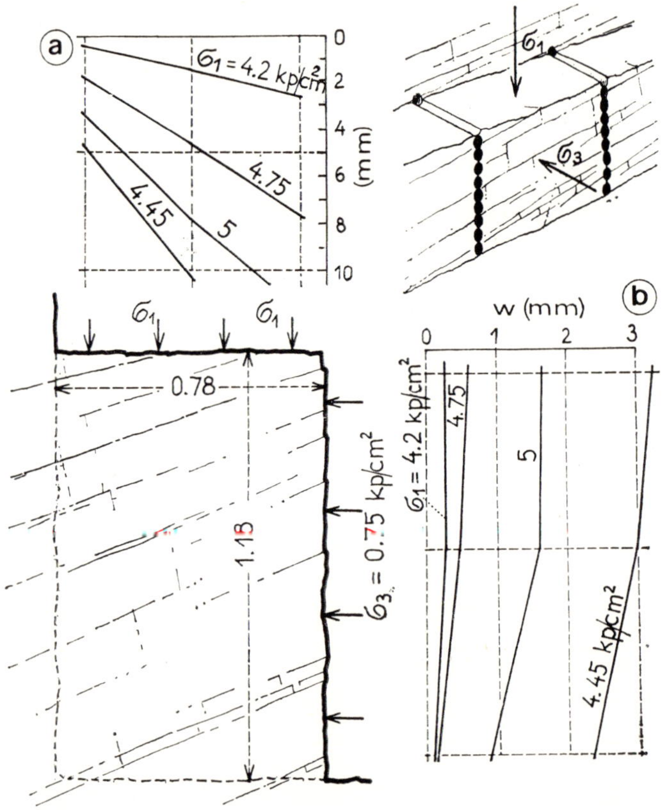

3-24. Arrangement of a plane shear test on a rock block which is delimited by drilling dense holes. a — Deformation of a surface loaded by σ_1, b — deformation of a surface loaded by σ_3.

Again the question arises of how the parameters of deformation with this behaviour should be established, and how they should be introduced into the statical analyses. Preference is to be given to the tests with which the entire sample or block can be strained as uniformly as possible. With rocks, *plane shear tests* (Fig. 3-24) are satisfactory. After drilling dense holes in the circumference of the block it can be trimmed for testing. One face of the block is loaded by σ_3 and another by a force representing σ_1. The arrangement of the test should satisfy two requirements: The loaded faces should be allowed to deform freely and $\sigma_1 > \sigma_2 > \sigma_3$. The first condition can be satisfied by inserting several Freyssinet's flat jacks between the block faces and the loading jacks, and the second by making the length of the block long enough in the direction normal to the plane made by the directions of σ_1 and σ_3.

As often the whole body, or at least a thick slip zone, is affected by plastic deformations, the pseudo-elastic solution is more acceptable than with the dilatant behaviour. The phenomenon of contractancy can be simulated by reducing the magnitude of ν. The variations of the modulus of deformation with $\sigma_1 + \sigma_3$ and $\sigma_1 - \sigma_3$ should be introduced according to the results of tests. As the portion of elastic deformation is small they can be neglected.

3.5. Strength of rocks. First it should be remembered that the strength of a rock does not represent anything other than the maximum resistance that the rock is capable of developing under the particular system of loading. Therefore the strength does not represent the stress with which the rock structure begins to be damaged by loading. As *shear failure* is the most widespread mode of rock failure in the practice of civil engineering let us concentrate on the question of *shear strength.*

As explained in Section 3.1, the resistance against shear loading originates in three components: *cohesion, shear friction* and *dilatancy.* As the resistance given by friction and dilatancy increase with increasing σ, so also the shear strength τ_f increases with σ. Therefore the strength can be represented in Mohr diagram by an inclined line which rises with the increasing σ. Its inclination towards the σ axis is denoted by Φ_f and is called the *angle of shear strength.* The term "angle of internal friction" does not therefore correspond to reality, as the increase due to dilatancy also participates in the Φ_f angle. In clayey soils cohesion also increases with σ as the particles come into closer contact. The point of intersection of the line with the axis of ordinates is defined as the *strength intercept* and is denoted by τ_{0f} (or more simply τ_0). The term *c, cohesion,* is often used in practice instead of τ_{0f}.

Questions associated with the problems of shear strength of rocks can be summarized as follows:

3.5.1 Components of strength. When dealing with the strength of a particular rock it is of practical interest to recognize in what quantities the individual

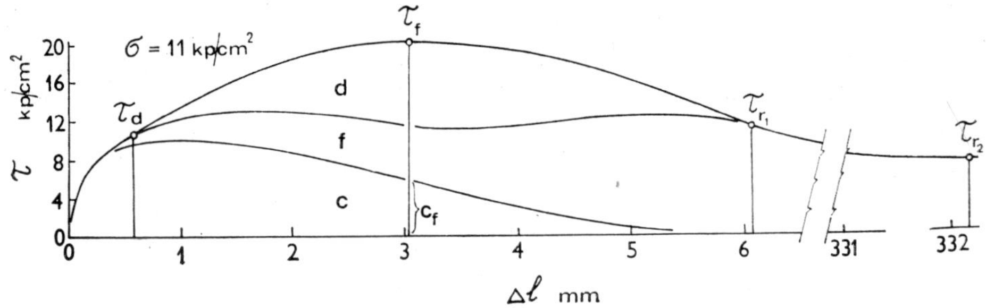

3-25. Diagram of a block shear test on gneiss (Jeseníky Mts.). Using the results indicated in Fig. 3-21c, the resistance component owing to dilatancy d has been defined and the remaining portion approximately divided into the components of cohesion c and friction f. The magnitude by which cohesion contributes to strength τ_f is indicated by c_f. The contributions of cohesion and dilatancy disappear when residual resistance τ_r has been achieved.

components participate in the strength. One of these can be established easily, namely the *dilatancy*. Let us denote it by d. As the point of application of the normal force is elevated in the course of dilatancy, the relation:

$$d = \sigma \tan (\vartheta + \omega)$$

can be written, according to which d can be determined for every point of the D-line (Fig. 3-19c). The result is plotted in Fig. 3-25. The division into the *cohesion* component, c, and *friction* component, f, is more difficult; it is based on the conception that with small deformation the resistance is produced rather by cohesion and with large deformation by friction. Cohesion disappears after the slip surface has developed entirely, while the shear resistance of rock diminishes to the value of *residual resistance*, τ_r.

Because the components d and f increase with increasing σ, we can write $\tau_f = \tau_{0f} + \sigma \tan \Phi_f$. Since the angle of shear resistance can be found by tests, τ_{0f} can be separated and this provides additional help for establishing the value

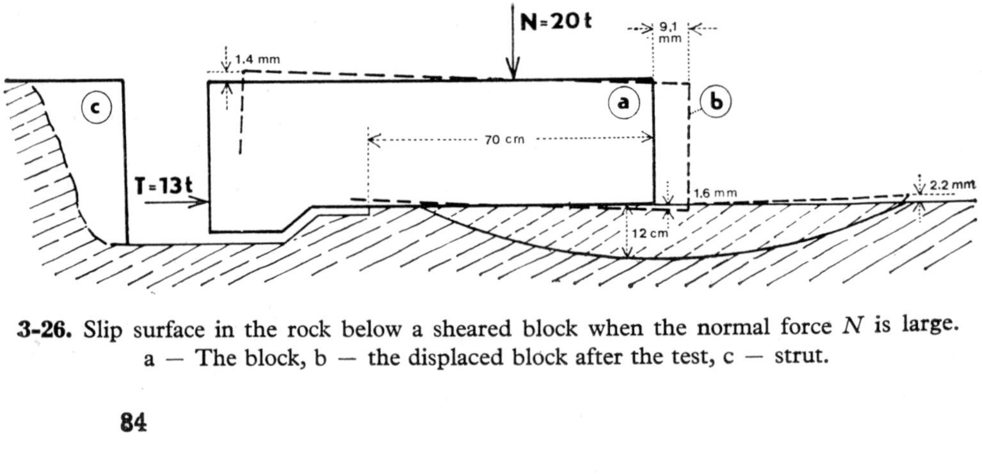

3-26. Slip surface in the rock below a sheared block when the normal force N is large. a — The block, b — the displaced block after the test, c — strut.

of *true cohesion* (except clayey soils and weak rocks with which cohesion also increases with σ).

It should not be forgotten that dilatancy is an indication that damage of the rock structure has started. As this damage is not desirable, the point of onset of dilatancy could serve as a good guide for practical design. This is supported by the fact that the onset of dilatancy can be fixed with a satisfactory exactness by tests. With *block shear tests* as presented in Fig. 3-21a the development of the variation in height is influenced by the moment generated by the T-force around the block base. In order to exclude this moment, the T-force is often set so as to act in an oblique way with a direction towards the centre of the base (Fig. 3-27). Otherwise the T-force is situated in the plane of the base (Fig. 3-26). Neither

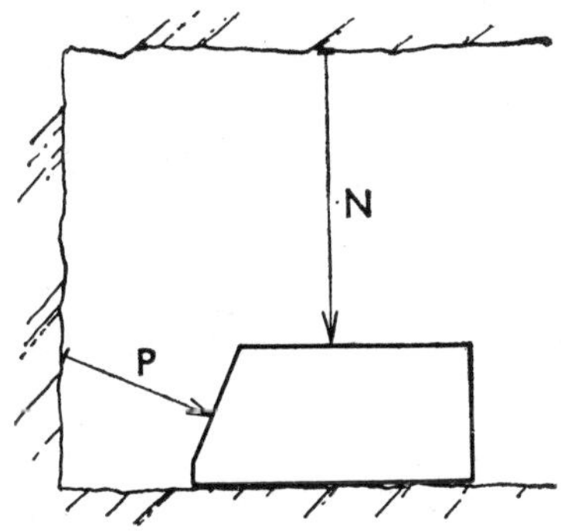

3-27. Block shear test with P-force acting in an oblique direction.

technique is free from imperfections since with a large normal force, even when directed to the centre of the base, the block inclines as the T-force increases.

As already known, dilatancy transforms into *contractancy* under conditions of large normal stress. Therefore the component of dilatancy is no longer active in the rock strength. Phenomenological manifestation of this is that the strength line in the Mohr diagram bends, exhibiting smaller Φ_f.

3.5.2. The orientation of the slip surface. Knowledge of the orientation of the slip surface is of importance not only with the analysis of the shear strength tests but also with the formulation of the equations simulating plastic flow in rocks.

Let us observe an isotropic homogeneous rock material with the strength given by the relation: $\tau_f = \tau_0 + \sigma \tan \Phi_f$ (Fig. 3-28c). What is the magnitude of the angle β_f made by the direction of σ_1 and the slip plane (Fig. 3-28a, b)? Fig. 3-28c shows that $2\beta_f = 90° - \Phi_f$ and therefore $\beta_f = 45° - \Phi_f/2$. This magnitude is often adopted in solutions of practical problems as well as with the analysis of plastic flow. More rare is the case that two conjugate slip surfaces develop (Fig. 3-29). Nevertheless the actual β_f is greater. This can be proven:

(a) By the results of tests.

(b) By the conception that the damage of the rock structure, and therefore also the development of the slip surface, originates at stressing smaller than the

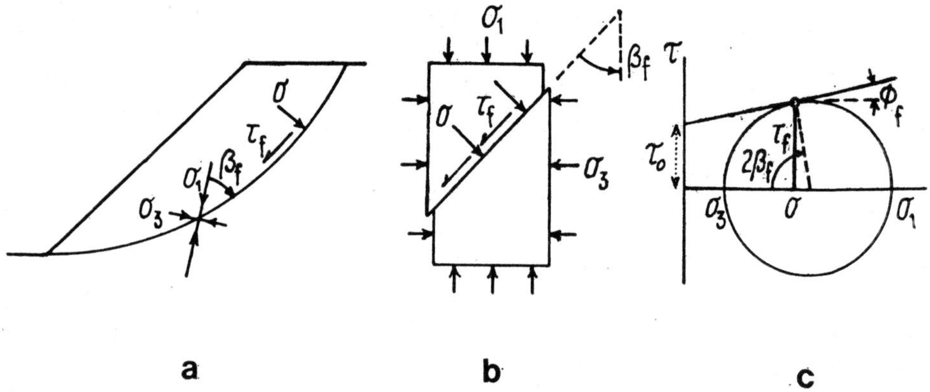

3-28. The direction of the maximum principal stress makes an angle β_f (a, b) with the slip surface. The theoretical magnitude of β_f can be determined by plotting the Mohr's circle (c).

strength. Therefore also the angle of resistance Φ is smaller than Φ_f. After all, we know that max τ acts along the plane inclined at 45°.

(c) By the consideration that the theory of plastic flow is concerned with plastic deformations, and its principles are based on the postulate of the maximum of dissipated energy. Therefore, only the magnitude of the Φ angle based on the friction component of the shear resistance should be included into the relation $\beta_f = 45° - \Phi/2$.

Before the slip surface has fully developed, i.e. until the stage of the residual resistance τ_r is reached, the component of dilatancy also participates in the plastic flow.

3.5.3. The influence of the pore-water pressure. As follows from above, the shear strength of rocks is defined by the relation $\tau_f = \tau_{0f} + \sigma \tan \Phi_f$. $\sigma \tan \Phi_f$ indicates how the strength increases with increasing σ, because this

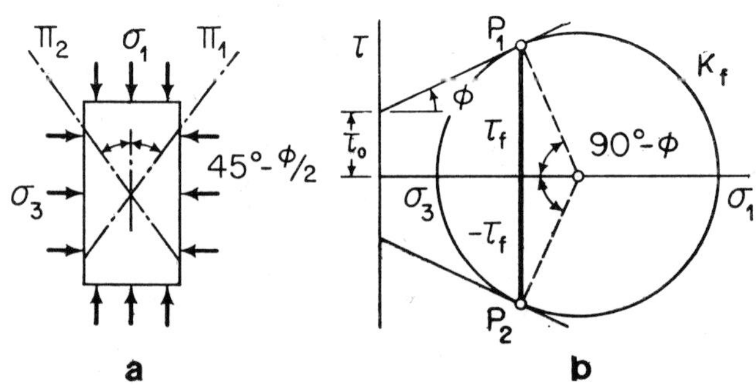

3-29. Two conjugate slip surfaces develop under several circumstances, e. g. when the loaded body can expand in one direction only, as it is the case in the bottom of cuttings or in the neighbourhood of faults.

makes the elements pack closer. But the pressure of water contained in pores or other voids of the rock mass acts in the opposite way. If pore-water pressure is denoted by u the remaining *effective stress* is equal to $\sigma' = \sigma - u$. As τ_f varies with σ', τ_{0f} and Φ_f also are affixed by a stroke and the relation:

$$\tau_f = \tau'_{of} + \sigma' \tan \Phi_f'$$

is written.

Due to the great importance of the water pressure (pore-water pressure, pressure of water in joints, uplift), it is necessary to establish the value of u for every studied point within the rock mass. With permeable rocks u equals the hydrostatic pressure of the ground water, but in less permeable, compressible soils, as with clay or silt, the pore-water pressure increases with the decrease in volume due to compression or contractancy. A long time is necessary before the excess water is pressed out from the pores. It may also be said that the *consolidation* of the soil is slow. The slowest consolidation proceeds in the middle of a layer where the

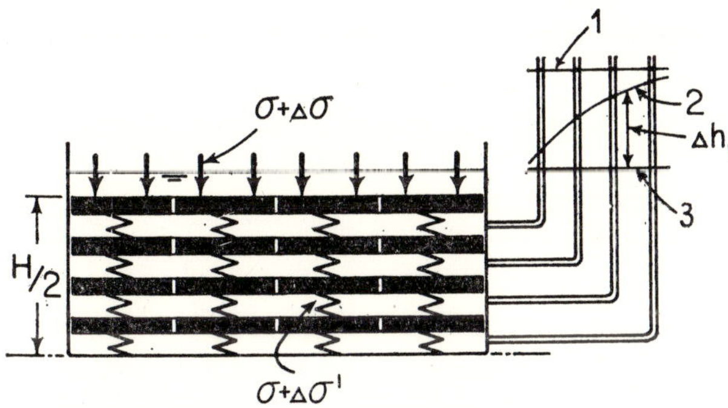

3-30. Symbolic model of a proceeding consolidation in the upper half of a layer of saturated soil or weak rock with low permeability. Immediately after the loading of the layer by $\Delta\sigma$, an increased pore water pressure develops (straight line 1, connecting the piezometric water levels). As consolidation proceeds, water flows out through the small openings of the slabs, the springs contract and the pore-water pressure is defined by the piezometer level Δh_w according to curve 2. After consolidation process has finished, the pore-water pressure decreases to the level shown by curve 3, which is given by the permanent ground-water level. Ground water generates the uplift acting on the soil particles according to Archimedes law.

distance to the surface is longest (Fig. 3-30). Therefore if the layer has consolidated (contracted) due to $\Delta\sigma$ to:

10 20 40 50 60 70 80 90 %

of the final contraction, σ' in the middle of the layer increased only by $\Delta\sigma'$ which equals:

0 2.5 10 23 40 53 70 87 % of $\Delta\sigma$.

In order to measure the *pore-water pressure*, hydrogeological observation boreholes are not suitable, as they withdraw pore-water pressure from a wide surrounding area before becoming filled with water. Specially constructed *piezometers* are necessary.

An effect similar to the uplift results from the forces exerted by the flow of percolating water on the soil particles, or in short from seepage forces (Section 7.6.3).

3.5.4. Numerical values of shear strength. Most important is the magnitude of shear *strength along the predestined surfaces* of weakness in rocks (bedding planes, schistosity planes, fissures). Small asperities allow dilatancy to take part also in the resistance. However, with normal stress greater than about 5—20 kp/cm² (lower with clayey shales, higher with hard crystalline rocks), *dilatancy* still disappears. Therefore the first section of the strength line bends towards the origin of co-ordinates in the Mohr diagram, thus demonstrating the considerable participation of dilatancy in Φ'_f. With greater σ' the line bends to the right so that Φ'_f decreases while τ'_{of} appears in the mathematical formulation. This is because the material of asperities begins to be crushed.

The τ'_{of} parameter varies widely according to the character of asperities. Where

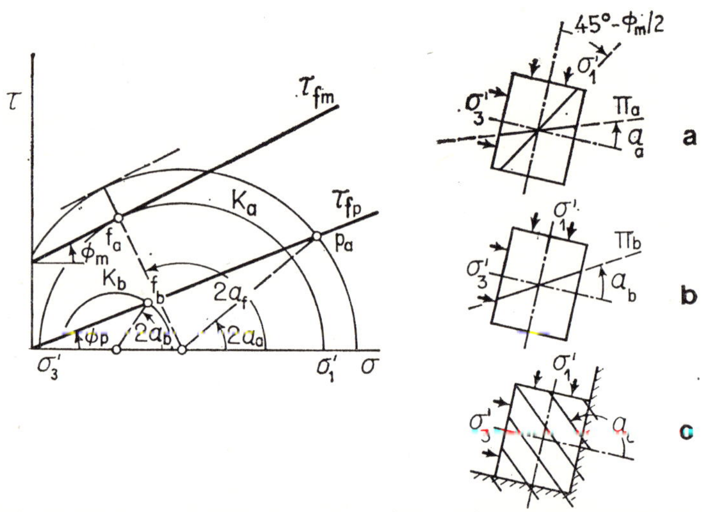

3-31. Rock is loaded by compressive stresses σ'_1 and σ'_3. The shear strength τ_{fm} of the rock material and shear strength τ_{fp} along an existing plane of weakness π are known. The question whether slip failure will develop along plane π can be analysed. Plane π cannot serve as a slip surface in case (a), since the Mohr circle K_a touches the envelope τ_{fm} before reaching point p_a, when σ'_1 increases. Therefore a new slip surface develops, making the angle 45° — $\Phi/2$ with the direction of σ'_1. The slip surface makes use of the existing plane of weakness in case (b). The circle K_b reaches point f_b before touching the envelope τ_{fm}. In case (c), the slip surface cannot make use of the existing planes of weakness since the boundary conditions do not allow it.

the infilling of the open joints is clayey, τ'_{of} is very small and Φ'_f equals that of clays, being of the order of 20 degrees. The presence of a small quantity of montmorillonite is sufficient to decrease this value to 12 degrees. The magnitude of Φ'_f along the smooth surfaces of joints in granite is often only 26 degrees and with rough surfaces about 32 degrees. Higher magnitudes of Φ'_f are usual with carbonates (more than 28 degrees) and sandstones (up to 35 degress). The crystalline bonds on the planes of foliation also promote higher values of shear strength; e.g. amphibolite (damsite of Dalešice, Moravia, Drozd, 1966) exhibited τ'_{of} of almost 10 kp/cm² and Φ'_f of 38°. But τ'_{of} diminishes rapidly when the rock is only slightly weathered and Φ'_f decreases to about 26° with a partly weathered rock.

With regard to the considerable scatter of values, measurement of shear strength on the predestined planes is necessary. Field shear tests on blocks are preferred in Europe while field torsion tests on large-diameter drilling cores are frequent in the U.S.A.

3.5.5. Influence of the surfaces of weakness on the shear strength of the rock mass. First it is necessary to investigate to what degree these planes can participate in the development of slip surfaces in the rock mass. An example of the analysis of this problem is shown in Fig. 3-31. In case (a) the slip surface cannot follow the inherited surfaces of weakness. But at greater α the failure can proceed along them (case b). In the same way, the problem can also be studied at $\alpha > 90$ degrees (case c), when supposing that failure is possible by sliding towards one side only, because of the boundary conditions. But the results obtained in this way often do not fit reality, as the thickness of the layers or the *interlocking* of blocks during the first tilting also intervenes. However, the comparison of results of analysis with those of field shear tests is always instructive.

Since field shear tests are expensive, Mencl (1968) compiled the results according to the most common types of macrostructure (Table 3-1). In several cases, in addition to the magnitudes of parameters τ'_{of} and Φ'_f, the normal stress σ is also registered indicating the borderline between the dilatant and contractant plastic deformation.

3.5.6. Influence of the medium principal stress. Questions concerning the strength of rocks were analysed above as plane problems, by supposing $\sigma_1 > \sigma_2 > \sigma_3$. But other situations also exist, an outstanding example being that of the classical *tri-axial test* on cylinders under the conditions of $\sigma_1 > \sigma_2 = \sigma_3$. A similar situation exists under circular foundations and, to some degree, in convex slopes of cuttings also. A smaller strength is associated with the latter stress situation as particles, grains, or blocks are displaced one from another in the directions of both σ_2 and σ_3 under the influence of σ_1. In contrast, the stress state $\sigma_1 = \sigma_2 > \sigma_3$ permits a higher strength. Such is the case of concave slopes, of circular construction pits, of shafts, etc.

These differences are most obvious with dense sand. If a test in a cubical

Tab. 3.1

Values of c (kp/cm²), Φ and σ (kp/cm²)[1]) related to rock structure and rock material

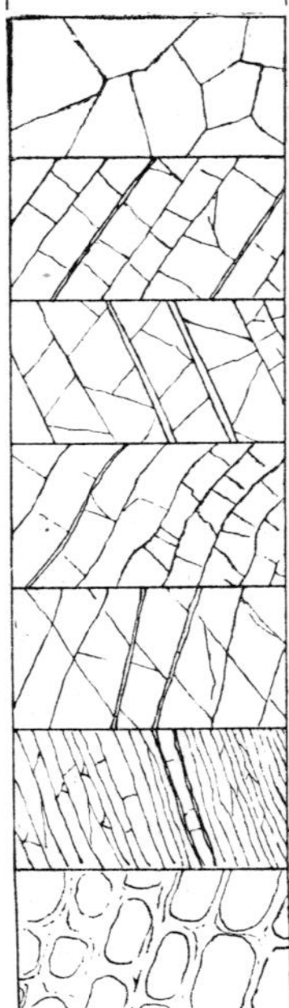

Rock structure	Type	Joints	Infilling[2])	100 kp/cm² 0°	100 kp/cm² 30°	100 kp/cm² 60° [3])	100 kp/cm² 90°	100 kp/cm² 120°	100 kp/cm² 150°	500 kp/cm² 0°	500 kp/cm² 30°	500 kp/cm² 60°	500 kp/cm² 90°	500 kp/cm² 120°	500 kp/cm² 150°
				Simple compression strength of rock material (on the edges of blocks) — Inclination of discontinuity planes from the principal stress axis σ_3											
	Y	closed	S C				2 53° 1 42°						7.5 57° 3.5 53°		
		open	S C				0.65 42° 0.3 30°						1 48° 0.35 33°		
	H	closed	S C	3 53° 2.6 48°	3.4 43° 3 36°	0.1 35° 0 25°	2.5 50° 1.8 47°	2.2 46° 1.4 39°	2 49° 1.7 41°	4.8 56° 4.6 52°	4.6 46° 4.6 38°	0.2 39° 0 29°	4 54° 3 50°	2.5 49 2.3 42°	3 52° 2 44°
		open	S C	2.4 46° 1.3 40°	2 40° 1.1 30°	0 29° 0 21°	1.6 43° 1 38°	1 38° 0.6 31°	1.1 38° 0.7 32°	3.5 50° 1.6 44°	3 42° 1.5 33°	0 35° 0 23°	1.8 45° 1.5 40°	1.5 39° 1 35°	1.8 40° 0.9 37°
	K	closed	S C	2.5 40° 2 34°	1.7 37°5.5 1.6 34°	0.05 35° 0 25°	0.7 37° 0.6 29°	0.7 38° 0.6 27°	0.7 40° 0.6 32°	3 41° 2.5 36°	3 39° 2 35°	0.1 39° 0 28°	1.6 40° 0.9 30°	1.5 41° 0.8 31°	1.6 43° 0.8 33°
		open	S C	2.5 36° 0.7 30°	2 35° 0.6 29°	0 29° 0 21°	0.5 31° 0.4 21°	0.5 32° 0.4 25°	0.5 37° 0.4 32°	2.6 37° 1 34°	2.3 36° 1 33°	0 34° 0 21°	0.9 32° 0.7 25°	0.9 33° 0.7 28°	0.9 39° 0.7 36°
	SH	closed	S C	2.8 48° 2.5 44°	1.8 42° 1.4 36°	0.1 38° 0 28°	1.5 40° 1 34°	1.3 42° 1 36°	1.2 50° 1.1 42°	4.3 53° 4 45°	4.2 45° 4 37°	0.15 40° 0 32°	3.5 53° 2.7 49°	2 43° 1.8 37°	2.6 53° 1.8 45°
		open	S C	2.3 40° 1.2 39°	1.1 37° 1 28°	0 30° 0 23°	1 30° 0.6 28°	0.8 32 ° 0.5 30°	1 39° 0.6 34°	3.5 42° 1.3 40°	3 38° 1.2 29°	0 36° 0 24°,	2 46° 1 36°	1.4 35° 0.8 32°	1.7 40° 0.9 35°
	X	closed	S C	0.25 42° 0.2 30°	0.15 38° 0.1 28°	0 35° 0 25°	0 35° 0 28°	0.15 35° 0.1 30°	0.15 36° 0.1 32°	0.28 43° 0.25 32°	0.2 39° 0.15 30°	0.1 38° 0 29°	0.15 36° 0.1 33°	0.2 36° 0.15 32°	0.2 37° 0.15 32°
		open	S C	0.2 30° 0.15 22°	0.1 29° 0.05 21°	0 29° 0 21°	0 30° 0 23°	0.1 32° 0.05 25°	0.1 33° 0.05 26°	0.25 32° 0.15 23°	0.15 30° 0.05 22°	0 34° 0 22°	0 34° 0 24°	0.15 33° 0.1 27°	0.15 34° 0.1 28°
	I	closed	S C	1.7 39°3 1.3 34°	0.9 36° 0.6 33°	0.1 30° 0 25°	1 35°4.5 0.6 28°	1.3 38° 0.7 27°	1.4 39° 0.8 31°	2.8 40° 2.2 35°	2.8 38° 1.8 34°	0.2 38° 0 27°	1.6 36° 0.8 29°	1.4 39° 0.8 30°	1.5 40° 0.9 32°
		open	S C	1.2 34° 0.4 29°	0.9 32° 0.3 28°	0 29° 0 21°	0.6 28° 0.25 21°	0.6 33° 0.3 24°	0.7 34° 0.3 31°	2 35° 0.7 33°	1.9 34° 0.5 32°	0 34° 0 23°	0.7 30° 0.4 24°	0.7 31° 0.5 28°	0.8 37° 0.6 36°
	O	closed	S C				0.35 37° 0.22 35°						0.45 40° 0.3 36°		
		open	S C				0.1 26° 25°						0.13 33° 28°		

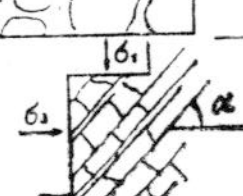

[1]) σ — in several cases the normal stress is registered indicating the beginning of contractant (ductile) behaviour.
[2]) Infilling: S — none or sandy, C — clayey
[3]) The resistance with $\alpha = 60°$ is given directly by the strength along the planes and should be checked by tests. The minimum of resistance may occur with $\alpha < 60°$.

tri-axial apparatus gives the strength angle as 38 degrees under the conditions of $\sigma_1 > \sigma_2 = \sigma_3$, a value of 42 degrees is obtained at $\sigma_1 = \sigma_2 > \sigma_3$.

The Huber—Hencky—von Mises theory of strength treats the preceding questions in a phenomenological way. We are interested in the results because of their practical importance, and in the physical basis of the phenomenon.

3.5.7. Time factor. Several natural slopes exhibit slow displacement, termed "*creep*". Caution is necessary when using this term in engineering geology as its meaning is different in mechanics. E.g., the slowness of deformation of a clay can be caused by a slow consolidation during the decrease in volume due to contractancy. Similarly, a particular time is necessary before the dilatant increase in volume takes place. Therefore clay has to absorb the necessary amount of water from environment before the slip surface can develop. Thermal variations, influence of frost and periodic effects of precipitation can also induce slow slope movements. But these changes cannot be considered as creep in general mechanical terminology where constant environmental conditions are assumed.

Up to a certain degree of loading, smaller than the strength, the velocity of the transient creep decreases and the process tends to equilibrium. The material is, therefore, capable of resisting the loading, only it exhibits a *transient, retarding creep* for a certain time after the loading has been imposed. This problem can be also formulated as follows: what is the magnitude of the *long-term strength* as compared with the strength obtained from tests conducted over a limited time period? First, a necessary warning is that laboratory strength tests are frequently made on thin samples, yielding misleading results. The long-term strength found from these tests is too small. But long-term tests arranged in such a way that the barrel-shaped deformation of cylindrical samples was reduced (and therefore, the extension fracture was limited), produced a *permanent creep* when load was over 85 % of the standard strength. Therefore, with the exception of situations allowing development of extension fracture (high and steep rock faces, high walls of underground halls) the reduction of the standard strength by 10 to 15 % is sufficient. For clay Skempton (1969) instances the strength decrease of 3.5 % with each successive log cycle of time. For example, if a test conducted for one hour gives 100 % of strength, the test conducted for 10 hours would yield a strength of 96.5 %, for 100 hours 93 % and for 10^6 hours (144 years) 79 %. Skempton recommends that 24-hours tests be performed and then the reduction corresponding to the expected duration of the structure, e.g. 100 years, would be 14 %.

With dilatant behaviour creep is almost immeasurable. But it does continue, as can be detected by using *relaxation tests* (Mencl and Trávníček, 1964). With them the compression force is measured acting against a fixed obstacle. Creep deformations associated with a *ductile* behaviour are remarkable as deformation increases due to the thickness of the slip zone (Mahút, 1971).

3.5.8. Residual resistance. After the strength has been surpassed the shear resistance of rocks decreases to the magnitude of the residual resistance τ_{r1}

(Fig. 3-25). The magnitude of τ_{r1} is of the order of more than half of the peak strength and its physical nature is principally frictional. Nevertheless, it is not yet the minimum of resistance, which appears only after a final smoothing of the slip surface (τ_{r2}); this requires a shear displacement of several decimetres. The magnitude of the angle of minimum residual resistance was 18 degrees in Jurassic marlstone on the Vaiont slide. The tectonically damaged granite of the rockslide at Ružín (East Slovakia) presented a value of about 37 degrees. Marls of Brno (Neogene) gave a value of 13 to 16 degrees. Much lower values can be obtained when higher clay or montmorillonite content is present.

Besides other consequences of the decrease in shear resistance after a certain displacement has developed, the urgency for stabilization of a rockslide before the displacements reach magnitudes of several decimetres should be kept in mind.

3.6. Natural stress state in rocks. Investigation of the stress state existing in rocks in natural conditions is of great interest in engineering geological studies for several reasons: The existing stress state at a particular site helps in the recognition of processes which modelled the ground surface in the past. It is also assumed as the initial state for statical analysis. If neglected, the existing stress state can bring about important destructive processes during excavation of underground works or large cuttings.

Vertical stress in a rock mass is generally determined by the overburden pressure, γD, where D is the depth below the ground surface. Discrepancies in this value are common under an uneven surface. Under valley slopes and floor the actual magnitude is commonly greater than that defined by the overburden, since the principal stress produced by the weight of the upper parts of the valley slopes does not have a vertical direction. A large concentration of stress is often present near the toe of a slope as tendency of the valley floor to lift after the erosion of the valley was resisted by the slopes.

Large discrepancies exist in the magnitude of *horizontal stress*. When discussing its magnitude, the *pressure at rest* is generally taken as a reference value. This is given by the condition that the rock is loaded vertically (e.g. by its own weight) and can deform in the vertical direction only, as horizontal deformation is prevented by the surrounding rock. This condition, i.e. the condition of zero normal strain in the horizontal direction, yields:

$$\sigma_h = K_0 \sigma_v$$

σ_h and σ_v being the horizontal and vertical stresses and $K_0 = \nu/(1 - \nu)$. Assuming for a granite $\nu = 1/7$, $K_0 = 1/6$ is obtained. With sand $\nu = 0.33$ and $K_0 = 1/2$ and with saturated clay $\nu = 0.45$ and $K_0 = 0.82$ are approximate values.

As noted previously, the actual horizontal stress commonly differs from the magnitude of the pressure at rest. An increased stress is often encountered in *tectonic depressions* where the displacement along an inclined fault may promote great horizontal forces. Large horizontal stress is often present in flat areas as a remnant of the ancient state where the overburden was up to several hundred metres high but has been denudated until now. This case is not rare in Neogene clayey deposits, e.g. in the Carpathian Foredeep, where horizontal stresses of the magnitude of 3 to 5 kp/cm^2 were registered. The weight of the continental ice sheet also left horizontal stress in rocks, as is the case of the almost horizontally bedded deposits of Palaeozoic age in North America. In folded mountains large horizontal stress is also often present. Here and there this stress may by of recent origin owing to neotectonic processes. Remarkable concentrations of residual stress can be encountered under valley floors, since the valley slopes moved towards one another during the erosion of the valley and the rock under the floor contracted in a horizontal direction.

Large horizontal stress induces deformation in rocks; this in turn reduces the stress but does not extinguish it. Horizontally bedded layers under valley floors often bend upwards and horizontal voids appear in them, filled by secondary material. The surface of weak deposits may be undulated. Partial slip surfaces bending downwards and vanishing at depth are frequent in slopes.

From the point of view of engineering geology such a state may sometimes be favourable but its bad effects prevail. Favourable is the fact that the structures founded on a deposit which has been prestressed in the horizontal direction exhibit smaller settlements; or the presence of horizontal compression contributes to the reduction of tension in the rock in a tunnel roof (Section 12.6). The bad effects are more common: Horizontal displacements of the lower parts of the valley slopes produce tensions in the upper parts and this advances deep *weathering of rocks*. Rigid rocks are damaged by tension cracks when they are supported by weak rocks which increase in volume after being exposed to stress relief due to erosion. Cracks in the river bed near the crown of the American part of Niagara Falls (Fig. 3-32) give rise to a gradual destruction of this waterfall. Walls of the large intake canal of the Niagara power station moved about 9 cm towards one another and the bottom heaved up. Similar features can be encountered during deep excavations in Tertiary or Cretaceous deposits (Section 11.4). A large stress concentration in hard rocks gives rise to *spalling* and *rock bursts* during tunnel excavations or in mining works.

In valley slopes, especially in their upper positions, in the meander necks or in tectonic elevations, rocks are free from horizontal stress. This is a disadvantage in tunnel excavation (Fig. 12-18). Such a location should be avoided especially when designing underground *caverns* of water power stations (Section 12.8). Rock flanks free from horizontal compression are often impaired by deep continuous *tension cracks* which promote large gravitational movements, e.g. when provoked by seismic forces, as in the Andes.

Therefore, a need for *measuring the stress* existing in rock masses appears. For the design of large underground openings this investigation is categorical

3-32. Tension cracks in Lockport dolomite (Silurian) near the crest of the American part of the Niagara Falls. Cracks are parallel with the crest and develop as a result of free energy stored in the underlying clayey shale. The view is taken showing the artificially drained river bed 20 m from the crest of the fall. (Photograph by Mencl.)

because its results are decisive for the possibility of realization of the opening. Techniques for measurement of the stress state in rocks have developed into a large branch of rock mechanics.

(a) *Removal of material,* a technique initiated by Sachs (in Dow, 1956) for metals. After a circular opening is excavated in the rock, (e.g. a circular drill hole or a circular gallery) the displacements of the points set proximal to the opening develop. When a general relationship between the stress components and the displacements is known, the magnitude of the former can be determined by measuring the latter.

E.g. for the simplest case of $\sigma_1 = \sigma_2 = \sigma$, the elastic analysis yields

$$\sigma = \Delta r \,.\, r \,.\, E/(1 + \nu)\, a^2$$

where Δr is the displacement of the observed point, r is the distance of the point from the centre of the opening and a is the radius of the circular opening. The calculated stress σ is compression when Δr occurs in a direction towards the centre of the opening. The instantaneous deformation modulus should be applied for E (e.g. free of volume changes produced by consolidation). In order that the pseudo-elastic solution yields reasonable results the points should not be set within the rock zone (Section 12.6) damaged by the stress concentration around the opening. On the other hand Δr decreases when r increases and, therefore, r should not much exceed $2a$. One of the important practical requisites of the measurement is not to injure the measured points during the drilling of the opening. An example of this technique adapted for soils is shown in Záruba, Mencl, 1969, Fig. 6-15.

The technique described above requires that the measuring of displacements can be performed on an exposed surface or on a rock face. But stress state existing on a rock face does not usually coincide with that which is present within the rock mass. Therefore measuring techniques in boreholes have been developed. Several methods exist, but usually the variations in diameter of a drill hole are measured after the surrounding rock has been relieved of stress. This is done by *overdrilling* the borehole by a diamond bit of larger diameter. Measurement of diameter variations is performed by means of special pressure cells inserted into the original drill hole; this requires sensitive measuring devices.

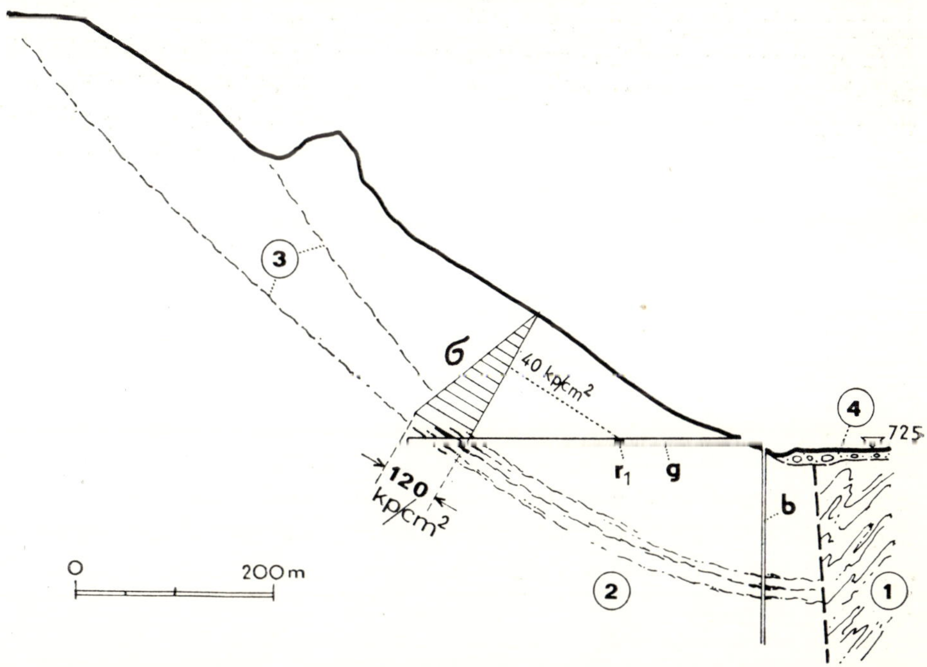

3-33. Schematic illustration of the statical situation near the cross heading r_1, in which the stress present in the rock was measured at the damsite of Čierny Váh, Slovakia. 1 — Werfen shale and quartzite, tectonically disturbed, 2 — dolomite, 3 — assumed planes of weakness, 4 — valley terrace, g — exploratory gallery, b — borehole, σ — assumed diagram of stress acting parallel with the slope.

(b) *Strain cancellation method.* The basic idea of this method is reflected in its name. The surface of a drill hole is subjected to a compression so large that the walls return to their original position. The stress necessary to effect this is considered to be equal to the pre-existing stress in the rock. A flat slot is mostly used instead of a cylindrical hole. Freyssinet's flat jack is inserted into the slot and this induces the necessary compression on the slot walls.

Comparison of results of different measuring methods conducted in the exploration gallery at the dam site Černý Váh (Slovakia) may be of interest. The rock is a large block of dolomite, separated from another rock mass by a fault, and based upon a rather weak breccia layer (Fig. 3-33). Both simple statical analysis as well as the finite element method (Mejzlík, 1972) yielded a magnitude of stress of the order of 50 kp/cm^2 at a particular point within the rock. The measured magnitude of stress was (Groma, 1972):

— 50 to 60 kp/cm^2 on the basis of the displacements of observation points set in the vicinity of a cross-cut (Trávníček and Mencl, 1966).

— 30 to 32 kp/cm^2 on the basis of the displacements of observation points set in the vicinity of an exploration shaft (Nevický and Kraus, 1971).

— 29 to 39 kp/cm^2 on the basis of the strain cancellation method (Stavební geologie, n. p., 1966).

In order to determine the stress state at a particular point, i.e. to establish the magnitudes and directions of principal stresses, investigations of *strains* ε in three different directions are necessary, if a two-dimensional problem is considered. Equations exist for the calculations of principal strains, but simple graphical procedure is also sufficient with the problems of geotechniques by treating strains in a Mohr circle in the same way as with stresses. From the magnitude of principal strains the values of principal stresses can be calculated, using the relationships discussed in Section 3.4.2.

3.7. Statical solutions in engineering geology. The aim of statical analyses in engineering geology is to establish whether and under which conditions the proposed structure can be constructed. As necessary statical analyses are often intricate, they pertain to the sphere of activity of civil engineers, but the co-operation of engineering geologists should not be underestimated. Several approaches to the solution of statical problems in engineering geology exist:

3.7.1. Method of analogy. The engineering geologist is capable of contributing to the solution because of his capability of anticipating the behaviour of rocks under the conditions of the proposed project and of his power of estimation which is produced by perpetually facing problems of geotechnical character. In addition, simple, engineering solutions are of help to him if he is precautious enough not to use them without criticism. Finally, the wealth of experience gained by long practical activity as well as by comparison with problems solved elsewhere cannot be over-emphasized. Such an approach has been denoted as the "method of analogy" in several Soviet books. Since the engineering geologist is not capable of supporting his estimations with results of 'exact' or 'academic' solutions, his report may be found to be less reliable. Such an approach is wrong; the engineering geologist should be asked to submit his opinion on the design on the basis of the "method of analogy". When his report is at variance with the results of other analyses, the attention attracted to the particular problems ensures sound corrections not only of the presumptions of the method of analogy, but also often of those of the mathematical analyses.

3.7.2. Engineering solutions. Since shapes of bodies and properties of materials treated in geotechnical problems are intricate, simplified analyses have been developed. Their aim is to respect theoretical relations as much as possible and to complement them on the basis of experience, model measurements or estimation. This is the reason why the range of validity of the formulae or of the analyses is limited, and serious problems may arise when the formulae are employed by unexperienced or theoretically poorly trained workers.

Due to mathematical limitations most engineering calculations concentrate on the analysis of several distinct and well-defined stages of the behaviour of the proposed structure or rock body. These selected analyses are often termed "*limit-design solutions*" and the most important stages are:

(a) *Limit state of stability* or *ultimate load state*. The principle of the analysis consists of calculating the maximum forces which the studied body can resist. The ratio of the sum of these forces to the sum of actually acting forces represents the *factor of safety*. Stability analyses of slopes (Section 7.8) or of foundations (Section 10.2.4) are typical examples of this approach. An advantage of this approach is that the stability of the body in bulk is studied, making the solution relatively simple. The most important deficiency is that the problem of so-called *progressive failure* is not considered. Yet one portion of the body often fails while the others still resist and yield later. This process is respected by increasing the allowable factor of safety; the engineering solution is turned into the method of analogy in this way.

(b) *Limit deformation state*. The applicability of the design is often examined with regard to the values of calculated displacements (*settlements*) of several points of the proposed structure. This approach, usual with buildings (Section 10.2), has several merits: the assumed magnitude of the loading of the foundation rock or soil is limited by the actual magnitude of loading. This makes the use of pseudo-elastic solutions more applicable (Section 3.4.3). Moreover, results of this solution can be compared with the *actual settlements* measured after the structure has been constructed. Owing to these comparisons the particularities of statical analyses can be further improved. Unfortunately, methods of this type have not yet been developed for slopes of excavations.

(c) *Limit state of incipient failure* or the *critical state*. The danger of progressive failure was mentioned in (a). Progressive failure can occur even when the factor of safety is considerably greater than one. For example, cracks appear on the slopes of cuttings, which makes the penetration of water possible and the entire collapse of the slope is only a question of time. But no engineering solution to this problem exists. Stability calculations in which the resistance at the yield limit or that at the *onset of dilatancy* are introduced instead of shear strength, should be regarded as a provisory approach to this problem. The application of a *factor of safety* not much greater than one can be associated with this solution.

3.7.3. More exact solutions. The observation of the behaviour of the studied body from first deformations to first fissures, and further until failure is almost reached, with the possibility to choose a particular state as critical, led to the application of two methods:

(a) *Models of equivalent materials.* One of the most important merits of this technique is that the in situ process is duplicated as realistically as possible. The term "equivalent" requires that relationships between the geometrical quantities (dimension, displacements) as well as between the physical properties of the *model* should conform to those of the prototype. The dimension scale $\alpha_L = L_m/L_p$ (L are the lengths, m = model, p = prototype) is mostly of the order 1/20 to 1/100. The scale of unit weights is governed by the properties of the model material, $\alpha_\gamma = \gamma_m/\gamma_p$. When both scales are respected, the scale for the quantities having the dimensions of kp/cm^2 (stress, strength, modulus of deformation) is given by $\alpha_\sigma = \sigma_m/\sigma_p = \alpha_\gamma \cdot \alpha_L$. The magnitude of Φ, the shear strength angle, remains equal in the model and in the prototype.

The *model* is erected in a box, and generally has dimensions representing a cross-section of the prototype. In order that this model is not influenced by friction on the walls the thickness of the model should be at least 50 cm. Owing to the fact that the displacements are α_L times smaller in the model than in reality, and the stresses are α_σ times smaller, very sensitive measuring devices are necessary. Moreover, the measuring installations must not affect the development of the stress and strain fields in the model. The construction of the measuring devices and their practical performance present the most difficult problems in model methods.

The *model material* should exhibit Φ equal to that of the prototype and its deformation and strength characteristics should obey the relationships α_L and α_σ. Both conditions require model materials with a small but exactly controlled value of τ_0 ("cohesion") on one hand, but with a high modulus of deformation on the other. Several materials are used: ferrosilicon (like hard steel balls) or ballotini (glass balls), mostly 0.06 to 0.1 mm in diameter. These materials are mixed with a small quantity of glycerol, bentonite, or synthetic resin, or they are simply wetted, to obtain a small cohesion, but the shear strength angle is then only 15 to 30 degrees. Therefore materials with a greater angle are often used, e.g. sand, or it is necessary to compose the model from small blocks.

The model method has been applied successfully for the analysis of mining

problems (Mining Institute of the Czechoslovak Academy of Sciences). It is a unique method capable of simulating the processes, with which the modes of disintegration or collapse can be investigated and controlled.

Nevertheless, the construction of models is cumbersome and no materials exist which would maintain the above ratio of α in all situations of straining: in elastic properties, in the mechanism of plastic flow, in the mode of failure, and in the strength values.

Application of models using equivalent materials is of advantage for engineering geologists as they can construct and use them within the limits of their knowledge of mechanics and as they can compare the structural changes obtained with those observed in situ.

(b) *Mathematical solutions.* The revolutionary advance in computation techniques providing the possibility to solve thousands of equations makes possible the analysis of structures composed of a great number of elements. Amongst several methods of this kind, the *finite element method* has been most widespread. The studied rock mass (e.g. a valley cross-section, as an example of a plane problem) is divided into a net of elements, mostly triangular. A dense net of elements is chosen for areas where frequent differences in stress state are expected. Each element has the mechanical properties which belong to the rock in the particular area. Uniform or uniformly defined stress and strain fields are assumed to develop over the entire area of the element during loading. The components of external forces (loading, gravity forces) are assumed to act only in the corners of the elements. Therefore the formulation of relations between the displacements of the corners and the components of forces acting in the corners can be made. But the corners of an element are also nodes where the element is interconnected with other elements and where it shares common acting forces as well as common displacements. In this way a linkage between all the elements originates which can be mathematically formulated into a large set of linear algebraic equations.

When applying this method, care should be taken that the most important stress-strain characteristics for the studied rock are not neglected. The subsections 3.4 and 3.5 were written with respect to this requirement.

This method develops into a practical tool in engineering geology. Results are near reality when the mechanical properties have been well established and where the relations were not simplified in an improper way. Regarding these conditions it is obvious that comparison of the results with those of other methods is desirable.

3.8. Classification of rocks. Although this is not the aim of engineering geological activity, classification of rocks can be of assistance. Communication of information and exchange of experience are impossible without it. For engineering-geological purposes applied classifications have been developed with the objective to compare the studied rock with a rock type whose properties have already been established. This makes it possible to consider which properties of the studied rock are the most important, which tests can be used, which values of parameters can be expected, and how the test results can be applied when analyzing the particular problem.

3.8.1 The applied classification has been most elaborated with *soils* and identification as proposed by Casagrande (1948) is the most widespread. In Czechoslovakia this method has been incorporated into the standards for dams. This makes possible the exploitation of the wealth of *experience* gained by others; moreover it is unnecessary to burden the brain of engineers by learning any singular system.

Casagrande's procedure starts in removing particles larger than 3 inches (7.5 cm) from the soil. The soil is then classified as being coarse-grained or fine-grained using the No. 200 sieve (0.074 mm, which is the size that can be distinguished by the unaided eye) and the 50 % criterion.

(a) *Coarse-grained soil* is classified as *gravel* (G) if it contains more than 50 % (by weight) of grains larger than 4.76 mm. In the opposite case, it is *sand* (S).

Gravel is to be classified as:

— *Clean* if it contains less than 12 % of fines (0.074 mm, No. 200 sieve). Final classification is made on the basis of the *coefficients of uniformity* and of *curvature*. When the former coefficient is greater than 4 and the latter one between 1 and 3, clean gravel is classified as *well graded* (GW). Otherwise, it is *poorly graded* (GP).

— *Dirty* gravel contains more than 12 % of fines. It is classified as *silty gravel* (GM) or *clayey gravel* (GC), by using the same criterion as will be outlined with fine-grained soils.

Similarly, *sand* can be:

— *Clean, well-graded* (SW), when satisfying identical criteria, as with gravel, only that the coefficient of uniformity should be greater than 6.

— *Clean, poorly graded* (SP),

— *Silty sand* (SM)

— *Clayey sand* (SC).

The effect of gradation is not without importance, even with "dirty" soils: gravel containing 10 % of fines and exhibiting a coefficient of uniformity of 20, that of curvature of 2, and a plasticity index of 6, can be classified as GW — GM rather than GW — GC.

(b) *Fine-grained soils* contain more than 50% of fines (smaller than 0.074 mm). Further classification is made from the results of *consistency limits* (Section 3.3.2) and *silts* (M), *clays* (C), and *organic soils* (O) are distinguished. Another criterion is whether the values of consistency limits are *low* (L), as with "lean" clay, or *high* (H), as with "fat" clay. This classification is based on the diagram plotted in Fig. 3-34. Probably, the fact that silty soil can also show high values of consistency limits may be surprising, but soils with these properties do exist. For example, the soils of the East Slovakian lowland contain a large amount of micaceous and diatomaceous fines. They are assigned to group (H) on account of a small specific height but in spite of this classification the soil possesses only weak cohesion.

This classification scheme allows several questions of practical importance to be answered. Examples are given in Section 13.10. On the other hand, many other problems require that the *state of soil* is also known, e.g. the density of sands.

3.8.2. Unfortunately, a comparable applied classification does not exist for rocks, especially because rocks, as compared with soils, possess in addition a complicated *macrostructure.* Several applied classification systems have only been established for individual types of structures, e.g. for the purpose of tunnelling (Section 12.7). Table 3-1 also represents an attempt to typify rocks with the aim of estimating their strength using the method of analogy.

The most general classification of rocks is based on the spacing of *discontinuity planes.* For example, Deere (1968) recommends that planes spaced less than 5 cm apart should be distinguished as very close, when 5—30 cm as close, when 30 cm — 1 m as moderately close, when 1—3 m as wide, and when greater than 3 m as very wide. A counterpart to this is the classification of rocks based on block size.

Advanced applied classifications of rocks (see, e.g., Coates, 1964) principally consist of the examination of the most important *index properties* (Section 3.3) and the definition of a grouping based on these properties. However, Coates made progress by attempting to use mechanical behaviour as another important classification feature. According to him, the rock may be classified according to whether it fails in a brittle or in a ductile way. Objection may be raised that the mode of failure does not depend on the rock only, but also on the amount of hydrostatic stress, i.e. on the boundary conditions (Sections 3.4.4 and 3.4.5).

3.8.3. Weak rocks. Several questions involving the classification of this rock kind have been outlined in Section 3.3.4. From the point of view of engineering-

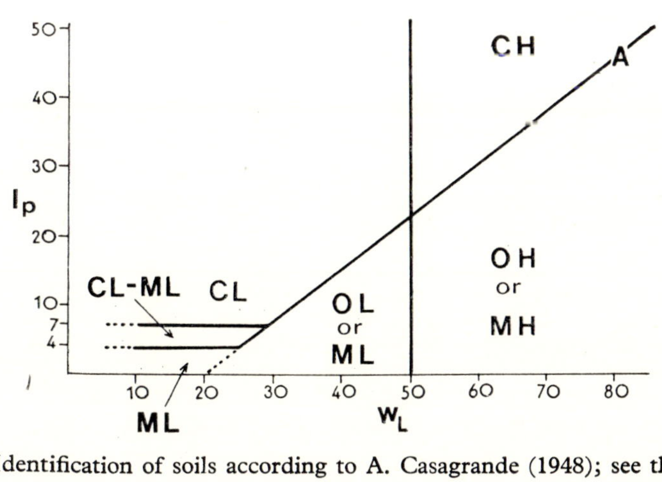

3-34. Identification of soils according to A. Casagrande (1948); see the text.

geological classification, the definition of a weak rock is of basic importance. Are they rocks which disintegrate after being submerged in water (Peshkovskii and Pereskovova, 1971), or those which do not disintegrate in water but after wetting and freezing, or are they simply all rocks with a compression strength smaller than an arbitrary limit (e.g. 500 kp/cm²)? In many countries engineering geologists are spared of the consideration of disintegration due to frost. But in Central Europe this feature is an important factor, which can cause failure, e.g. of slopes.

The tendency prevails to consider changes in structural cohesion after repeated saturation and desiccation as an important feature of fine-grained sedimentary rocks as well as of low-grade metamorphic rocks. According to this criterion weak rocks may be classified as:

(a) *Claystone* or *siltstone*, which disintegrate in water and often exhibit great swelling pressure. Their porosity is rather less than 30 %, as compared with 40 % of preconsolidated clay. Kazda (1968) showed that for the characterization of rock the decrease in strength after saturation as well as the swelling pressure are more useful than a direct examining of disintegration in water. The consistency limits supply only limited information for the given purpose. Claystones and siltstones often possess, due to their large condensation pressures, a *fissile structure*, and then they are called *shales*. Porosity of shales is often less than 20 %. Shales containing less than 50 % of particles smaller than 0.002 mm are called *silty shales*, otherwise they are *clayey shales*. Often very thin layers of silt alternate with those of clay. When a considerable amount of sand is present (about 25 to 45 %), they are called *sandy shales*. The tendency to disintegration is encouraged by the presence of organic matter.

(b) *Argillite* and *aleurolite*, which have experienced a small incipient recrystallization owing to large pressure. Their porosity is often less than 10 %. Due to a slight metamorphism, particles are aligned, but weak crystalline bonds resist disintegration after wetting. However, they cannot prevent disintegration due to frost.

(c) Rocks with crystalline bonds, which are borderline cases between hard and weak rocks. Regarding the character of bonds two groups exist:

— *Cementation types* in which bonds are produced by solid materials. *Calcareous shales*, containing 25 to 35 % of calcium carbonate, is a typical example. Nevertheless, from the point of view of behaviour, the distribution, and not only the quantity, of calcium carbonate, is important. Calcareous shales may therefore also disintegrate in water and then they are rather classified as *clayey* (or *silty*) *shales* in engineering geology.

— *Crystalline types*, which experienced more intensive recrystallization. Their porosity is usually less than 7 %; rock does not disintegrate as a result of wetting, and disintegration due to frost action is only a long-term process. *Chlorite slates* and several *phyllites* belong to this category.

Chapter 4

SUBSURFACE EXPLORATION

For the refinement of geological investigation various exploratory methods are available, the number and effectiveness of which steadily increase. The most common procedure in the engineering geological investigation of a building site is to drill or excavate holes, pits or trenches. It is therefore necessary for the engineering geologist to have at least a background knowledge of the principal methods, their advantages and drawbacks, of the kinds of samples obtainable, etc. so that he can choose the most economic and suitable procedure. The use of an inadequate exploratory method caused many-a-time the loss of money and time on unproductive investigation of the building site. The subsurface investigation for engineering constructions is made with the use of (1) test pits and trenches, (2) exploratory drifts, (3) sounding and (4) boreholes.

4.1. Test pits and trenches. Test pits are open pits with vertical walls, which are dug by hand and provided with sheeting if necessary. The cross-sectional

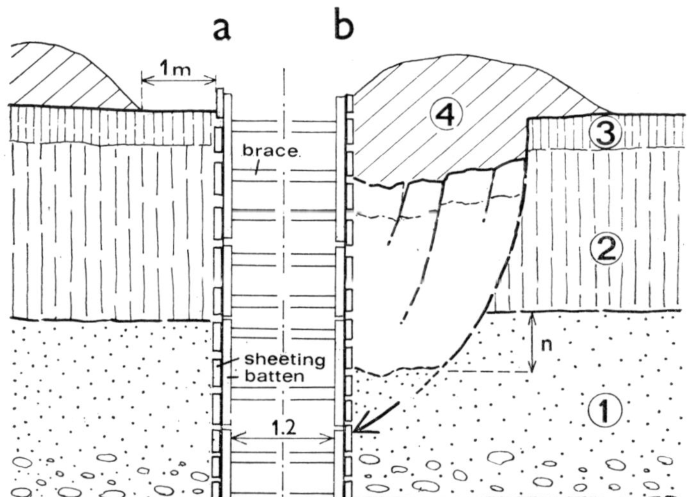

4-1. Test pit with horizontal sheeting (a); when the sheeting is not sunk simultaneously with excavation, the vertical section plotted may be incorrect (b). 1 — Sand and gravel, 2 — loess, 3 — humic soil, 4 — man-made ground.

dimensions of pits up to 6 m depth are 1.8 × 1.2 m. One dimension should not be smaller than 1.8 m so that working tools of normal length can be used. For test pits 10—15 m deep, the size 2×1.5 m is recommended, or even larger when water pumping is to be expected. Since the test pits must remain accessible for several days or even weeks, they should be provided with sheeting.

The test pits are timbered either with horizontal or vertical sheeting boards. Horizontal sheeting (Fig. 4.1) is used in rocks which remain unsupported in a 1 m high wall for at least several hours. The pit is excavated in stages 0.60 to 1 m deep and the timbering is inserted and braced against earth pressure. This sheeting method is not very suitable for geological investigation, as there is always the possibility that some of the boards may conceal an important boundary between two different rocks. Another drawback is the difficulty of removing one board without loosening the whole sheeting.

It is more suitable for geological investigation to timber a test pit with vertical boards which are braced with wedges against frames (Fig. 4-2). If gaps can be left free between the boards, the succession of strata can be observed on all walls of the pit. When necessary, the boards can be drawn out one after the other and replaced after the exposure has been measured.

In loose soils, particularly in water-bearing sands, it is necessary first to drive the boards into the ground and then excavate the material from the space thus delimited. When the pit must be fully timbered, the inspection of walls is im-

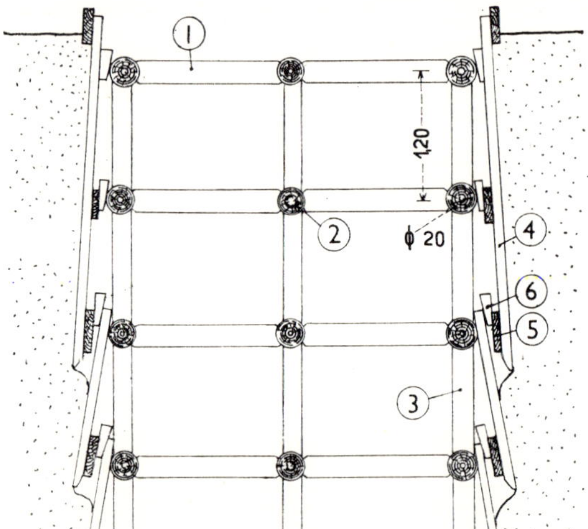

4-2. Test pit with vertical sheeting. 1 — Frame, 2 — braces, 3 — struts, 4 — sheeting boards, 5 — "tucking" pieces, 6 — wedge.

practicable and only samples taken during excavation are available. A well-made borehole gives the same results and usually at a far smaller cost.

Exploratory trenches are open excavations with vertical walls, which usually have to be timbered. A trench is made 0.8—1.20 m wide; broader trenches are

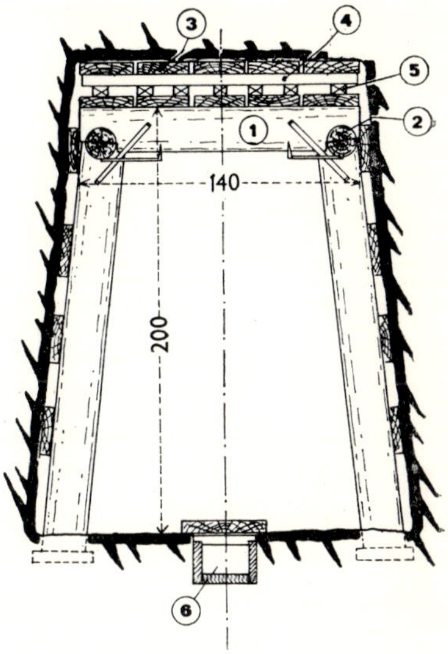

4-3. Timbering of exploratory drift. 1 — Cap piece, 2 — braces, 3 — sheeting boards, 4 — tucking pieces, 5 — wedges, 6 — drainage trough.

difficult to timber and the amount of excavation is unnecessarily increased. The exploratory trenches are excavated mainly on slopes, since they provide a continuous section through slope deposits. If their bottom is arranged at a moderate angle downslope, the water need not be pumped and they are easily accessible. They are therefore used mainly for laying out the communication lines on mountain slopes.

In investigating a site for the engineering work of particular importance, such as a dam or large quarry, exploratory trenches are excavated to the whole height of the slope or the bedrock is exposed within a large continuous area.

4.2. Exploratory drifts are excavated where accessible exposures should be available to a greater depth, e.g. for the investigations of damsites, tunnels and exploration of mineral deposits. A drift should be at least 1.40 m wide and 2 m high. For the sake of stability the drift widens downwards (Fig. 4-3). If necessary, the drift is provided with timber supports; in particular, the drift opening must be ensured when the rocks are weathered. The drift excavated at the site of a designed tunnel must have larger dimensions, depending on the working procedure to be used.

For the investigation of the bedrock at a damsite, drifts have often to be excavated under the valley bottom. They are as a rule accessible by vertical or inclined shafts. The drifts are excavated in the firm bedrock and the boreholes are made upwards from them to find the contact of the bedrock with the overburden. Figure 4-4 shows the exploration of the bedrock surface below the Drance-river valley in Savoy (France). At first the drift G 1 was driven from the shaft in firm Jurassic

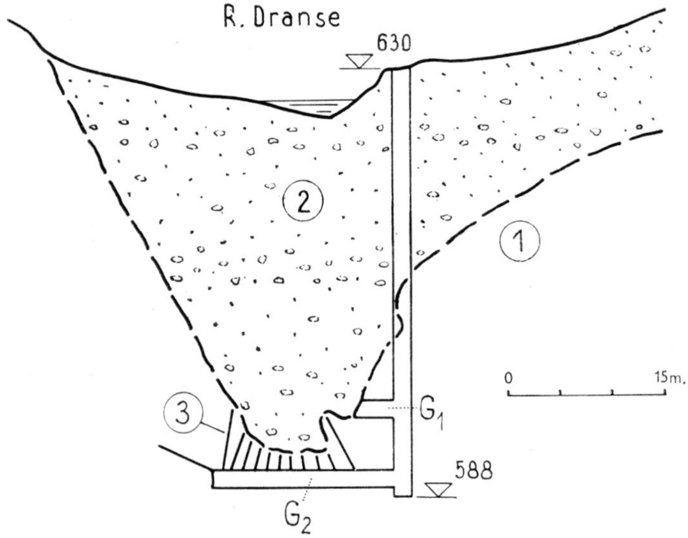

4-4. Determination of the bedrock surface below the floor of the Dranse river valley using galleries and boreholes. 1 — Jurassic limestones, 2 — river deposits, 3 — boreholes. (Gignoux, Barbier, 1955.)

limestones and when it encountered the valley alluvium, the shaft was deepened and drift G 2 was excavated across the valley. The accurate depth of the bedrock was established by boreholes drilled from the roof of this drift (Gignoux, Barbier 1955).

The exploration of a damsite with the use of exploratory drifts should be executed carefully so that the bedrock is not unnecessarily disturbed. The use of drifts is justified when the depth of loosening and weathering of rocks is studied, as these will be in any case removed during foundation excavations. Below the foundation surface the drifts are made only when necessary and without explosives or with only small charges. Their dimensions should be as small as possible. Large charges and long boreholes may disturb friable rocks so that the original fracturing cannot be distinguished from the cracks formed by blasting.

In investigating a damsite, exploratory drifts should not be driven parallel to the valley, below the foundation surface, since they would provide paths for water seepage. Their subsequent filling with concrete and grouting is always expensive.

4.3. Sounding is used for getting preliminary information of the thickness of deposits in the flood plain, of peat and other little cohesive soils on the firm rock. A steel rod is jacked or driven by a hammer into the ground whose character is inferred from its penetration resistance. It is a rather primitive procedure but in some cases it was used with success to complement the geological sections between an open network of boreholes.

This method is applied, for example, for establishing the necessary length of piles to be driven into sands and gravels of different density. From the diagrams showing the progressive penetration of rods the depth at which the penetration resistance changes is apparent but no information about the composition of soil can be obtained. The forced-in rods can thus serve to delimit the extent of individual beds, the composition and physical properties of which must be examined by boring.

Sounding with hand-operated device is often used in mapping the surficial deposits. The device consists of a steel pipe 30 mm in diameter and 1 m long and of a steel rod 2 m long and provided with an oblong groove. The pipe is hammered into the ground by a wooden mallet to the depth of 1 m under simultaneous turning. On pulling out the auger, a continuous profile 1 m long is obtained. The 2 m long rod is dropped into the bored hole and hammered 1 m deeper. The soil samples from the depth of 1 to 2 m are retained in the groove of the rod. This method is a suitable complementary tool in geological mapping but it is not adequate in investigation of deeper foundation ground. The samples are also too small to be reliably determined and evaluated.

Vibration sounding can be used in superficial deposits to a depth of 20 m and gives good results in loose and cohesive soils with the exception of tough clays, coarse gravel and stony detritus. This method is based on pushing the pipes by weight and vibration of special vibrators. Two working procedures are used.

(1) A steel pipe 2—3 m long and 6 cm across, provided with a V-shaped groove with a central angle of 60—120° is attached to the rod set, through which the effect of vibration is transmitted to the bottom of the hole. In cohesionless soils or at sounding to a greater depth casing must be used, which is also forced into the ground by vibration.

(2) The other method is usable for sounding to a depth below 10 m. Pipes consisting of parts 2.5 m long and 7.5 cm in diameter are provided with a V-shaped groove with a central angle of 100° for the whole length (except the sleeves at the end). They are gradually pushed into the ground by vibration. On pulling out the pipes a continuous profile of the hole is obtained. In measuring the thickness of the individual beds, the compaction of the soil by vibration must be taken into consideration. It is expressed by the difference between the height of the soil column and the depth of the hole, which must be measured during the pushing down of the pipes. The samples furnished by this procedure are of satisfactory

quality. The working procedure is relatively rapid, but it can be used only in rocks of suitable character.

4.4. Boreholes. Boreholes used for engineering geological investigation are made using various methods and devices ranging from simple hand-operated augers to ingenious drilling machines, which permit the drilling of holes several thousand metres deep. Auger boring is used for shallow holes of about 20 m depth. Percussion and core drilling is used for holes to 300 m depth. Drilling to greater depths is rarely used in engineering geological investigations.

4.4.1. Auger boring. The boring rig consists of several augers of various shapes, a drill rod set, casing, a steel or wooden tripod, a crane or winch, and auxiliary tools. For drilling shallow holes in soft rocks augers of spoon or barrel type, helical and bailer types are used (Fig. 4-5).

A *spoon-shaped auger* is a hollow cylinder made of steel, cut lengthwise, open at the bottom and provided with a sharp bit. The bit cuts the rock and holds it in the spoon, which is attached to a drill rod. Additional rods are screwed on as the hole is deepened. The spoon-shaped auger is suitable for soft homogeneous rocks, such as loams, clayey sands, fine-grained detritus, weathered materials, granite and gneiss eluvium, etc.

Depending on the character of the rock an auger with a wider or narrower slot is chosen so that the loosened rock will not fall out. Rock samples obtained by augering are disturbed as the rock is cut away by a bit and pressed into a cylinder. In cohesive soils the foremen often try to attain a quicker output by pouring water into the hole. As the consistency of the rock is thus changed this procedure is not allowed in some countries. Wetting does not usually reach the lower end of the spoon and the sample from this end is most representative of the state of soil, but it cannot be regarded as an undisturbed sample.

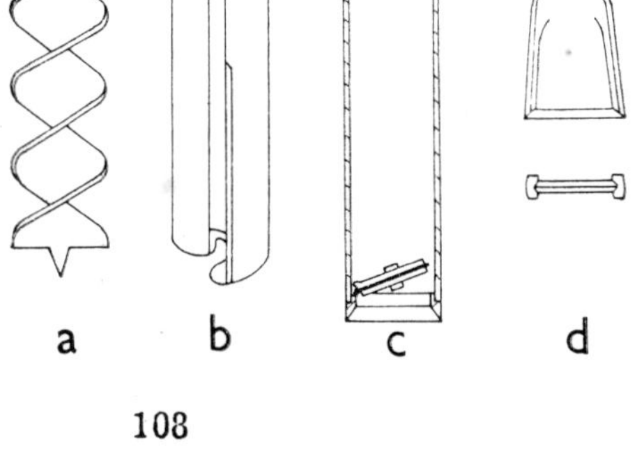

4-5. Characteristic auger tips. a — Helical (worm) auger, b — spoon auger, c — bailer, d — chisel.

The *helical auger* is made of steel and has a spiral form with coils of different steepness. This tool is used mostly in stiff clays and marls, clayey shales, and for loosening dense sand and gravel or stony detritus. In sands and gravels a helical auger with a cylindrical sleeve has proved suitable since pebbles caught on coils do not fall away.

The *bailer* or *valve* sludge pump is a cylinder made of steel, provided with a hardened bit on the lower end. Inside the bit there is a leather flap or spherical valve. This tool is used in loose water-bearing soils, such as sands and gravels. The bailer is dropped abruptly to the bottom, and the material penetrates into the cylinder through the half-opened valve. The movement is repeated several times until the larger part of the cylinder is filled. The hole must be secured by casing. The bailer is also used for bailing the drilling fluid at percussion drilling.

The *drill rod* set consists either of steel pipes or steel rods of square section. It is screwed together from pieces 2—8 m long. Care must be taken not to damage the threads during transport or drilling. For lifting the rods or other tools which may have fallen into the hole various special instruments are used. The lifting is always difficult, especially in deep boreholes, and sometimes it has been necessary to abandon the unfinished borehole.

Drilling tools and rods are attached to a steel or wooden derrick over the head of which a cable is suspended. The tool is lifted from the hole with a hand crane or a motor-driven winch. (Fig 4-6).

The *casing* consists of steel pipes which are used to protect the walls of a borehole in non-cohesive soils from caving in. The diameter of casing should be 10—15 mm larger than that of the boring tool. Exploratory boreholes should always be protected with casing, as rock fragments loosened from the wall and fallen on the bottom may lead to incorrect geological interpretation. Casing pipes are sunk and lengthened gradually as the hole is deepened. The first pipe is provided with a steel shoe at the lower end. If the borehole cannot be completed with the initial diameter, a new set of pipes of smaller diameter must be sunk into the borehole. Casing pipes should be pushed to the very bottom of the hole but not deeper than is necessary for drilling. The pipes sunk deep below the bottom of the hole could close an aquifer which would thus escape recording. In some countries a prescription holds that casing should not be driven deeper than 0.50—0.70 m below the bottom of a borehole.

Boring diameter. Shallow boreholes usually have a diameter of 150—300 mm. If undisturbed samples should be taken, the diameter of casing pipes must be at least 200 mm, since a normal core sampler is 160 mm across. In cohesive homogeneous soils such as loams, loess, clays, a smaller diameter may be used, as no heavy tools are needed for drilling.

A larger diameter must be chosen in coarse gravel and stony detritus since heavier tools necessary for breaking hard pebbles are of larger size. In coarse gravel a diameter of at least 300 mm allows the removal of even fairly large pebbles by a bailer without breaking them. Another advantage of the large diameter is the fact that a larger amount of soil is obtained so that the ratio of sand, gravel and pebbles can be estimated with greater precision.

When it is to be expected that a borehole cannot be completed with the initial diameter, boring should be started with a larger diameter, so that a narrower pipe can be inserted. This is the case, for example, when near-surface aquifer should be closed.

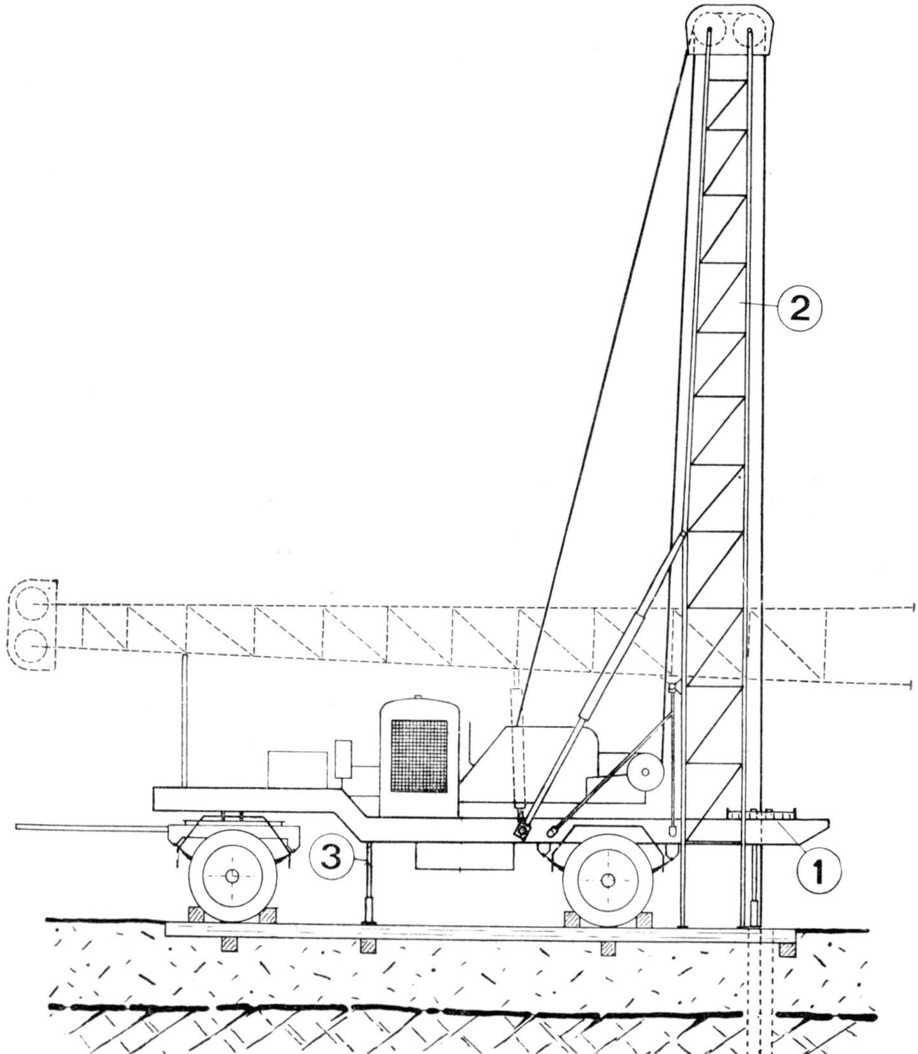

4-6. Truck-mounted rig for auger and percussion drilling (B-120 M). 1 — Revolving table, 2 — derrick, 3 — supports.

Samples obtained by auger boring are disturbed (loamy and clayey soils are kneaded) but the original moisture content and consistency of soils can be appraised from them except when ground water penetrated into the hole or water was poured in. It must not be forgotten that in boring in water-bearing soils by a bailer the

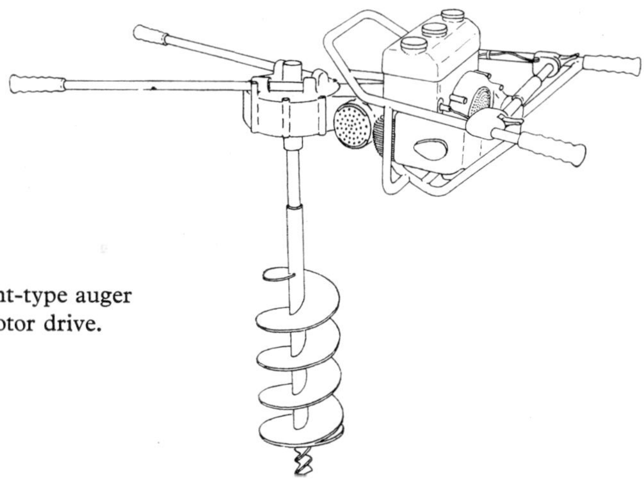

4-7. A light-type auger with motor drive.

sample is washed during boring. Clay and silt particles are poured out together with water so that the sample consists mostly of relatively pure coarser fractions, and is not representative. Cases are known when pure sandy gravel was recorded in the bore logs as a suitable concrete aggregate but when excavated, it proved to be contaminated by clay admixture so that it was unusable without treatment.

As mentioned above, some boring crews use various procedures to make boring easier and more rapid, such as pouring water into holes when boring in clayey soils or adding clay when boring in loose sand or friable sandstone. All these depreciate boring results, distorting the true picture of the soil and may cause an erroneous evaluation of geological conditions.

Auger boring is effective in soft cohesive soils such as loams, clays, marls, soft clayey shales and non-cohesive sands, gravel and detritus to a depth of several tens of metres. Where they are interlaid by layers of harder rocks or of coarser gravel, auger drilling should be combined with percussion drilling.

A light auger type driven by a small engine (2.5 HP) has been developed for geological mapping and investigation of surficial deposits in areas with difficult access (Fig. 4-7). The engine is mounted on a truck or trailer (Fig. 4-8) or may be carried by a worker on his back; rotation is transmitted by a flexible shaft. Augers of normal types are used; the recently developed helical auger of 2.5-in diameter gives good results in clayey rocks (Fig. 4-9). Auger rigs of similar construction are a useful tool of a mapping geologist thanks to their easy manageability and rapid working.

4.4.2. Percussion (churn) drilling. Percussion drilling consists of breaking the rock on the bottom of the hole by a steel chisel (Fig. 4-5d). After lifting the chisel it falls by its weight on the ground. After each blow the chisel is turned a little so as to bore a circular hole. Chisels are made of hard steel and are of various forms. In firm rocks of medium hardness, not strongly jointed, flat straight-edged chisels are generally used. In the rocks traversed by open joints, chisels with a cross bit or Z-bit are applied, as they are less likely to stick in the joints. The form of the bit depends on the hardness and character of the rock.

Previously, a chisel with rods was suspended on a brake-staff which enabled the chisel to be lifted regularly. In shallow boreholes the tool was lifted by hand and it had to be worked by four to six men. Today this cumbersome labour is replaced by machine drive.

A drilling jar is inserted between boring rods and the chisel so as to protect the rods from the effects of blows. At the moment of impact this device loosens the

4-8. Drilling equipment with transportable motor. (Photograph by Testlab Co. Chicago.)

chisel from the rods. In hard rocks the weight of the chisel is increased by a bar which is inserted between the jar and the chisel.

The cuttings and slurry must be removed regularly from the borehole so that percussive blows are not damped. In order to clear the hole the rods with the chisel are withdrawn to the surface and a bailer is lowered into the hole to remove rock fragments and slurry. The slurry should be sufficiently thin for easy bailing and therefore some water is poured into the borehole. Percussion drilling is difficult and slow in firm rocks. In boreholes of great depth, extraction of the chisel, dismantling of boring rods and lowering of a bailer delay the work. This method does not comply with the present-day requirements.

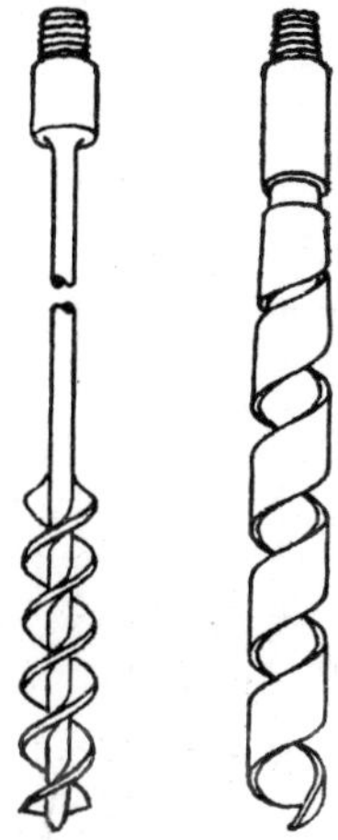

4-9. New types of augers for drilling small-diameter holes in clayey rocks.

Percussion drilling is not suitable for engineering geological investigation. It is rather difficult to identify the rock accurately and determine its technical properties from the rock detritus and slurry.

4.4.3. Wet percussion drilling. In order to accelerate drilling to a greater depth the wet percussion method has been introduced. The detritus obtained by boring is washed out by water forced into the borehole. The velocity of water must be high enough for the rock detritus to be brought to the surface by water flow. This procedure is more rapid as drilling need not be interrupted and the rods to be lifted when crushed rock is removed. Water under pressure keeps the bottom of the borehole clean, which also increases the output of the chisel. A dense liquid, such as a mixture of clay and water, is sometimes used for flushing. This drilling fluid (or „driller's mud") carries rock fragments more easily and consolidates the borehole walls in poorly cohesive rocks.

Cable drilling differs from normal percussion drilling in that a chisel is suspended on a rope instead of on boring rods. Slow dismantling and extracting of rods is replaced by winding the rope on a drum.

High-speed drilling is based on the principle that drilling is faster if a chisel makes a larger number of blows from a small height than a few blows from a larger height within a given interval. Rapid movement of the chisel is produced by the oscillation of a lever on which the set of rods is suspended. Strong springs inserted between the suspender and rods increase the length of oscillation and thus also the strength of blows. Rock detritus is generally withdrawn by flushing. This method is used for exploration for oil and deep-lying coal deposits. It is not used for engineering geological investigation as the rock is totally broken and pulverized and brought to the surface in the form of thin sludge.

4.4.4. Rotary drilling. The rotary system is used when high rate and output is demanded as, for example, in deep oil wells. In this system the whole bottom of

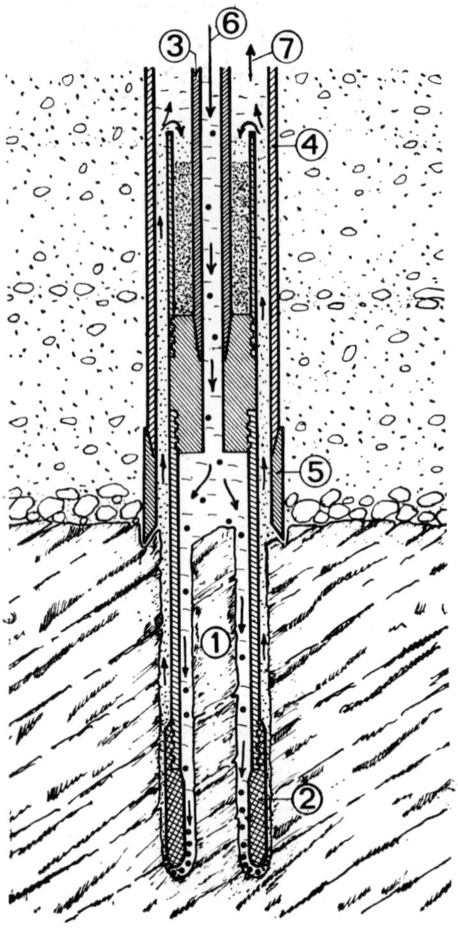

4-10. Shot core boring. 1 – Core, 2 – coring bit, 3 – string of drilling rods, 4 – casing, 5 – drive shoe, 6 – pressure water and shot, 7 – water rising to the surface.

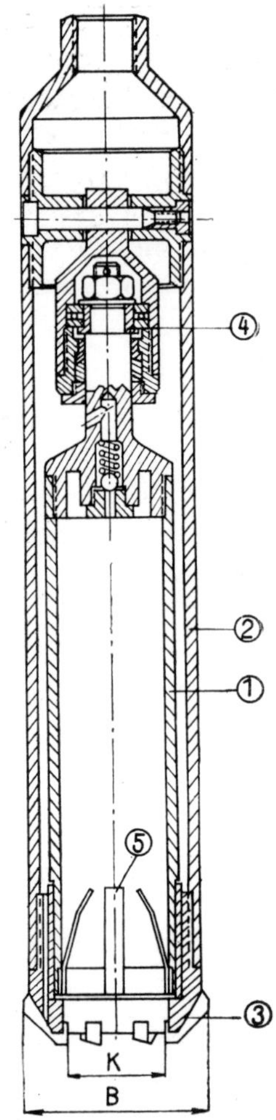

4-11. Double-tube core barrel with rotating inner tube. 1 – Inner tube, 2 – outer tube, 3 – drill bit, 4 – swivel-type head, 5 – core catcher. (Salzgitter A. G.)

4-12. Diamond bit for core boring.

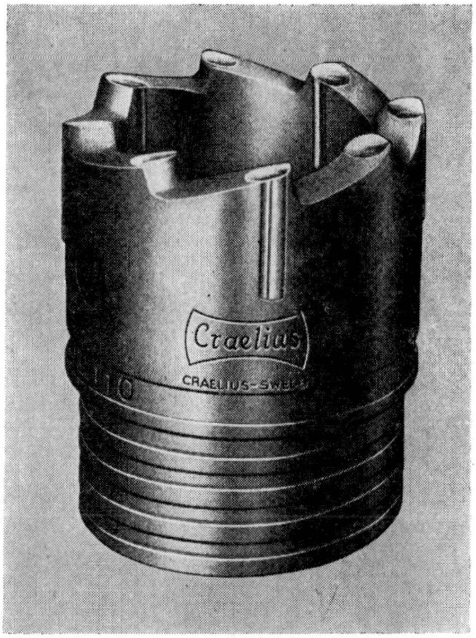

4-13. Steel bit provided with hard-metal inserts.

the borehole is ground by a rotating bit so that only crushed rock is obtained, which is washed out by drilling fluid.

4.4.5. Core boring. By this method the rock is ground only at the periphery of the borehole, and a continuous rock sample — the core — remains in the central part. It is the most suitable procedure for investigation of hard rocks. Boring is carried out with the use of an annular bit. The bit is screwed to a steel cylinder of the same dimensions, called the core barrel, into which the core passes gradually (Fig. 4-10). The core barrel is usually 2—3 m long. The core generally breaks off on a bedding plane or joint. A steel ring inserted between the bit and core barrel is used to grip the core. This core lifter has the form of a truncated cone and consists of springs which let the core pass into the core barrel but not back from it. After deepening the hole by 0.50—1.50 m, at the most by the length of the core barrel, the rods with the bit and core barrel are extracted to the surface. The bit is screwed off and the core removed from the core barrel, carefully marked and deposited in a special core box. Water under pressure constantly brought into the hole during boring carries away the cuttings.

In soft and jointed rocks double-tube barrels should be used. The inner tube protects the core from being disturbed by drill water. The water flows between the outer and inner core barrels and comes into the contact with the rock only under the bit. The original double-tube barrels were tightly connected but at present barrels with a revolving inner core barrel are suspended on a ball bearing. With the use of this device the inner tube enclosing the core may remain stationary while the outer tube is revolving (Fig. 4-11).

Bits used for core boring are of various types (Fig. 4-12). *Diamond bits* have proved to be very effective. Originally, boring was made with black diamonds (so-called carbonados) which are found only in Brazil. They are composed of tiny cemented crystals and virtually represent the hardest rock on the earth's surface. Later on, fragments of white diamonds (bortz) were also used but they are not so tough and split along cleavage planes. The fragments are cast in metal so that the mass is penetrated by diamond spalls. Boring with diamond bit gives smooth and fairly long cores but it is relatively expensive. In softer or disturbed rocks *steel bits* or bits with teeth toughened by *hard metal* are used (Fig. 4-13). In boring with these bits care must be taken not to loosen a tooth by impact, as it could within a few minutes destroy the bit or, when shifted upward, cut off part of the core barrel.

In exceptionally hard rocks shot (or calyx) core boring is used (Fig. 4-15). Steel shot, 0.5—2 mm across, is fed together with water to the crown provided with a narrow slot cut. The shot is brought into fast rotation (150—200 turns in a minute) thus grinding the rock under the bit. This method is more difficult than diamond drilling as the cores have rough surfaces and can easily be broken or crushed during drilling.

The modern core boring systems make it possible to drill in vertical, horizontal and diagonal directions. The equipments are produced today in several types developed for different depths and borehole diameters. A diameter of 76—150 mm is commonly used for exploratory boreholes several tens of metres deep.

For an easy and fast transport the drilling equipment, together with the tilting

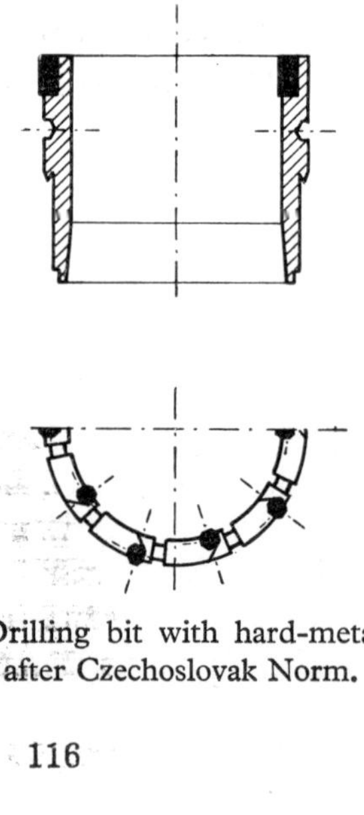

4-14. Drilling bit with hard-metal teeth after Czechoslovak Norm.

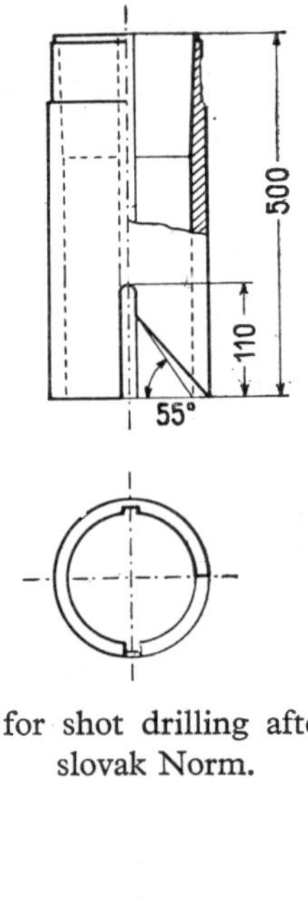

4-15. Bit for shot drilling after Czechoslovak Norm.

4-16. Exploratory boring for the Prague underground in the Vltava river bed; the rig is mounted on a raft.

rig, is mounted on a truck or crawler. For drilling through water, e. g. in a rive bed, the drilling equipment is placed on boats (pontons) (Fig. 4-16) or on driven in-barges.

In Craelius Atlas-Copco drilling rigs the drive is transmitted mechanically from the power unit through a clutch and a gearbox to a friction wheel in the spindle head of the drilling equipment (Figs. 4-17, 4-18).

The equipments of Wirth type (Fig. 4-19) have a hydraulic drive mounted directly on the spindle head; this arrangement facilitates a relatively high lift and longer core runs. In drilling rigs used for large-dimension boreholes and deep hydrogeological holes, a rotary table is driven by a mechanical gear and turns itself the first rod. The heavy Salzgitter rigs and some Soviet systems are of this type.

The Craelius-Duplex equipment has been developed for boring in non-cohesive soils (e. g. for grout holes, anchoring, drainage boreholes). This method is based on the principle of a double drill rod string; the holes are drilled simultaneously by an external string of strong tubing, which is also provided with a drill bit, and by the internal drill rod string. A great advantage of this system is that the inner rod string with the core barrel can be pulled out quickly through the external string to remove the core. When a borehole is sunk into the bedrock, the external rod string can be left in the hole and drilling continues only by the inner tube with a core barrel. The Craelius-Duplex method works with diameters of 86 and 66 mm.

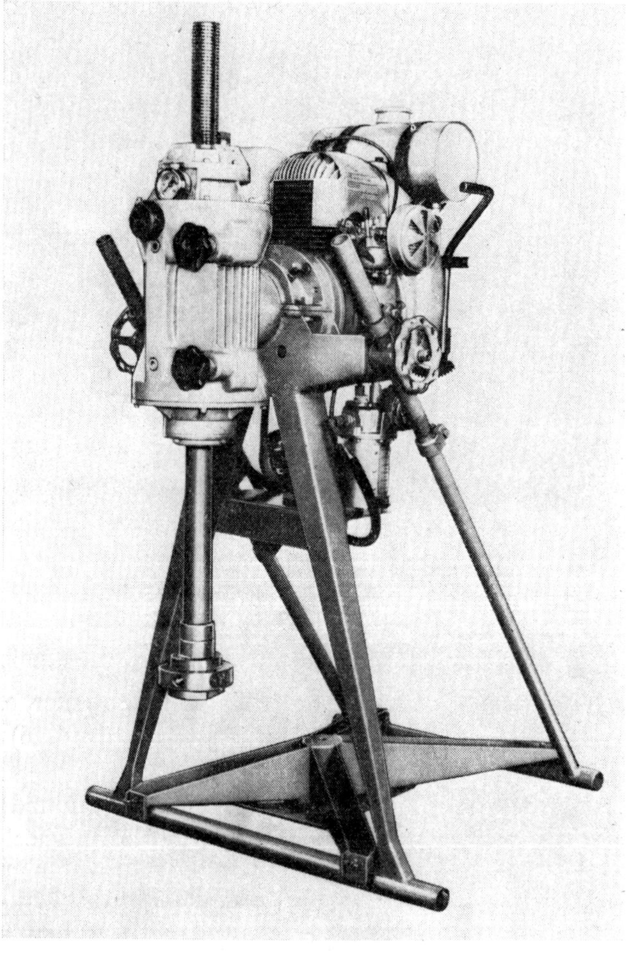

4-17. Atlas-Copco rig for drilling to 75 m depth.

4-18. Atlas-Copco rig D 750 for core boring to 750 m depth. 1 — Motor, 2 — winch, 3 — chucking head, 4 — hydraulic system

In recent years the *Wire-line* core boring system, which was originally used in oil prospection, has also been introduced in engineering geological investigation. It is essentially a double-tube barrel, the inner tube of which can be lifted by means of a wire rope without pulling out the drill rods. During drilling the inner tube is connected with the outer by jaws; on their opening the inner tube is released and can be pulled out with the core by a rope, on which an assembly serving to open the jaws and grip the inner tube is suspended (Fig. 4-20). The Wire-line system is suitable mainly for boreholes 100—200 m deep, but can also be used for depths of 40—50 m in disrupted rock, where the core must be often withdrawn. In this case the application of this method saves time, as the string of rods and the outer tube need not be pulled out.

For special purposes devices enabling core boring with a diameter of 750—1200 mm were developed. Boring is performed by a rotating cylinder, either with steel shot or with a steel crown provided with teeth of hard metals, which can be used in softer rocks. In hard rocks the core must be broken from the bottom with the use of a small explosive charge. Figure 4-21 shows the Calyx system for large-diameter boring without a drill rod string. The engine with a core barrel and crown are sunk into the borehole on a rope and fastened to the walls of the bore hole With this drilling equipment holes 90—180 cm across and as much as 100 m deep can be drilled.

Large diameter boreholes have been used advantageously for investigations of damsites in cavernous limestones, for reconstructions of dams and for water supply. The yield of wells can be increased by driving short galleries below the lowered water table from boreholes 900 mm in diameter (Fig. 4-22).

Large-diameter boreholes provide the best opportunity to study the properties and the mode of deposition of rocks, as a geologist can be let down into the hole to examine the whole succession of beds on the walls of the borehole. He can measure the orientation and density of joints or fissures, their habit and type of filling, as well as the thickness and nature of interbeds, etc. Large-diameter boreholes are usable only when the inflow of ground water can be mastered either by pumping, or by grouting the rock before drilling.

4-19. Drilling rig of the firm Wirth et Co.

Boreholes made for engineering geological investigations, particularly for grout curtains and tunnels, should maintain the desired direction as accurately as possible. In order to check the direction during drilling, deflection of the hole, i. e. deviations from the vertical or desired direction, are measured by various

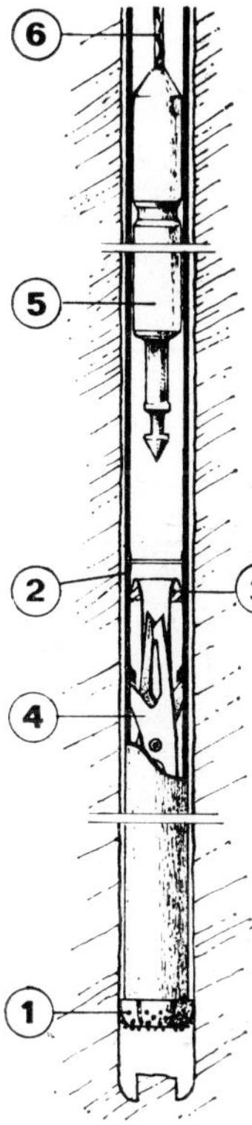

4-20. Wire line core boring (Atlas-Copco). 1 — Drilling bit, 2 — outer core barrel, 3 — inner core barrel, 4 — jaws, 5 — overshot assembly, 6 — wire.

devices, from very simple to ingenious instruments called *inclinometers*. The direction of a vertical hole can be checked quite simply, by the occasional insertion of a bottle partly filled with thin plaster. The deviation from the vertical is given by the angle made by the surface of hardened plaster with the side of the bottle. Melted gelatine or acid, which etch a line on the glass cylinder, are also used for this purpose. Some more complicated instruments are based on the same principle but are combined, for example, with a compass so that the dip and direction of the borehole can be measured at the same time. Some inclinometers use a freely suspended pendulum whose projection on the compass is photographed. Electric inclinometers applied for very deep boreholes measure the deflection of the borehole continuously.

Core boring must be carried out expertly and carefully so as to ensure the highest possible core recovery. The length of the core obtained does not always correspond to the depth drilled; we speak about the *loss of core*. The percentage recovered

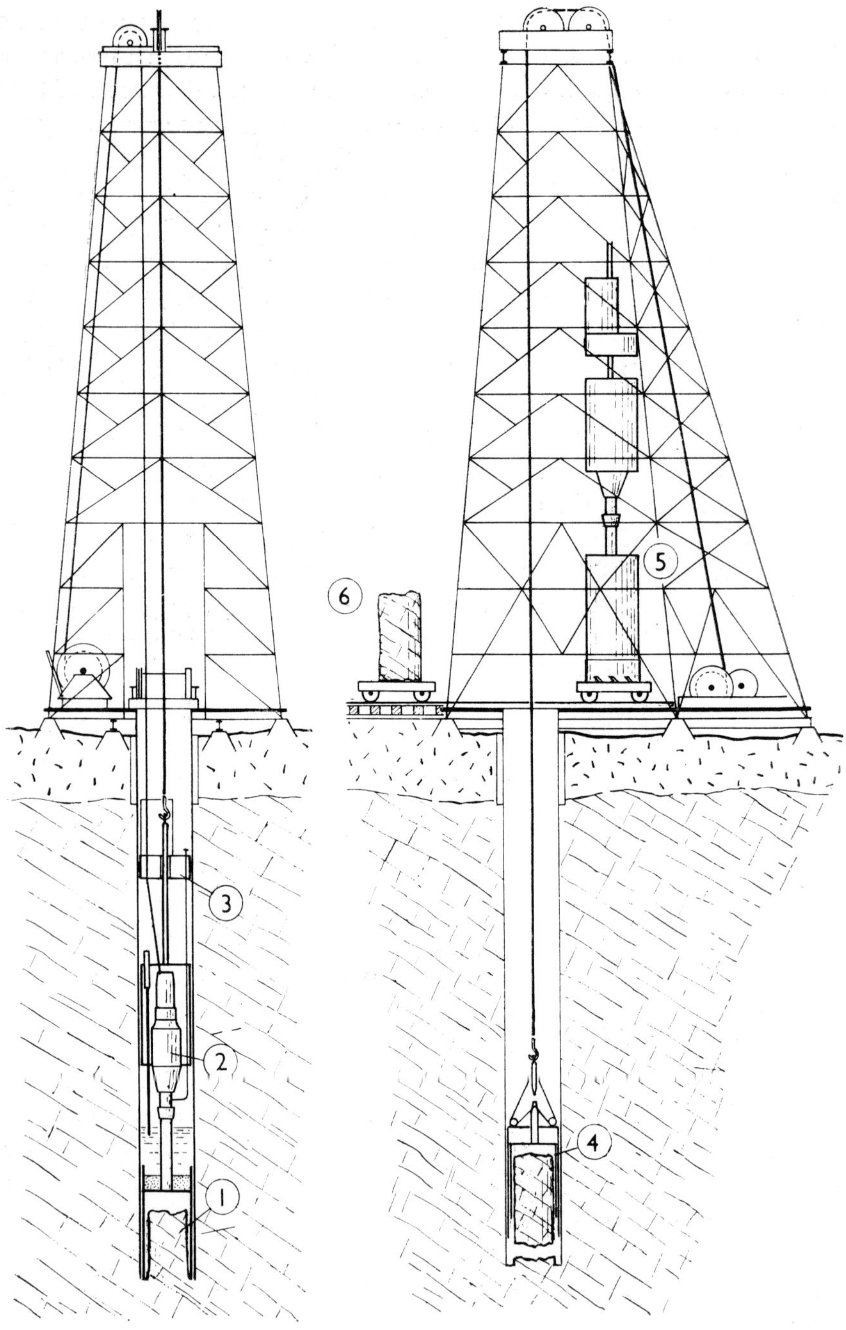

4-21. Calyx rig for large-diameter boring with motor and core barrel suspended on a cable. 1 — Core barrel with a bit, 2 — electric motor, 3 — anchored rod, 4 — lifting of the core, 5 — core barrel and motor set aside when the core is being lifted, 6 — core.

4-22. Core recovered from a 90 cm borehole. (Photograph by Záruba.)

from each core run should be recorded. A full core recovery can be obtained in firm homogeneous rocks not traversed by joints. In granite and other plutonic rocks or in thick-bedded limestones and sandstones 90—100 per cent of core is recoverable. Core recovery in thin-bedded and schistose rocks amounts to at least 70 per cent. Densely jointed rocks, especially when the joints dip steeply, furnish low recovery; generally only rock fragments are brought up in the core barrel, while the prevalent part of cuttings is flushed by drilling fluid. In this case the cuttings are also preserved so that the character of the rock can be at least roughly estimated. The quality of boring in disturbed rocks depends greatly on the correct choice of boring diameter and on the skill and experience of the driller foreman. The larger the diameter and the shorter the core runs, the greater is the possibility of obtaining high core recovery. Also of importance is the pressure under which wash water is forced into the borehole. The pressure should be the lesser the softer and more disturbed is the rock, as the rock would be still more disturbed by a strong flow of pressure water.

With the present progress made in boring techniques, 70 % core recovery is desirable even under difficult conditions. In establishing the efficiency standards, not only the hardness of rock and the length of the borehole, but also the core recovery, should be taken into consideration.

A continuous core, however, cannot always be obtained even if boring is carried out with outmost care, particularly in soft and strongly jointed rocks. Since in these cases a direct examination of the borehole walls is desirable also in holes of small diameter, various optical instruments on the *periscope basis* have been constructed. The inspection of an illuminated borehole is thus enabled by means of a system of mirrors and telescopes. These apparatuses did not furnish satisfactory results.

Photography of the borehole walls by special cameras proved more advantageous. The apparatus is mounted in a cylinder of stainless steel, 80 cm long and 7 cm in diameter and provided with a window in its mantle. An oil-damped compass is placed on the bottom of the cylinder and above it a conical mirror, which reflects the picture of the wall into an objective. Illumination, the shutter and drift of the film are synchronized with the device for extracting the cylinder from the borehole so as to obtain a continuous photographic record. The walls of the boreholes can also be photographed under water provided that water is not too turbid. This method gives very good results particularly with the use of colour film.

A similar method was used in the German Democratic Republic for geological investigation of areas with extensive leaching of gypsum complexes. The size of karst cavities was photographed by a simple device constructed for this purpose.

The application of *television* for borehole logging has been a great improvement on the previous methods. The first apparatus was developed in 1957 in the Federal Republic of Germany by combining a miniature television camera with an appropriate optical device. The instrument is mounted in a steel cylinder, 800 mm long and 62 mm in diameter, which is suspended on a television self-supporting cable up to 400 m long. An image of the borehole side is projected on the screen, being reflected by an inclined mirror. On either side of the mirror there is a row of bulbs which are switched on alternately, since it can be determined from the shadows whether the joints are open or filled. A compass and a level in the lower part of the cylinder make it possible to determine the direction of the axis of the instrument. The range of a television camera is limited by the length of the cable. A depth of 400 m is quite sufficient for engineering geological purposes. In Czechoslovakia, the industrial television camera of the firm Tesla has recently been modified for engineering geological investigation. It was tested at the construction of the Orlík Dam and proved successful in checking the results of grouting. The TV camera is combined with a photographic apparatus so that the desired parts of the borehole can be photographed.

4.5. Evaluation of subsurface exploration. The investigation of a building site is not completed by carrying out several boreholes and pits. To fulfil the task it is necessary to describe and measure the holes, take rock samples and evaluate the work from the engineering geological point of view. Otherwise the entire exploration would be waste of time and money. It is also of utmost importance to submit the final report in time and in a comprehensible form for both the designer and construction engineer.

The progress of work should be recorded systematically in the field and instructions for further investigation should be drawn immediately from the results. Unfortunately, this is not always the case in practice. The records and descriptions of rocks are often in the charge of the foreman and only occasionally inspected by the engineer or geologist. Even if the foreman is conscientious and experienced, his foremost interest is in continuous operation, repair of equipments, high output and timely completion of work, which might be at variance with the requirements of engineering geological investigation. Drilling to a greater depth, more frequent sampling or taking more undisturbed samples, all delay the work and reduce the output of the crew. Moreover, a specialist in drilling may be far from being an experienced petrographer.

Since the results of boring depend on the accuracy and objectivity of records made during the work, a technician having basic knowledge of geology should be present permanently at the building site. He will supervise the working procedure, describe the sequence of strata, carry out necessary measurements, take samples and will be responsible for correctness of records, cooperating thus with the geologist who is charged with the management of exploration.

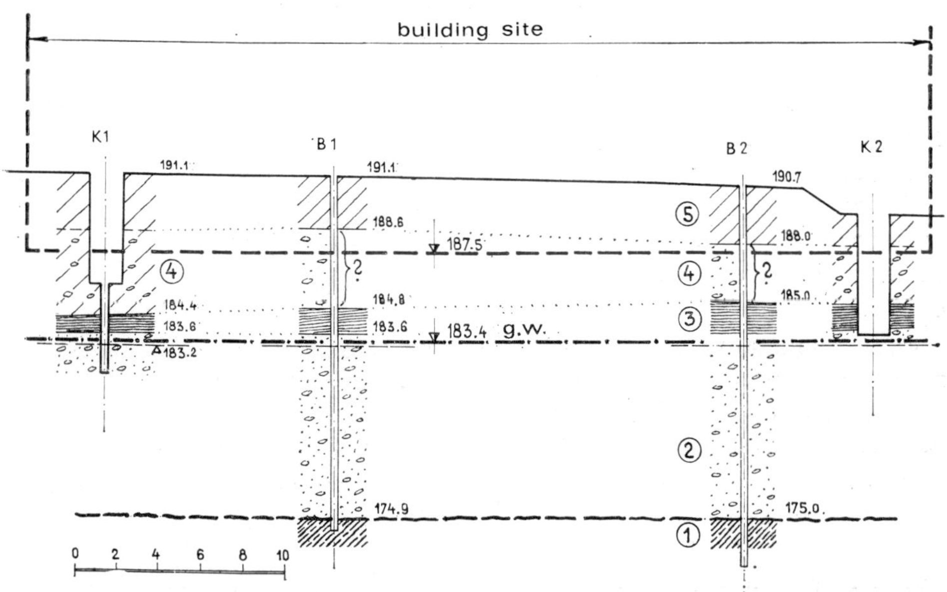

1-23. False interpretation of boring results. K 1 and K 2 control pits have shown that bed 4 is man-made ground and not gravel in natural position. 1 — Ordovician shales, 2 — sand and gravel, 3 — clayey-sandy alluvium, 4 — sandy gravel (man-made ground), 5 — young fill.

In auger boring, for example, undisturbed samples for laboratory tests cannot be taken from all beds. It is therefore important to describe and designate cohesive soils (loams, clays, marls) right on their extraction from the borehole. The original state of these sediments cannot be determined from the dried-out and hard samples, as they are found in a sample box after several weeks. The natural state should also be recorded in loose soils, the samples of which do not virtually change with time.

The investigation of one building site on the bank of the river Vltava in Prague demonstrates the importance of expert evaluation of boring results (Fig. 4-23). A bed of sand and gravel, 13—14 m thick, was encountered in several boreholes. Its surface was recorded at a depth of 2.50 m and a thin clayey-sandy interlayer was recorded at 6 m depth. On the basis of the bore logs the building was to be founded on a continous footing on the surface of sand and gravel bed (el. 187.50 m). In excavating the foundation, the gravel proved to be loosened and highly compressible with randomly distributed layers of loamy fills. In test pits carried out additionally it was ascertained that the gravel bed is not in an original position but that it is man-made ground deposited on clayey alluvial deposits. The foundation had to be deepened and the building was founded on a true gravel bed (elev. 183.40 m). The engineering geological investigation was not carried out with sufficient care. The upper layer of gravel and sand could be recognized as a fill or at least designated as loose and little consolidated. Moreover, fragments of bricks and Medieval ceramics were found on the surface of clayey alluvium. As has been later ascertained, they were also found in the boreholes but the foreman thought they got there from the surface layer through inadvertency and did not record their finding, neither did he deposit them into a sample box. This is one of many cases when boring technically well carried out was depreciated by its wrong evaluation.

Boring log. The progress and results of boring should be recorded daily in a copying log book. In addition to normal records on the number of workers and working hours, on the kind of work, weather and temperature, the records should contain:

a) Number and dimensions of boreholes being drilled.

b) Description of rocks, particularly of their properties which change with time (by drying out, rapid weathering or disintegration). As a definitive determination of rocks and their properties is possible only after laboratory examination of samples, the rock will be identified from visual inspection, according to the experience and knowledge of the technician. It is desirable that he has a working knowledge of the geological and petrographical terminology and of soil classification, but he may even use local names.

The characters described are: The colour of the rock in the natural state before drying, both the lightness and hue (e. g. dark greyish-green), the surface of rock fragments (lustrous, dull, smooth, etc.), smell of the rock, and mode of deposition, if possible.

Since the strength and compressibility of clayey soils are controlled mainly by their consistency, this must be described accurately. The terms for consistency of clayey soils agreed upon must be observed to avoid misunderstanding. Soil should be called hard or solid when it is so dry that its fragments can be broken by a hammer. Semisolid or very stiff soil is usually dark in colour, moist but does not contain enough water to become plastic; it breaks and crumbles when rolled in the hand. Stiff soil is difficult to mould, but it can be rolled in the hand into cylinders 3 mm in diameter. Soft soil can be kneaded without great pressure. Pasty soil can be squeezed out through the fingers of a clenched hand. The foremen should be instructed in this terminology, as they often denote every clayey soil as plastic regardless of its consistency.

c) Daily output, i. e. the depth of a borehole attained during one shift. Every interruption of work and its cause should be mentioned, e. g. repair of sheeting in a test pit or sticking of a boring bit in drilled holes.

d) The working procedure (in test pits the use of a pick, jack, blasting, etc., in boreholes augering, percussion; in core boring the type of bit and other particulars) should be recorded. In sandy and gravelly soils, the penetration resistance of the material is to be noted. As undisturbed samples cannot be taken from these soils, every record of their density is important, since it indicates their technical behaviour.

e) The kind of sheeting in test pits and the dimensions and depths of casing in boreholes.

f) The height of ground-water table, if found, should be recorded at the beginning of work and in the following days. If pumping is necessary, the efficiency of the pump should be checked and the time of pumping and its length should be recorded accurately, so that inflow can be determined. In boreholes every change in wash water (e. g. loss) must be mentioned.

g) Sampling, especially the depth and manner of taking disturbed and undisturbed samples. In core boring the loss of core and its cause should be noted (the falling out of the core from the core barrel or excessive fracturing of rock). In the boring log only actually established facts and measurements should be recorded. The progress of exploration should be described objectively, without personal conjectures. In addition to the written log it is recommended to draw a section of every borehole. Instead of a drawn section a printed form can be used, where the records are entered in prepared columns.

4.5.3. Surveying of boreholes. The position of all boreholes should be surveyed accurately and plotted on a plan. The position of the boreholes is given by co-ordinates drawn from fixed points well marked in both the plan and on the site (crossroads, buildings, bridges, etc.) The altitude of the land surface at the borehole is determined by levelling linked to the nearest datum point. If it is not possible, a suitable fixed point is chosen which could be linked up additionally with the level network. It should be noted that many boreholes became valueless because their position has been incorrectly recorded.

The pits and trenches are measured in detail and in case various rocks crop out in the walls, all sides of the pit should be drawn. The outcrops of beds, their strike and dip, course of joints, their spacing and filling, depth of weathering and other observations are plotted graphically. It is possible to recognize, even on a small outcrop, whether the rock is in situ or has been displaced (e. g. by slope movements) from the mode of deposition of the beds, from the course of principal joint systems, or from schistosity and foliation.

When ground water appears in the hole, the depth of water table from the ground surface is measured. Simultaneously, the water level in the nearest water-course, pond, well and neighbouring boreholes is determined in order to ascertain the mutual relationship between the ground and surface waters and to observe the oscillation of water table due to subaerial agents (e. g. heavy rainfall, changes in temperature and air pressure). The water table should be recorded daily, at best before the beginning of work. If the water table in a hole rose after the first measurement, the original and subsequent depths should be recorded.

This is of special importance for identifying aquifers with confined (artesian) water. In little permeable soils boring may continue below the ground-water table without any perceptible inflow, as small permeability prevents water from percolating into the hole. Water appears only after a certain time or at a certain depth. If several aquifers are encountered in a borehole they must be closed carefully by casing so that their yield and/or overpressure can be recorded.

On the other hand, water can flow into a borehole with such a rate that it also drags along fine sand; this feature must likewise be recorded in the boring log.

It should be remarked that even a simple test pit can be carried out inadequately, and lead to erroneous conclusions. Pits that were not timbered in time may cave in. As a result the ground surface close to the hole will sink (Fig. 4-1) and if such a pit is surveyed, the profile will not correspond to reality, since the beds are not in their original position. The error will be the more serious if an „undisturbed" sample is taken from such a loosened rock.

In evaluating test pits one must bear in mind that they are only temporary exposures and that no omission can be rectified after they have been back-filled.

4.5.4. Sampling and preservation of samples. Samples for detailed examination are taken from all rocks which occur in each borehole. They are taken from the bottom or wall of a test pit or from a borehole, and not from a heap of excavated material. The samples should not be long exposed to sunshine or rain and they should not be washed or otherwise treated. Immediately after extracting they should be deposited in boxes and designated by the number of borehole or pit, ordinary number, and the depth of their derivation.

For preservation of samples, wooden boxes divided into compartments with the dimension 10 × 10 × 10 cm are used. The lid should closely contact the sides of

4-24. Boring cores deposited in a core box.

compartments to prevent mixing of samples during transport. Samples can also be deposited separately into boxes of waxed cardboard. Samples of soils in which consistency and water content are of major concern are best preserved in glass jars with air-tight stoppers.

The cores are preserved in boxes 1—2 m long and every core run and depth of extraction are noted (Fig. 4-24). Rock samples should be described conscientiously, immediately after the core has been extracted, as some rocks such as clayey shales, claystones and marlstones rapidly alter by drying, weather and disintegrate. If a rock fragment is sent for laboratory testing, a wooden block is put into the box instead of it, with the record stating when and why the core was taken away. As it is very important to know the correct orientation of core fragments, they should be denoted, for example, by an arrow showing the upper end of the core.

The core boxes should be deposited in a dry store room so that the samples are easily accessible. Until subsurface explorations are definitively evaluated and the construction is completed, the rock samples should be preserved in perfect order as important documentary material of economical value. Unfortunately, on many a large construction, little care is devoted to preservation of samples; the boxes are deposited haphazardly on unsuitable places, so that the samples are often unattainable, particularly when boring was extensive. The costly exploration work is thus useless and in vain.

The reliability of statical analyses depends on the correct estimate of rock properties. It is therefore desirable to secure *undisturbed samples* of cohesive soils (mainly of clays) having the original water content, porosity, compressibility and strength; samples of sand and loose soils should be taken so as to preserve the original pore volume.

Taking undisturbed samples requires special instruments and procedure. It is difficult, and even impossible, in loose soils. Although sampling of cohesive soils is easier, samples are generally more or less damaged. The degree of their disturbance depends on the technical arrangement of the sampling procedure.

Normally, a thin-walled cylinder with removable tubes is forced into the soil (Fig. 4-25) or driven into it by hammering. In order to avoid the overtopping of the cylinder, a guide cylinder is sometimes used; this rests on the soil surface with its widened base and is anchored or held in place by hand. In taking samples from boreholes, samplers are pushed or forced into the ground by blows of a bar.

The following principles should be observed, if the least disturbance of samples is to be ensured:

a) To avoid further treatment of a sample before its deposition into an oedometer, the sampler should be provided with removable containers.

b) To decrease friction between the sample and the cylinder, which disturbs the soil, the inner diameter of the opening of a sampler should be 0.5—1.5 per cent smaller than the inner diameter of the sampler.

c) When the walls of a sampler are too thick, the soil is compressed along the perimeter and bulges in the centre. The area of the base including the thickness of the mantle should therefore exceed the surface of the opening by 15 per cent at the most. If the diameter of a sampler is 100 mm and its wall is 14 mm thick, the difference between the two surfaces amounts to 16 per cent, which nearly surpasses the limit of admissibility. It is thus more advantageous to choose samplers with larger diameters.

d) The structure of a sample, especially of fine-grained soils of small cohesion

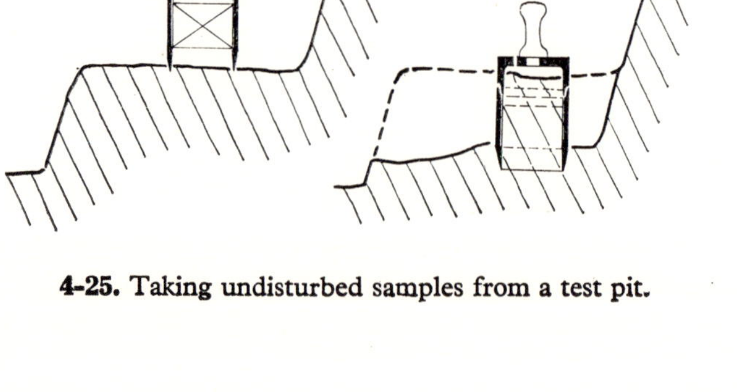

4-25. Taking undisturbed samples from a test pit.

is damaged by vibration produced by hammering. It is therefore preferable to push the sampler into the ground. The sampler should be withdrawn from the borehole some fifteen minutes after its being pushed in. In test pits the penetration of a sampler may be facilitated by careful removal of the soil around its cutting edge.

e) In normal cases the length of a sample should not surpass 2.5 times its diameter. A sampler with an inner piston is used for samples of greater length.

As the sample may slip out of the sampler and remain stuck in the borehole, samplers with pistons have been developed (Fig. 4-26). In taking a sample, the

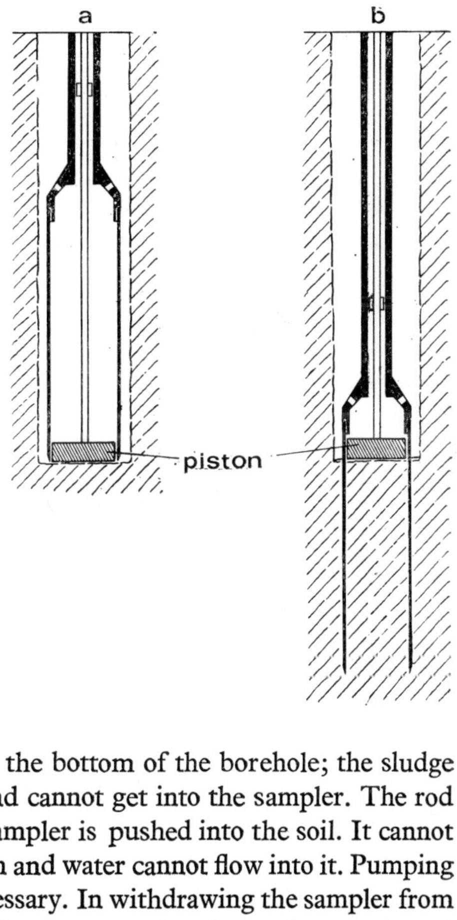

4-26. Sampler with a piston. a — Sampler set on the bottom of a hole, b — sampler pushed into soil.

cylinder with a piston inside is pressed to the bottom of the borehole; the sludge with water is thus pushed to the sides and cannot get into the sampler. The rod with a piston is then held firmly and the sampler is pushed into the soil. It cannot penetrate deeper than the level of the piston and water cannot flow into it. Pumping which would disturb the soil is thus unnecessary. In withdrawing the sampler from the borehole, the piston prevents the air from penetrating above the sample and consequently, the falling out of the sample.

Undisturbed samples cannot be taken from sands, gravels and stony soils; the original structure of sands is disturbed by pushing in the sampler and in stony soils penetration of the sampler is hindered by firm fragments and pebbles. To prevent the disturbance of sandy soils, they may be consolidated by bituminous solutions or by freezing before sampling. These methods are, of course, rather complicated and expensive.

Since the foundation engineer needs to know how the properties of soils change with depth, undisturbed samples are taken every one to two metres. In the boreholes, samples should be taken simultaneously with drilling. In test pits they are taken either during digging or all at once after the pit has been completed. In the latter case the pit is excavated in a step-like fashion, but its length must be somewhat greater. An undisturbed sample is taken from every step and the sampler is hammered vertically. In a pit excavated earlier, the surface layer, which may be dried out, should be removed before sampling. When the pit is provided with vertical sheeting, a board may be loosened and samples taken from the wall of the hole.

The samples after extracting from the ground are relieved of the pressure of surrounding soil on the upper and lower surfaces. The effects of the change in the state of stress, however, are reduced greatly by capillary forces in clayey soils. To prevent loss of moisture the sample containers are sealed at the ends with hot paraffin. During transportation care must be taken not to disturb the structure of the sample, especially in little cohesive, fine-grained soils.

4.6. Construction of geological sections from bore logs. It ought to be emphasized that in some cases subsurface exploration, even if carried out conscientiously, may provide only an approximate picture of natural conditions. In areas with intricate geological structure and irregular depositional conditions, a borehole shows the geology of the nearest vicinity only. If the geological structure of the wider area is inferred from several boreholes, the probable course of beds

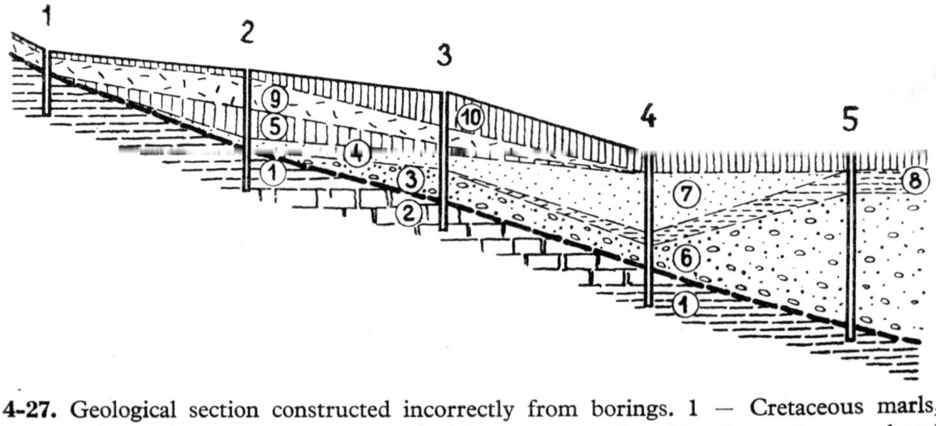

4-27. Geological section constructed incorrectly from borings. 1 — Cretaceous marls, 2 — sandstones, 3, 4 — sand and gravel of lower terrace, 5 — loess loam, 6 — sand and gravel of the valley terrace, 7 — sand, 8 — Holocene clayey alluvium, 9 — slope detritus, 10 — recent soil profile, 11 — flood loam, 12 — filling.

between the individual boreholes must be reconstructed. This should be done with great care and judgement. In evaluating the results of exploration the geologist should mention in his report to what degree his conclusions are reliable, which questions were not settled unequivocally by investigations and about which

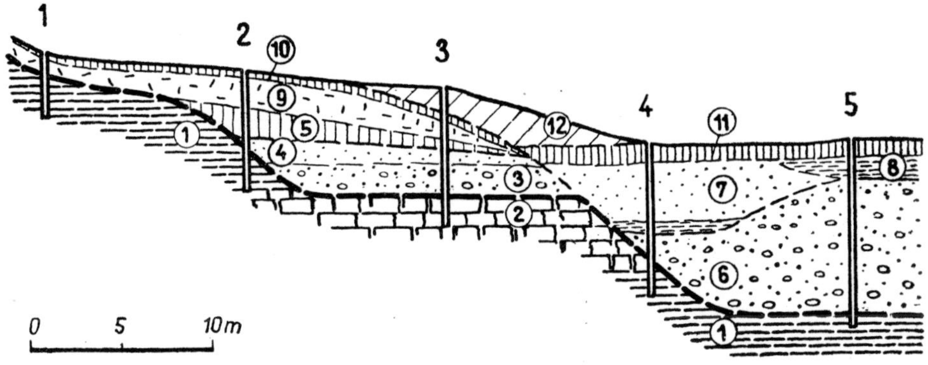

4-28. Correct geological section constructed from the same boreholes.

points he is in doubt. Both in the reports and graphs the established facts should be strictly separated from presumptions; for example, beds and their boundaries established by boring are drawn with a solid line, and their postulated course between the holes with a dotted line. An example of unreasonable and mechanical joining of boring records is shown in Fig. 4-27. A correct section constructed from the same boreholes is shown in Fig. 4-28.

If the geological structure is complicated, it is recommended to use an additional method, such as block diagrams, for presenting a more accurate picture of the geological situation. For illustration, a block diagram of the building site of a factory near Prague is shown in Fig. 4-29: six geological cros-sections are plotted in axonometric projection and linked by one longitudinal section. The cross-sections were constructed from boreholes drilled under each footing. In areas with particularly diversified relief and complicated geology it is of advantage to construct a model of boreholes from sticks, cardboard bands or glass on a base which represents a datum plane.

4.7. Plan and lay-out of subsurface exploration. In preparing a reasonable and economic plan of subsurface exploration, the engineering geologist must take into consideration the importance of the construction, the intricacy of geological conditions of the area and the questions which should be solved by exploration. The larger and more important is the construction and the more complicated are geological conditions, the denser is the network of boreholes which must be planned. In planning reconnaissance investigations for a large industrial plant or a hydraulic work with a dam and power plant, problems which are decisive

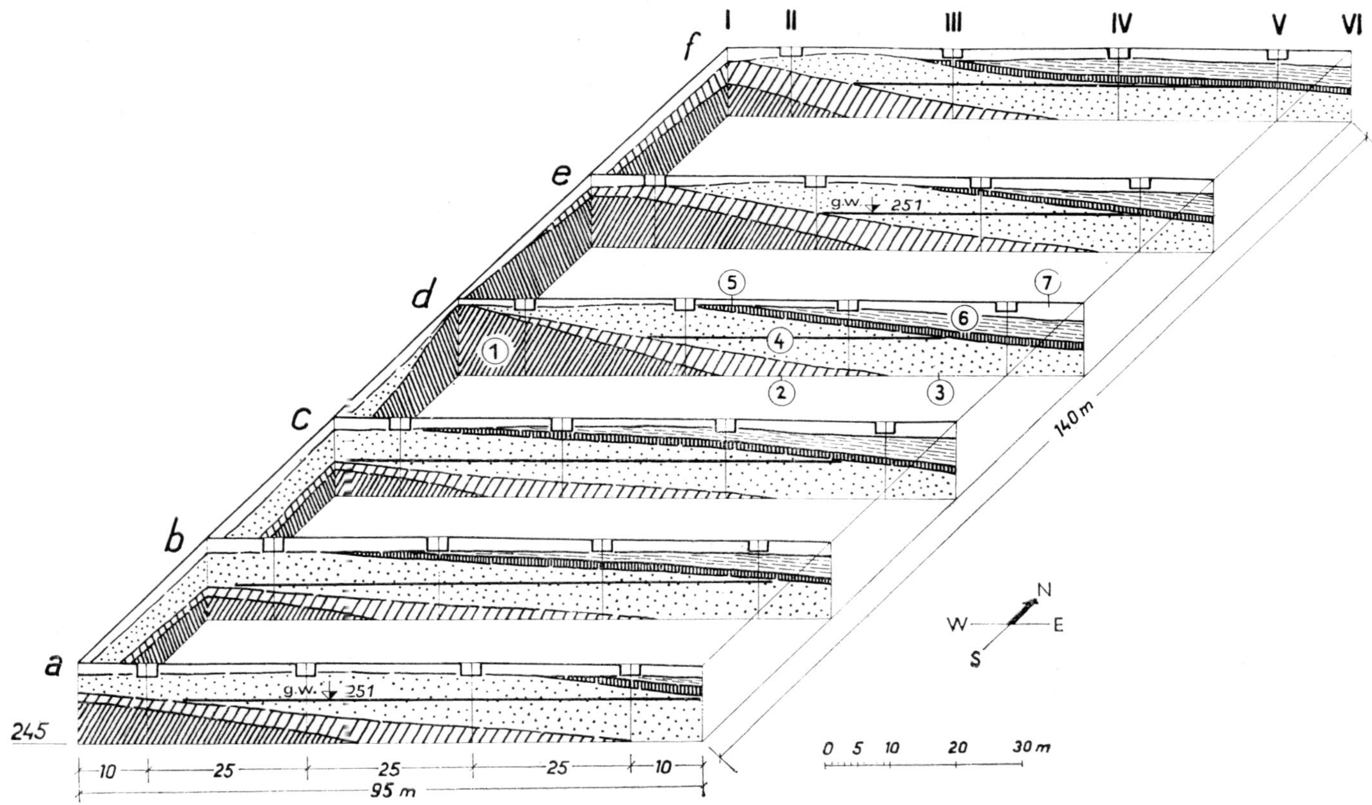

4-29. Geological conditions of a building site in axonometric projection. 1 — Algonkian slates, 2 — kaolinitic clays (decomposed slates), 3 — Cretaceous sandstones, 4 — ground-water table, 5 — glauconitic sandstones, 6 — Cretaceous marls, 7 — loess.

for the realization of the project must first be studied. Only afterwards, when the general concept of the work has been solved, can detailed investigations be planned.

In compiling the plan of detailed investigations we have thus a general idea in which rocks the boreholes will be drilled, at what depth the bedrock surface or the ground-water table may be expected, etc.

First to be decided is the method of subsurface exploration. In order to choose the most convenient procedure, advantages and drawbacks of the individual methods should be known.

The advantage of test pits, trenches and galleries is that they are directly accessible so that the rocks can be inspected in the walls and bottom of excavations. It is possible to study the deposition of rocks, the strike and dip of beds, tectonic features, jointing and the type and depth of weathering. In digging the pits, technical properties of rocks and their resistance to excavation may be well estimated. Samples can also be taken directly, so that their mixing, disturbance or contamination is almost excluded. The drawbacks of test pits are (1) a large amount of hand labour and, consequently, a relatively long time when test pits are numerous, and a high cost when they are excavated to greater depths; (2) increased cost when ground water must be pumped; (3) sheeting, which is necessary in most cases, raises the price and makes direct inspection and measurement in the walls difficult. Test pits and trenches cannot be recommended where excavation works can loosen and disturb the surrounding area, e. g. on slopes prone to sliding or in fine water-bearing sands.

The advantages of boreholes are a rapid working procedure and a relatively small cost per 1-m drilling; neither depth nor ground water hinder the progress of work. The disadvantage of boreholes is that they mostly furnish only disturbed samples so that they do not afford sufficiently reliable data for geological assessment of the bore profile. In particular, percussion drilling gives only crushed detritus, from which the original state of the rock or the dip and strike of beds cannot be estimated satisfactorily. The strike and dip of beds may be deduced from several boreholes located side by side, but in some cases a confusion of individual beds cannot be precluded.

Well-performed core boring is a very efficient method for investigation in firm rocks. From the rock core we can determine not only the petrographical character and weathering of the rock, but also the dip of beds, jointing and fracturing of the rock mass. Another advantage of core boring is that the hole can be drilled in vertical, inclined or even horizontal directions so that the most suitable direction, for example perpendicular to the strike of beds, can be chosen (Fig. 4-30). Core boring, however, does not afford satisfactory results in densely jointed rocks.

In choosing the working method it must be borne in mind that reliable results should be obtained at a reasonable cost. If the kind and character of rocks in a deep cutting should be determined for the assessment of their workability, percussion drilling should not be used. A sample of crushed rock would hardly help to answer these questions. A test pit or core boring would be suitable in this circumstance. On the other hand, a test pit should not be used for determining the surface and character of bedrock below water-bearing alluvium on the bottom of a flood plain. Excavation of test pits in the alluvium is very difficult, intensive pumping is necessary and the pits must be well sheeted. In this case auger boring will be convenient, complemented by core boring, if needed, for accurate determination of the bedrock surface.

Whenever a direct inspection of the rock is desirable, or the suitability of the rock as a building stone is to be assessed, or the stability of rocky slopes should be verified, test pits or exploratory drifts (in slopes) should be used. When exploration should be carried out to greater depths, they can be replaced by core boring.

4.7.1. The lay-out and number of boreholes. The location and number of boreholes are controlled by the size and importance of the planned structure and by geological conditions of the building site. Generally, they are established according to the results of reconnaissance investigation and the disposition of the principal objects of the structure. The boreholes are placed in several lines so that several geological sections across the building site and the neighbouring area can be constructed. In general, it is more reasonable to examine several sections by means of a large number of boreholes than to situate the same number of boreholes over the whole building site, depending on the lay-out of the individual buildings.

The spacing of the sections studied and of single boreholes cannot be established with universal validity. In some area a few boreholes afford satisfactory information of the geological structure of the site and the character of rocks, whereas in another area with more complicated geological conditions a far larger number of boreholes is needed. The depth and surface of the bedrock, for example, are dependent largely on the geological history of the area. The surface of bedrock is generally diversified where there are rocks of various strength and weatherability and where the weathering material has not been removed by erosion. The bedrock may also be strongly disturbed along fault and fractured zones. In these cases such a simple problem as

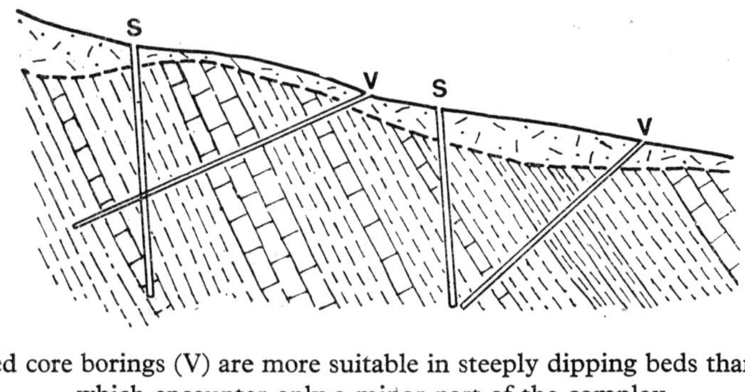

4-30. Inclined core borings (V) are more suitable in steeply dipping beds than vertical (S), which encounter only a minor part of the complex.

determination of the depth of the bedrock surface necessitates a dense network of boreholes (Fig. 4-31). In contrast, below valley terraces the weathering material is generally removed and the bedrock is levelled by river erosion. A few holes are sufficient to obtain necessary information (Fig. 4-32).

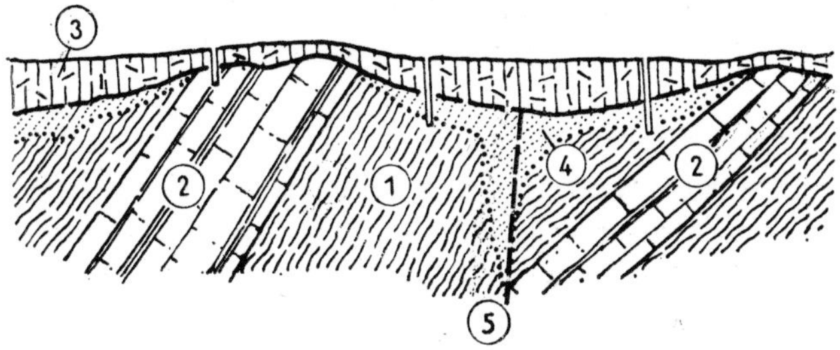

4-31. Deep and irregular weathering of rocks on ancient peneplains makes it necessary to carry out a dense network of boreholes for locating the unweathered surface of bedrock. 1 — Gneiss, 2 — quartzite, 3 — eluvial loam, 4 — weathered gneiss, 5 — fault.

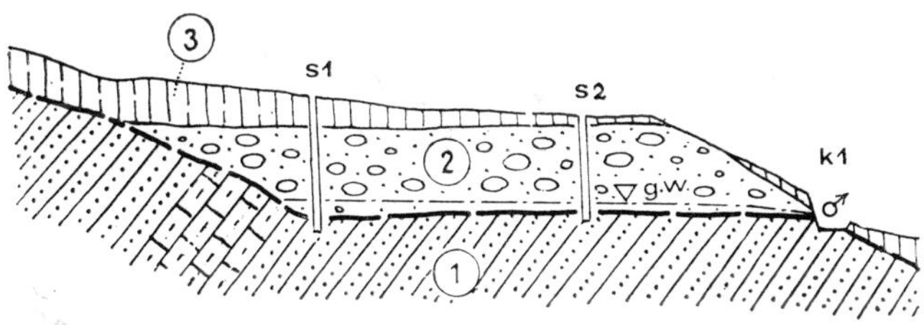

4-32. A few boreholes are sufficient for locating the surface of bedrock below valley terraces, since it is levelled by river erosion. 1 — Bedrock, 2 — sand and gravel, 3 — loess loam.

Careful subsurface exploration is a prerequisite in the areas of disused quarries or in the proximity of old fortifications, where the land surface was changed appreciably by man's action. In these places even a dense pattern of boreholes does not preclude unpleasant surprises.

Apparently simple are geological conditions in wide river plains. Levelling of the valley floor by flood loams and a considerable thickness of sediments lead to the erroneous assumption of a uniform character of the alluvium. In reality, however, these youngest sediments may be far from uniform even within a building site of limited extent. Sands and gravels are often eroded by river meanders, which after the shift of the river channel are gradually grown over by vegetation and silted by flood sediments with a high organic admixture. In view of these complicated relations, a comparatively large number of boreholes should be used in flood plains created by a meandering stream.

In investigating a building site it is recommended to make exploration also of its nearest surroundings. If boreholes are located only within the extent of a building site, they do not give a complete picture of the geological conditions, since these may change at its periphery. Figure 4-33 shows a generalized section of a building site in Prague. As it was surrounded by built-up areas, two test pits sunk to the bedrock were located within the site. During excavation works, the bedrock proved to descend abruptly under the southern wall (sector a-b), more steeply than in sector b-c, where the slope of the bedrock surface was determined by two test pits. Since the southern wall had to be founded at a greater depth, the cost of foundation constructions rose appreciably on account of strong inflow of corrosive ground water at that level.

At some building sites where the ground below the foundation should not be unnecessarily disturbed, the test pits are placed outside the ground plans of foundation constructions. Those near the foundation should be backfilled immediately after their evaluation, and the fill properly compacted, or they should be filled with lean concrete.

4.7.3. The depth of test pits and boreholes is determined according to the requirements of the planned structure and with the objective of verifying the presumed geological structure of the area. If, for example, the workability of rocks in deep cuttings should be determined, the test pits are excavated to the unweathered layers of the bedrock, occasionally down to the level of the communication route. Test pits for the study of foundation beds are put down to the firm bedrock or, in case it is too deep, to a depth which might be affected by foundation constructions (see Chapter 10). Boreholes for the exploration of damsites are sunken to the depth at which water seepage might still occur; they must pierce the weathered zone and reach the sound bedrock. Explorations for tunnel works generally require very deep boreholes. If a tunnel runs deep below the mountain ridges and geological conditions are complicated, boreholes are often drilled to the level of the tunnel to check the postulated geological sections.

For building purposes the pits and boreholes are generally deepened to the firm

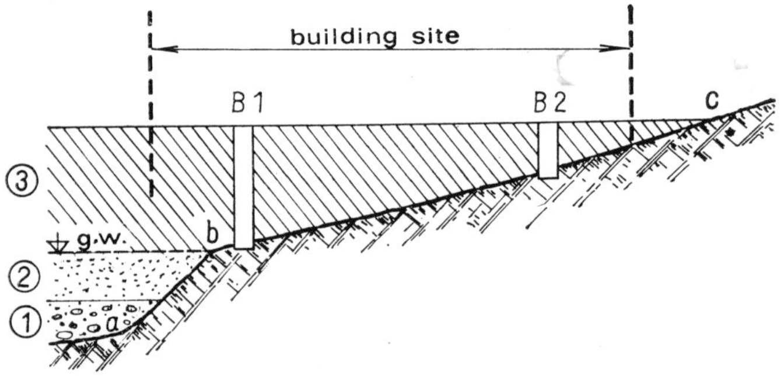

4-33. Unsuitable distribution of boreholes on the building site. A borehole placed at the margin would reveal the unexpected steep slope of the bedrock surface in a-b section. 1 — Gravel, 2 — sand, 3 — filling.

bedrock. It is not difficult to decide at what depth a test pit should be terminated, as the surface of bedrock is easy to establish in an open pit. In boreholes, however, the identification of the bedrock surface requires experience and knowledge of the local geology. Boring should not be ended until the nature of the hard bottom

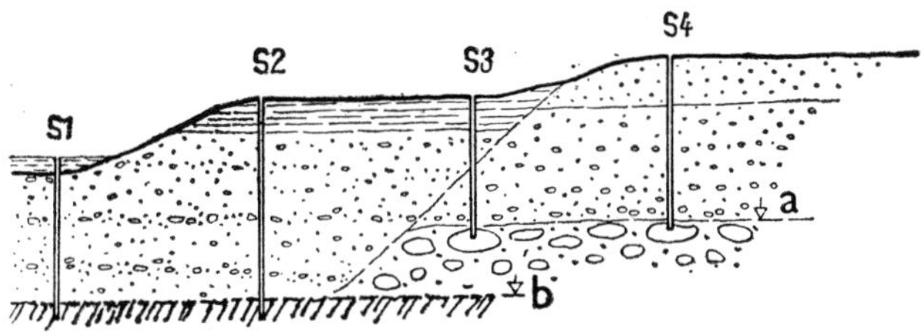

4-34. False determination of the bedrock by boring. S 3 and S 4 boreholes ended in boulders. a — Erroneously determined bedrock surface, b — true bedrock surface.

of the hole has been recognized. This implies that core boring or percussion drilling should continue at least 1 m into the hard rock to obtain enough fragments for a reliable determination of the rock. The determination of bore samples is not always sufficient to decide whether or not a borehole has reached the rock in situ, and it must be checked against the geological conditions of the wider area. Several examples can be cited when the depth of the bedrock has been incorrectly deduced from boring results. In establishing the depth of valley alluvium (Fig. 4-34), boreholes were terminated allegediy in granite and the building programme was elaborated on this assumption. The excavation, however, showed that the boreholes were sunk only to a continuous layer of large granite boulders in the valley terrace. This error could easily be rectified if the engineer had at least a background knowledge of geology, as both the river banks were made up of unaltered Algonkian shales, so that the presence of granite below the river bed was next to impossible.

Chapter 5

GEOPHYSICAL METHODS

Engineering geological investigation by means of borings is time consuming and expensive, and it provides information of the geological situation at the point of the borehole only. Therefore the tendency prevails to establish the geological picture of the site by less expensive and more extensive methods. In principle, these methods register the characteristics of a natural or artificial physical field, influenced by the physical properties of rocks. These changes can be measured by sensitive devices with a satisfactory precision over relatively long distances. After establishing the characteristics of the field at more points, the distribution of the characteristics in the studied area can be plotted and the boundary lines between rocks of different properties drawn. Several of these methods have been widely used and accomplished. They constitute a field of an individual science, geophysics. Of primary interest is for which purpose any of the methods can be employed in engineering geological investigation and which is the chance of the reliability of results. In principle, geophysical methods are of use during the investigation of the geology of a site (Sec. 5.1), during the detailed investigation of the properties of rocks (Sec. 5.2), and in the hydrogeological exploration (Sec. 5.3).

5.1. Geophysical methods used in the geological investigation of a site. A geophysical survey is made at the ground surface in the area which has already been studied by a net of borings or pits. These serve as support points for the geophysical survey. Good results were obtained with methods as follow:

(a) *Electrical resistivity method.* This entails measuring the parameters of a direct current field, that is generated in the ground by means of electrodes E_1 and E_2 (Fig. 5-1). When setting the collecting electrodes S_1 and S_2 by turns at variable spacings, the potential gradient between S_1 and S_2 can be measured by a sensitive potentiometer. From the potential gradient and from the intensity of the measured current the average or apparent resistivity of the rock can be calculated. Denoting the spacing between S_1 and S_2 a(m) and assuming that rocks between S_1 and S_2 possess a homogeneous resistivity ϱ (ohmmetres), the formula for ϱ gives:

$$\varrho = 2\pi a \Delta V / J$$

where ΔV is the potential fall between S_1 and S_2 (volts) and J is the measured current intensity (amperes). The corresponding electric field extends to a depth of about one third of the distance between E_1 and E_2. By increasing the spacing a, as is done with the so-called depth probe resistivity method, the area of the electric field extends to a greater depth and may reach different rocks. Since variations in ϱ of different rocks are considerable (e. g. ϱ is about 10^{-1} to 10^2 ohmmetres for wet sand, 1 to 10 for clay, and 10^3 to 10^5 ohmmetres for granite), the measured quantities are sufficient to indicate these differences. The depth where the resistivity changes and where, therefore, a boundary between rocks is to be expected, is obtained from additional calculation.

The resistivity method is especially suitable where horizontally bedded rocks are concerned. However, the water content in rocks and the water table influence the results, as the conductivity of water is many orders of magnitude greater than that of rocks. But the conductivity of water depends on temperature and on the content of dissolved salts. These features obscure the influence of the geological conditions upon the electric field. The true cause of the variations in resistivity can be determined by borings, which thus function as key points.

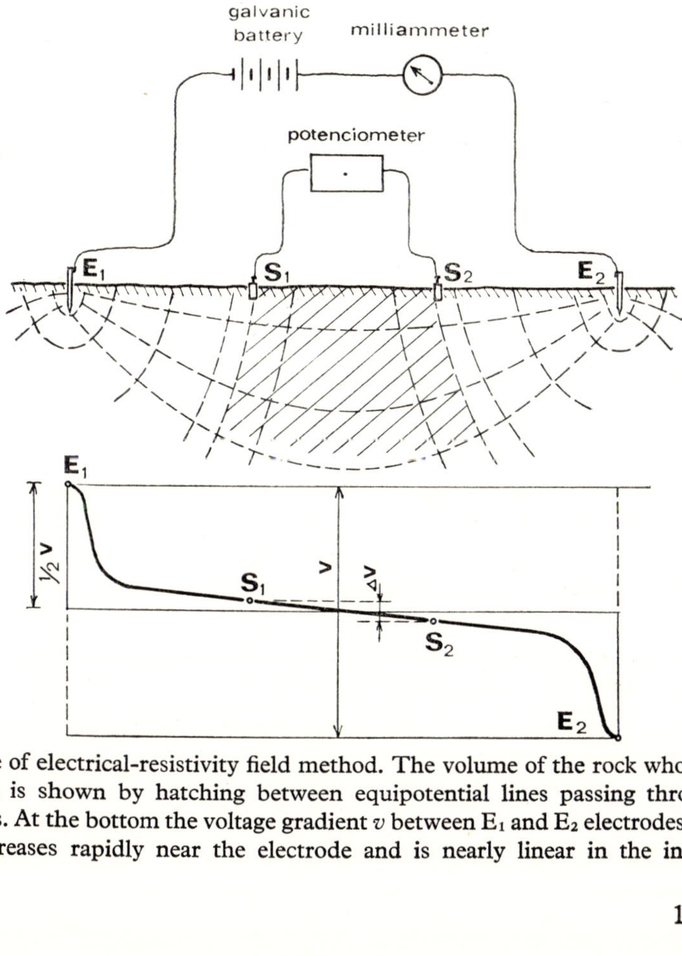

5-1. Scheme of electrical-resistivity field method. The volume of the rock whose resistivity is measured is shown by hatching between equipotential lines passing through S_1 and S_2 electrodes. At the bottom the voltage gradient v between E_1 and E_2 electrodes is indicated. Voltage decreases rapidly near the electrode and is nearly linear in the internal third.

The resistivity method is most applicable for the study of the surface of a compact bedrock under a weak cover, even if a water table is present. On the other hand, when the rock surface is weathered to a considerable depth, the results are less reliable. This is why the resistivity method works better below the valley plain than below the higher valley terraces. On rock slopes, the thickness of partly weathered rock can also be measured by this method.

The resistivity method was employed, for example, to determine the thickness of alluvial deposits under the dikes of the Mississippi River for a length of about 50 km. The surface of the clay of Tertiary age lies at a depth of about 30 m and was determined with an average error of 3 per cent. Borings were drilled at 1500 m intervals instead of 300 m as presumed. The investigation term was a fifth and the cost 40 per cent of the presumed ones (Perret, 1948).

(b) *Constant separation resistivity method.* This is employed when the depth of the bedrock surface varies over a short distance, as, for example, with buried river channels, or when the boundaries between rocks of different character are almost vertical. This method is often used in combination with the depth probe method. The distance a remains unchanged but all electrodes are shifted as a unit to a new station after each reading. Lateral variations in the geological conditions can be investigated in this way and the presence of vertical faults under the overburden can be detected.

From the point of view of operation the simplicity of the resistivity method is of advantage.

(c) *Seismic methods.* Their application requires more complicated procedures and devices. Seismic methods are based on the phenomenon that elastic waves generated by an explosion or by a weight incidence propagate at a different velocity depending on the elastic properties and the density of the given rock. The velocity of the longitudinal (P or compression) waves, which reach the recorder first, is about 250 to 1000 m/s for dry sand, 750 to 1250 m/s for saturated sand, but 4500 to 6000 m/s for granite. Wave-refraction processes are also interpreted in the results (Fig. 5-2).

The velocity of waves is measured at different distances of the seismometer from the source. When the spacing is small, the velocity of direct waves (V_1) is obtained corresponding to the wave propagation in the cover. But greater velocity (V_2) is obtained with greater spacings, as it originates from the waves travelling along the boundary between the cover and a denser underlying rock. The spacing which corresponds to the evidence of greater velocity is called the critical spacing, L_c. The depth of the boundary can be calculated from:

$$D = L_c(V_2 - V_1)^{1/2} \; / \; 2(V_1 + V_2)^{1/2}$$

The seismic method is very helpful in engineering geological investigation. For example, a considerable depth of weathered rock could be detected by this method, although only a small depth of weathering was evident in the face of a quarry in the vicinity. Weathered rock manifested itself by an extremely low velocity of seismic waves generated by explosions. As with electrical resistivity methods, caution is necessary when the surface of a weak bedrock is found under the Quaternary deposits; for example, the surface of the Palaeogene sandstone under a cover of 20 m of alluvial gravel had to be determined at the damsite on the Bystrica river in the East Carpathians. The seismic survey indicated the existence of a 7-m depression in the bedrock surface, what was difficult to interpret from the point of view of the geological development of the valley. Assumption of the existence of this depression complicated the location of the dam. Nevertheless, subsequent borings disproved the existence of the depression. Results of seismic survey were probably influenced by a dense fracturing of sandstone along a set of faults. It is to be considered that the velocity of seismic waves is of the order of 1500 m/s in dense gravel, of 1700 m/s in hard sandstone, but only about 1000 m/s in partly weathered zones of sandstone.

(d) The study of *the earth's magnetic field*. This may be useful at certain locations. The presence of rocks which have special magnetic properties (susceptibility) due to their nature, can be determined. For example, steeply dipping layers of basic rocks in the gneiss environment can be discerned in this way. On the contrary, when an increase in the intensity of radium emanation was measured (radiometric method), the presence of rather acidic rocks or of faults can be assumed.

(e) *Gravimetric survey*. This deals with the anomalies in the gravity field and was successfully employed in finding the cavities in carbonate rocks (Enslin, 1953).

5.2. Determination of rock properties. Electric resistivity and seismic methods are of particular value for a detailed investigation of the differences in rock properties.

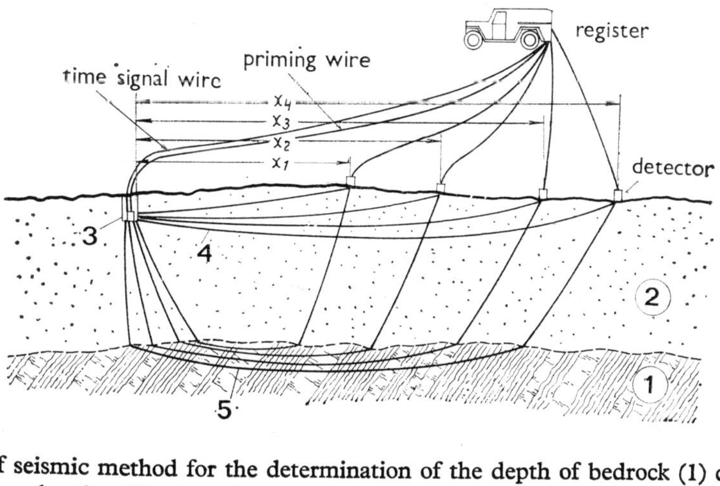

5-2. Use of seismic method for the determination of the depth of bedrock (1) covered by overburden (2); 3 — charge, 4 — direct waves, 5 — refracted waves.

(a) *Constant separation resistivity method.* This is of the same character as shown in Fig. 5-1, only the three electrodes are set into a borehole in a vertical direction. One electrode is set on the ground surface. Collecting electrodes are moved along the borehole; they measure the potential gradient at different depths. The most important effect of this method is that the geology of neighbouring borings can be correlated and the geological sections plotted. When the borehole is dry, the depth to which the drilling fluid has penetrated can also be evidenced, and the assumptions of the rock permeability can be made.

(b) *Seismic measurement.* The aim is often to establish the seismic modulus of deformation (Sec. 3.4). The detailed investigation of other physical properties is also based on the recording of longitudinal waves. The velocities of seismic waves when passing either from one boring to another (Fig. 3-14), or from the surface to a point inside the boring, are measured.

When the boring wall is to be investigated over the entire depth, *ultrasound impulses* are also used. These are transmitted from the upper end of the cylindrical travelling device, and recorded on its lower end. Quantitative information concerning the density of joints, the deformation properties of the given rock, and even the stress intensity can be obtained. The latter case is exemplified by Fig. 12-19, where zones of rock weakening due to tunnelling and of stress concentration around the tunnel opening are shown. Very informative are the results of measurements from one boring to another with the aim of determining the homogeneity of rock. Bertacchi et al. (1967) presented the results of measurements which yielded a uniform velocity of 5700 m/s for seismic waves and that of 6000 m/s for ultrasound waves in the diorite to a depth of 40 m below the dam foundation. Farther from the dam the seismic value was 4850 m/s on average, but the ultrasound value varied between 3500 and 6000 m/s.

Recent investigations deal not only with the wave velocity, but also with absorption characteristics of rocks, manifested by a pronounced decrease in amplitudes and frequency. They indicate the presence of joints or minor faults, which do not manifest themselves by differences in velocity (Dvořák, 1969). The tendency arises to determine the density number of rock from the three named parameters.

(c) *Gamma radiation measurement.* Because the determination of the density of soils, especially of sand, is expensive, this has found a broad scope of interest. The source of radiation is mounted on a probe, situated either on the ground surface or on prospecting rods. Intensity of the radiation dissipation decreases with increasing density of soil. For example, densities of soils have been measured in this way over large areas of aerodromes since 1953 (Lane, Torchinsky, Spinka, 1953), with an error of 2 to 3 %. This method has also been widely employed in boreholes. As the reflection of the emitted radiation is measured, the term gamma-gamma measurement has also been adopted.

5.3. Geophysical methods used in hydrogeological research. The question at which depth underground water penetrates into a borehole is of primary importance, especially when dealing with stability of slopes. Often the inflow is very small, but water exhibits uplift. With the exception of temperature measurements several geophysical measurements are made: clean water, penetrating into the borehole, dilutes the water inside, which has been artificially enriched with a soluble compound. The degree of dilution can be detected by the changes in either intensity of luminous flux (radiator and photo-electric cell are necessary), or in conductivity and radioactivity of water. The level at which water escapes can also be detected by measuring the movement of the layer of diluted water.

It is apparent that geophysical methods provide information which cannot be obtained by classical methods of enginering geological investigation.

Chapter 6

WEATHERING OF ROCKS

In all foundation excavations carried to the bedrock a surface layer of weathered rock is first encountered; undisturbed fresh rocks very rarely occur at or near the earth's surface. It is therefore necessary for the engineer to know the properties of weathered rocks, which usually differ from the character of fresh rock, and the processes that caused physical and chemical alteration of the rock mass.

The thickness and nature of the weathered zone differ, depending on the kind of rock, the weathering agents, duration of weathering processes, the slope of the terrain, and on the rate and mode of displacement of weathered material.

The term 'weathering' generally denotes processes of mechanical disintegration or chemical decomposition of rocks caused by natural agents, man included. Alterations of rocks are produced mainly by temperature changes, insolation and rapid cooling of the rock, frost action, chemical action of percolating water, effects of plant roots and smoke gas, etc. The type of weathering is controlled mainly by climatic conditions. In arid regions, for example, mechanical (or physical) weathering effected by temperature changes predominates and fine-weathered material is removed chiefly by wind. In polar regions and high mountains freeze-and-thaw is the main weathering agent. In temperate and particularly tropical zones, a warm and humid climate supports deep-reaching chemical weathering, which results in chemical decomposition of rock components.

If climatic conditions of a given area remain unchanged for a long time, a typical profile through the weathered layer develops near the surface. The composition and depth of this profile correspond to climatic conditions and the kind of parent rocks. In some favourable places old weathering crusts covered by younger sediments have been preserved. From these fossil-weathering crusts the climate of the respective geological period can be inferred.

According to the character of principal agents, physical (or mechanical) and chemical weathering are distinguished, but it must not be forgotten that the two processes generally act together in intimate co-operation. Depending on climatic conditions, one of them may predominate in a given area.

6.1. Physical weathering. Physical weathering first produces fracturing of rocks, which gradually proceeds to their disintegration into minute particles, but without any chemical change. Physical weathering essentially results from the change of stress in the near-surface layers. When the stress increases so as to exceed the strength of the rock, this disintegrates. Changes in stress conditions are caused by the following processes:

1. Unloading on removal of overlying rocks by erosion;
2. freezing of water in cracks and joints;
3. growth of crystals;
4. volume changes of rock due to changes in temperature;
5. action of plants, mainly tree roots.

Under climatic conditions of the temperate zone the effects of physical weathering are relatively small. In deep erosional valleys pressure release by unloading is the dominant factor; in high mountains where there is a great temperature fluctuation, changes in the rock volume and forces produced by freezing of water in joints cause shattering of rocks. Physical disturbance of rocks due to neoformation of minerals occurs rather rarely, for example, in pyrite-rich rocks, where the decomposition of pyrite may produce sulphates. If gypsum is formed, which has a larger volume, its crystals wedge open the cracks and bedding planes. Mechanical effects of tree roots are limited to the surface layers of the bedrock.

6.1.1. Fracturing of rocks by unloading. The stress release on unloading a rock mass is an important factor especially because it is often the first step in a continuous series of weathering processes. The destructive action of air, water and plant roots is possible only when these agents can penetrate the rock along release joints.

Both sedimentary and igneous rocks originated at a greater pressure than is the pressure they were exposed to after having been tectonically uplifted or laid bare by erosion. Atmospheric pressure at the earth's surface is far lower than the pressure even at a small depth below the land surface or the sea level. Every ten metres

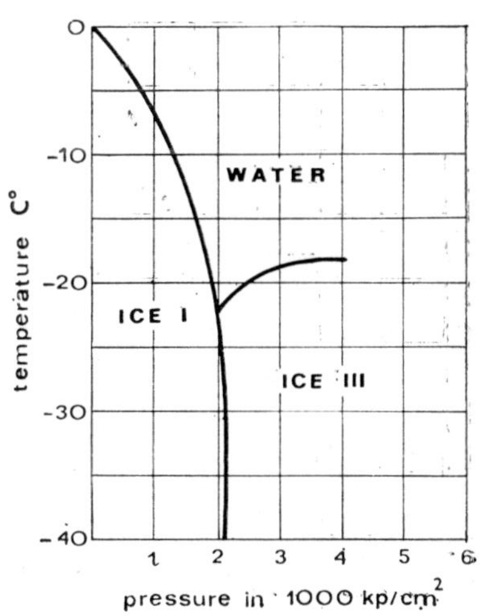

6-1. Under increasing pressure water remains in the liquid state up to −22 °C and 2,100 kp/cm². Under higher pressures the ice III modification develops, which is of higher density than ice I. (After E. C. Dapples, 1959.)

of sea water or four to five metres of rock mass increase the pressure by about 1 kp/cm^2.

Unloading of a rock mass lowers the confining pressure within it and the internal stress is relieved by expansion of rock. Slight expansion manifests itself as minute shifts along the original planes of weakness, i. e. bedding planes in sedimentary rocks and joint systems in homogeneous igneous rocks. Primary open joints are not very frequent in some massive rocks, such as sandstones and granites. Prominent joints (so-called sheet joints), however, form as the original overhead load is gradually stripped off by erosion. They are roughly parallel to the land surface and develop concentric shells or layers up to several decimetres thick. Jointing extends gradually downwards with the proceeding unloading.

The amount of residual stress in massive rocks is surprisingly large. It has been reported that excavated granite blocks may expand 0.1 %. This also explains uparching and sudden up-springing of thin rock sheets, which sometimes occurs in a freshly excavated quarry or excavations in valley floors.

Fracturing of rocks due to unloading rarely extends deeper than a few tens of metres below the surface, since at a greater depth expansion is restrained by the weight of the overlying strata. In deep-lying tunnels and mines the pressure relief often shows itself by rock bursts, i. e. popping of slabs from the walls or floor of the newly opened spaces. Special precautions are therefore taken to protect workmen and equipment, and blasting is usually performed at the end of the working day, so that pressure equilibrium can be re-established overnight.

The breaking-off of rock slabs due to unloading may be a serious problem during dam construction in a deep erosion valley. The excavation of dam foundations in the valley floor leads to unloading of bedrock and hence to further breaking off and uplift of rock blocks. To excavate the foundation down to the "undisturbed" portion of the rock mass would be a waste of money, as it would produce only further disruption. This phenomenon was observed, for example, in the granite foundation for the Grand Coulee Dam on the Columbia River and in andesite excavated for the construction of the Boulder Dam on the Colorado River. Newly formed joints must be sealed by contact grouting immediately after placing the first concrete layers.

6.1.2. Freezing of water in joints. Water freezing under atmospheric conditions increases its volume by about nine per cent, which causes a rise of pressure and liberation of latent heat. The higher the increase in pressure, the greater must be the decrease in temperature for freezing to continue. Figure 6-1 shows the amount of pressure necessary for lowering the freezing point of water by only a few degrees: a confining pressure of 200 atmospheres will lower the freezing point only by ca. 2 °C.

Frost wedging is effective in humid climates, especially in high mountains, where the daily temperature oscillates around the freezing point. On a sunny day melt water fills fissures in the surface layer and freezes at night. The recurrent freeze-and-thaw action shatters the rocks on mountain slopes. Loosened rock fragments fall to the foot of the slope, mainly during snow melting in spring, and accumulate in talus fans and cones. Transport routes in mountainous areas can be threatened by stone falling; protective walls or benches are therefore set up in hazardous sections.

The slopes of talus fans depend on the shape and surface of rock fragments. The larger and rougher the stones, the steeper the slope of the fan. Detritus of flat smooth fragments of shales and marlstones develop a slope of 25—30°, angular granite fragments a slope of 35—40°. Water-saturated talus is stable at an appreciably smaller slope, so that a sudden torrential rain can disturb even old talus fans. They may also be disturbed during earthquake.

6.1.3. Crystallization of neoformed minerals. Rainwater entering the rock always contains trace amounts of dissolved substances, which increase as the water percolates through pores and cracks of the rock. When it evaporates or freezes, the solution becomes oversaturated and salts precipitate. The growing crystals may generate a considerable pressure which disintegrates the rock similarly to the pressure of growing ice crystals. In case a rock contains a sufficient amount of water, some anhydrous minerals (anhydrite $CaSO_4$) can be converted into hydrates ($CaSO_2 . 2\,H_2O$) by which their volume is increased.

In cities where the air is polluted by sulphurous coal smoke, rainwater is actually a dilute sulphuric acid, which strongly corrodes limestones and marbles. Gypsum

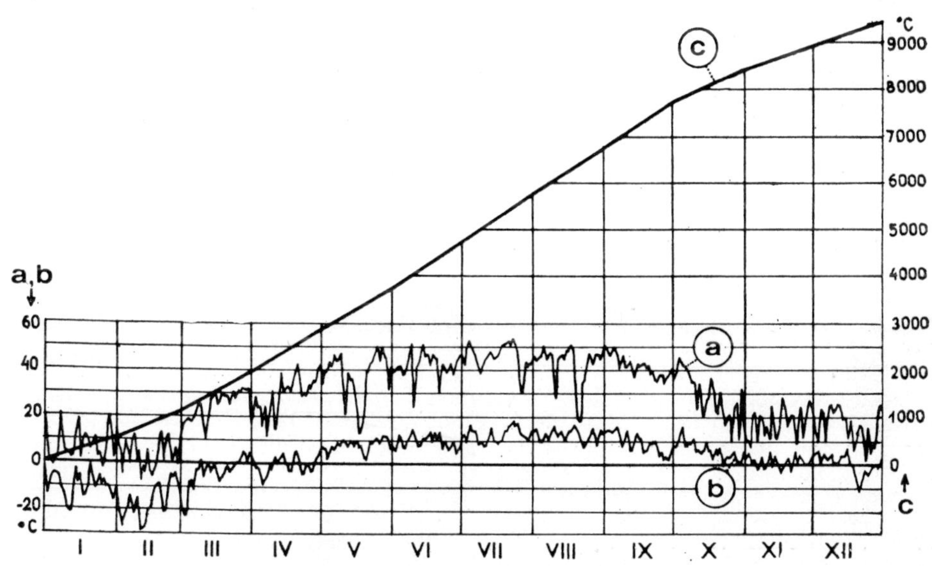

6-2. Graphic representation of varying daily and nocturnal temperature on the ground surface in Prague during 1929. a — Maximum day temperature on the sun, b — minimum night temperature, c — mass curve of greatest daily differences in temperature (Q. Záruba, 1931.)

that crystallizes in the cracks of building stones causes peeling off of thin flakes, which hastens the weathering process. Many ancient monuments and buildings bear traces of combined detrimental action of moist, smoke-polluted air and of gypsum precipitation.

The wedging activity of forming salt crystals in rocks is regarded as a mechanical weathering process, but it is closely related to chemical weathering.

6.1.4. Disintegration of rocks by temperature changes. Rocks at the earth's surface are exposed to changes of temperature, which also contribute to their disruption. During the day the rocks are heated by insolation and are cooled at night. Since they are not good conductors of heat, the temperature of their surficial parts is considerably higher than that of the deeper layers. The stress thus generated within the rock produced a system of cracks along which thin scales break away.

The disturbance of rocks by alternating heating and cooling is supported by different thermal dilatancy of minerals, their different colour (dark minerals absorb more heat than the light coloured) and by the difference in grain size. Although even in the temperate zone the daily differences in temperature are considerable (Fig. 6-2), this mode of physical weathering has very small effects there. It is more effective in deserts where the changes between day and night temperatures are far greater.

6.1.5. Action of vegetation. The participation of vegetation and other organisms in rock weathering depends on climatic conditions. In humid tropical zone with abundant flora their activity is comparatively strong. In arid regions (deserts) the scarcity of plants reduces their effects. The roots of shrubs and trees penetrate thin cracks and in growing they exert pressure on their walls. Cracks widen gradually and thus enable rainwater to penetrate deeper into the rock. It has been found that the roots of beech and larch are more powerful agents than those of oak and pine. Humic acids and other organic substances intensify the process of chemical weathering.

6.2. Products of Pleistocene physical weathering. In Central Europe physical weathering was most intensive in the Pleistocene, when the climate of periglacial areas was affected by the proximity of the vast ice sheet. The front of the ice reached the boundary mountains of Bohemia, and the Alpine glaciers which were more extensive than they are today contributed to the adversity of climatic conditions. In glacial periods the climate of Central Europe was similar to the present conditions in Subarctic regions, for example, in Siberia, Spitsbergen and Alaska; under very cold and dry climate physical weathering was intense and the ground was perpetually frozen to a great depth, with thawing occurring only within a thin surface layer, during the summer months. At the surface marshy tundra developed, as the melt water could not soak into the impervious frozen ground. The rain and melt water accumulated in the surface layer which became nearly liquid, and water-logged soil flowed downslope, even if the slope gradient was small (solifluction).

Intensive physical weathering due particularly to frost activity is evidenced by stone detritus and block fields, which have been preserved on some slopes formed of solid rocks with block jointing. The perenially frozen ground is evidenced by surficial loosening of rocks and by wedge-like widened joints, which during glacial periods were filled with ice, replaced later by loam or sand. These frost cracks also originated in the weathered surface layers of the bedrock, which show a characteristic loosening and plication (Fig. 6-3). These phenomena occur mainly on

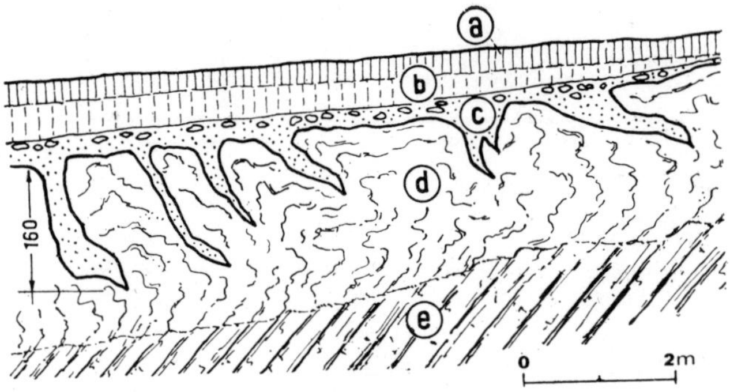

6-3. Frost wedges filled with sand in Ordovician shales; excavation in eastern part of Prague. a — Degraded chernozem, b — loess loam, c — fine sand, d — weathered shales, e — firm shales.

plateaus and in the upper part of slopes, where bedrock was exposed to frost action. They are less marked or altogether lacking in the lower part of the slopes or in the valley floors, where bedrock is generally protected by detritus or gravel accumulation.

All the above phenomena characteristic of periglacial areas were first recognized and described as recent features of the polar regions, and only afterwards has their fossil form been identified in the Pleistocene deposits of periglacial areas.

6.2.1. Disturbance of bedrock by freezing. In periglacial areas the surface layer of bedrock may be disturbed to a depth of about 2—6 m below the original land surface. In this surface layer, which may be markedly developed chiefly in shales and thin-bedded rocks, the rock is densely jointed or plicated into small folds (Fig. 6-4). Rock fragments are upheaved and tilted and cavities between them are free or filled with fine debris and clayey loam. The material showing the character of eluvium passes downwards gradually into solid parent rock. On sloping ground the disturbed strata are bent and dragged out parallel to the slope surface. The form of weathering is controlled mainly by the character of rocks (Figs. 6-5, 6-6).

Loosening and contortion of surface layers is produced by alternating freezing

6-4. Periglacial disturbance of Cretaceous sandy marlstone; Přední Kopanina quarry near Prague, height of quarry face about 6 m. (Photograph by Záruba.)

and thawing, in particular where ground water flowed out. Thin ice layers forming in shaly or thin-bedded rocks were fed by capillar water rising from the lower unfrozen zone. As freezing continued downwards, a considerable stress developed and caused irregular rising and undulation of the surface, and in places the formation of surface folds. After the ice had thawed, the ground surface partly subsided but the rock fragments often remained erected and shifted, so that many cavities occur in this active layer. Figure 6-7 shows the loosening and contortion of Algonkian shales at a building site in the north-eastern part of Prague. The rock is shattered to a depth of 2.80 m, but the disturbed layer was not laterally moved, as is seen from a continuous passage of unweathered beds into the disturbed zone. The Pleistocene age of this feature is confirmed by two beds of loess-loam covering the disturbed bedrock. The lower loam was contorted consistently with the surface layer of the bedrock, whereas the younger light-brown loess levelling the ground into a uniform plain has been preserved in the original undisturbed position.

Compressibility of the surface bedrock layer shattered by periglacial weathering is very high, as has been established at many building sites. It therefore should be examined in the first stage of investigation, as changing the lay-out of foundations after the detailed research has been accomplished will cause unnecessary expense.

High compressibility of disturbed shales has been determined by loading tests,

6-5. Weathered basalt columns; quarry near Konstantinovy Lázně, north-western Bohemia. (Photograph by Záruba.)

6-6. Predominant mechanical weathering of stressed granite forming the summit of Králova Hola, Low Tatra Mts. in Slovakia. (Photograph by Záruba.)

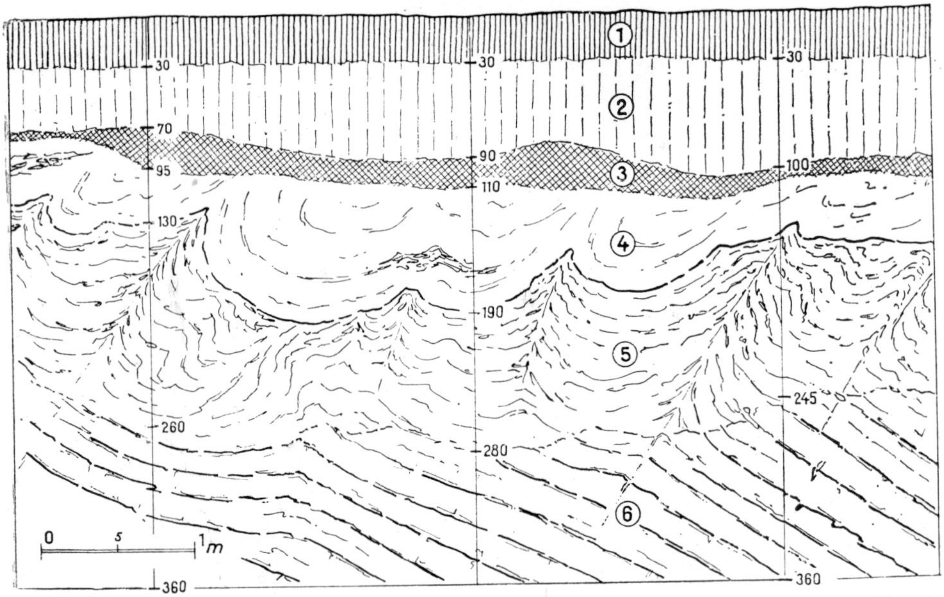

6-7. Loosening and crumpling of Algonkian shales by periglacial freezing. 1 — Humic soil, 2 — younger loess, 3 — relict of fossil soil, 4 — lower loess loam affected by cryoturbation, 5 — loosened weathered shales, 6 — firm Algonkian shales.

particularly when elastic and permanent compressibility was measured. From these tests, in which the load is removed after every increase in loading, the share of permanent and of elastic compression is inferable. The result of a test made on shattered Ordovician shales in Prague is shown in Fig. 6-8. The disturbed shales have small elasticity (P_2) and high permanent compression (T_2), hence also a high total compression, which at 5 kp/cm² attains as much as 8 mm.

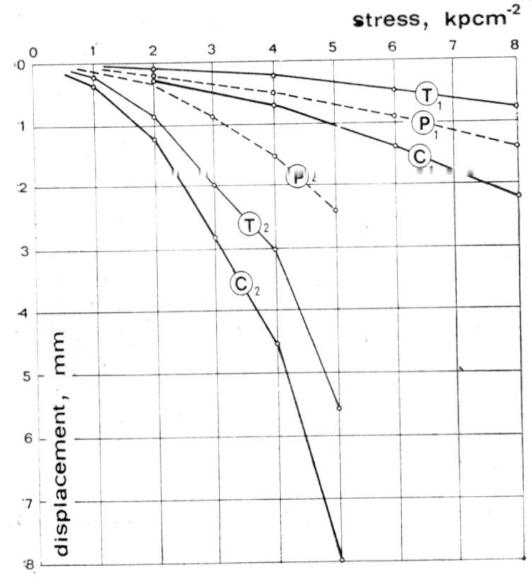

6-8. Results of loading tests on Ordovician shales; in undisturbed shales elastic deformation P_1 exceeds permanent deformation T_1 and the total deformation C_1 is small. In weathered shales the permanent deformation T_2 is greater than the elastic deformation P_2.

In contrast, in unweathered shales elastic compression (P_1) predominates over permanent (T_1), and total compression (C_1) is only 1 mm for the same loading.

Figure 6-9 demonstrates the importance of periglacial weathering for foundation design. In the excavation for a factory near Prague terrace gravel was found with the exception of the northern part, where a slope formed of Ordovician shales was exposed. In the surface layer, 2—3 m thick, the shales were disintegrated and penetrated by frost cracks, about 0.6—1 m wide at the upper end. They extended 2 m below the rock surface, but their original depth was very likely greater, since the present surface of the shale complex does not correspond to the ground surface of Pleistocene time, when they originated. The wedge-shaped cracks are filled with sand washed or blown out from the neighbouring terrace. They were originally filled with ice. Progressive freezing of thaw water, which penetrated into weathered shales, caused widening of the cracks. Along the wedges, shales are compressed, contorted and in places deformed into small-scale folds.

Load tests have been performed at the level of the foundation base, both on sandy gravel and weathered shales. The shattered shales proved to be more compressible than the terrace gravel. The application of 6 kp/cm^2 produced a settlement of 1.91 mm in gravel and 5.66 mm in shales. Since even a small loading would produce a relatively great settlement of the construction, pillars of the frame structure could not be founded on weathered shales; the foundation in the northern part was therefore excavated down to the unweathered rock.

Figure 6-10 shows an instructive section through frost disturbed shales at another building site in Prague. In the excavation, sand and gravel of the river terrace were encountered and the basal part of a slope made up of Ordovician shales was exposed. When unweathered, the shales are firm, dark grey, and dip into the slope (60—65° NNW), but their surface layer is disintegrated and loosened by periglacial weathering to a depth of 2—4 m and penetrated by several deep cracks. During excavation, five such wedge-shaped cracks were found; they were 50—60 cm wide near the surface, narrowing gradually to 20—5 cm and disappearing 4 m below

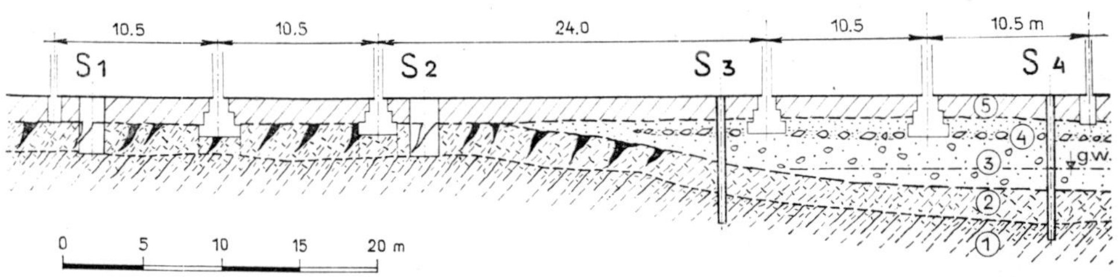

6-9. Section across a building site in eastern Prague. Weathered shales disturbed by frost wedges are more compressible than terrace gravels. Footing between S 1 and S 2 holes had to be founded on unweathered shales. 1 — Ordovician shales, 2 — surface shale layer disturbed by freezing, 3 — sand and gravel, 4 — layer of dreikanters, 5 — man-made ground.

the surface. The cracks are filled with fine sand, which was transported by wind from the near terrace. The weathered shales penetrated by frost cracks had a compressibility ten times greater than firm unweathered shales and twice as great as the sand and gravel. The building has to be founded on unweathered shales, two metres deeper than was planned originally.

6.3. Chemical weathering. Chemical weathering involves processes by which chemical reactions change rock components into secondary minerals.

Chemical alteration is of two kinds: it occurs either at or near the ground surface, or at depth. The former, representing chemical weathering proper, is controlled by exogenous agents, the depth of subsurface water and the resistance of rock. Their intensity generally decreases with depth. Chemical alteration at depth is connected with processes that act at different depths below the surface and is independent of subaerial factors. Their intensity is not in direct relation to the surface relief. These deep-seated alterations include, for example, saussuritization of plagioclases in basic, mainly gabbroic rocks, serpentinization of ultrabasic rocks (peridotites), uralitization of gabbro, propylitization of andesite, i. e. alterations resulting from internal conditions.

Surficial chemical weathering must not be mistaken for these deep-seated chemical alterations which may affect large rock masses and cannot be accounted for by the near-surface position of these bodies. If, for example, chemically altered andesite is encountered near the surface, there is little hope that a sound rock will

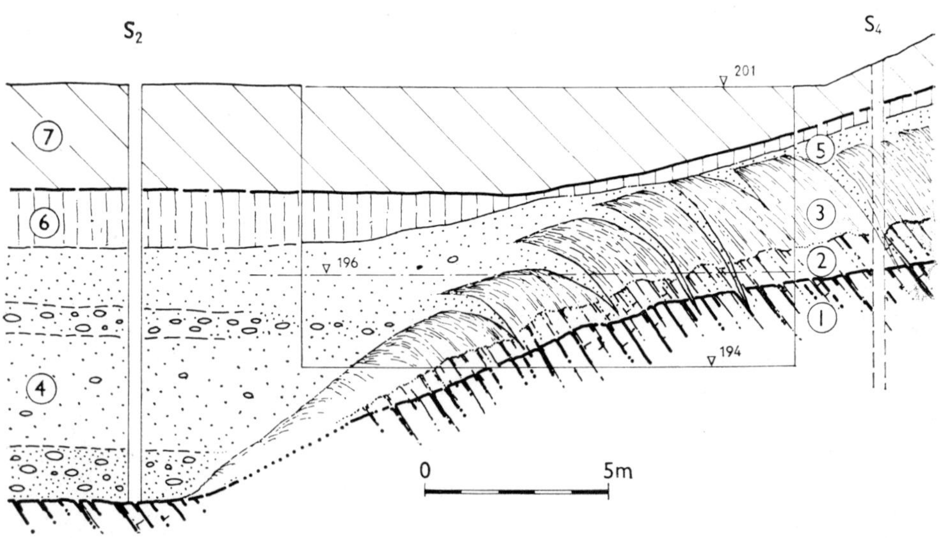

6-10. Section through frost cracks in a foundation pit in Prague. 1 — Ordovician shales, 2 — weathered shales, 3 — surface bending of strongly weathered shales, 4 — sand and gravel, 5 — wind-blown sand, 6 — sandy loam, 7 — man-made ground.

be found at a greater depth. An engineering geologist should keep this fact in mind when determining the depth of weathering in granitoid rocks.

Superficial chemical weathering must also not be confused with hydrothermal alteration, which often affects rock zones of a great thickness. Hydrothermal alterations are the result of chemical activity of hot water solutions, which are generally of magmatic origin. The solutions generated by condensation of water vapours and other gases escaping from the magma are forced into pores and fractures of the rock and react with its components. If a hydrothermally decomposed zone is encountered at a damsite, it is most likely that decomposition persists to a great depth. There is no use in removing such a clayey zone from the foundation pit, since it may extend both laterally and downwards.

6.3.1. Chemical weathering proper. The principal agents of chemical weathering are rain water, oxygen, carbon dioxide and organic acids. Chemical activity of rainwater is due to a large content of oxygen and carbon dioxide. Various salts and acids dissolved in water, humic acids and other organic substances support chemical weathering of rocks.

The depth of surficial chemical weathering varies with the depth of the groundwater level and water percolation which is controlled by the permeability of rocks. The rate and intensity of chemical weathering depends on climatic conditions, in particular on the temperature and the amount of precipitation. The petrography of rocks, their texture and structure, and mode of deposition also influence its intensity.

Particular attention should be paid to seasonal fluctuations of water in the zone of aeration, where the pores are partly filled with water and partly with air. Through this zone rainwater percolates to the lower layers and rises by capillarity to the surface during dry spells, and evaporates.

In determining the depth of the zone of weathering, the length of time for which this process was active should be taken into consideration. Weathering usually extends to greater depths on elevated plateaus, which were long exposed to the action of subaerial agents, than on the slopes of erosion valleys, where the rocks have been laid bare only "recently", geologically speaking.

On some ancient plateaus there are relics of *fossil weathering*, which may extend to a depth of several tens of metres and are occasionally covered by younger sediments. For example, in central Bohemia, a fossil zone of weathering has been preserved at many places under Cretaceous sediments, which indicates a very warm to tropical climate.

The principal processes of chemical weathering are oxidation, hydration and hydrolysis, carbonation and solution.

Oxidation. Most of the rocks originated in an environment of higher pressure and higher temperature than is now present on the earth's surface. Chemical changes caused by weathering are thus predominantly of exothermic nature, i. e. secondary minerals are formed with a liberation of heat. They have a larger volume but a lower density. Such a process is oxidation. It is frequent in minerals containing bivalent iron (Fe^{2+}), which is oxidized to Fe^{3+} and gathers OH^- ions to form hydroxide $Fe(OH)_3$. Hence the rusty or yellow-brown colour of many weathered rocks. Sulphides are also readily oxidized; pyrite and marcasite (FeS_2) are converted first to iron sulphate and then to limonite. During this weathering process sulphuric acid is liberated, which dissolves minerals of the weathered zone, among others calcium carbonate. Crystallization of newly formed sulphates, mainly gypsum ($CaSO_4 . 2 H_2O$) and epsomite ($MgSO_4 . 7 H_2O$), on bedding planes and joints of pyrite-bearing shales contributed to further disintegration of the rock. Therefore, rocks containing pyrite are not suitable building stones; they cannot be used for decoration purposes either, because their appearance becomes marred by rusty spots with time.

During oxidation of sulphides a large amount of heat is released. Decomposition of pyrite and marcasite, which form an abundant admixture in bituminous and brown coal, often causes self-ignition of near-surface coal seams. Burning of coal tips can also be the result of heat liberation at sulphide weathering.

Hydrolysis. This weathering process mainly effects decomposition of complex silicates. Chemical activity of water is increased by its ability to partly dissociate into free ions — H^+ and OH^-. The amount of dissociated water increases with the increase in temperature and in the content of CO_2 and acids, because the acids raise the concentration of H^+ ions. On chemical decomposition of silicates various clay minerals are formed; the nature of the environment and the intensity of weathering control the type of neoformed clay minerals. Less intensive chemical weathering in a slightly alkaline environment causes partial alteration of feldspar into clay minerals of the illite group. In the alkaline environment bentonite weathering gives rise to clay minerals of the montmorillonite group. In a slightly acidic environment potassium feldspar changes into kaolinite (kaolinization) in a warm and humid climate.

Under high temperature and high moisture conditions in tropical regions a very intensive weathering process occurs; most of the silicic acid is removed in the form of gel and the insoluble iron and aluminum hydroxides concentrate in the upper layers. This process is known as lateritization.

Hydration is essentially a process in which water is incorporated in the chemical composition of anhydrous minerals, with a subsequent increase in volume. When anhydrite is changed into gypsum ($CaSO_4 . 2 H_2O$), its volume increases by 60 per cent. Swelling of Triassic rocks containing anhydrite made the driving of some Alpine tunnels very difficult. Other minerals also undergo hydration, e. g. hematite changes into limonite. This alteration is again accompanied by an increase in volume.

Carbonation is a weathering process in which secondary carbonates and bicarbonates are produced by the compounding of carbon dioxide with some weathering products, such as oxides of calcium, magnesium, potassium and other elements.

All surface waters contain dissolved CO_2 of atmospheric origin. From water percolating through weathering rocks limestone most frequently precipitates, crystallizing on bedding and joint planes.

Solution of rocks and minerals is also a chemical reaction of weathering. The ability of water to dissolve rocks depends on its purity, temperature and the time of reaction. Rain and surface water contains oxygen, carbon dioxide and other substances, by which it is enriched in the atmosphere. Percolating through the ground, this water absorbs additional carbon dioxide formed in the soil by the activity of organisms and by the decomposition of organic substances. Minerals do not show the same degree of solubility in water. Chlorides (halite) and gypsum are most soluble, limestone dissolves readily, somewhat better than dolomite. Quartz and muscovite are virtually insoluble and are therefore frequent components of clastic sediments.

Water containing dissolved carbon dioxide changes calcium carbonate ($CaCO_3$) into calcium bicarbonate ($Ca(HCO_3)_2$), the solubility of which is ten times greater than that of the normal carbonate. The origin of karst phenomena in limestone areas is the result of this process. When limestones contain insoluble impurities such as quartz grains and clay, these accumulate as residual soil on the surface of dissolved limestone. The residual soil, generally coloured red by oxidized iron compounds, is known as *terra rossa*. From the thickness of the residual soil can be estimated how much the surface of limestones was lowered. The amount of lowering depends on the proportion of soluble and insoluble components. In the chalk area of England the land surface was lowered about 50 cm in 4000 years. This value has been obtained by comparing the original level of limestones beneath prehistoric mounds, which protected the surface from dissolution, and the altitude of the limestone surface in the neighbourhood. The age of the mounds has been determined by archaeological finds.

The solubility of salt and gypsum in the areas of salt deposits is a very serious phenomenon from the engineering geological point of view. Underground cavities formed by dissolution of gypsum may be of enormous dimensions and increase fairly rapidly when situated within the range of ground-water fluctuation. Sinkholes may originate as a result of the collapse of cave roofs. In the neighbourhood of Halle, Germany, extensive deformations of the ground surface and subsidence threaten the existing structures and complicate the planning of the new ones.

6.4. Investigation of the weathered zone. One of the important tasks of engineering geological investigation is a correct determination of the depth and degree of weathering. The depth of the weathered zone varies over a wide range depending on the resistance of rocks against weathering and on their jointing. Weathering extends to great depths in rocks containing pyrite or marcasite. The greatest depth has been determined in densely jointed rocks and around the issues of ground water.

To express the degree of weathering objectively, it is necessary to introduce suitable tests, the results of which may be interpreted numerically. Weathering of rocks and its classification for engineering purposes was studied by P. G. Fookes, W. R. Dearman and J. A. Franklin (1971). These authors differentiated six grades based on the fracture spacing index, the results of strength tests performed on irregular lumps of rock, and on rock porosity. This method has been developed on the basis of the study of various rocks of the south-western part of England. Figure 6-11 gives the geotechnical assessment of Carboniferous sandstones and

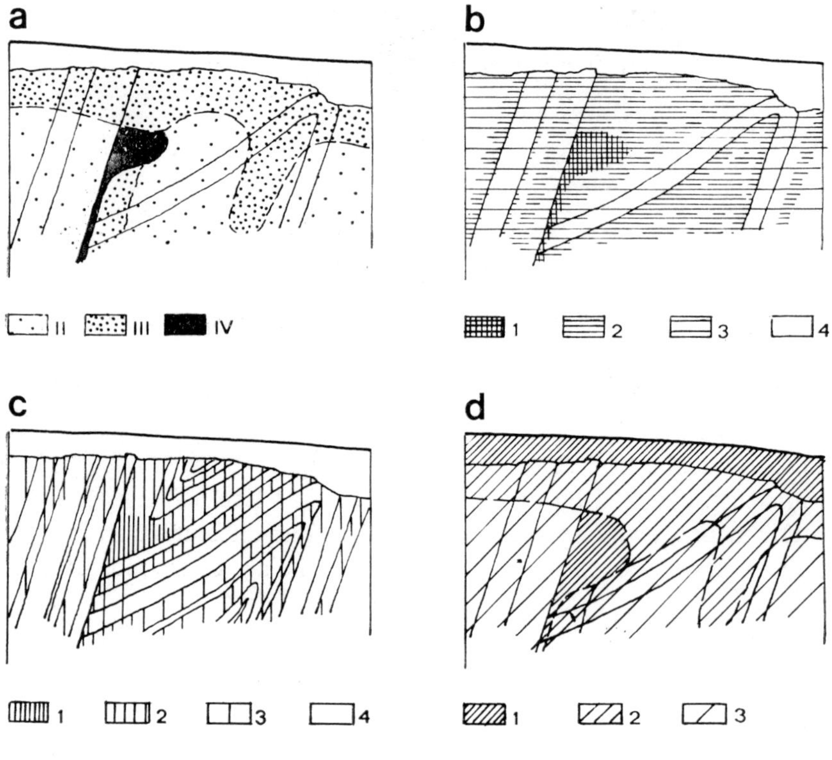

6-11. Geotechnical assessment of a quarry face in Carboniferous sandstones and shales. (Fookes, Dearman et al., 1971.) a — Grades of weathering: II — slightly weathered, III — moderately weathered, IV — highly weathered; b — fracture spacing: 1 — very low < 2 cm, 2 — low (2—6 cm), 3 — medium (6—20 cm), 4 — high (20—60 cm); c — point load strength; d — mode of excavation: 1 — scraping, 2 — ripping, 3 — blasting.

shales in a quarry face, together with the classification of their weathering grades. Since this method requires rather extensive field works and laboratory tests, it can be applied only in detailed research for quarries and large excavations.

J. S. Zolotarev (1971) discriminates three zones of weathering. The upper, dispersion zone of weathering (I) is distinguished by complete alteration of the mineral and chemical composition and of the character of the primary rock. Clayey and sandy soils produced by weathering are readily removed by denudation and are scarcely preserved in situ.

In the middle (II), fragmentary zone of weathering, physical wathering predominates, accompanied by chemical decomposition of varying intensity. This zone is generally divided into several horizons (A, B, C, D). Eluvium of this zone is characterized by very heterogeneous composition and physical properties. The quantity of secondary oxides of iron, gypsum and carbonates varies greatly. The strength and elastic properties of weathered rocks differ from those of unweathered rocks.

The lower, fractured zone of weathering (III) is controlled by tectonic disturbance (faults and fractures) which enables the downward penetration of subaerial agents. The thickness of this zone depends on the configuration of the land surface and on the stress release in the massif.

During the preliminary engineering-geological investigation the grade of weathering must be determined on the basis of simple macroscopic criteria. The division into four grades, proposed by R. Ondrášik (1968), proves convenient for this purpose:

(1) Zone of complete weathering, in which processes of physical and chemical weathering have advanced so that the rock is completely disintegrated and decomposed. The weathered rock consists of secondary minerals and finely disintegrated resistant minerals.

(2) Zone of intensive weathering. The rock is disintegrated into loose small fragments and single minerals. Interstices between fragments are commonly filled with loamy material. Chemical weathering is so strong that all less resistant minerals are decomposed but the original macrostructure of the rock may be preserved.

(3) Zone of slight weathering. The rock is mechanically weathered into large fragments and blocks in primary position. Spaces between fragments and blocks are formed by widening of primary joints or tectonic fractures and are filled with loam or minute rock fragments. Chemical weathering is limited to the sruface layers of blocks and fragments.

(4) Zone of partial weathering. Physical weathering is manifested by the widening of primary and tectonic fractures and by loosening of coherence along planes of weakness, which facilitates disintegration of rocks along them. Chemical weathering is effective along the walls of fractures. Steep wide fractures may be filled with loam or small rock fragments.

In assessing the grade of weathering it should also be noticed whether the weathering products have remained in place or whether they have been transported by water or wind. Some gneiss layers can be disintegrated to sand, while others, being silicified, remain as solid blocks in place. When the sand is removed by rainwash, an inspection of the land surface can give quite an incorrect idea of the state of the rock. The easily weatherable portions may escape attention and the proportion of sound rock may be overrated.

Test pits and galleries are suitable for the determination of the depth and grade of weathering, as they allow a direct inspection of the rock, sampling and the performance of in-situ tests, if necessary. It is difficult to discern the depth of the weathered zone in boreholes. In disturbed zones the core recovery is generally small; only small fragments of solid unweathered rock components are obtained, and it is not easy to decide whether the loss is due to unsuitable drilling technique or to dense fracturing of the rock. A suitable drilling technology and careful extraction of the core are necessary for a correct interpretation of the drilling results.

The depth of weathering cannot be readily determined from the boring samples of some sandstones, because sand and sandy detritus have the same appearance as the parent rock, although the material has been redeposited. The depth of the weathered zone cannot be estimated from a small boring sample, if the rock underwent periglacial weathering, which is dominantly mechanical. The rock fragment can be sound and firm even if it is derived from a congeliturbate layer. In this case the character of the weathered zone can be determined only from the loosening and fracturing of the rock and not from the petrographic properties of a small fragment. Surprisingly, the weathered zone in some rocks can be of a greater strength than the original rock. For example, in claystones minute cracks may form on freezing. The rock has the character of fissured clay and is in part permeable. After a secondary cementation by, e. g. limonite, the weathered zone shows a greater strength than the original claystones.

Geophysical measurements are also used for the determination of the depth of a weathered zone. The seismic refraction method has been applied successfully to the determination of changes in the physical properties of rocks caused by weathering. The resistivity method proved useful in some cases, since the resistivity is related directly to the grade of weathering. Resistivity generally drops appreciably in the weathered zone, because the voids of loosened rock are generally filled with water enriched in alkaline ions. Since the results of this method cannot yet be interpreted with sufficient precision, they should be checked by drilling a few boreholes.

Chapter 7

SLOPE MOVEMENTS, LANDSLIDES

Hill slopes undergo a permanent development owing to various processes which mould their shape. Of major importance for all engineering works are earth movements caused either by natural forces or by human interference. They are of very different kinds, depending on the type and number of factors and their interaction.

7.1. Economic significance of slope movements. In some areas slope movements may produce great damage, both directly and indirectly. Landslides or rockfalls have been known to destroy whole towns and cause the death of hundreds of people. Railroads and highroads crossing slopes prone to sliding are often interrupted, because the stability of the slope has been disturbed by excavations (Fig. 7-1) or surplus loading with high embankments. Some railway lines have to be relinquished, owing to sliding hazards and high maintenance costs. Slope movements may also cause great difficulties in the construction of tunnels and dams (see Chapter 13). Landslides make difficult and endanger work in quarries and, on the other hand, quarries wrongly opened and operated can disturb the stability of the whole slope.

Slope movements indirectly entail great damage when the downslipped material dams a river valley giving rise to a temporary lake. The collapse of such a natural dam causes disastrous floods below it. Extremely dangerous are sudden landslips and rockfalls on the shores of seas or reservoirs. In the Norwegian fjords rockfalls produce waves up to several tens of metres high, which seriously damage the settled coast. The huge rock slide into the Vaiont reservoir (Italy) in 1963 raised a water wave almost 100 m high, which overflowed the dam and devastated the valley below it.

This brief review clearly indicates the economic importance of the study of slope movements and their prevention and control for all engineering works, the compilation of regional and settlement plans, the lay-out of transport routes and location of hydraulic structures.

7.2. Factors producing earth movements. It is of primary importance to recognize the conditions that cause the slope to be hazardous for sliding and the

7-1. Highway interrupted by landslide caused by excavation for dam abutment, eastern Slovakia. (Photograph by Záruba.)

factors that trigger the movement. Only an accurate diagnosis makes it possible to appreciate the extent of the danger and to propose adequate remedial measures. The great variety of slope movements reflects the diversity of factors that may disturb the slope stability. The most important are as follows:

(1) *Change in the slope gradient*: This may be caused by natural or artificial interference (e. g. by undermining of the foot of a slope by stream erosion, or by excavations). Exceptionally, the angle of the slope is steepened as a result of tectonic processes, such as subsidence or uplift. The increase in slope gradient produces a change in internal stress of the rock mass and the equilibrium conditions are disturbed by the increase of shear stress.

(2) *Surplus loading* by embankments, fills and spoil heaps. This produces an increase in shear stress and in the pore-water pressure in clayey soils, which decreases their shear strength. The more rapid the loading, the more dangerous it is.

(3) *Shocks* and *vibrations*. Tremors produced by earthquakes, large-scale explosions and machine vibrations also affect the equilibrium of slopes by evoking a temporary change of stress due to oscillations of different frequencies. In loess and loose sand shocks may disturb intergranular bonds and thus decrease cohesion.

In water-saturated fine sand and sensitive clays, the displacement or rotation of grains can result in a sudden liquefaction of the soil.

(4) *Changes in water content*:

(a) Rain and melt water penetrate into joints producing hydrostatic pressure. In soils the pore-water pressure increases and, consequently, the shear resistance decreases. Measurements of rainfall have confirmed that recurrent slope movements occur in years of exceptionally high rainfall.

(b) It has been found that two beds at the contact of which a sliding plane was formed, show different values of electric potential. The increase in the water content leading to slope movement is explained by electro-osmotic processes.

(c) In clayey rocks, the deleterious effect of atmospheric water is stronger when rains come after a long dry period: clayey soils are desiccated and shrunken and water readily percolates deep through the fissures.

(5) *Effects of ground water*:

(a) Flowing ground water exerts pressure on soil particles, which impairs the stability of slopes. Abrupt changes of water level, for example, in reservoirs, cause the increase of pore-water pressure in the slopes and this in turn may lead to liquefaction of sandy soils.

(b) In fine sand and silt the flowing ground water flushes out fine particles and the strength of the slope is weakened by the cavities that are formed.

(c) Confined ground water acts on the overlying beds as uplift.

(6) *Frost effects.* Water freezing in rock fissures increases in volume and thus tends to open them. Rock penetrated by fissures consequently has a lower cohesion. In clays and clayey sandy soils ice laminae are formed which, on melting, enlarge the water content in the slackening surface layer. Freezing of water on the surface

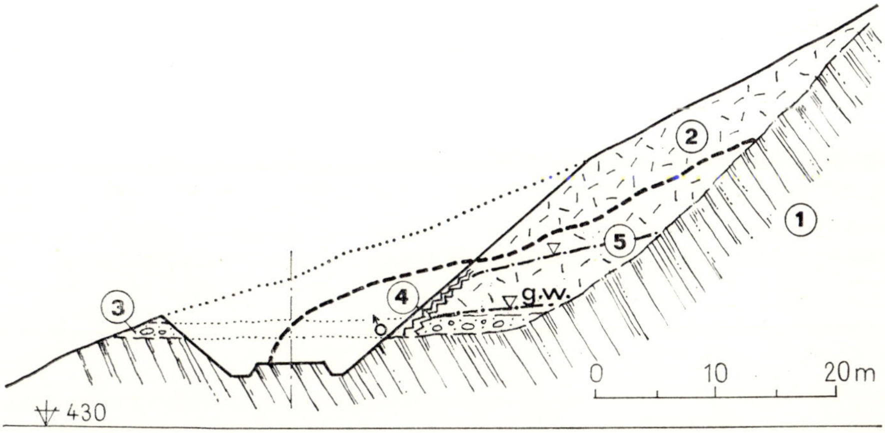

7-2. Stability of cutting disturbed by freezing of water at the slope surface (4), which caused the rise of groundwater table (5). 1 — Permian arkosic sandstones, 2 — slope detritus, 3 — terrace sand and gravel.

7-3. View of the slope shown in Fig. 7-2; freezing springs cause impounding of ground water. (Photograph by Záruba.)

impedes drainage of the slope, so that the ground-water table rises and equilibrium is eventually disturbed (Figs. 7-2, 7-3).

(7) *Weathering*, both mechanical and chemical, gradually disturbs the cohesion of rocks. In many landslides, chemical alterations, such as hydration, ion exchange in clays, are thought to contribute to the origin of landslides. The liability of slopes to sliding has been established, for example, in areas made up of glauconitic sands and clays.

(8) *Effects of vegetation.* The roots of trees maintain the stability of slopes by their mechanical effects and contribute to the drying of slopes by absorbing part of the ground water. Deforestation of slopes interferes unfavourably with the water regime in subsurface layers.

7.3. Division of slope movements. The great variety of slope movements provides many possibilities for their classification. They can be classified according to the mode and rate of movement, the shape of the sliding plane, the type of slipped material, and a number of other criteria. Many authors have endeavoured to develop a suitable classification of slope movements. Terzaghi's (1925) division based on physical properties of the rocks involved has proved adequate for engineering purposes. In the Soviet literature, Savarenski's division into asequent, consequent and insequent landslides according to the shape of sliding planes, is frequently used. Asequent landslides develop in homogeneous cohesive soils along curved, approximately cylindrical surfaces. Consequent landslides occur along bedding planes, joints or planes of schistosity, dipping downslope. Insequent landslides run across bedding and are generally of large dimensions; sliding planes extend deep into the slope.

Popov (1951) recommends taking regional conditions into consideration also, which owing to the diversity of controlling factors modify the individual landslide types.

In the American literature Varnes's classification is generally used; its essentials have been accepted by the Swiss Association for Rock Mechanics (1964).

In Czechoslovakia, Nemčok, Pašek, Rybář (1973) have endeavoured to refine the classification and terminology of slope movements. They proposed a division according to mechanism and rate of movement, into four basic groups: creeping, flow, sliding and fall. Creeping involve a wide range of slow movements (in cm/year) from talus creep to long-term gravitational deformations of mountain slopes. When the sliding mass contains so much water that the movement shows the character of flow, earth- or mudflows will originate. Relatively rapid movements along a definite sliding surface belong to the third group. Abrupt movements of solid rocks, in which free fall plays the main role, are placed in the fourth group.

The division of slope movements into creeping and sliding has several drawbacks, which were pointed out by Skempton and Hutchinson (1969). The rate of movement may change within one landslide and very slow movement may occur even along a distinct sliding plane.

The most frequent factor resulting in creeping is the contractant behaviour of the rock at depth (Fig. 11-10). This phenomenon is perceptible in cuttings of small depth, in large long-term gravitational deformations of mountain slopes and in squeezing of soft rocks from beneath the overlying hard rocks. At depth a shear zone of great thickness originates in these movements, while a well-defined sliding surface forms where the normal stress is small (e. g. in the upper part of the slope).

The adequate classification for engineering geology should enable some general directions for investigation and preventive or remedial measures to be determined for each established group of slope movements. The authors think it most convenient to class sliding phenomena with regard to regional conditions of the country. Since the major part of landslides in Czechoslovakia involve Quaternary cover deposits, they place them in a separate group. Slope movements in the pre-Quaternary bedrock are divided according to the character of rocks and to the type of movement.

Using these principles the following scheme has been proposed (Záruba, Mencl, 1969):

7-4. Pine wood on an old landslide, Dneboh, northern Bohemia. (Photograph by Záruba.)

A. *Slope movements of superficial deposits* (slope detritus, weathering material), produced mainly by subaerial agents:

(a) Talus creep, effecting terminal bending of strata.
(b) Sheet slides.
(c) Earth flows.
(d) Debris flows, muren, liquefaction of sand.

B. *Slides in pelitic rocks* (clays, marls, claystones, clayey shales, etc.):

(a) Along cylindrical surfaces.
(b) Along composite sliding surfaces.
(c) Caused by squeezing out of soft underlying rocks.

C. *Slope movements of solid rocks*:

(a) On predisposed planes (bedding, schistosity, jointing or fault planes).
(b) Long-term deformations of mountain slopes.
(c) Rockfalls.

D. *Special kinds of slope movements* which are an important geological phenomenon in particular countries:

(a) Solifluction.
(b) Slides in sensitive clays (quickclays).
(c) Subaqueous slides.

The age of landslides and their geomorphological development. The time factor plays an important role in the formation and development of slope movements. As the factors change with the course of time, the movement passes through several phases of development. At first, indications of the disturbance of equilibrium will appear, i. e. fissures form in the upper part of the slope. Afterwards the loosened mass is propelled into motion, travels downslope and is gradually deposited at the foot of the slope. According to the development we differentiate the initial, advanced and final stages of slope movements.

On the basis of age, *contemporary (recent)* and ancient landslides are distinguished. *Ancient* landslides that cannot revive under the present climatic and morphological conditions are denoted as '*fossil*'. When they have been subsequently covered by loess loam or other deposits they are termed *buried* landslides.

For engineering purposes the division based on the degree of stabilization into active, dormant (potential) and stabilized landslides has proved convenient.

Active landslides are easy to recognize by their configuration, because their surface forms are fresh, spectacular, and as yet not effaced by rainwash and erosion. Trees are diverted from their original position, roads and alleys traversing the slide area are interrupted and buildings are often damaged (Figs. 7-4, 7-5, 7-6).

7-5. Stretched tree roots in fissures indicate active movement. (Photograph by Záruba.)

7-6. Road dislocated along a crack suggests recent movement. (Photograph by Záruba.)

Dormant (potential) slides are usually overgrown or disturbed by erosion, so that traces of the last movement are difficult to discern. Since the slide-provoking factors are still present, the movement may be renewed.

Stabilized landslides originated under morphological and climatic conditions that cannot occur today.

7.4. Slope movements of surface deposits. Controlling factors of these slope movements are the character of slope deposits, morphology of the slope and, chiefly, the activity of subaerial agents.

7.4.1. Talus creep and terminal bending of beds. Talus creep is the result of various minute processes. In winter loosening of rock fragments and the upheaval of surface layers are produced by freezing. On spring thaw, the particles do not fall back to their original position but, under the influence of gravity, move slightly downhill. The movement of loose stone debris is caused mainly by temperature changes (expansion by heat and shrinkage on cooling).

The surface layers of clayey material move downhill as a result of slow plastic deformation (creep). This movement does not develop a discrete slide surface but a broader zone, within which minute partial movements occur. It is restricted to a surface layer, the thickness of which does not exceed the depth of seasonal variations in temperature and humidity. The shifts are of centimetre order per annum but they may produce diverting of telephone and electric poles, and damage to retaining walls, buildings and subsurface lines.

The creep of detritus layers leads to the terminal bending of strata. Friction active between the creeping detritus and the surface of the bedrock produces a gradual bending of beds. The dragged-out and disrupted layers of the bedrock are incorporated into slope deposits, thus increasing their thickness.

Dragging out and bending of beds could not develop on moderate slopes under present climatic conditions. These features date mostly from the Pleistocene, when even hard rocks were so disturbed and shattered by recurrent deep freeze and thaw that they were dragged downhill.

Terminal bending of strata is widespread on slopes made up of shales, thin-bedded sandstones or limestones, weathered gneiss, granite and other rocks. Figure 7-7 shows a typical bending of beds from the outskirts of Prague, where it is very frequent in Ordovician shales.

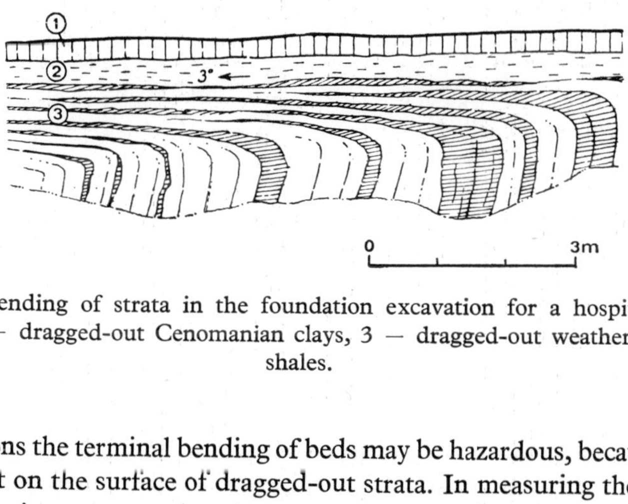

7-7. Surface bending of strata in the foundation excavation for a hospital in Prague. 1 — Loess, 2 — dragged-out Cenomanian clays, 3 — dragged-out weathered Ordovician shales.

In excavations the terminal bending of beds may be hazardous, because a liability to sliding exist on the surface of dragged-out strata. In measuring the dip of beds in shallow test pits, care must be taken not to measure the false dip of near-surface beds affected by terminal bending.

7.4.2. Sheet (slab) slides of slope detritus and weathering material may be of large dimensions, but their thickness is generally small, not exceeding several metres. The disturbed slopes show several stages of sliding, ranging from initial fissuring of the surface layer up to advanced forms with several generations piled on top of one another. Sheet slides are generally at rest in dry weather, the movement being renewed after heavy rainfall and during spring thaw, in particular after a long frosty period. The surface layers of clayey-sandy and loamy detritus are enriched in water during freezing of the ground. The water rising by capillarity

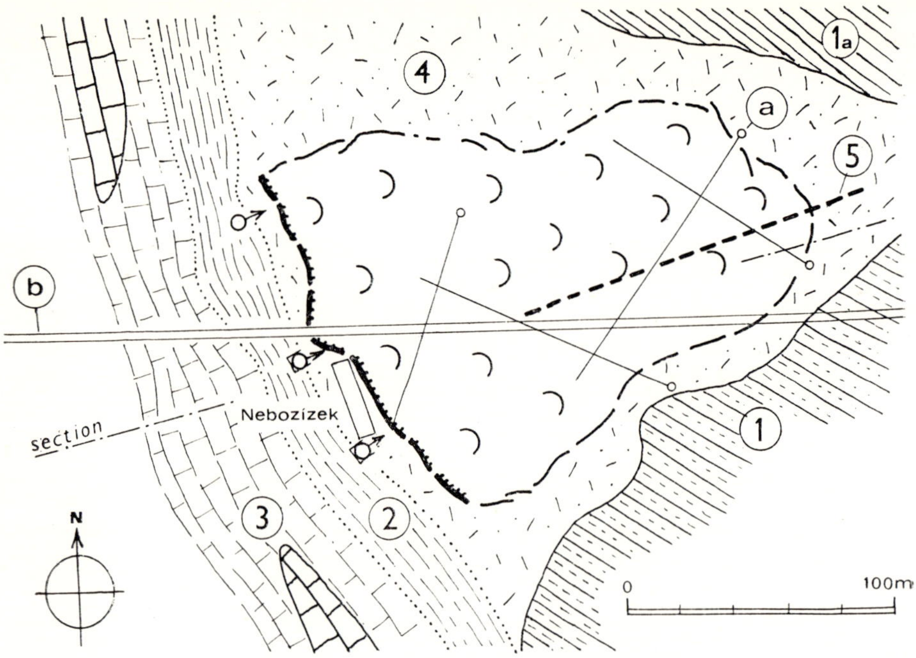

7-8. Sheet slide of slope detritus on the slope of Petřín in Prague. 1 — Ordovician shales, 2 — Cenomanian claystones, 3 — Cenomanian sandstones, 4 — slope detritus, 5 — tentative course of the Prague Fault, a — drainage boreholes, b — embankment of the funicular.

towards the surface from the lower unfrozen beds successively freezes into thin ice laminae, which on thawing cause slaking of the surface layer.

Sheet slides often occur in Cretaceous and Tertiary clayey sediments and commonly in flysch areas.

A sheet slide that disturbed the slope of the hill Petřín in Prague in 1965 is shown in Fig. 7-8. The movement, which destroyed the embankment of a funicular railway, involved slope debris and weathering material of Cretaceous sandstones and claystones filling a depression in Ordovician shales, which formed along the Prague Fault. The sliding surface follows in part the dragged-out Cretaceous claystones and in part the surface of weathered underlying shales. The slide is about 200 m long, and in the upper part 130 m wide and 4—8 m deep. The head scarp originated below the spring line, where several springs appear, draining the groundwater accumulated above the Cretaceous claystones (Fig. 7-9). The movement was triggered by increased rainfall during 1964—1965 and by the increased discharge of all springs in the upper part of the slope, due to insufficient tightness of water and sewer lines in the plateau above.

The first task of corrective treatment was to divert the springs beyond the slide area; subsequently the slide itself and its substratum was drained by oblique drillholes. Permanent stabilization of the area has been achieved by a drainage gallery, which intercepted an aquifer feeding the springs in the slope.

7.4.3. Earth flows. Sheet slides and slides along cylindrical slide surfaces can pass into earth flows, when the masses accumulated in an erosion gully become so saturated by water that they move downhill as a narrow flow, occasionally to a considerable distance. According to the material and its consistency, debris, earth or mud flows are differentiated.

The course and shape of earth flows adapt to the surface relief; the earth flows usually follow erosive gorges or channels of brooks. The movement is generally provoked by exceptional rainfall. Atmospheric water soaking into the detritus greatly increases its weight and reduces its shear strength. The earth flows therefore move more rapidly than sheet slide. There are, of course, also transitional types; rocks may tear loose in the head scarp and slide down curved surfaces to the lower part of the slope, where the water-soaked mass moves as a flow.

Earth flows are abundant in the Flysch Belt of the Carpathians, from which a number of slope movements of this type have been described both from the Slovakian and Polish side of the mountain range.

A typical earth flow occurred in 1962 in the Slovak Beskides Mts., in the Riečnica river valley (Fig. 7-10). The area is made up of the Tertiary Magura Flysch, consisting dominantly of marly and clayey shales with intercalations of glauconitic sandstone. The shales weather readily, so that the slopes are covered by a thick layer of clayey-sandy detritus, which is liable to sliding. The earth flow originated in the upper part of a hill slope, in a depression drained by a small brook. This depression was filled by slope debris up to 15 m thick, slid down during older movements. After a heavy downpour, fissures appeared on the houses of Liesková village, situated in the upper part of the landslide area. Several days later, the water-soaked detritus poured into the brook valley, through which it travelled as a narrow flow. The maximum rate of motion was measured in the middle part of the flow (25 m in 1 hour). The graph showing the rate of motion is reproduced in Fig. 7-11.

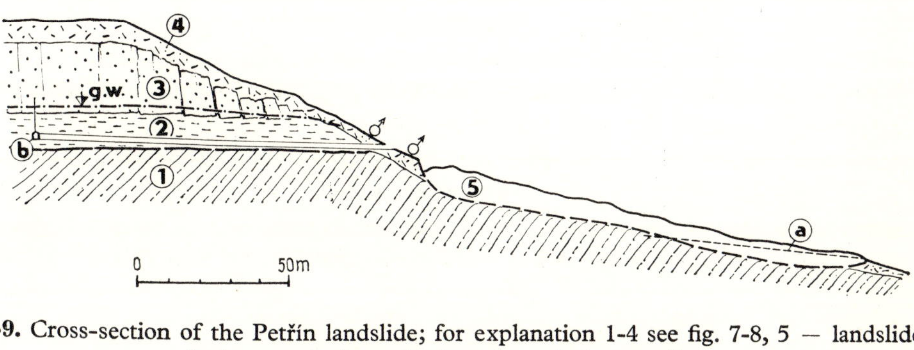

7-9. Cross-section of the Petřín landslide; for explanation 1-4 see fig. 7-8, 5 — landslide body, a — drainage borings, b — drainage gallery.

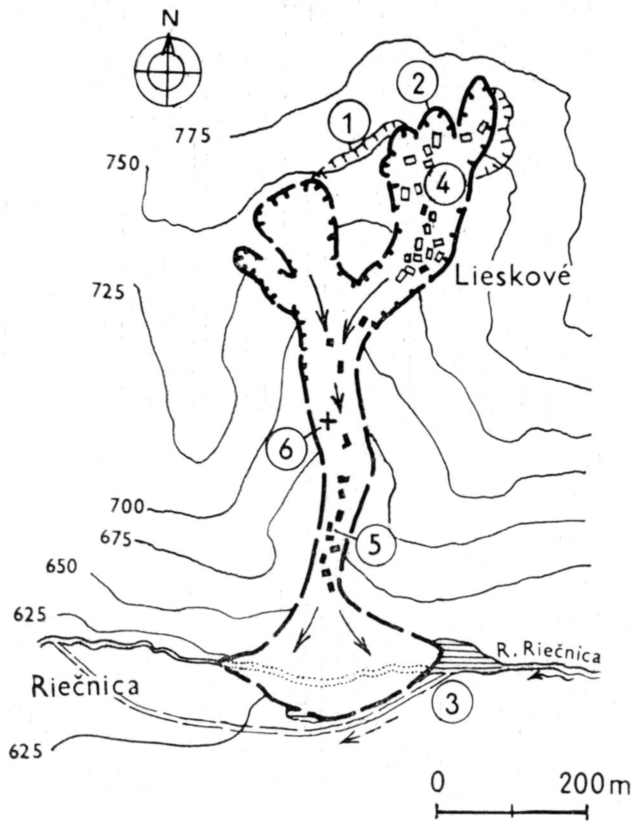

7-10. Earthflow near Riečnica (Slovakia) from 1962. 1 — Ancient head scarp, 2 — newly originated head scarp, 3 — dammed lake, 4 — original site of houses in Liesková, 5 — site of houses after sliding, 6 — place where the rate of movement was measured. (Řepka, 1963.)

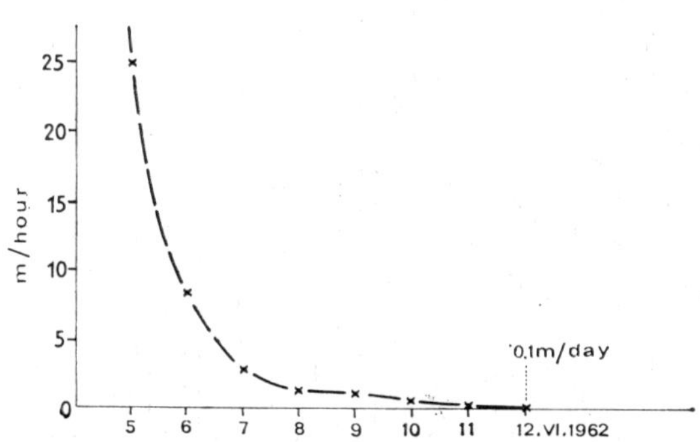

7-11. Rate of movement measured in the middle part of the earthflow.

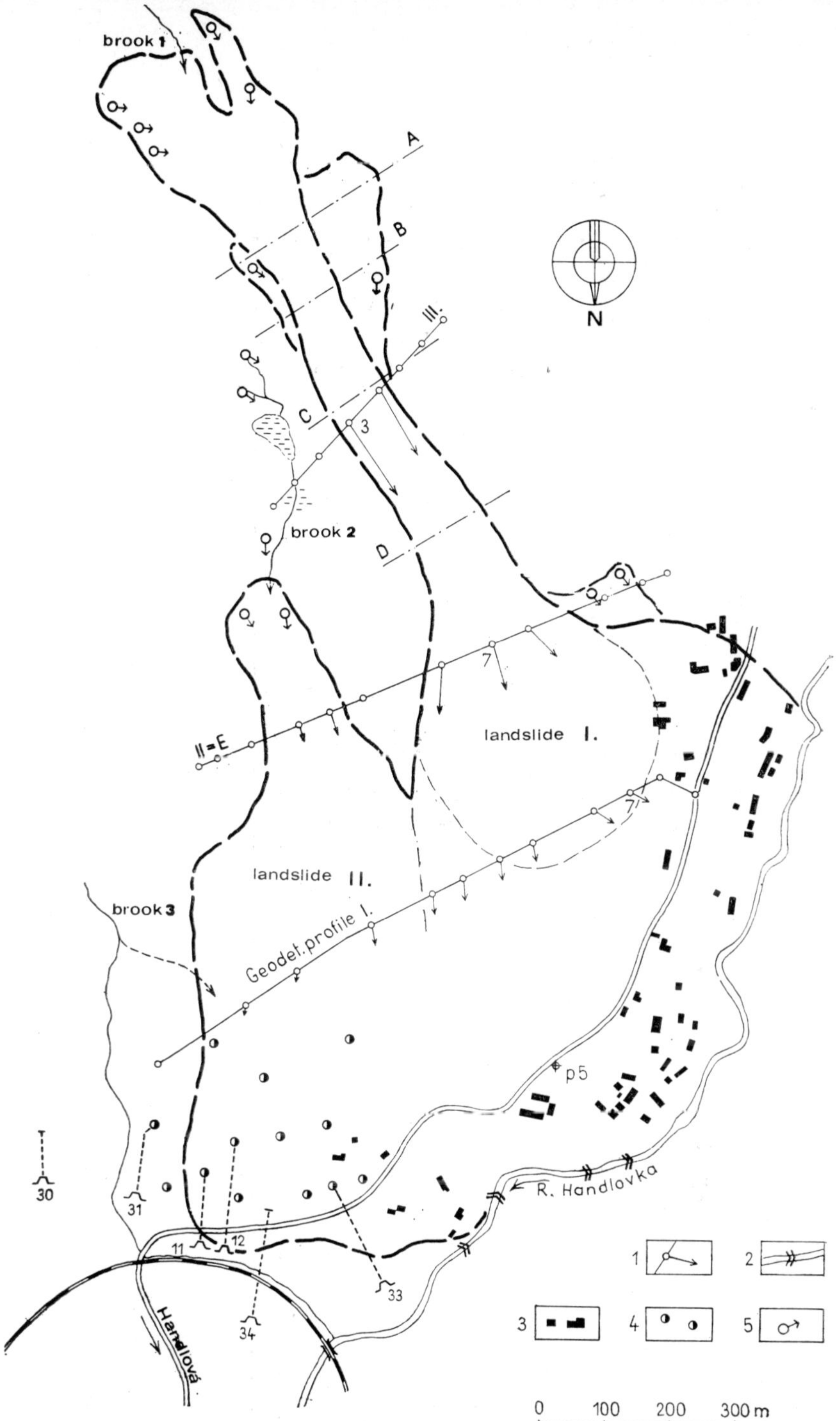

7-12. Situation of earthflows near Handlová in 1960–1961. 1 – Vectors of points surveyed from January 1 to May 31, 1961, 2 – rubble work barriers in the Handlovka bed, 3 – destroyed buildings, 4 – hydrogeological boreholes from which water was pumped by submersible pumps, 5 – large springs in the root area. (Záruba, Mencl, 1969.)

The earth flow was 950 m long and the volume of the debris involved was estimated at 900,000 m^3. The clayey-sandy detritus was kneaded through, strongly water-soaked and mushy so that in the first days the flow was inaccessible. The village Liesková was completely destroyed and two more villages in the river valley were demolished. The earth flow dammed the ancient channel of the river Riečnica and the impounded water threatened several other villages.

This disastrous slide was caused by unusually heavy rainfall during spring months, which, according to the records of the ombrometric stations, attained 237 per cent of the fifty-years' average. The accumulation of clayey-sandy detritus, its liability to slaking and suitable morphology of the slope were contributive factors.

One of the largest earth flows occurred near Handlová in Slovakia in 1960. During several weeks it attained an enormous size: the total length was 1800 m

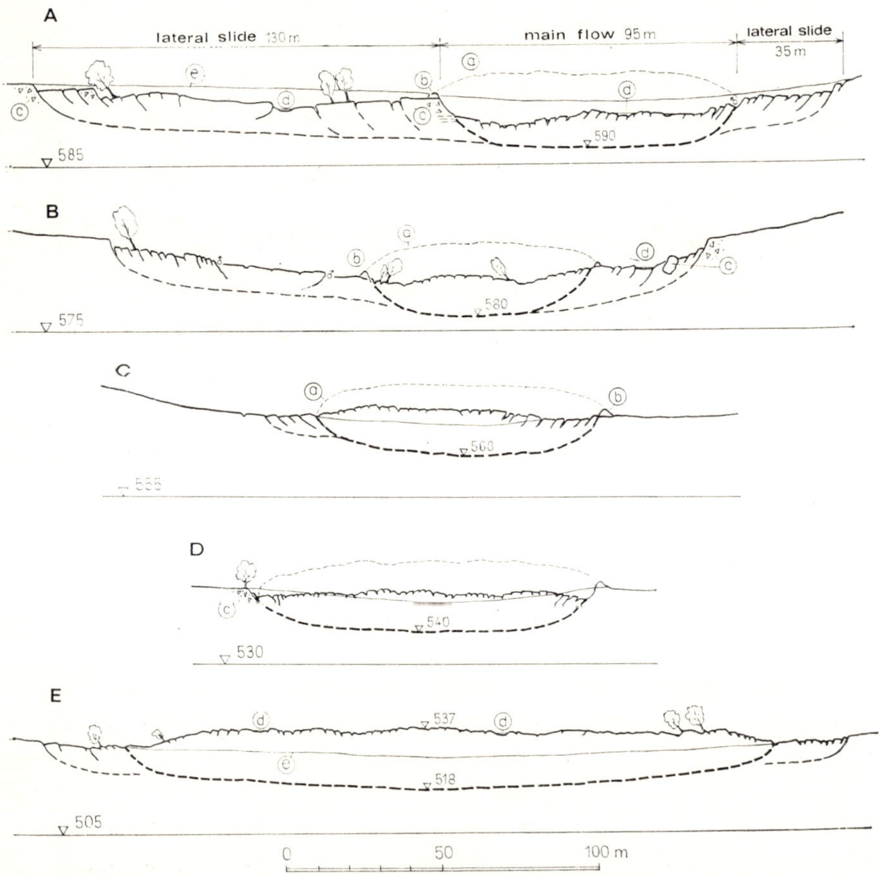

7-13. Cross-sections of earthflow I near Handlová. a — The highest level of the earthflow, b — squeezed lateral ridges, c — andesite blocks, d — lakes in the depressions of the surface, e — original ground surface.

7-14. The middle part of earth flow I near Handlová (courtesy Geological Survey, Prague).

and the volume of the downslipped mass exceeded 20 millions m³ (Figs. 7-12, 7-13). The earth flow destroyed 150 houses, disturbed the higway, water lines, etc. The head scarp formed high on the slope in Neogene clays and tuffites, at the foot of andesite sheets. The loosened weathering material and andesite detritus moved as a narrow flow into the river valley, which was dammed and constricted at several places. For details we refer to Záruba and Mencl, 1969 (Figs. 7-14, 7-15).

Earth flows on mountain slopes, the root areas of which are filled with loose detritus, are a serious obstacle to all transportation routes, because their correction is difficult and expensive.

7.4.4. Debris flows, muren. The term debris flow or talus flow is used for rapid movement of slope debris provoked by sudden floods. According to Skempton (1969) they seem to partake more of mass transport than mass movement. They originate in loose, little consistent rocks, into which water seeps readily. Mountainous debris flows are called by a local Alpine name, *muren*, or *debris avalanches*. They commonly originate above the timber line in gorges filled with rock fragments. During torrential rains, the unsorted material ranging from fine sandy detritus to large boulders can be swept downhill. The ratio of solid particles to water is usually 1 : 1. The muren move very swiftly; moving trains have been

7-15. The root area of earth flow I near Handlová. (Photograph by Záruba.)

known to be caught and buried by such a flow. In temperate zones their occurrences are confined to high-mountain areas. Yet they can be produced by careless interference on any slope covered with loose detritus, for example, by clearance or disturbance of the vegetation cover, or on the slopes of deep road and railway cuttings left ungrassed for a long time.

Debris flows are detrimental in arid and semi-arid regions, where the rock detritus, loam and mud are swept down the valley during intermittent torrential rains. In these regions, the ground surface is not protected sufficiently by vegetation cover from strong rainwash.

Volcanic mudflows are of similar character. Volcanic explosions are generally accompanied by heavy rains, which wash down loose ash and minute ejecta as a mushy mass towards the foot of the volcano. Such a volcanic mudflow buried the town of Herculaneum upon the explosion of Vesuvius in A. D. 79. Debris flows also form on the slopes of dormant volcanoes covered by mountain glaciers, when volcanic activity is rejuvenated. The glaciers melt suddenly due to the heat of eruption, and the flood waters carry down the slope debris and morainic material.

Slope disturbances produced by liquefaction of sand can be classed with this group. Running sand is not a special kind of soil, because every loose soil can become liquid under certain conditions. Liquefaction of sand is caused by forces developed by water flow in the sand. Liquefaction of sand may be provoked by abrupt lowering of the reservoir level, or on puncturing the impermeable cover of a sandy aquifer. Failures of this type are relatively rare in natural slopes.

Loose, highly porous sands may also be liquefied by vibrations, which result in the re-arrangement of sand grains; the density of the mass is greater and porosity smaller. The superfluous water is expelled. Since the water cannot escape immediately, the pore-water pressure increases, reducing friction between grains, and the soil becomes liquid for a short interval.

This type of slope failures is comparatively frequent on seashores (e. g. in The Netherlands) or on lake banks. Analogous slides may occur on artificial slopes, such as poorly compacted sandy embankments, along reservoirs, upstream faces of earth dams, etc. As mentioned above, these slope accidents are caused by the high porosity of sand and, therefore, fills composed of sand must be thoroughly compacted.

Running sand is very dangerous in coal basins, because it may cause caving-in of mine spaces and sudden subsidence of the ground surface in undermined areas.

7.5. Landslides in clayey rocks

7.5.1. L a n d s l i d e s a l o n g c y l i n d r i c a l s l i d e s u r f a c e s (s l u m p s). If the shear strength of unconsolidated or poorly consolidated clayey rocks is surpassed, deep-reaching slides form along curved slide surfaces in the slope. Slides of this kind are called slumps. In homogeneous clayey rocks (clays, claystones, argillaceous shales) the shape of the slide surface may be regarded as cylindrical. Using theoretical computations and measurements a somewhat different shape of the slide surface might be assessed, but the difference in stability calculations is small, in any case smaller than the reliability of the method applied for the determination of the strength of soil.

The slumps have a characteristic form (Fig. 7-16). Since the slide surface is curved, the movement involves rotation and the surface of the slumped mass is usually inclined towards the slope. The root area shows a typical concave form and the material is accumulated at the foot of the slope. Transverse cracks, which open in the tongue, are gradually filled with water. The equilibrium of the slope is thus further disturbed. The slumped mass is often so waterlogged that the tongue takes on the character of an earth flow.

Slumps in clayey rocks are of various sizes; those in artificial cuttings are relatively small, but those developed on high river banks or on a sea coast may attain enormous dimensions. Occasionaly many millions cubic metres may be in motion.

The slumps can gradually extend by headward caving of the scarp. The partial movements occur along curved slide surfaces, and the whole slide area is usually randomly undulated. The resulting slide surface is then not cylindrical. The depth and shape of the slide adapt themselves to the geological structure of the slope and to the local mode of deposition.

Steep concave slopes of river valleys built of clayey rocks may be disturbed by a whole series of slumps, the alcoves of which are close to each other. The slumped masses often extend into the river channel, but their removal by floods causes a new disturbance of the slope equilibrium.

7.5.2. Landslides along composite slide surfaces. Large landslides in clayey rocks generally develop a new curved slide surface in the upper part, but in the lower part use a stratum of smaller strength and of suitable position. These movements are transitional to the slides along predetermined surfaces.

A large landslide of this type has been examined in the Váh-river valley (Slovakia) near Súčany, where a power plant was designed. The river Váh flows there along the southern foot of the Malá Fatra Mountains, in a broad valley cut in Neogene marls and silts with interbeds of friable sandstone and conglomerate. West of Súčany, marls and silts crop out in a steep concave bank, 400 m long and up to 24 m high (Fig. 7-17). This concave bank is the front of a slide, the head scarp of which is 900 m from the right bank of the river. The surface of the slide area is partly covered by an ancient alluvial cone. The deeply dissected ground surface is suggestive of the recurrence of sliding processes in the geological past. The outcrop of the slide surface has been established on the left bank or the Váh, 60 m from the river channel. Neogene sediments were raised along this slide surface several metres above the Holocene floodplain, and the bent willow trunks indicate that the slide has been in motion until recently. An interesting feature was that the Váh flowed over the lower part of the slide, undermining its toe, while the slide surface outcropped on the other bank (Fig. 7-18). River erosion and the slide

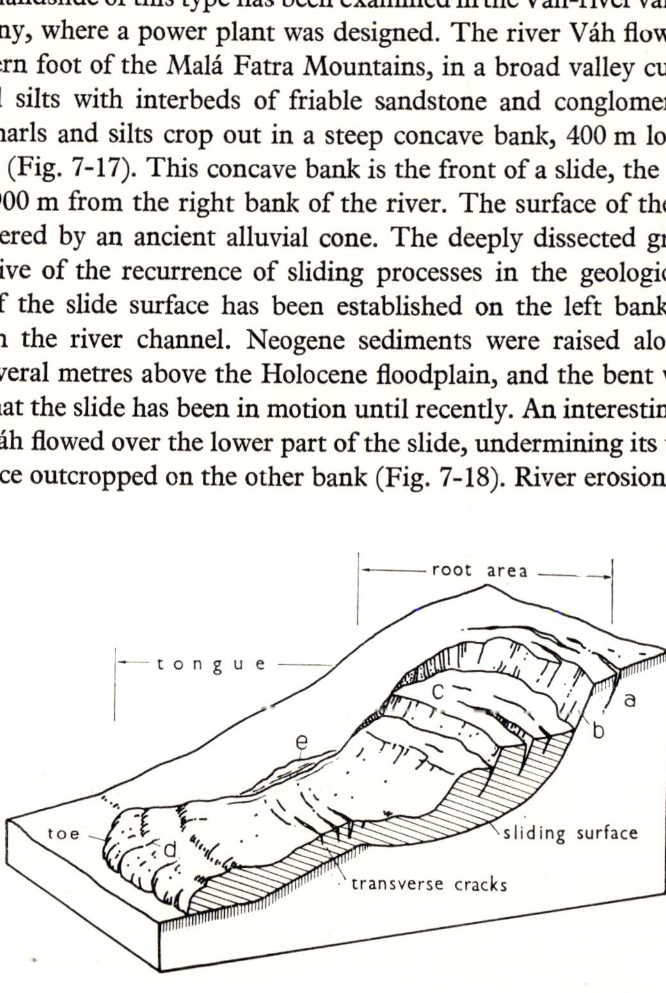

7-16. Principal parts of a landslide and major cracks (modified after Varnes). a — Lunar cracks, b — head scarp, c — transverse cracks, d — radial cracks, e — lateral ridges.

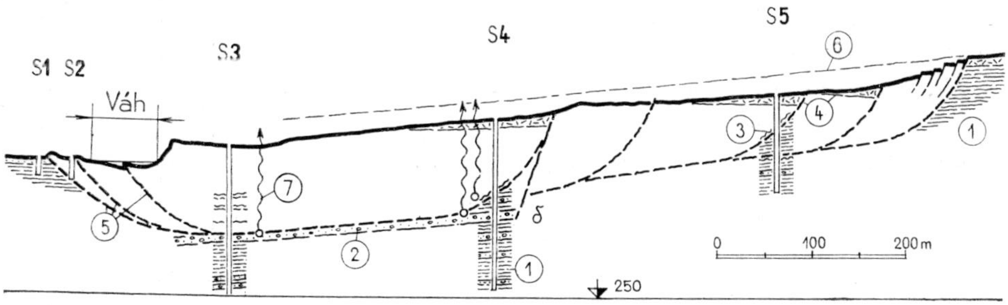

7-17. Section of a deep landslide with a compound slide surface near Sučany at the foot of the Malá Fatra Mts., Slovakia. 1 — Neogene marls and sandy marlstones, 2 — Neogene sandy gravels, 3 — kneaded layers, 4 — granite debris, 5 — partial slip surfaces, 6 — original ground surface, 7 — confined ground water in Neogene gravel. (Záruba, Mencl, 1956.)

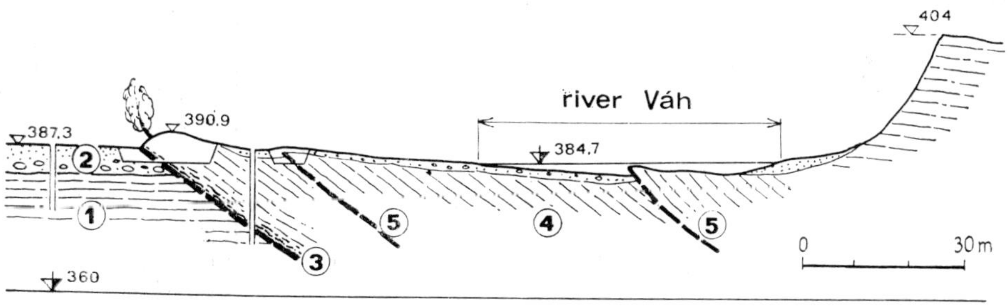

7-18. Section of the frontal part of the Sučany landslide. 1 — Neogene marls in original position, 2 — floodplain sand and gravel, 3 — kneaded layer squeezed above the flood plain, 4 — raised Neogene marls eroded by the Váh, 5 — partial slip surfaces. (Záruba, Mencl, 1956.)

movement thus acted in opposite directions. As the movement was relatively slow and river erosion strong, the two factors were almost in equilibrium. In the length profile of the thalweg, a distinct knickpoint exists at the site of the landslide. Such steps originate in stream channels by selective erosion in hard rocks, but in this particular case it formed in soft marlstones which were crammed incessantly into the river channel by sliding movement.

The slide area was surveyed and the rate of movement was measured by geodetic methods (Fig. 7-19). During 1947—1957 the horizontal displacement averaged 8—10 cm in a year. The shape of slide surfaces and moulded zones was studied in test-pits and boreholes. The shear zone dips 34—40° into the slope on the left bank, and was found at a depth of 80 m on the right bank. The outcrop of a partial slide surface was established in the river bed. The volume of the mass involved is estimated to be 40 milion m³ (Záruba, Mencl, 1958).

The landslide developed in a relatively flat terrain by lateral erosion of the Váh, and as a result of hydrogeological and tectonic conditions. Basal conglomerates contained an artesian water horizon with 13 atm overpressure, which produced the uplift of the moving mass. The movements parly occurred along ancient fault planes, which predetermined the shape of the landslide.

The stability analysis has shown that the results of laboratory tests do not conform fully to reality. Hence, it follows that when a large mass is in motion and the slide

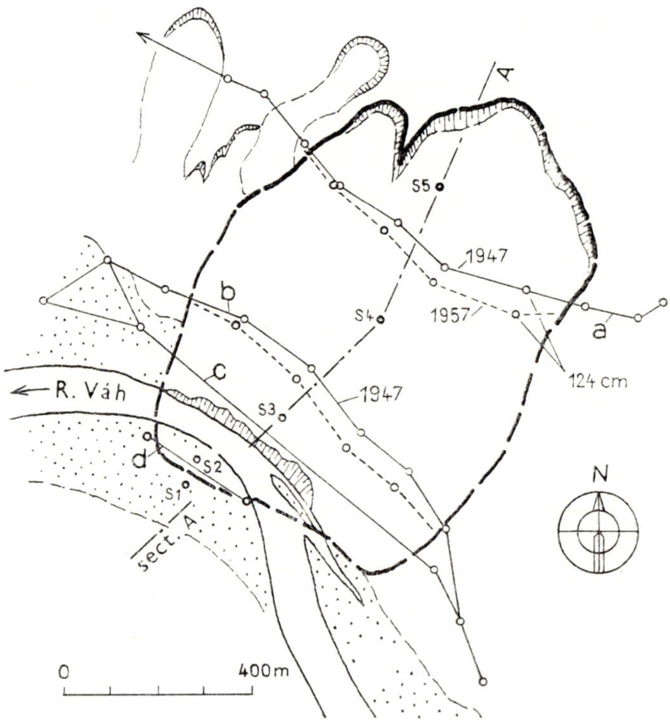

7-19. Plan of the Sučany landslide, lay-out of observation lines (c, d) and polygons (a, b). (Petrášek, Záruba, 1959.)

surface runs deep under the land surface, the results of shear strength tests performed only on a small scale cannot be relied upon. In these tests only one clear-cut slide surface develops, but at great depths and thus also at high pressures (in the case studied up to 21 kg/cm²), the movement produces thick remoulded zones, in which the structure of the mass is totally altered. The change involves even the finest particles. This fact was already mentioned by Denisov (1951) who found that pressures exceeding 10 kg/cm² cause disturbance of the internal structure of clay and reduction of the angle of shear strength.

Owing to the Súčany landslide the Krpelany-Súčany-Lipovec water scheme had to be redesigned. The diversion canal and the Váh channel were relocated so as to avoid damaging by the movement. The movement ceased as soon as the toe of the slide was not undermined by river erosion. The slope was stabilized by a loading fill (about 60,000 m³) placed on the valley floor where the squeezed-out ridges of Neogene deposits indicated the outcrops of slide surfaces.

In dimensions and character the landslides in the Váh valley resemble slope movements described in the Soviet literature from the valleys of the Volga, Moskva, Dnieper and other major rivers. Sliding is most extensive in the middle and lower courses of the river Volga, between the towns of Gorkii and Volgograd and particularly affects the steep western banks. Slumping mainly involves Jurassic and Cretaceous clays. Fossil slumps buried by young river deposits provide evidence that the conditions favourable to slope movements were repeated several times in the geological past.

Many landslides disturbing sea coasts may also be included in this group. An example of extensive sliding on the sea coast is the well-known landslide nar Folkestone in southern England. Upper Cretaceous rocks overlie the Gault clays on a steep bank. During the last two centuries the Dover-Folkestone railway line has been interrupted several times by recurrent sliding. The motivating force is marine abrasion of clays at sea level, which results in unloading of the foot of the steep bank and, consequently, in its slumping along the curved slide surfaces. The stability is decreased by hydrostatic water pressure in the joints cutting across the Cretaceous rocks. Tests on samples taken in borings have revealed that the strength of clays close to the slide surface is far smaller than that of the undisturbed clays. The reduction of strength is likely to be due to the unloading of clays and the seepage of water from glauconitic rocks, which contributes to slaking of clays.

The origin of these slumps mainly depends on the consistency and shear strength of the argillaceous rocks, and to a certain extent on hydrogeological conditions. The relationship between renewed sliding movements and excessive rainfall is well demonstrated by a deep landslide in Portland (Oregon), which affected the area of the local Zoo. The slide formed in decomposed basalt (stiff clayey silt) (Radley-Squier, Versteeg, 1971). In the upper part of the slope a curved detachment surface developed, which in the lower part follows the surface of dense weathered basalt, nearly flat (Fig. 7-20). The landslide is about 200 m long, 300 m wide and the layer

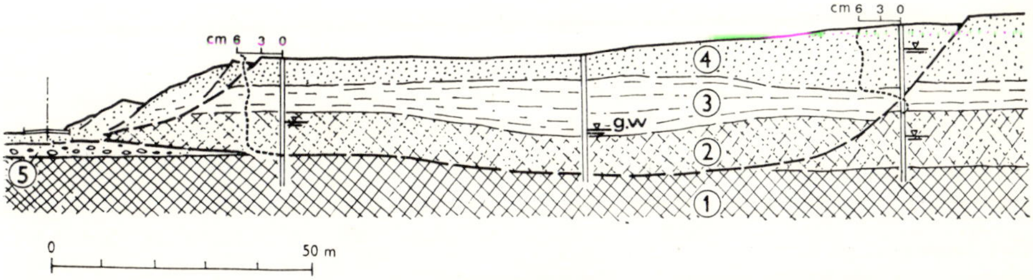

7-20. Cross section of Zoo landslide in Portland (U. S.). 1 — Weathered basalt, 2 — decomposed basalt, 3 — stiff brown clay with sand interbeds, 4 — loess loam, 5 — silt with pebbles, 6 — displacement in cm, April 1969 — February 1970. (Radley-Squier, Versteeg, 1971.)

of clayey material involved in the slide is approximately 25 m thick. It is an old slide area, the stability of which was disturbed by excavation for a new road in 1957. In 1958—1959 increased rainfal caused a new movement, during which a distinct marginal crack delimiting the head scarp was formed. During 1963—1970 the landslide was studied in detail, particularly the relationship between recurrent movements and rainfall. From Fig. 7-21 it is apparent that the movement was

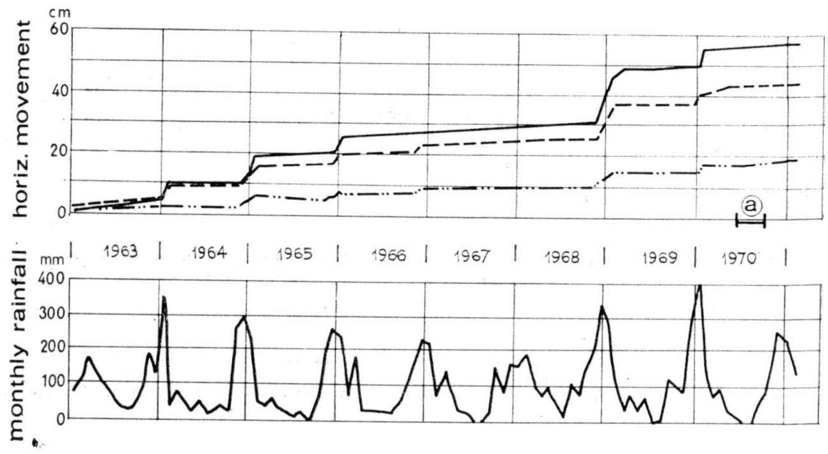

7-21. Relation of horizontal movement of Zoo slide (Fig. 7-20) to rainfall intensity; a — period of buttress construction.

reactivated when the winter rainfall exceeded 200 mm per month. The landslide was stabilized by a system of relief wells and drainage borings and by a rip-rap buttress in the front of the landslide.

Large-scale fossil landslides of this type were recorded by Solle (1967) from the Nile valley in Upper Egypt. A thick complex of limestone and marly shale of Eocene age slid there along curved slide surfaces. Near the valley floor the shear surface runs within subhorizontal clayey shales, rich in montmorillonite (Esna Shales). The landslides occurred in some of the Pleistocene pluvials so that under present arid climatic conditions they are a fossil phenomenon.

7.5.3. Slope movement caused by squeezing of soft rocks. This group comprises block slides, slope deformations caused by squeezing of soft clays in erosion valleys or cuttings, and failures of embankments erected on a substratum of small bearing capacity.

The process is common in nature but because of its extremely small velocity it escapes attention. The instability of the slope is perceptible after a longer time, when the continuous minute deformations have reached measurable values.

Squeezing of soft rocks in the valley floors, so called *bulging*, has been studied and described by Hollingworth (1944) in the iron-ore open mines near Northampton in central England. Relatively deep valleys are cut in solid Jurassic limestones and pelitic shales and soft Lias clays are exposed in the valley bottoms. Although the complex lies nearly horizontally, the beds are strongly deformed in the valley sides and under the valley floor. Limestone beds dip towards the slope and are disrupted by open fissures. The marginal blocks of solid rocks were sinking simultaneously with the bulging of clays in the bottom (Fig. 7-22). The stress that

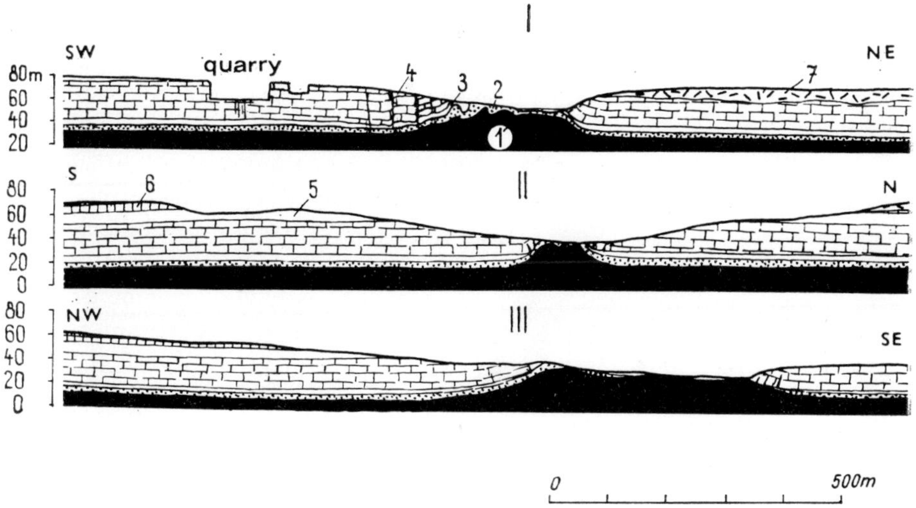

7-22. Bulging of the Lias clays in the valley floor, Lincolnshire, England. I—II — Section of the Bytham brook valley, III — section of the Guash river valley; 1 — Lias clays, 2 — Northampton Sandstone, 3 — Upper Estuarian Clay, 4 — Lincolnshire Limestone, 5 — clay, 6 — Main Oolite Shale, 7 — glacial sediments. (Hollingworth et al., 1944.)

caused the upheaval of clays resulted from the difference in the loading of clays in the valley floor and below the valley sides. The deformations can be interpreted as squeezing-up of plastic substance from the loaded medium into the unloaded one.

A similar phenomenon has been observed in the Lucina-river valley near Ostrava

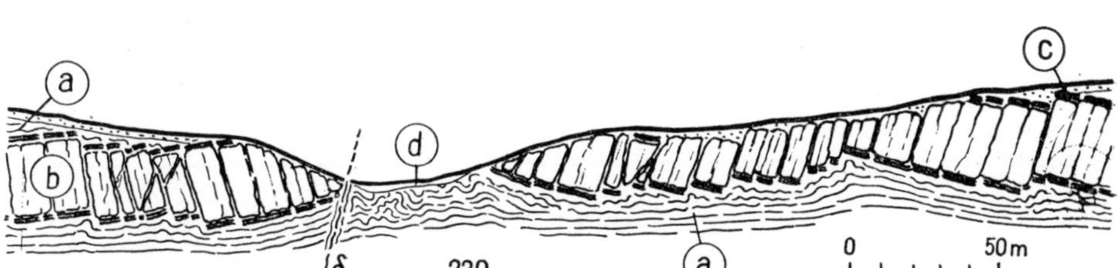

7-23. Squeezing of marly shales in the floor of the Lucina r. valley near Ostrava. a — Lower Cretaceous marly shales, b — teschenite, c — contact metamorphosed slates, d — disturbed shales in the valley floor. (Záruba, 1956.)

(Czechoslovakia). The section in Fig. 7-23 shows that the river cut its channel in marly shales of early Cretaceous age, which are transected by thick concordant bodies of teschenite. Teschenite is disrupted into blocks by faults parallel with the valley. Heavy teschenite blocks gradually sank into soft shales and moved slowly downslope. The squeezed-out shales in the valley floor were carried away by the stream. The inclination of the sinking blocks is clearly seen from the relics of contact-metamorphosed slates above and below the teschenite blocks. The major deformations date from the Pleistocene, as is evidenced by the inter-block steps filled with slope debris and loess loam.

Squeezing of soft rocks in the floor of erosive valleys is also known from Rumania, where these features are denoted as "valley anticlines". Rumanian geologists studied the diapir-like squeezing of salt domes after their unloading by stream erosion. Plastic deformations are frequent in Tertiary clayey sediments in the basins of the rivers Oltul, Tarnava Mare, Arges, etc. Anticlinal bends develop below the valley floors, while beds are nearly horizontal in the slopes (Ilie, 1955; Záruba, 1958). The identical orientation of valley-anticline axes with that of drainage pattern suggests that these features are young and caused by stream erosion.

These surface rock deformations and their accompanying phenomena call for great attention in civil-engineering constructions. Similar phenomena can occur in excavations, where soft rocks are encountered below hard rocks.

Block slides develop on slopes, where soft clayey beds underlie steep-walled blocks of hard rocks. The marginal blocks sink gradually into the soft substratum and travel downwards. Tension produced in the overlying layers causes the marginal blocks to be torn away and moved, being eventually separated by gaping fissures from the main mass. Generally, the bottom part of a block moves outwards and the upper surface of it inclines into the slope.

The processes that give rise to block slides are similar to those leading to the squeezing up of soft rocks in valley bottoms. The plastic layer underlying heavy rock blocks, however, is of relatively small thickness. It facilitates the motion and tilting of the blocks but it is too thin for bulging to develop in the valley floor.

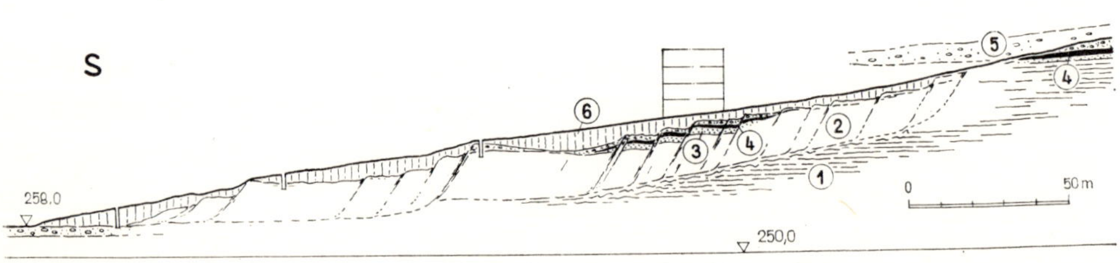

7-24. Fossil block slide near Turnov (Bohemia). 1 — sandy marls, 2 — moved blocks, 3 — friable sandstones, 4 — glauconitic sandstones, 5 — Pleistocene terrace gravel, 6 — loess loam.

Hazardous slopes for failures of this type occur, for example, where rigid sandstones forming sheer walls lie on soft marls (Fig. 7-24) or where basalt sheets rest on tuffs or marls (Fig. 9-10). In Slovakia, some block slides at the edges of andesite sheets show a similar character.

Large-scale block slides occurred in the area flooded by the Bratsk reservoir near Baikal Lake in Siberia. Extensive block slides have been described from the valleys of the rivers Angara and Ilima. The slopes there are formed of Ordovician quartzites and sandstones lying on Cambrian claystones, into which the marginal quartzite blocks are sinking (Fig. 7-25).

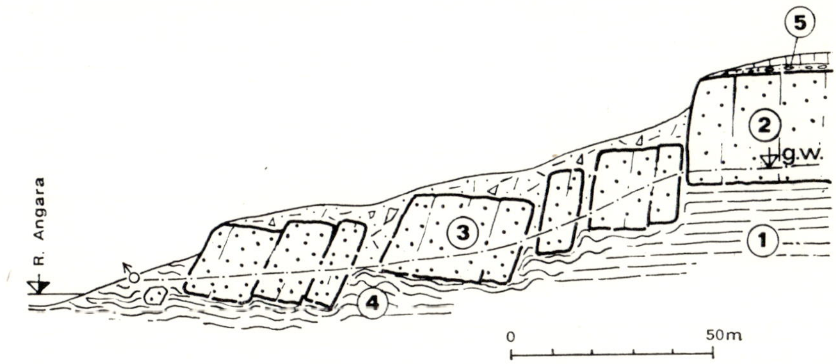

7-25. Block slide on the left bank of the river Angara near the Bratsk Dam, Siberia. 1 — Upper Cambrian claystones, 2 — Ordovician sandstones, 3 — sunken sandstone blocks, 4 — weathered claystones disturbed by plastic deformation, 5 — terrace gravel. (After Palshin et al., 1963.)

The velocity of block slides is mostly so small that they can be justifiably ranked among creep. They belong to this category because the movement does not occur on a definite slide surface, but the blocks sink gradually into their soft substratum, which undergoes plastic deformation and "creeps" downslope. Exceptional rainfall may disturb the consistency of the substratum to such an extent that a sudden subsidence and shift of blocks takes place. Part of the city of Algiers was destroyed by such disastrous block movements (Agard, 1948; Záruba, Mencl, 1969). An inter-disciplinary research carried out after the last disaster has shown that the failure of the slope was facilitated by extremely unfavourable lithology of the slope. The strongly jointed limestones are underlain by glauconitic marls, and when the percolating rainwater came into contact with them, it was enriched with potassium compounds released during the weathering of glauconite. The alkalized water promoted dispersion of marls so that the limestones rested on a virtually semiliquid mass.

7.6. Sliding movements of solid rocks

7.6.1. Rock slides along predetermined surfaces, bedding, joint or fault planes commonly originate when the planes of separation dip downslope and their continuity is disturbed at the foot of the slope.

In stratified rocks with even and smooth bedding planes, the dip of strata is usually the maximum inclination at which a slope is permanently stable. If the beds are undercut by stream erosion or by excavation, they maintain their position only by friction. The coefficient of friction increases with the roughness and unevenness of the bedding planes. Friction can be reduced by climatic factors (freezing and thawing of interstitial water) or by hydrostatic pressure of water in joints if a free outflow of water is obstructed. Failure may also be provoked by the increase of slope angle as a result of tectonic uplift.

Rock slides on bedding planes or other surfaces of separation may be disastrous in high mountains where the movement, due to great height differences, may attain an acceleration nearly equal to that of rockfall. In young mountains, slopes are prone to sliding because streams with steep gradients readily cut into the bedrock and the adjustment of the slopes cannot keep pace with erosion.

One of the largest slope failures of this type occurred on April 25, 1974 in the Mantaro river valley in Peru (Fig. 7-26). A huge complex of Permian sandstone with marl interbeds slid along bedding planes from the top part of a steep slope. A debris flow over 8000 m long blocked the canyon forming a temporary lake in

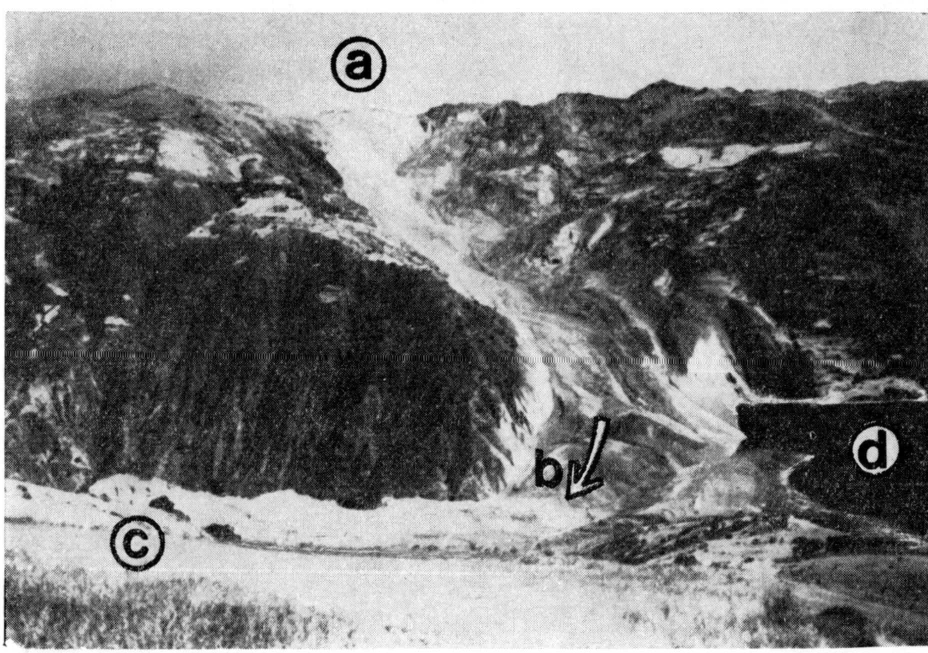

7-26. Rock slide and debris flow on the Mantaro river in the Andes Mts. (Peru). a — Head scarp, b,c — debris flow, d — temporary lake. (Photograph by Kojan.)

which about 670 million m³ water was retained. The volume of rock mass is estimated at 1000 million m³. Because of the height difference of about 1500 m the movement reached a velocity of 120—140 km/hour, being accomplished within 3—4 min. The disastrous slide destroyed several villages and 450 people lost their lives.

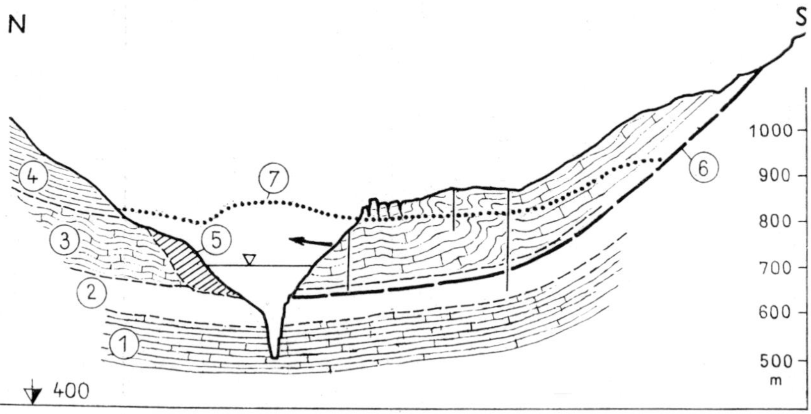

7-27. Huge block of Jurassic and Cretaceous limestones slipped into the Vaiont reservoir in the Italian Alps and caused disastrous flood in the Piave-river valley in 1963 (Selli, Trevisan et al. 1964). 1 — Dogger limestone, 2 — thin-bedded limestone with clay interlayers (Malm), 3 — thick-bedded limestone with chert (Cretaceous), 4 — marly limestone (Cretaceous), 5 — relict of an old landslide, 6 — sliding surface, 7 — valley blocked by landslide.

After 44 days the debris dam was overtopped and in two days washed away. The flood wave was up to 35 m high but thanks to a timely evacuation of the valley no life was taken.

Six months before the slide, cracks were observed in the upper part of the slope, but no earth tremors or increased rainfall preceded immediately the movement. Since, however, the valley lies in a seismic area, previous vibrations are likely to have induced the loosening of sandstone beds. A contributing factor is a strong river erosion which resulted in a very deep valley with steep slopes (Kojan and Hutchinson, 1974).

Landslides are not rare in the Mantaro river valley. Snow (1964) described a similar slide from 1945, when the valley was obstructed by a large slide in jointed granodiorite, and recorded remnants of several earlier slope failures, which form conspicuous features in the topography of the area.

The slide of Jurassic limestones along a bedding plane caused the clogging of the Vaiont reservoir in the Italian Alps in 1963. An immense limestone mass (more than 240,000,000 m³) filled the reservoir and put out of action one of the largest arch dams in the world. The greatest damage was produced by a water wave 100 m high, which overflowed the dam, destroyed the town Longarone and devastated the Piave-river valley below the dam (Figs. 7-27, 7-28).

7-28. Vaiont landslide from 1963. (From picture postcard). 1 — Sliding surface, 2 — slid masses, 3 — arch dam, 4 — limestone laid bare by flood wave.

Rock slides along bedding planes are frequent in the flysch areas of the Carpathian Mountains. Their origin is due to the character of rocks and to the topographical configuration. In the Carpathians, streams with a high gradient cut readily into the soft flysch deposits and the angle of slope lags behind in adjusting itself to the rate of erosion. Consequently, the slopes become very steep, often steeper than the dip of strata and, if this is towards the valley, conditions are created for the development of rock slides (Fig. 7-29).

Sliding along bedding planes may also be induced by excavation for road, railway

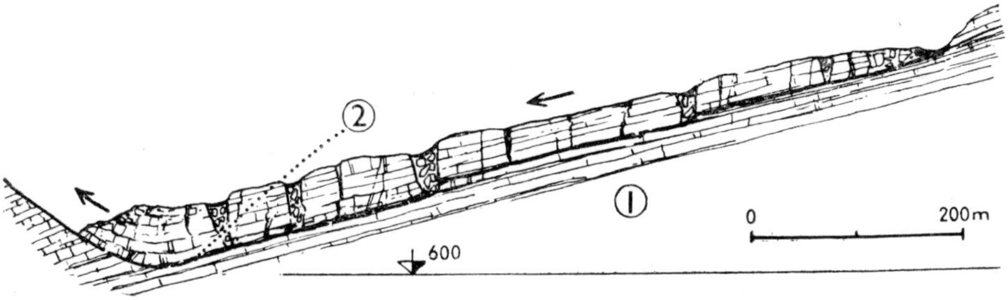

7-29. Landslide along a bedding plane. 1 — Cretaceous sandstones and claystones, 2 — presumed form of the valley before slope failure. (Novosad, 1966.)

and other constructions or by exploitation of mineral resources. Figure 7-30 shows a cutting through the Bohdalec elevation in Prague, excavated for the enlargement of a railway track, which should replace an old one-track tunnel. The slope is formed of Ordovician shales with quartzite intercalations, which are openly folded and inclined 15°-20° into the cutting. When the rocks set into motion during excavation, the slope was graded at 1 : 1 in the lower part and at 1 : 1.5 in the upper part, with three horizontal benches. Despite these controlling measures, a block of rock, about 8,000 m³ in size, slipped into the cutting on a bedding plane during the winter months. The bedding plane functioning as a slide surface had a dip of about 15°.

7.6.2. Long-term deformations of mountain slopes (gravitational slides). Some long-term deformations of mountain slopes may be included in this group although their motion is of creep character. They are termed gravitational folding or gravitational slides. They occur in rocks capable of plastic deformation as a result of componental movements along planes of stratification, schistosity or jointing without the formation of a continuous slide surface (Fig. 7-31). Many such phenomena have been observed in young mountain ranges, such as the Alps or Carpathians (Zischinsky, 1966; Nemčok, 1973). They most frequently occur on the slopes formed of phyllites, mica-schists, paragneisses, or chlorite schists that can be deformed by differential movements.

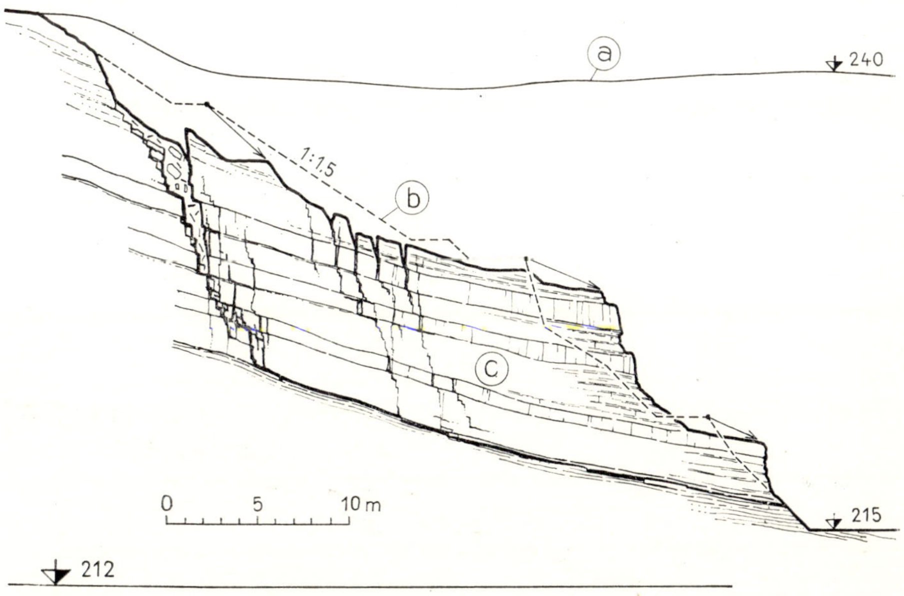

7-30. Failure of a railway cutting at Bohdalec in Prague. a — Original ground surface, b — graded slope of the excavation, c — shales slipped along bedding planes. (Záruba, 1968.)

The long-term loosening and slipping of rocks along moderately inclined bedding planes can originate when the joints and fissures open gradually due to climatic effects (mainly by freezing of water in joints) or when the residual stress is released as a result of valley deepening. In the galleries driven into the rock abutments of a dam on the Morávka river (in the Beskydy Mts., Moravia) loosening of subhorizontal sandstone beds extended as far as 50 m from the slope surface. Long-term slipping of sandstone blocks on moderately inclined clayey shales has also been observed on other slopes of this mountain range (Novosad, 1966).

The gravitational slides should be thoroughly registered, as they occur more frequently than is generally thought. Many steep slopes thus loosened are hidden under younger slope deposits. The timely recognition of these phenomena can save difficulties in dam or tunnel construction, particularly in designing pump-storage reservoirs in high mountains.

7.6.3. Rockfalls. The term rockfall refers to abrupt movements of loosened blocks or complexes of solid rocks detached from rock walls or roof ceilings. Rockfalls are distinguished by a very high velocity resulting from free fall. Their size ranges from isolated stones to enormous masses of rock.

Stones and blocks which have fallen down build up slope talus or debris fans at the foot of the mountain slopes. In places they coalesce into extensive aprons. Their slopes range from 25° to 40°, depending on the shape and size of the rock fragments. The fall of large rock masses separated high from a mountain wall may attain a velocity of up to 200 km per hour. In this way some rock streams originated in the Alps and other high mountains.

Rockfalls are relatively frequent on steep mountain slopes, especially in valleys overdeepened by mountain glaciers. They are also abundant on rocky coasts of lakes and seas, as well as on concave banks of deep erosive valleys (Fig. 13-3).

Hundreds of rockfalls have been recorded from young mountain ranges, such

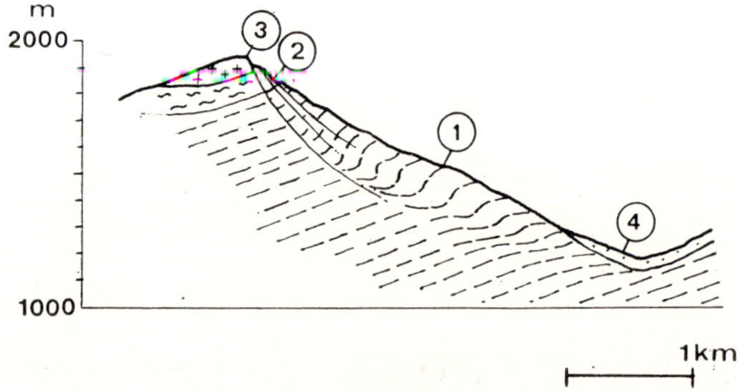

7-31. Section of the Ráztoky mountain slope affected by gravitational deformation (West Tatra Mts.). 1 — Biotite gneiss, 2 — migmatite, 3 — granite, 4 — glaciofluvial and slope deposits. (Nemčok, 1972.)

as the Alps, Carpathians, Himalayas, Andes and Rocky Mountains. One of the largest was the rockfall in the valley of Bartanga river in the Pamir Mts. in 1911. A rock mass of about 4,800 million m³ fell down and blocked the valley creating a lake 75 km long and 262 m deep. In 1893 a gigantic rockfall occurred in the upper course of the river Ganges in the Himalayas, forming a 7 km lake by damming the valley. In the following year the temporary dam collapsed and a disastrous flood devastated a large part of the Ganges valley.

According to Soviet authors, these extremely large rockfalls and landslides are generally triggered by earthquakes. Solonenko (1972) reports that the 1957 earthquake in the Baikal graben induced rockfalls in the Mujiskii khrebet at a distance of 220—230 km from the epicentre, over an area of more than 150,000 km². He designates these phenomena as seismo-gravitational and mentions other examples, chiefly from Central Asia, from the Murgab-river valley in the Pamir (1887, 1911) and the Narym-river valley (1946), which was blocked by rockfalls at several places. The earthquake in the southern part of the Tyan-Shan in 1949 also provoked large rockfalls and landslides which produced greater damage than the earthquake itself.

Of the rockfalls in the Alps let us mention the rockfall near Elm in Switzerland in 1881, which has been studied in detail by Swiss geologists. Figure 7-32 shows

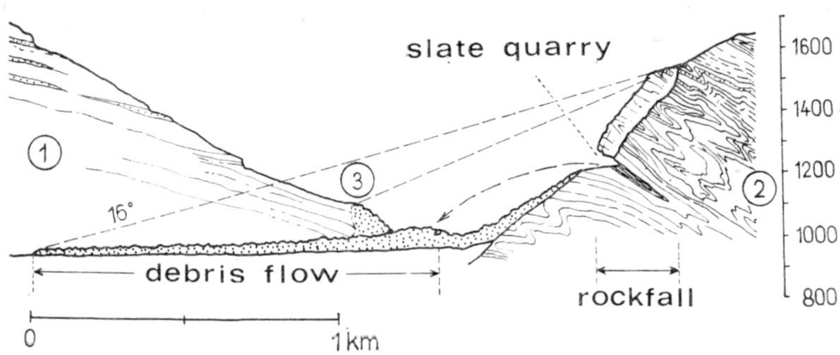

7-32. The rockfall near Elm in Switzerland from 1881 (after Heim). 1 — Flysch sandstones and shales, 2 — folded Blattengrat Beds, 3 — rock debris pushed up the opposite slope.

a cross-section of it. The slope is made up of folded shales and sandstones of the Flysch Zone, which dip into the slope. The failure was triggered by undercutting the foot of a steep slope in several quarries on roofing slates. The fall occurred in three phases, following after one another at 17 and 4 minutes intervals, respectively. During a few minutes more than 10 million m³ of rock filled a 90-hectare area of the valley to a height of 10 to 20 m. Eighty three houses were buried and 115 lives were taken.

In the Carpathians, a large rockfall occurred east of Georgheni in Rumania in

1828. The limestone blocks dammed the valley creating a lake (Lacul Rosu), which still exists. Numerous tree trunks protrude above the lake level.

Rockfalls from steep walls of the Norwegian fjords are disastrous mainly because they may give rise to huge swash waves (up to several tens of metres high) which flood the coast.

Rockfalls are especially dangerous because they occur quite suddenly and unexpectedly. The stabilization of hazardous rock walls against rock falling is difficult and expensive. Today bolting with the use of prestressed steel or cable is used for this purpose.

7.7. Specific types of slope movements. This group comprises slope movements specific for certain environments, where they are an important factor in the modelling of the relief of the area and, occasionally, are also of economic significance.

7.7.1. Solifluction is a combined flow and slip movement involving the surface layer on a frozen substratum. Solifluction phenomena are known mainly in subarctic regions and, to a lesser extent, in high mountains. The deeply frozen ground thaws there only to a small depth (about 0.5 m) during a short summer. Melted water accumulates in the soil because it cannot penetrate into the frozen and thus impermeable substratum, and the waterlogged bed flows downslope as a dense sludge, even on rather moderate slopes.

During the Pleistocene, areas lying within the periglacial zone were subject to an extent of solifluction similar to that found today in the subarctic regions. On many slopes it left behind compacted slope detritus composed of small and coarser fragments of solid rocks stuck into a sandy or clayey matrix. The material generally constitutes several layers, which level all depressions of the bedrock into a continuous slope.

7.7.2. Sliding of sensitive clays belongs to rapid slope movements of a flowing nature. Sensitive clays, called quick clays in Norway and Leda clay in Canada, are sediments of marine origin. After sea regression they have formed flattish areas which are situated today at altitudes ranging to several hundred metres above sea level. The strength of these sediments decreases progressively because of the reduction of salt content in the pore water. The water percolating through permeable silt laminae in the clay complex removes salt from the soil by osmotic processes. Rainwater has the same effect. Bjerrum (1955) describes a case when NaCl content in pore water dropped from 20—30 g/lit. to 1.2—2.9 g/lit. The decrease in salt concentration is simultaneous with the loosening of bonds between clay particles and the bond water and, consequently, with the reduction of strength of clays. In addition the liquid limit decreases and the sensitivity of clay, i. e. the proneness to the striking loss of strength by moulding, increases.

It is characteristic that the water content of clays does not change during this process. The loss of strength results in sliding movement of clays, which behave

7-33. Air photo of a landslide in quick clays near Skjelstadmark, Norway, in 1962 (Courtesy Norwegian Geotechnical Institute).

as a viscous fluid. The sliding of quick clay slopes is very treacherous, as it may also affect areas that are nearly flat (with the angle of slope smaller than 5°) and occur at an unusually great speed.

One of the largest slope failures of this type occurred near Vaerdal, north of Trondheim in Norway in 1893 (Holmsen 1953). A layer of sensitive clay was laid bare by stream erosion. The liquefied clay with a volume of 55 million m^3, flowed down in 30 minutes and covered an area 8.5 km^2 large. The temporary lake created by damming the valley flooded 3.2 km^2. The flat terrace in the broad valley of the river Vaerdalselven is today suggestive of absolute safety, but the monument bearing names of 111 people who lost their lives on this occasion testifies to the catastrophic possibilities.

Figure 7-33 shows a landslide in quick clays which occurred near Skjelstadmark, north of Trondheim (Norway) in 1962. The movement was initiated by stream erosion which caused at first a small slump on the bank. Sensitive clays that were thus laid bare flowed rapidly away, filling an ancient valley to a height of 10 m. The landslide is 2.8 km long, the head scarp is 12—15 m deep and the movement involved about 2.1 million m^3 clay.

Landslides in sensitive clays are also frequent in the eastern part of Canada.

Sudden liquefaction of sensitive clays is sometimes caused by vibrations and shocks. The greatest damage produced in the town Anchorage in Alaska by the earthquake in March 1964 was not inflicted by seismic shocks but by sliding movements triggered by them. The town Anchorage is located on a littoral plain, made up of a 20 m thick bed of sand and gravel, which lie on a thick layer of clay. Earthquake tremors caused liquefaction of a bed of sensitive clay 7—10 m thick, which extends to the city area at sea level. The upper beds of stiff clay and the overlying sand and gravel literally "floated" on the liquid clay seawards. Towards the end of March the ground was frozen to a depth of several feet so that the moving rocks did not disintegrate and moved in large blocks. Many houses travelled several metres distance without great damage, although all subsurface service lines were torn. The blocks moved almost horizontally, the sensitive clay being squeezed in front of the slide into ridges. At the head of the slide deep grabens limited by antithetic fissures originated (Fig. 7-34).

Subhorizontal slope movements of similar type are known from the coast of the Black Sea in Bulgaria (Pašek, Rybář, 1969).

7.7.3. Subaqueous slides form by the slipping of unconsolidated sediments on an inclined sea bottom, consisting mainly of clayey or silty and calcareous muds and to a lesser degree of fine sandy deposits. The slides may be of various dimensions and the resulting structures range from simple bending of beds to complicated slump structures, composed of pseudo-conglomerates with fragments and galls of disturbed rocks.

Favourable conditions for the origin of subaqueous slides exist in deltas where deposition proceeds at a relatively high rate and the slope is fairly steep. The movement may be stimulated by seismic or other shocks. In the underconsolidated clayey sediments, excess pore-water pressure will develop and on account of it slides may result even on very gentle slopes, without being triggered by any external event.

Subaqueous slides may initiate turbidity currents, which carry stirred-up sediments mixed with water far into the sedimentary basin. Subaqueous slides have been identified in almost all ancient geological formations, especially in the flysch sedimentary facies.

Subaqueous slides and turbidity currents may cause the rupture of submarine cables. A large turbidity current near New Foundland, provoked by an earthquake in 1929, progressively broke off submarine cables. Its velocity (up to 4.4 m/sec) has been established from the time intervals between successive cable failures.

7.8. Stabilization of slopes in slide areas. The main attention of engineering geologist should be directed to the prevention and/or stabilization of slope movements. If a potential slide area cannot be avoided, interference with the hazardous slope should be limited to a minimum extent, an assiduous investigation should be conducted and a suitable treatment suggested.

7.8.1. The investigation of slidc areas is governed by the same principles as investigations for other engineering works, but special stress is laid on some conditions and factors.

First of all, the engineering geologist must develop a working hypothesis of the type of the existing or potential slope movement, the position and shape of the

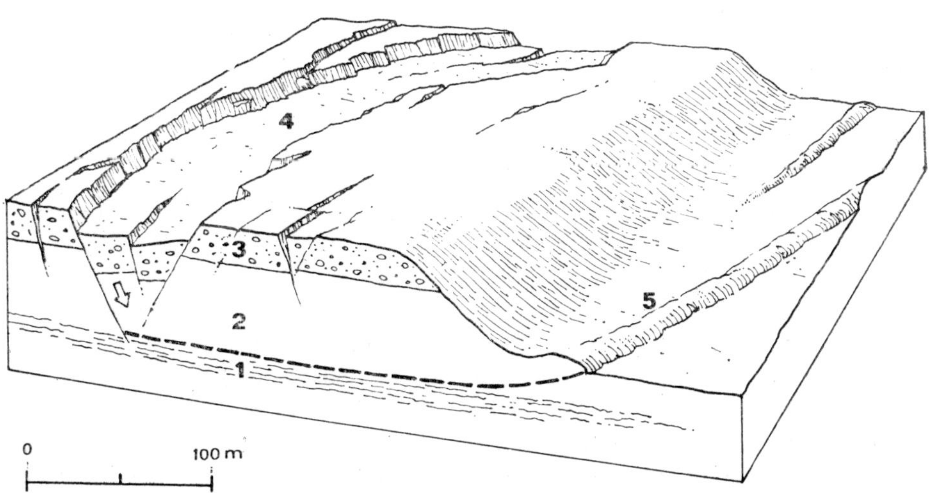

7-34. Block diagram of a translational landslide caused by liquefying of sensitive clays during Alaska earthquake in 1964 (Hansen, 1969). 1 — Sensitive marine clays, 2 — stiff clays, 3 — sand and gravel, 4 — graben in the root area, 5 — squeezed-out clays.

slide surface or zone, and the factors which could promote a slope failure. All these assumptions must be evidenced by field investigation and statical analysis. To this purpose it is necessary to establish, in addition to the geological conditions, the following parameters:

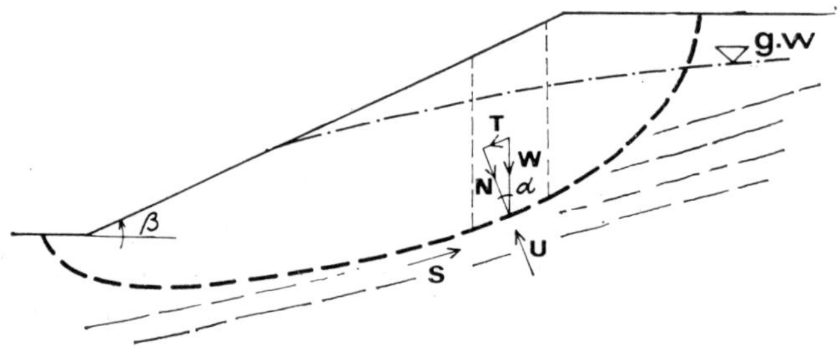

7-35. Diagrammatic representation of forces acting on the existing or anticipated slide surface (see the text).

— All surfaces of weakness which could promote slope movement, such as bedding planes, faults, continuous joints and slide surfaces of ancient dormant landslides. The study of topographical features of the area and the examination of steps in the slope profile and at its foot are also advisable.

— In existing landslides the position of a slide surface or zone by the study of samples from boreholes. The position of the slide surface in boreholes may be determined from the deformation of the borehole. In serious cases, an exploration gallery should be driven, which will also serve as a drainage gallery.

— In existing landslides lines of stakes should be set up and vertical and lateral offsets of these stakes should be recorded at suitable time intervals. The depth of the slide surface can be estimated from the vectors of displacement.

— In existing slides the root area of the slide should be studied in detail. A broad zone penetrated by cracks inclined into the slope is indicative of a deep-reaching slide.

— In hazardous slopes it is necessary to ascertain the paths along which the slope is fed by atmospheric water or subsurface water of aquifers. For this purpose observation tubes should be set at various depths, depending on the presumed hydrogeological conditions. Tubes perforated over the whole length are not recommended, because they may link several horizons of subsurface water. In little permeable soils it is advisable to set up piezometers for the measurement of pore-water pressure.

— The stress should be measured at several points on the ground surface.

— In examining the mechanical properties of rocks the parameters of residual shear strength should also be determined.

7.8.2. As far as possible, a statical analysis should be attempted. From the previous sections and Chapter 11 it is apparent how complicated the origin and development of slope failures may be. As in complex cases statical analysis does not always furnish reliable results, it is necessary to decide whether its application is justifiable, so as to not discredit this useful method. A statical analysis can at least contribute to the elucidation of a given problem, as it will show how the change of factors which is suggested for the treatment of the slope or for the stabilization of the slide would improve the existing state.

The applied statical methods have the character of stability analyses (Sec. 3.7). In the study of slope movements a factor of safety has to be determined, i e. the number equal to the ratio of forces *S* resisting the movement to driving forces *T* (Fig. 7-35) representing the tangential component of gravity *W*, which provokes sliding. Resistance *S* is given by the strength of rock and this increases with the rising normal stress. Therefore, the normal component *N* of gravity *W* also appears in the statical analysis. Under the ground-water table or at the pore-water pressure the component *N* is decreased by uplift *U*. It may also be said that the factor of safety *F*, strictly speaking its inverse value, expresses the proportion of rock strength needed for the stability of the slope.

Further computation procedure is today subject to discussion and demands a wide knowledge of soil and rock mechanics. For the first information the diagram

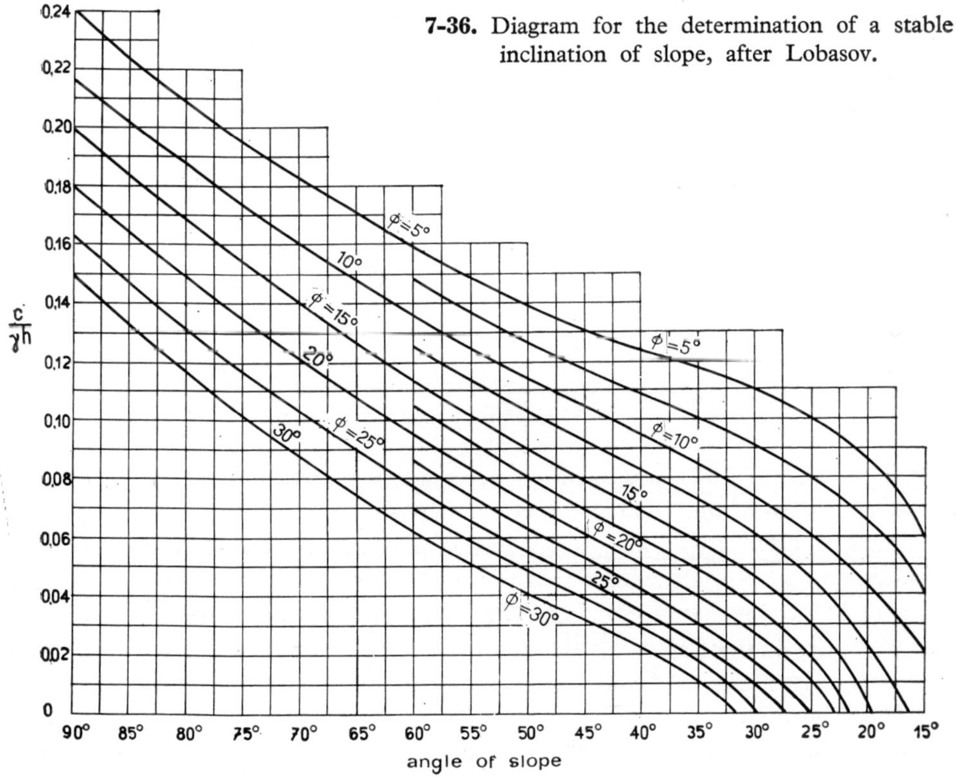

7-36. Diagram for the determination of a stable inclination of slope, after Lobasov.

after Lobasov (Fig. 7-36) is satisfactory; it is valid for $F = 1$ and only for $U = 0$. Since a definite F must be introduced (e. g. about 1.7 for the slopes of clayey soils), the c and Φ values must be reduced.

The deficiencies of stability analyses have been discussed in essentials in paragraph 3.7. As the progressive disturbance of slopes is frequent, an analysis taking into consideration the limiting state of local disturbance would be more convenient. Unfortunately, such analyses have not yet been elaborated.

In existing landslides, where the residual resistance of the rock (Sec. 3.5.7) is determined by reverse calculation of slope stability at $F = 1$, a relatively small increase of F (e. g. 1.15) is sufficient for stabilization of the slope.

It is evident that the study of slope safety deserves great attention, even from a theoretical aspect.

7.8.3. Stabilization treatment. The measures to be applied are mentioned in the previous sections and in Chapter 11. Only a survey and general evaluation of them will be given here.

(a) *Modification of slope shape.* The stability of a hazardous slope can be improved by unloading the active part of the slide and loading its toe. This is valid particularly for slopes where the rock mass is not relieved by uplift. The displacement of rocks is today enabled by modern effective machines for earth and rock excavations. Care must be taken, however, not to reduce the stability of the slope above the slide by unloading the existing or tentative root area. When the toe of a slope is surcharged by a fill, it is advisable to drain the underlying bed.

(b) *Drainage of the slide area.* An important factor controlling slope movements is the hydrostatic pressure of water, which can act as lateral pressure in pores and fissures of the rock or as uplift acting on the impermeable roof of permeable beds. These can be fed by water from the surface or from neighbouring aquifers.

It is therefore necessary to divert rainwater and brooks from the slide area by the shortest way possible. The lengthening of flow by horizontal ditches is to be avoided. Where the drained water could infiltrate the slope, it should be carried over through pipes. The whole drainage system must be permanently maintained, particularly during winter months.

Subsurface drainage is one of the most adequate measures for the treatment of landslides. If the downslipped layer is only a few metres thick, flat trenches will be deepened by a bulldozer and filled with permeable material. Water from a greater depth is drained by borings, which are today carried out up to a length of over 200 m. In North Bohemia a drainage boring 231 m long was carried out, for the major part through a coal seam. The drawback of drainage borings is that in liquefied sand they can rarely be driven to more than 60 m length and that they often become clogged. It is of advantage to combine a horizontal drill with vertical drainage borings.

In flat areas a system of horizontal borings driven from a pit or a system of drilled wells may be used. In sandy soils pumping wells are more advantageous, as they can be provided with suitable filtres. In active landslides, however, the wells do not serve as long as horizontal drainage borings.

(c) *Supporting and anchorage* of a *sliding slope*. Frequently, the configuration of the terrain does not allow for a decrease of the angle of slope and drainage; if it is possible at all, it does not suffice to stabilize the slope. In these cases various walls are constructed to complement the forces necessary for restoring equilibrium. The statical analysis of a supporting wall must be made very carefully. Excavations for a continuous retaining wall very often cause renewal of the movement. Of advantage are supports made up of cast-in-place piles. Since embedded piles have a low resistance to shear stress, their bearing value may be increased by anchoring their heads into the rock.

If the ground water is not drained from the slope behind the retaining wall, the surface layers would become waterlogged on spring thawing, slip downhill and often overflow the head of the retaining wall. Consequently, stabilization of a moving slope without its drainage is far from being satisfactory. All the corrective measures mentioned above should be supplemented by covering the slope with

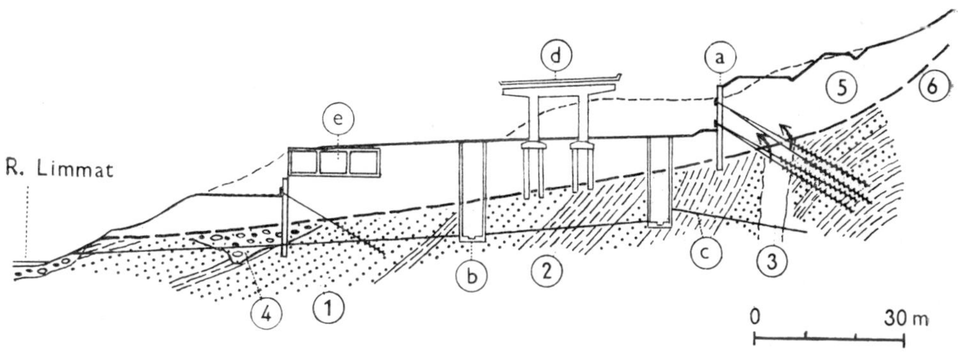

7-37. Complicated stabilization of a sliding area on the route of motorway near Zurich (according to Gautschi and Schindler). 1 — Sandstones, 2 — marls of Lower Molasse, 3 — fault zone, 4 — sand and gravel of the abandoned Limmat channel, 5 — clayey-sandy material of ancient landslide, 6 — slide surface; a — anchored piling wall, b — pumping wells, c — drainage boreholes, d — bridge construction, e — box construction for unloading of slope.

a permeable bed to prevent the water in the near-surface layer from freezing.

Figure 7-37 shows an example of a complicated stabilization of an ancient landslide. The motorway on the concave bank of the river Limmat near Zurich (Switzerland) should be constructed over the existing transportation routes. The marls, silts and sandstones of the Lower Molasse are covered by an extensive landslide dating from the last Würm interstadial. Detailed research has shown that any interference with the landslide would provoke partial movements, unless thorough precautions are taken. The road cuttings therefore had to be supported by cast-in-place piles anchored into the bedrock. Since the outflow of ground water from the sandstones and the fault zone were among the factors promoting slope movements, the permeable beds of the bedrock were drained by horizontal drainage borings and wells. The motorway was carried over on a bridge spanning the dangerous area. The piles of the bridge foundation were driven into the bedrock. The cutting for the lower higway was also secured by anchored piles and the upper part of the slope was relieved by a box-beam structure of reinforced concrete.

Chapter 8

EXCAVATION AND WORKABILITY OF ROCKS

In addition to other duties, the engineering geologist should appraise the likely excavation conditions and workability of rocks, as these properties greatly influence the economy of the construction work.

The time and cost of construction are controlled mainly by the workability of rocks. In transportation routes it may also affect the alignments of roads, as earthwork is the dominating part of building operations. The provision of suitable machine equipment also depends on the timely determination of rock excavability.

While individual items of the cost specification and other relations of economical character may change without serious consequences during construction, a false estimate of the probable excavation and haulage methods may cause heavy losses. For example, in constructing the water-supply tunnel 7.5 km long below the bay near the City of New York the gneiss, contrary to expectation, proved to be so hard and resistant to drilling that the expensive tunnelling machine had to be replaced by conventional excavation after only 135 m had been driven.

The engineering geologist may be invited to participate in discussion either by the owner and his designer (mainly for setting up the cost specification) or by the contractor. Besides the costs, the questions of excavation methods may also be considered.

8.1. Resistance of rock to excavation. In order to estimate the resistance of rocks to excavation, they are classified on the basis of either their resistance or the tools needed for excavation. The division into seven groups, as established by Rziha (1889), has been widely used. Rziha distinguished:

(1) Loose soils, (2) soils easy to spade, (3) soils easy to dig, (4) crumbling weak rocks, (5) rocks easy to blast, (6) rocks difficult to blast, (7) rocks very difficult to blast.

This and other classifications of soils and rocks have often been subject to criticism. Therefore the tendency arose to reduce the number of groups and to leave the estimation of the resistance to excavation of the rock in question to the contractor. If this is the case, the contractor should have the possibility to participate in the geological investigation of the building site and to consider its extent and usefulness. Nevertheless, this may lead to other controversies. In Czechoslovakia the problem was solved (in 1935, by the Standard ČSN No. 1175) by establishing four groups of rocks according to their workability (see following). The recently valid standard ČSN No. 75 3050 Earth Works, 1963, lists seven groups according to resistance to excavation.

Within the range of groups 1 to 4 the division is fairly clear. Even when the present standard does not require the identification by use of a tool, it can still be applied. Soils, which can be easily removed using a shovel, belong to the first group. When the shovel penetrates with difficulty and a spade is necessary, the soil is placed in group 2. When digging is necessary, the soil is ranked in group 3 and when digging is very difficult, in group 4.

The important boundary between a difficult soil (or weak rock) and the rock itself is given by the difference between groups 4 and 5, which is also reflected in a marked difference in cost. A typical representative of group 4 is a dense gravel containing more than 10 per cent of cobbles 10 to 25 cm in size, or a cohesive soil of hard consistency. Controversies often arise with fissured or partly weathered rocks. The criterion whether blasting is necessary to loosen them is often misleading since modern heavy hydraulic excavators are capable of excavating even a firm rock without blasting. Therefore the attribute of group 5 is the spacing and filling of discontinuities, which determine whether the loosening of rock after a shovel bite will spread into the surrounding rock or not. On this principle, Cretaceous sandy

8-1. Sheet jointing of granite. Quarry on Malmön Island in Sweden. (Photograph by Záruba.)

8-2. Rectangular jointing of granite. Zvírotice quarry in Bohemia. (Photograph by Záruba.)

marls have been classed in group 5. Rock debris also belongs into this group.

Within the range of groups 5 to 7, drillability (see following) and the density of discontinuities are of importance. Rocks and weak rocks forming beds up to 15 cm thick belong to group 5 and hard rocks (e. g. gneiss) with spacing of discontinuities up to 25 cm belong to group 6.

While studying the given problems of practical importance, the engineering geologist should respect the provisions of the standard; nevertheless, he is also expected to pay attention to the causes of differences in the drillability of rocks (Sec. 8.2), as well as to their discontinuities.

As is seen, the resistance to excavation is shown by both petrographic features of the rocks (mineral composition, structure and texture), and the fabric of the rock mass (bedding, thickness of beds, their strike and dip in relation to the face of excavation, intensity of tectonic disturbance, spacing of joints and other discontinuities and the degree of weathering).

In sedimentary and metamorphic rocks the planes of separation mostly coincide with the bedding and foliation planes. These rocks are either thin- or thick-bedded, slabby or tabular, and the layers are usually transversely jointed, so that their excavation is on the whole free from difficulties. Exceptional are coarse to bouldery, hard conglomerates and agglomerates, which are difficult to excavate and belong to group 6 or 7, even if the boulders are bound by clayey cement. Excavation in these is often more difficult than in a compact rock.

The excavability of igneous rocks varies depending on their jointing. Uniform

8-3. Spherical jointing of diabase. Disused quarry near Lodenice in Bohemia. (Photograph by Záruba.)

sheet (Fig. 8-1) or block jointing (Fig. 8-2), common with plutons, makes excavation easier than irregular or polyhedral blocks, which are closely interlocking. Spheroidal and irregular large blocks also make excavation difficult, since the blocks or corestones have to be blasted. Irregular jointing is encountered in some granites and spheroidal jointing is encountered in effusive rocks. (Fig. 8-3). Excavation of basalt is also difficult due to columnar jointing (Fig. 8-4), especially when the columns are of large size.

The character of bedding and joint surfaces is also important for the appraisal of likely excavation conditions. The joints may be plane, undulated, smooth or rugged, opened or secondarily filled. Filled joints are favourable for blasting, even when filled only with clayey or detritus material. Open joints decrease the intensity of blasting since the charge may partly escape through them. When the joints are healed by calcite, silica or limonite, the rock shows the character of a compact rock mass.

Excavation of rocks is also greatly influenced by climatic conditions. If the rock is worked during the winter, the blocks of even apparently dry rock may stick to one another due to the freezing of "quarry water". This increases the resistance of rock to excavation and should be respected in assessing the rock group.

If the workability of rocks is to be appraised after a longer period, especially after winter months, climatic conditions must be considered. The surface of excavation may be so weathered as to be misleading. Therefore cleaning of the surface is necessary in order to evaluate accurately the resistance of the rock to excavation.

8.2. Drillability of rocks. Drilling is a considerable item of excavation costs, particularly in underground works where the consumption of explosives is several times greater than in open cuttings. Nevertheless, drillability may also be used as a parameter for the evaluation of the excavation resistance in cuttings and foundation pits (Záruba, 1941). The Czechoslovak standard, as mentioned above, uses drillability as a subsidiary criterion. In rocks of group 5, with a bed thickness below 15 cm, the drilling of a hole 1 m deep with a pneumatic hammer possessing a hardened bit edge 30 mm in diameter at a positive air pressure of 4·5 bar should

8-4. Columnar jointing of basalt. Quarry near Kamenický Šenov in Bohemia. (Photograph by Záruba.)

take less than 10 minutes. The time taken is 10 to 20 minutes with rocks of group 6, and more than 20 minutes with those of group 7. The practicability of this rule is impeded by the fact that compressors of today are capable of furnishing the hammer with a higher air pressure.

The evaluation of the drillability of rocks affects the choice of drilling equipment; inadequate equipment may involve great losses when hard rocks are concerned.

A new laboratory method for evaluating the drillability of hard rocks has been described by Selmer-Olsen and Blindheim (1970). These authors introduced indices for the drilling rate, established by correlating drilling results and laboratory-measured properties of rock samples. The rate of drilling in cm/min (hammer drill, simple edge and carbide tip) is about 65 per cent of the "drillability index" value. The latter is estimated on the basis of two parameters: the rate of laboratory drilling (Sievers's test) and the result of the Swedish test of brittleness, which is essentially a shatter test.

Today, the engineering geologist is often asked to indicate parameters which may serve to assess the resistance of rocks to rotating tunnelling machines. Because the high cost of this equipment is to be amortized after only few kilometres of driving, and since the machines are custom-built, the question is of delicate nature. The best the engineering geologist can do is to contact the builder of the particular machine. Otherwise the rock should be characterized by the testing of laboratory samples for compression strength, by its shear strength, by its hardness (with respect to the mineral composition, the fabric, matrix, and cementation), by the density of prevailing surfaces of weakness, their filling, etc.

The engineering geologist is expected to welcome the use of a tunnelling machine, as the smooth circular opening obtained makes a better arching of the rock possible. Shocks due to blasting do not occur. On the other hand, the inspection of the rock face should be made much more regularly as often the only available working room is ahead of the tunnelling machine.

8.3. Workability of rocks. Although the most important item in excavation cost is the resistance to excavation, consideration should also be given to other factors, such as the haulage and the deposition of muck. A broken-up, clayey soil which sticks to the shovel and is difficult to dump belongs to a higher category than it would if classed according to its resistance to excavation only.

The workability of rocks should be appraised firstly during the design stage of the structure and the preparation of equipment and, secondly, during the operation works.

(a) Preliminary assessment of the workability of rocks. This task is not easy and in the course of excavation the preliminary grouping has sometimes proved to be wrong.

Boreholes and small-size pits are inadequate. On the other hand, exploration trenches give good results, as they include a larger area and offer a continuous profile. As long as fresh rock is established reliably at any depth, a further deepening of the trench is not necessary. After drilling several holes and excavating the rock to the depth of about one metre, the properties of the rock can be appreciated. The best approach is by comparing the workability of the rock in question with a rock of similar properties, which is known from other sites. Therefore mineralogic composition, strike and dip of beds, density and orientation of important discontinuities, etc. have to be investigated. When comparison with similar rocks is not possible, a trial excavation should be performed.

The results of investigation are plotted in a longitudinal profile and the excavation group, in addition to the geological data, is recorded. Nevertheless, when the investigation has not been sufficiently detailed, it is better not to plot a continuous profile, since it could be far from reality.

(b) Grouping of rocks in the course of excavation. Several approaches are possible. Usually the boundaries between rocks of different groups are plotted in the cross-sections and the volumes are calculated. When individual beds or complexes of beds are concerned, their volume can be easily calculated as they form a distinct body. When scattered, the volume may be expressed as a percentage of the total volume.

When the axis of the cutting is perpendicular to the strike of beds, cross-sections are not sufficient. The longitudinal face of the exposure should be investigated and used together with the cross-sections for the specification of the volumes of different rock types.

8.4. Bulking (increase in volume) of rocks during excavation. This question should also be considered in connection with excavations. In some cases the volume of excavation has been deduced from the volume of the fill. The harder and more

Table 8.1

Rock	Bulking, per cent by volume	
	transient	permanent
Sandy soil	8–17	1–2.5
Peat, topsoil	20–30	3–4
Clayey sandy soil	14–28	1.5–5
Clayey soil	24–30	3–7
Fat clay	26–32	6–9
Marl, sandy marl	33–37	11–15
Soils with rock fragments	30–45	10–20
Rocks	45–50	20–30

compact the rock, the greater is its bulking due to irregular angular blocks and fragments. Cohesionless soils show the smallest increase in volume, especially gravel and sand of various grain sizes, as small grains fill the voids.

Transient bulking is to be distinguished from permanent. The former appears, for example, during haulage. Its magnitude is of the order of only 10 to 15 % with sandy gravel but attains 60 to 90 % with schists of a splintery or rod-like disintegration. The volume of the excavated material is therefore much larger than the in-place volume of the rock.

Permanent bulking varies depending on the mode of deposition. Due to the use of heavy machines it is not rare to find cases where several soils (loose sand, loess) show no permanent bulking, or may even decrease in volume.

Where the excavated material is not compacted, the question of the necessary superelevation of the embankment arises.

Values of bulking according to Averin (1953) which were used for Soviet standards are presented in Table 8-1.

Chapter 9

GEOLOGICAL INVESTIGATION OF BUILDING-MATERIAL DEPOSITS

In many engineering structures, particularly in communication and hydraulic works, the primary task of geological investigation is to find deposits of construction material, in order to supply the structure with a continuous, sufficient amount of suitable stone, gravel, sand, aggregate, etc. The supply of construction material is of particular importance in structures demanding a large volume of masonry or concrete, as the availability of suitable material generally determines the most economic performance of the work. The engineering geologist who is charged with investigations for such a large structure should possess enough knowledge and experience to estimate the possibility of procuring suitable construction material near the building site, and organize the prospection work.

The task of engineering geological investigation is either to appraise the existing quarries and occurrences or to find new deposits to be developed. For structures of minor size it is usually more convenient to use stone from quarries that are in operation, even if they are at some distance from the building site, because the small consumption would not cover the cost of the opening and outfit of a new quarry. For large civil engineering works (tunnels, bridges, dams) the opening of a new quarry is generally more economic in cases where suitable deposits occur in the proximity. These quarries are sometimes only temporary, especially in remote areas, because their cost of maintenance after the completion of the structure is not covered by the local demand.

The methods used for the search of building materials are analogous to those applied to other engineering geological investigations. The exploration should be carried out in four stages:

(a) Prospection for mineral deposits
(b) Reconnaissance of the deposit
(c) Detailed exploration of the deposit
(d) Research into the working and operation of the deposit.

For economic reasons the stages should follow one another logically. Only when the preliminary report is favourable or at least promising, should technical-geological investigation and technological tests be performed. An expert engineering geologist who is acquainted with local conditions can estimate, on the basis of reconnaissance exploration, whether the stone or sand is usable for the intended purpose. If the preliminary report is negative, all further investigations including technological tests may be omitted; exploration of another, more promising locality instead will save money and time.

9.1. Reconnaissance of a deposit. Preliminary investigation of a deposit or quarry should:

(a) Provide preliminary information about the building materials used in the neighbourhood, include a study of the geological maps and reports and hence give a preliminary idea of which rocks will be suitable for the planned structure;

(b) Appraise the quality of the rock and its usability on the basis of local inspection and petrographic analysis of samples;

(c) Determine the hydrogeological features of the deposit and estimate the economic reserves from the shape of the deposit and the thickness of the overburden;

(d) Assess the position of the deposit or the quarry with respect to transportation facilities and the building site.

(a) The first general information should be drawn from a *geological map* which should be as detailed as possible. The drawback of many geological maps is that they contain only stratigraphic terms, which may cover rocks of various physical properties, but not even a direct designation of the rock (e. g. granite) provides information about its usability for building purposes. On the ancient elevated plateaus, for example, granite crops out at the surface but is disturbed and weathered to an appreciable depth. On the other hand, granite building up the slopes of young erosive valleys is far less weathered. A good geological map is thus a valuable tool to those who can 'read' it, but not even the most detailed map can substitute a direct inspection of the deposit in the field.

Very useful preliminary information can be obtained from publications dealing with the geological conditions and mineral resources of the area studied. In Czechoslovakia a Collection "Inventory of Quarries" has been published. The volumes of this publication are arranged according to the sheets of the General Map. Each volume contains basic data on geological and technical properties of rocks from individual quarries, the results of technological tests and chemical analyses and a synoptic map of the area; the types of quarries and their size are designated by special symbols. These inventories of quarries are very handy for the preliminary search of suitable building stone, because the field investigation may be restricted to a few localities selected from this list according to the demands of the engineer.

(b) The rock must be examined at the locality and its suitability for the given purpose should be assessed. When the rock crops out on the surface, a sample for petrographic analysis is easy to take. An expert geologist can estimate roughly the strength of the rock, its friability, toughness, cleavage, shape of cleavage planes, colour and structure, in trimming a hand specimen. From the microscopic

9-1. Unequal resistance of basalt pavement to weathering causes unequal wear of the roadway. (Photograph by Záruba.)

study of a thin section, the usability of the stone for the given purpose, particularly its durability, may be ascertained.

The *weatherability* of the rock and the mode of weathering may be estimated by examination of exposed surfaces. It is advisable to examine local buildings, staircases, milestones, tombstones on local cemeteries, made up of the stone in question. Old tombstones often bear the date, which shows how long the stone has been exposed to weathering.

Figure 9-1 shows the economic losses that can be caused by using a stone of unsuitable quality.

(c) The mode of deposition of the rock is inferable from the outcrops of beds in the wider neighbourhood of the deposit. In an area not fully covered by overburden, the strike and dip of rocks may be measured and the shape of the deposit, its thickness and the nature of the overburden may be broadly determined. Where natural outcrops are few, at least several test pits and boreholes must be made. From these data the total volume of the deposit may be computed, at least approximately, and the reserves available may be determined.

(d) Also of importance is the geographical position of the deposit, its distance from the building site and the transportation routes. It should be considered particularly, whether it will be possible to construct an access road or a narrow-gauge railway, or whether it will be necessary to construct an inclined or suspended cableway in the mountainous terrain.

From the results of preliminary investigations, a general plan of exploitation may be prepared and the economy of operation estimated. Reconnaissance studies are often made on several quarries and occurrences before the most suitable is chosen. If the results of reconnaissance investigations are quite unequivocal, no other investigations are necessary and the deposit is either opened by a quarry or relinquished. In an area with few outcrops, the reconnaissance is only approximate and its results must be verified by a detailed investigation.

9.2. The aim of detailed engineering geological investigation is to determine the precise shape, thickness and extent of the deposit, the kind and thickness of the overburden, and to take reliable samples for technological tests. In essentials, it includes the following procedures:

(1) Detailed survey of the situation and altitude of the deposit area,

(2) geological mapping of the area and the study of hydrogeological conditions,

(3) boring or trial quarrying,

(4) taking of samples and technological tests,

(5) compilation of the results, computation of material reserves, preparation of the plan of exploitation.

(1) *Survey of the area.* For detailed investigation a good topographic plan is necessary, in which the results will be plotted. Topographic maps at a scale of 1:25,000 are usually not satisfactory; as a rule, the wider area of the deposit must be surveyed and a contour plan at a scale of 1:5000 or 1:1000 constructed. Photogrammetric plans have proved convenient, especially for steep rocky walls of abandoned quarries which cannot be measured precisely by any other method.

(2) In detailed *geological mapping*, all outcrops, the strike and dip of outcropping beds, the main joint systems, faults and fractures, the extent of overburden, as well as the position of test pits, boreholes and other artificial exposures are plotted in the map. The map must be thoroughly documented, i.e. all outcrops and exposures must be precisely surveyed and described. Hydrogeological data to be determined relate to the depth of the ground-water table below the surface, possible inflow of water into the quarry, its origin, the position of aquifers and faults. Hydrogeological conditions often determine whether the deposit is workable.

(3) *Subsurface exploration.* The amount of further field investigation depends on the frequency of outcrops. In the areas overgrown with rich vegetation or covered with weathering material, slope debris or loam, detailed geological mapping is not sufficient, and subsurface exploration must be carried out. It is an unreasonable policy to save on this exploration, as only accurate information concerning the form and volume of the deposit and its overburden makes it possible to assess reliably the amount of workable reserves and to prepare good operational plans. Many important constructions have to be discontinued for a long time

when the reserves of building materials were estimated wrongly and another source of building stone had to be searched for during the work. The delay in the supply of building materials is particularly troublesome in tunnel construction, because the excavated sector must remain long on provisional timber supports, resulting in loosening of the rock and great pressures on the timbering and lining.

The choice of the location and number of exploratory holes must be based on sound knowledge of the geological conditions of the area if the extent and form of the deposit are to be reliably determined.

9-2. Rixö granit quarry in Sweden. (Photograph by Záruba.)

In order to assess technological properties of the stone, its jointing and the degree of weathering, a direct examination in an open pit is necessary. Therefore test pits, trenches, galleries or trial blasting are used. If core boring is used, care must be taken to recover a continuous core, from which specimens for technological tests can be prepared. Boring generally serves for the determination of the thickness and composition of the overburden and for the exploration of gravel and sand deposits.

The method of subsurface exploration used depends on local geological conditions. If, for example, a succession of alternating, steeply inclined beds of sandstone and shale is to be established, either a trench in the upper part of the slope may be excavated or a gallery may be driven at the level of the floor of the future

9-3. Basalt columns used for the pavement of dikes in The Netherlands, (courtesy R. van Oord).

quarry. The gallery can be replaced by inclined core boreholes drilled at right angles to the beds.

When the jointing of the rock and the size of the blocks quarried are to be examined, and the rock is not accessible in a disused quarry or a steep rock wall, a small trial quarry must be opened or, at least, part of the wall must be broken off. The size of workable blocks in sediments is defined by the thickness of the beds and the density of cross joints. In igneous rocks jointing varies greatly, as it

9-4. Coarse columnar jointing of basalt is less suitable for crushed stone quarrying. (Photograph by Záruba.)

is usually predisposed by contraction joints, formed at the cooling and congealing of the magma.

The requirements for the size and shape of quarried stone depend on the purpose for which it is to be used. For dimension stone, sheet or block jointing is of advantage, as it enables roughly rectangular blocks of required size to be obtained (Figs. 8-1, 8-2). Polyhedral or bouldery jointing is less suitable. Spherical or concentric jointing of some diabases is quite unsuitable (Fig. 8-3). Columnar jointing distinctive of basalts (Fig. 8-4) is unfavourable because the columns of polyangular section do not possess a suitable shape and are difficult to dress. They are useful, however, for other purposes. In The Netherlands they are used for protection of dikes, where their great weight is of advantage (Fig. 9-3).

If the rock is to be used as *crushstone*, the demands on its jointing are different. It is most convenient if the stone disintegrates by blasting into small pieces which can be fed directly into crushers. Therefore, rock with sheet or columnar jointing (columns of smaller size) is preferred for the production of crushed stone. Rock which breaks into large blocks or stout columns is less suitable, as every block must be blasted anew or labouriously broken into fragments of convenient size. This implies a considerable increase in the use of explosives and more difficult operation (Fig. 9-4).

The excavation of test pits in solid rocks is relatively expensive; should it be economic, the rocks must be precisely surveyed and described. The records should involve all data on the mode of deposition, such as the strike and dip of beds, their thickness, density and strikes of joints, depth and kind of weathering, presence of ground water, etc. The test pits should be inspected by an engineering geologist and, under his supervision, samples of the principal rock types should be taken for technological tests. Inaccurate and incompetent logging of the holes, wrong sampling and unintelligible description of samples can depreciate even extensive subsurface exploration. No less important is horizontal and height survey of the sites of test pits, as otherwise they cannot be plotted accurately on the geological map and cross-sections.

(4) *Sampling*. Samples for technological testing must be taken from the actual rock that will be quarried for the given construction. It is well known that in almost every quarry samples of very different properties can be taken from one rock. A rock sample from a thick sound bed will obviously give better results

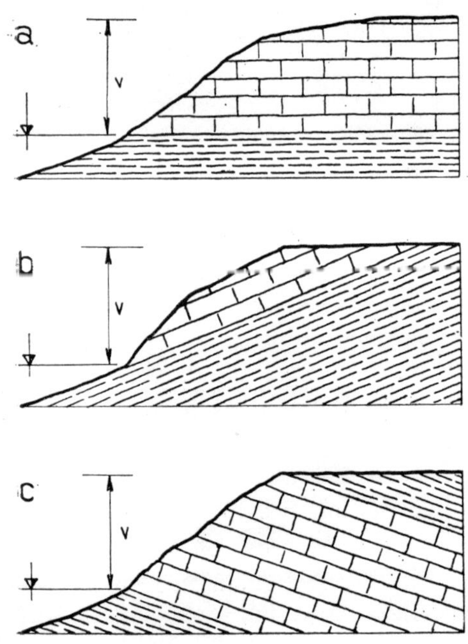

9-5. Importance of the dip of beds for the estimate of stone reserves. (For explanation see the text.)

than a sample taken from the neighbourhood of a fault zone. Such a stone can be penetrated by cracks invisible to the naked eye, but recognizable by testing. When the quarry is of large size, several samples are taken and each of them is tested separately. The different results of tests are then regarded as limiting values between which the average strength of the rock varies. The incompetent choice of samples or their confusion have many a time caused a wrong assessment of the usability of the rock or even a discrepancy between the results of technological tests and the geological account based on petrographical analysis and former experience. The selection of samples also requires great care because the preparation of testing material and technological tests are laborious and expensive, and if the samples are not reliable and representative, they are quite useless.

(5) *Evaluation of detailed investigations.* In order to determine the economic reserves and to open the quarry reasonably, the geological and hydrogeological features of the deposit must be examined. The geological conditions are most readily expressed in cross-sections, from which the total volume of stone as well as economic reserves may be computed.

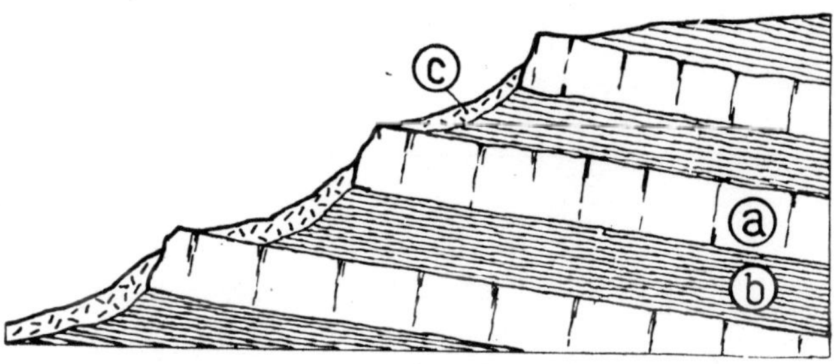

9-6. The slope is formed apparently only of firm rocks (a); soft rocks (b) are buried by detritus (c) of firm rocks.

In *sedimentary rocks,* unless they are strongly tectonically disturbed, the amount of stone is given by the thickness of the bed and its extent, which is deducible from a geological map. In a hilly area, only the stone supply in the valley slopes is generally taken into account, as quarrying below the valley floor is troublesome

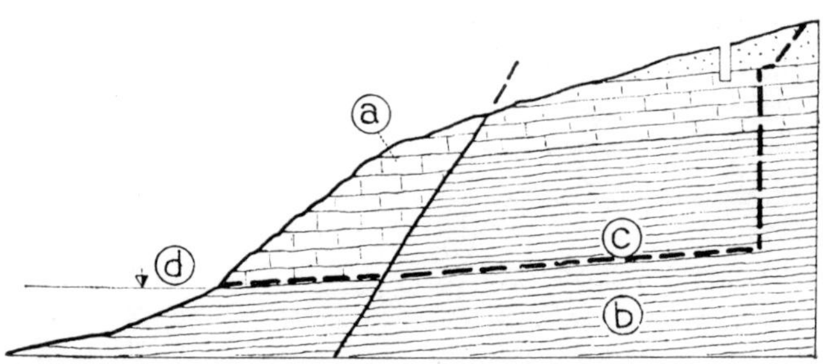

9-7. A fault disrupting the deposit gives a false picture of stone reserves. a — Limestone downthrown along a fault, b — shales, c — erroneous delineation of recoverable stone.

and often uneconomic (pumping of water, protection of the pit from flooding by surface water, lifting of stone blocks, etc.). Figure 9-5 depicts several simple examples showing outcrops of building stone in a slope with different dips of beds. Although the height of the outcrop is roughly the same in all three cases, the economic supply greatly differs. When the beds lie horizontally, the supply is much greater than when they dip downhill or into the hill; in the last case the economic supply decreases with the rising thickness of overburden. Care should be taken in estimating the stone reserves in a sequence of solid rocks (e.g. sandstone or igneous rocks) alternating with shales or other softer, non-utilizable beds.

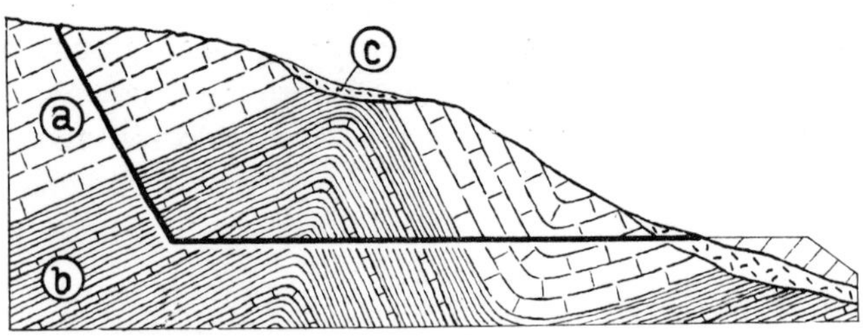

9-8. The slope in which a quarry was opened was thought to be composed only of sandstones, since the dip and strike of beds were disregarded. a — Sandstones, b — shales, c — detritus covering outcrops of shales.

If the interlayers of soft rocks are thin, they may facilitate the breaking of solid rocks, but when they are of greater thickness, they increase the waste amount and the cost of operation. A wrong estimate of the proportion of solid rock to the non-utilizable soft beds has often distorted the computation of economic supply (Fig. 9-6).

Mistakes in the computation of available reserves due to a false conception of the geological structure of the area are fairly frequent and result in appreciable economic losses. Figure 9-7 shows a shelf quarry opened in firm limestones for railway ballast. A visual inspection of the slope and a few boreholes sunken from the platform above the slope indicated that the amount of stone will be sufficient. An access road was constructed, as well as an effective crushing plant and a sorting plant, but soon after the opening of the quarry the estimate of the stone reserves proved to be incorrect. The limestone deposit was disrupted by a fault, subparallel with the slope foot, which separated the limestone from the underlying soft shales. An inspection of the wider environs of the quarry would easily have revealed the existence of the fault in transversal valleys. Another example of a wrong estimate of the economic supply is shown in Fig. 9-8. The quarry was opened in a folded sandstone complex. Owing to insufficient investigation, the presence of clayey shales below the sandstones in the anticlinal bend was not presumed. A thorough measurement of the dip of beds or a sole borehole placed suitably in the slope would clear up the geological setting of the slope.

Igneous rocks (granite, granodiorite, diorite, etc.) generally occur in such large amounts that the stone reserve is not limited by their mode of deposition. Granite may exceptionally be covered by relics of a gneiss mantle. The stone reserves may be reduced by a fault zone, by layers affected by hydrothermal decomposition

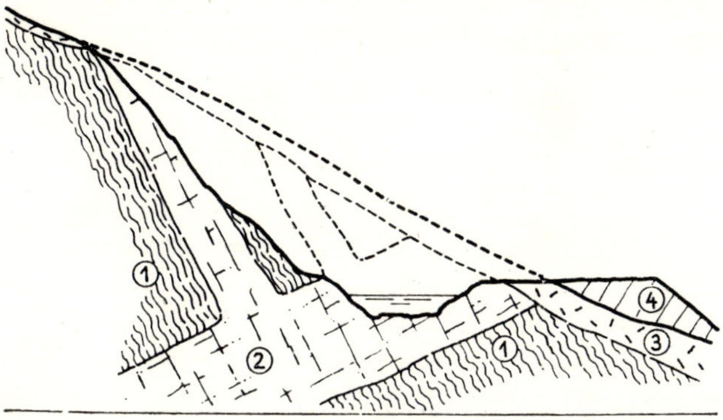

9-9. A quarry opened in porphyry dykes providing a small amount of exploitable stone. 1 — Gneiss and mica-schist, 2 — syenite-porphyry, 3 — slope detritus, 4 — filling.

or other alterations (presence of minute dykes and sills, inclusions or magmatic segregations), which depreciate the stone.

The estimate of economic reserves of hypabyssal rocks (porphyry, diabase, basalt) is sometimes difficult since the width of dykes and sills may vary over a short distance, even to disappearance. Figure 9-9 shows a deposit the volume of which was overestimated, even by an expert worker. The main body is a thick dyke of syenite porphyry piercing the biotite paragneiss. The dyke dips at about 20° into the hillside and sends out narrower sills, dipping at 65° downslope. One of these sills was already exhausted and the other is well exposed in the quarry face. Since it makes up the greater part of a high working face, the supply of porphyry appears to be appreciable, although the sill is in fact only 8 m thick, being over- and underlain by paragneiss not usable for the intended purpose.

In *effusive rocks* (basalt, andesite, melaphyre, liparite, etc.) extensive sheets

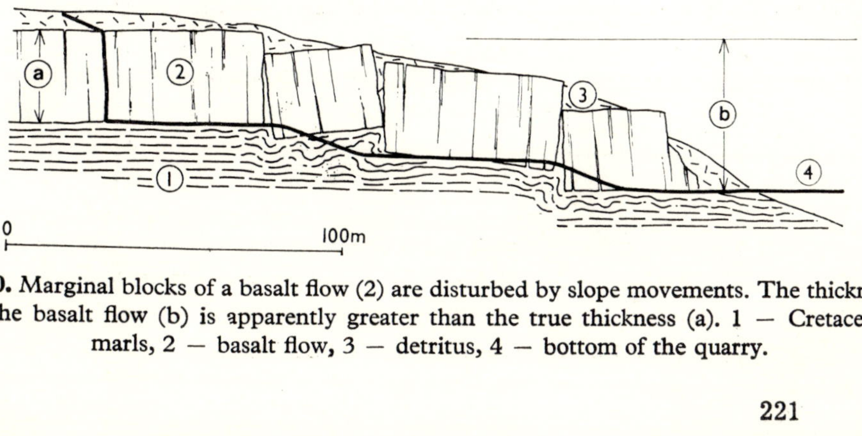

9-10. Marginal blocks of a basalt flow (2) are disturbed by slope movements. The thickness of the basalt flow (b) is apparently greater than the true thickness (a). 1 — Cretaceous marls, 2 — basalt flow, 3 — detritus, 4 — bottom of the quarry.

are commonly suitable for quarrying as the rock is relatively homogeneous and its structure is fairly regular. Along the surface of every sheet or flow, the rocks are porous to cavernous, which permits us to determine whether the body consists of one sheet or of several overlying flows. In the latter case, the rock may be inhomogeneous, especially if the flows are separated by tuff or agglomerate layers. In an abandoned quarry in northern Bohemia, three flows were exposed by quarrying, the largest of which was so weathered that it actually formed part of the overburden.

Marginal blocks of basalt flows lying on soft rocks (marlstones) are often disturbed by slope movements. The basalt blocks sunken to the foot of a slope suggest an apparently greater thickness of the basalt flow (Fig. 9-10). If the quarry is opened in a block at the foot of a slope, the floor must gradually be raised to a higher level. The height differences are unfavourable for the operation of a quarry and have to be overcome with the use of inclined cableways. Effusive rocks are often inhomogeneous around volcanic necks and conduits and enclose random layers of tuffs and agglomerates, which are virtually non-utilizable. Many quarries opened in similar conditions were soon abandoned because of uneconomical operation.

(6) *Overburden.* Sound rock exceptionally occurs directly on the ground surface. It is usually covered by weathering material or other unusable rocks, which are designated as overburden. The overburden may be of various character: it is either a weathered surface layer of the rock itself or a younger deposit of slope detritus, loess, terrace gravel, or waste tips and man-made ground.

The thickness and nature of the overburden often controls the economy of stone production; stripping of a thick overburden may make the operation economically prohibitive. It is not possible to establish universally valid directions determining the maximum admissible thickness of overburden from the economic point of view, as several other factors come into consideration: the composition of the overburden, the ratio of the volume of a deposit to the thickness of overburden, the value of the material exploited, etc. If, for example, the overburden consists of little-cohesive rocks, it can be easily stripped with the use of a shovel or a bulldozer, even if it is several metres thick. Stripping of a solid rock of a similar thick-

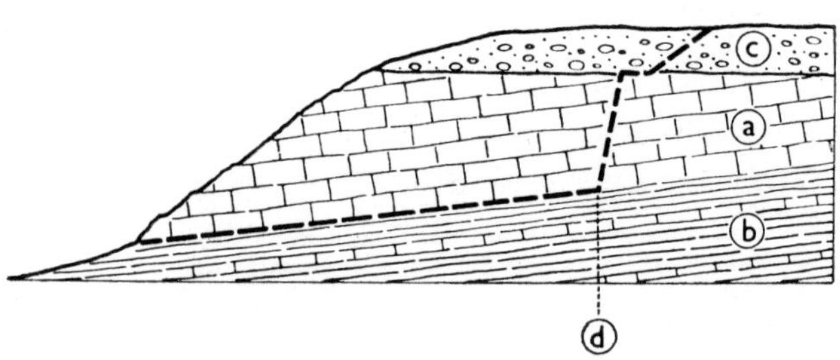

9-11. Limits of economic exploitation (d) are defined by the ratio of the thickness of deposit (a) to the overburden (c); b — unusable shales.

ness (e.g. shale) should be performed with the aid of blasting, and may be unremunerative for the quarry operation. The value of the quarried rock is another decisive factor. A thick seam of brown coal is mined in an open cast even when the thickness of the overburden is five times as great as that of the coal seam. On the other hand, in a quarry on common building stone stripping of an overburden only a few metres thick may be uneconomical.

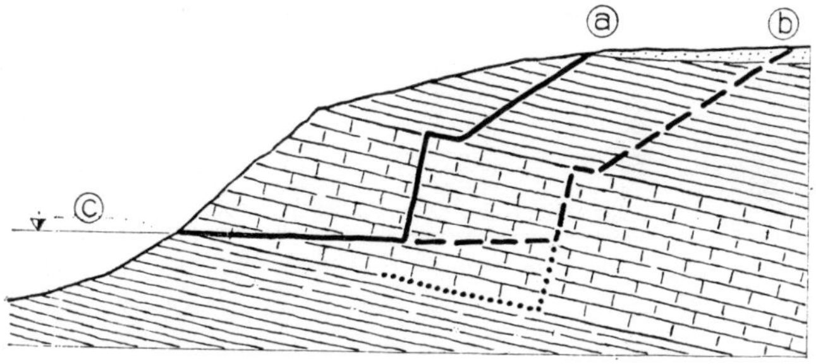

9-12. When the beds dip into the slope, the overburden increases rapidly relatively to the quarried rock. (For explanation see the text.)

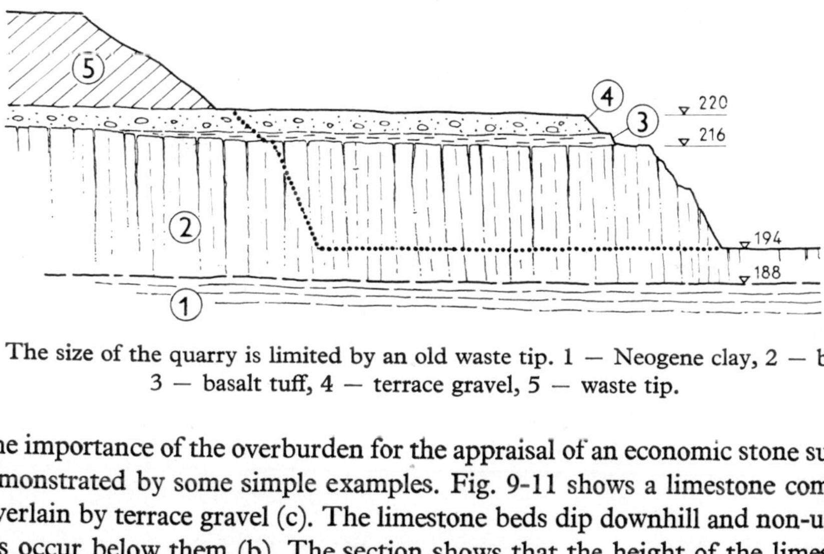

9-13. The size of the quarry is limited by an old waste tip. 1 — Neogene clay, 2 — basalt, 3 — basalt tuff, 4 — terrace gravel, 5 — waste tip.

The importance of the overburden for the appraisal of an economic stone supply is demonstrated by some simple examples. Fig. 9-11 shows a limestone complex (a) overlain by terrace gravel (c). The limestone beds dip downhill and non-usable shales occur below them (b). The section shows that the height of the limestone wall gradually decreases relative to the overburden. A dashed line delimits the economically workable part of the deposit. If the beds dip into the hill, the thickness of overburden increases rapidly in relation to the quarried rock (Fig. 9-12). When the quarry floor is horizontal, the overburden will amount to about 40 per cent in case (a) and 80 per cent in case (b). The overburden/stone ratio would improve somewhat, if the floor were lowering progressively according to the dip of beds but this alternative would be operationally disadvantageous since it would be necessary to lift the stone and pump the water.

The economic supply can also be restricted by man-made ground and waste tips. Figure 9-13 is a section of a basalt quarry in which the supply of high-grade stone is strongly reduced by an old waste tip. The quarry has been opened in a basalt flow covered by a thin bed of tuff and a 6 m-thick gravel terrace. Stripping of this natural overburden would not make the operation uneconomical, but the

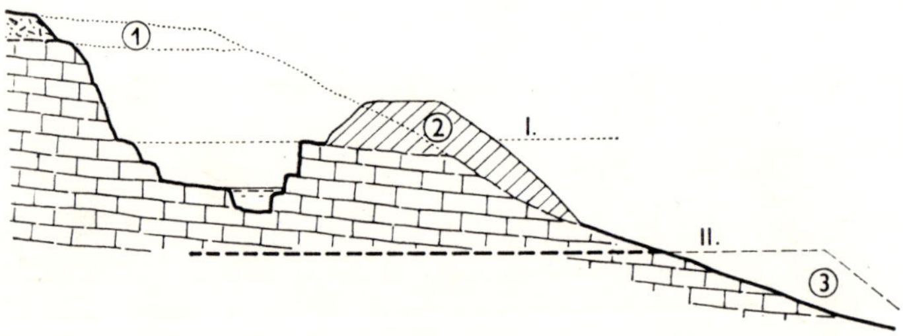

9-14. Quarry depreciated by depositing waste tip on the workable rock (2). I — Original quarry bottom, 3 — suitable place for a waste tip below quarry bottom II; 1 — overburden.

removal of an extensive waste bank deposited there from a coal mine would be all too expensive.

An important task of geological investigation is to find a suitable place for the disposal of overburden material. The spoil material should be deposited below the quarry floor, at the foot of the slope, where it will not impede operation and where it can be stored without lifting. It is a gross mistake to deposit the spoil material upon the stone reserves, which may be exploited in the future, and thus prevent the enlargement of the quarry. Such a depreciation of valuable stone is shown in Fig. 9-14. A quarry on sandstone suitable for building purposes was deteriorated by depositing the overburden on the surrounding slopes. If the floor of the quarry had been placed lower and the spoil material stored below the working level, the quarry could be extended laterally and operation would be easier.

9.3. Basic principles of quarry opening. Rocks are quarried to be used as

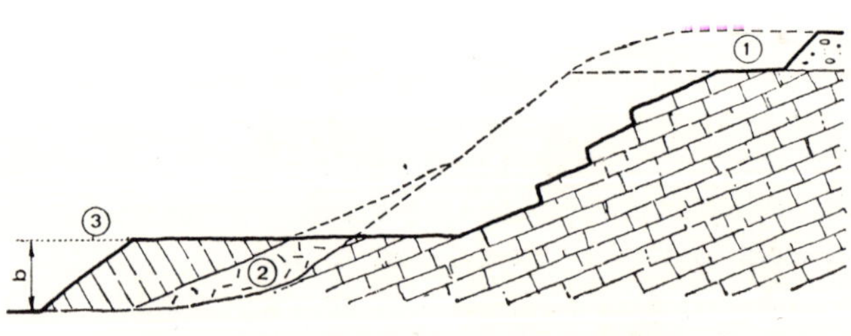

9-15. A shelf quarry, adequately located: bottom of the quarry is above the valley floor and the beds dip valleywards. 1 — Overburden, 2 — slope detritus, 3 — bottom of the quarry.

building stone, pavement and dimension stone, riprap, highroad and railroad ballast, concrete aggregate and in various industries. Each of these branches places other requirements on the properties of the rock and on the mode of exploitation, which is moreover controlled by the configuration of the landscape. Shelf quarries will be used in hilly areas and pit quarries in plains. Underground quarries are comparatively rare.

(1) *Shelf quarries*. Operation is usually easiest in shelf quarries, especially with regard to simple transportation conditions. In laying out a shelf quarry several principles should be observed relating to the position of the quarry floor and the direction of working. It is not advisable to place the quarry base at the valley floor, because the slope sediments accumulated there cover the deposit and their removal raises the price of operations. If the quarry floor is placed higher in the slope, the thickness of the cover is usually smaller (Fig. 9-15) and there is sufficient space left below the floor for the disposal of spoil material.

The strike and dip of beds should also be taken into consideration. A shelf quarry for building stone should be laid out so that the beds dip towards the floor of the future quarry. The stone may thus be extracted up-dip and transported readily along inclined bedding planes to the working place. The opposite dip, i.e. into the slope, is unfavourable for quarrying building stones, because the stuck blocks have to be laboriously loosened. Working is likewise difficult when the beds strike at right angles to the quarry face. In such a case it is advisable to excavate a cutting parallel to the strike of beds and then open the quarry so that the rock may be worked up-dip. These instructions are not valid universally. If the beds dip steeply to the quarry floor, the work may be hazardous for slipping of loosened blocks along bedding planes. If the slope is prone to sliding, the working face should be turned a little obliquely to the strike of beds. The dip of beds at right angles to the working face is thus reduced and one side remains braced up on the intact part of the slope, so that slope failure is not so imminent.

The dip of beds is not so important in quarries for crushed stone. The working face in these quarries is generally high and is blasted gradually by large charges or by a system of vertical holes. Some difficulties may arise from unfavourable jointing of the rock. In basalt quarries the working is easier when the columns are inclined towards the quarry floor or when they are vertical. If the columns dip into the quarry face, dangerous overhangs may form above the working place after a blast. Basalt flows adapt themselves to the shape of the land surface on which they congealed and since the contraction cracks extend inwards at right angles to the cooling surface, the columns may be arranged haphazardly. Figure 9-16 shows the disposition of a basalt quarry, which was well opened with columns inclined steeply towards the quarry floor. During quarry operations the columns were found to turn into another direction so that they had a dip of about 40° into the hillside in the upper parts. As a result, the columns after blasting remained partly overhanging on the steep face and were hazardous for the quarrymen.

Therefore, a new working place was opened from another side and at a higher level (II in Fig. 9-16).

In opening shelf quarries, one must not forget that at rocky edges and knobs the rock is generally harder than in gorges and depressions, which develop where the rock is softer, more weathered or tectonically disturbed. Quarries should be opened in the upper parts of mountain slopes and on mountain ridges only in most cogent cases, as they are disadvantageous in many respects. As a rule, the transport is difficult because of great height differences. A relatively long period of rough weather and a snow cover lying for several months shorten the working season and make difficult the quarry operation. Moreover, the rocks on mountain ridges are exposed to strong action of subaerial agents and are commonly weathered to a greater depth than in the lower part of the slope.

In laying out a shelf quarry, its position with respect to the cardinal points should also be taken into consideration. A north-facing slope is more suitable than a southern slope for opening a quarry for building and pavement stone. On a northern slope, the stone does not dry out so quickly; the rock moisture is preserved longer in the excavated blocks which are then easier to dress. At the higher altitudes, on the other hand, where the snow remains lying long into the spring, a south-facing slope is more advantageous, as the work can begin earlier.

(2) *Pit quarries* are used on ancient elevated plateaus, where even firm rocks may be disintegrated and decomposed to an appreciable depth, because the land surface has been exposed to mechanical and chemical weathering for long geological periods. In south-eastern Bohemia, for example, weathered granite extends to a depth of 12 m, but the deeper one proceeds, the sounder the granite. In these circumstances pit quarries have proved suitable and are deepened down to more than 50 m. R. Legget (1973) mentions an abandoned granite quarry at Aberdeen (Scotland), which reaches a depth of 146 m.

Pit quarries are sometimes opened in order to evade the removal of a thick

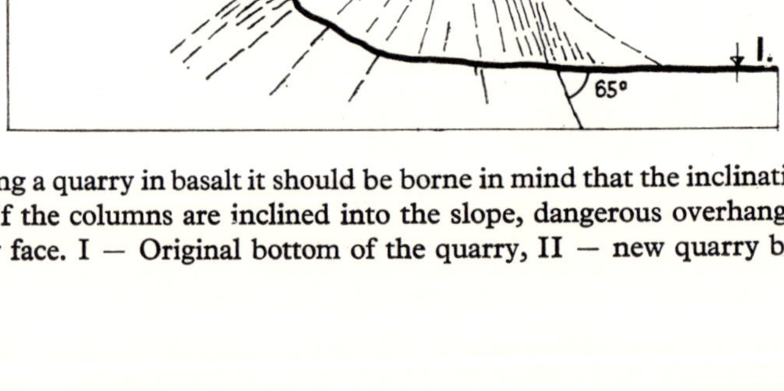

9-16. In opening a quarry in basalt it should be borne in mind that the inclination of columns may change. If the columns are inclined into the slope, dangerous overhangs form on the quarry face. I — Original bottom of the quarry, II — new quarry bottom.

overburden which separates the stone to be quarried from a valley. A gallery may be then driven from the quarry floor into the valley for the transport of stone and water drainage. Figure 9-17 shows the cross-section of such a quarry for crystalline limestone. The limestone deposit is lens-shaped and infolded in gneisses. The limestone beds, dipping at a steep angle, are separated from a valley by a belt of gneiss, about 150 m wide. The stone from the upper part of the deposit was

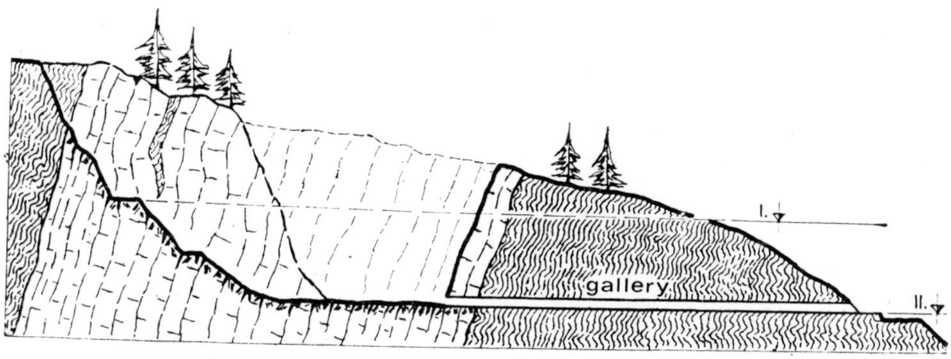

9-17. Section of a limestone quarry; the gallery is used for the transport of stone and for drainage. I — Original quarry bottom, II — haulage and drainage gallery.

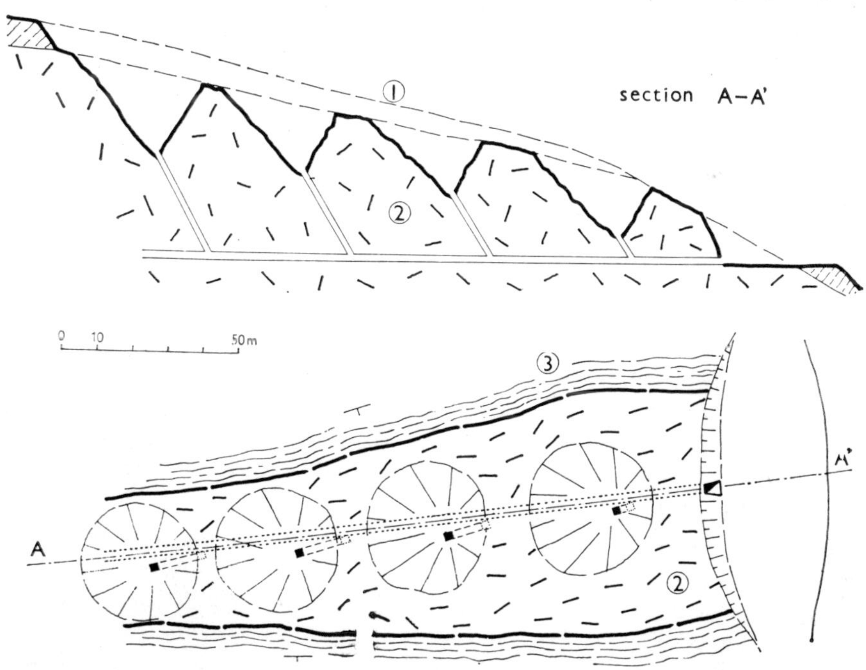

9-18. Amphibolite lenses in biotite gneiss are worked by a special procedure. The stone is transported by a gallery into which it falls through chutes. 1 — Overburden, 2 — amphibolite, 3 — biotite gneiss.

transported through a deep cutting. When the quarry reached a greater depth, a gallery was driven into the neighbouring valley for rock haulage.

A special working procedure in a pit quarry is shown in Fig. 9-18. A deposit of amphibolite has been selected for the construction of a dam in Moravia. Amphibolite makes up a lens-shaped body in gneisses, which extends perpendicular to the slope and is of insufficient thickness for being worked in a shelf quarry. Therefore, a gallery was driven at the level of the quarry floor, and several chutes were broken through into the gallery. The stone blasted on the surface fell directly into these chutes, which were widened funnel-like towards the surface, and through them into trucks.

(3) *Underground quarries.* When the overburden is of excessive thickness, the rock is sometimes worked in underground quarries and galleries. In Czechoslovakia, underground quarries are rarely used, as their operation is far more expensive than operation of shelf and pit quarries. Large underground quarries for building stone existed in ancient Rome, being used later as catacombes. In the underground of Paris large spaces excavated by limestone quarrying caused great difficulties in founding large structures and in construction of the underground. In the surroundings of the town of Bath in England Jurassic oolitic limestone has been worked since Roman times. Large underground spaces many kilometres long were used as air-raid shelters during World War II. Triassic limestone was worked in underground quarries in Kerch peninsula of the Crimea.

Safety measures in quarrying. If possible, quarries should not be opened in settled areas, as stone working, particularly when blasting is used, would endanger the surrounding structures and traffic on nearby roads.

Stability of quarry faces. Shelf and pit quarries are generally bounded by high steep walls, which very often are not permanently stable. A steep working face of an abandoned quarry can disturb the stability of a slope to a considerable distance, even after a long time. The neighbourhood of a quarry does not provide a safe foundation and should not be built upon. The floor of a quarry at the foot of a steep working face may be threatened by the sliding of loosened strata, falling of stones and blocks, or squeezing of soft rocks from beneath a high quarry face. All these conditions should be taken into consideration before opening a quarry, so that its operation does not cause damage which would exceed the value of the rock excavated.

9.4. Deposits of sand and gravel. The demand for sand and gravel steadily increases, as it is used not only for mortar and concrete, but large amounts are needed for the permeable subbase of highways and railways and in building foundations for the mats ensuring a uniform settlement of structures.

River terraces furnish large amounts of sand and gravel, which often occur at several levels above the valley bottom. They make up horizontal beds and are partially sorted. Pebbles are generally well worn by water transport. The quality of sand and gravel depends on the age of terraces and on the length of transport. In Tertiary and Early Pleistocene river terraces the sand often has a considerable loam and clay admixture and the pebbles of e.g. granite and crystalline schists are partly weathered. Younger terraces of large streams contain pure material which, as a result of long transport, is well sorted and firm.

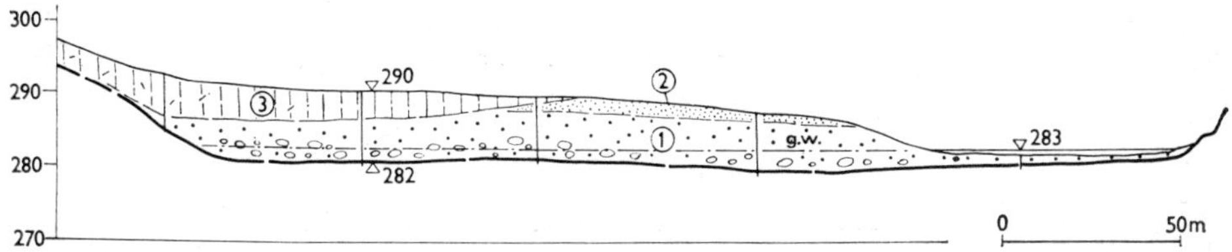

9-19. Sand and gravel of a valley terrace depreciated by the overburden; 1 — sand and gravel, 2 — fine micaceous sand, 3 — slope loam.

In Czechoslovakia, an extensive investigation of terrace deposits was conducted on the river Vltava in connection with the construction of several dams. A section of a Late Pleistocene river terrace is shown in Fig. 9-19. According to boring results, the sand and gravel deposit proved to be 7—9 m thick and its volume was estimated at 420,000 m^3. The material, however, was not usable for concrete, because the gravel contained about seven per cent pebbles of weathered granite. Technological tests have shown that they reduce the strength of concrete. It was not economical to sort out the granite pebbles in mass excavation. Moreover, it would have been necessary to strip a continuous cover of fine micaceous sand, and slope debris and loam at the foot of the valley side. For these reasons this large deposit was abandoned.

A similar Vltava terrace is shown in Fig. 9-20. The material was used for the construction of the Kamýk Dam, since the proportion of weathered pebbles was appreciably smaller and the deposit was not covered by unusable overburden. Exploitation was difficult because of large granite boulders at the base of the terrace and because of subsurface water, which reached above the lower part of the deposit (Fig. 9-21).

Sand and gravel often fill the valley bottom and make up the flood plain. This may be up to several kilometres wide when the river is meandering, and provides a large supply of material. In Central Europe the sand and gravel are commonly

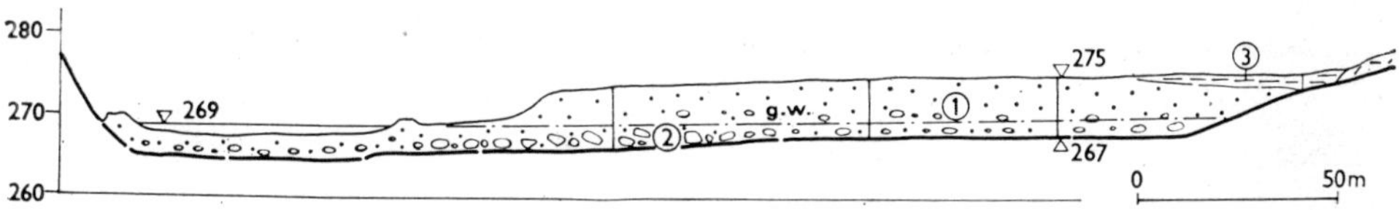

9-20. Material of the valley terrace used for the construction of Kamýk Dam on the river Vltava. 1 — Sand and gravel, 2 — coarse boulders, 3 — Holocene clayey deposits.

9-21. Granite boulders make the working of aggregate difficult. (Photograph by Záruba).

covered by loams, deposited as flood sediments in the Holocene. Man contributed to their formation by forest clearing and systematic ploughing, as the surface layers deprived of vegetation were washed down into the streams by torrential rains. In broad valley plains, sand deposits may be depreciated by humic loamy-clayey deposits filling the abandoned meanders. All the loamy sediments must be removed before the beginning of mass exploitation. It should also be taken into consideration that the greater part of sand and gravel of a valley plain generally occurs below the ground-water table.

In submountainous areas, sand and gravel usable in the building industry form alluvial cones in the lower parts of slopes. In some places these deposits constitute continuous layers of great thickness. The top parts of the cones consist of very coarse to bouldery gravel, whereas fine sandy material predominates at the lower margin. The gravel is poorly rounded, as the pebbles were generally transported only a small distance.

Glacial and glaciofluvial sediments were deposited by water flowing out from melting glaciers. They show various forms; in front of the continental ice sheets they constitute extensive plains and filling of deep valleys. The material is very inhomogeneous, comprising sand, gravel, boulders and very fine silt. The grains are mostly angular and pebbles are poorly worn. In some areas, the moraines of mountain glaciers and ice sheets are an important source of sand and gravel. With regard to the inhomogeneous material ranging from coarse boulders to fine silt, a thorough sorting must be carried out. The moraines of continental ice sheets have been preserved either as long ridges or as isolated mounds; their material is finer and of sandy or loamy nature. Deposits of mountain glaciers usually make up ridges running across valleys and mountain slopes; they are composed mainly of coarse gravel.

In some areas, aeolian deposits provide a large supply of sand. They occur along broad valleys of large streams, where the wind blew sand out of fluviatile sediments, or along lake or sea shores. Aeolian sediments in the form of sand dunes are most frequent in arid areas, where the land surface is not protected by vegetation. Aeolian sand is fine to medium grained and its usability is limited by the grain-size distribution. The sand is even grained, the grain size depending on the wind strength. It can generally be used only as one fraction for the preparation of concrete.

Sandy weathering material originated as a result of weathering of granite, sandstone, arkose, and other rocks is usable for the permeable subbase of highways. Since it contains a large proportion of weathered grains, it does not provide a suitable admixture for concrete. The layers of sandy weathering material are of varying thickness, which is unfavourable for exploitation. They are often partly redeposited and grade into sandy slope debris, providing permeable material of inferior quality.

The investigation of sand and gravel deposits is governed by similar principles, as are given in paragraphs 9.1 and 9.2. For the assessment of sand properties it is necessary to recognize its geological derivation, i.e. whether the sand is of fluviatile (terrace), lacustrine, glaciofluvial, or aeolian origin.

The economic reserves and the amount of overburden can be computed from several sections drawn across the deposit, which are constructed on the basis of boreholes or test pits sunk down to the substratum. When the deposit consists of several beds differing in grain size and quality, samples must be taken from each of the beds and the sampling localities plotted in the sections. In addition to the determination of grain size and clay admixture, petrographical analysis of gravel should be made to ascertain the degree of weathering and the possible content of deleterious particles, such as clay galls, flat fragments of shales, etc. If the sand or gravel contains loamy or clayey admixture, it should be determined whether the loamy material is dispersed or whether it envelops the grains. In the latter case, the sand and gravel are unacceptable for concrete.

In cases where alkaline cement is used for concrete production, it must be stated whether the gravel contains particles which when combined with this cement, would cause *expansion of the concrete*. From this point of view, gravel with chert, flint, cherty limestone, chalcedony, opal, volcanic glass, liparite, andesite, phyllite, etc. are unsuitable.

In some countries where alkaline cement was used, heavy defects in the concrete (in bridges, dams) occurred in the past decades, for which the above-mentioned rock particles were responsible. Detailed tests provided relevant results. Figure 9-22 shows the expansion within 24 months of mortar prepared from alkaline cement and some active rocks. For comparison, the behaviour of mortar from quartz sand and basalt is shown (D. McConnell, 1950).

In the sections of the deposit, the ground-water table and its annual fluctuation should be plotted, as the depth of the ground-water table controls the working procedure.

If the gravel is loamified or contains another unwelcome admixture (e.g. organic matter) which should be removed by washing, one of the questions that should be decided by geological investigation is the placing of settling basins, where the turbid water from washing plants will be let out.

9.5. Working of building material and protection of the environment. In exploitation of building materials we must observe the protection and economic use of mineral resources on the one hand and nature conservancy on the other. Mineral resources should be exploited and treated so as to utilize their economic value to the best advantage (Chapter 14).

From the point of view of environmental protection, the location and operation of quarries should disturb the character of the landscape as little as possible and should not endanger a valuable natural feature or historical monument. The economic standpoint will often be in discrepancy with the interests of nature conservancy, but in many cases it is possible to reduce the spoiling of the landscape,

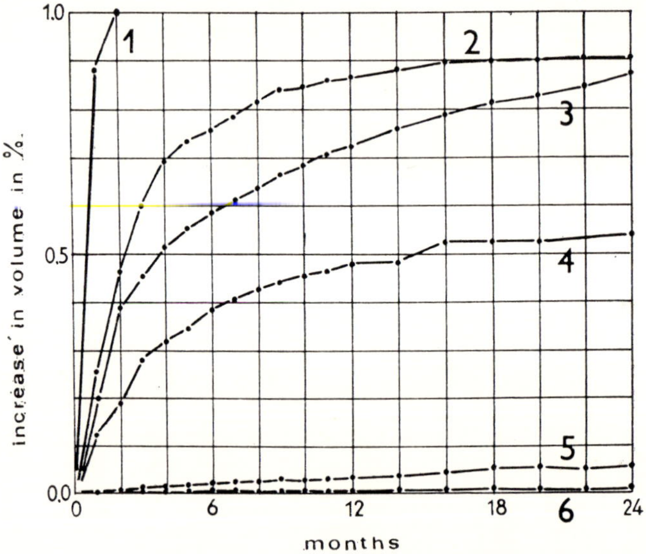

9-22. Increase in the volume of mortar prepared from alkalic cement and a mixture of the following rocks: 1 — opal, 2 — chert, 3 — rhyolite, 4 — andesite, 5 — standard quartz sand, 6 — basalt. (According to McConnell et al., 1950.)

for example, by depositing the stripped overburden into lateral valleys. In any case, a reasonably placed quarry which serves public needs has a justifiable place in the landscape, but when exhausted, it must be reclaimed so as not to mar permanently the face of the slope. The steep slopes should be arranged with an adequate inclination, covered by humus and planted.

In past decades quarries were opened unsystematically. At a time of increased demand, numerous minor quarries were founded, only to be abandoned when the demand decreased. As a result, in regions with deposits of required building material, large wood and agricultural areas were changed into barren pits surrounded by waste tips and garbage heaps from the neighbouring communities. Adjacent slopes often deprived of vegetation suffered from increased erosion and denudation of soil.

In the modern epoch of planned economy, small non-economic quarries are superseded by large regional quarries, well equipped with machinery. The operation in these quarries is directed to the full utilization of all exploited stone supply. This tendency is in accordance with the requirement of nature conservancy that the neighbouring area should not be spoiled with large tips of overburden material and non-utilized stone waste.

In planning the exploitation of a building stone deposit, the economic advantage should be overridden when the opening of a quarry would endanger a remarkable historical monument. Some regrettable examples may be cited from Czechoslovakia, where old castles, generally standing on high rocky hills for the sake of safety were threatened in this way.

Chapter 10

FOUNDATION OF BUILDINGS AND INDUSTRIAL STRUCTURES

The geological conditions of the site influence the design and the costs of foundations of buildings and of industrial structures to a large extent. The collaboration of the engineering geologist with the engineer therefore has become increasingly important, and large designing bureaus employ engineering geologists for this purpose. Since the structure and the subsoil constitute an interacting complex, the designer also has to observe the mechanical behaviour of foundation soils. The engineering geologist in his turn should understand the behaviour of different structures founded on various grounds so that he could supply the designer with all data necessary for statical calculations. Foundation problems, however, are not only of a statical nature; other aspects, such as the presence of ground water or the shrinkage of soils should be considered, and the engineering geologist is expected to be familiar with them.

All geological factors which can influence the structure at a particular site should be determined and submitted to the designer in order that he can produce the best design from the point of view of both structure and economy. In this stage of preparation works disagreement with the ideas of the architect may arise. The earlier the controversial questions are settled the better. A very difficult situation can develop when the architect has already fixed the location of the buildings as regards the plan and level, as well as the spatial arrangement of the structure; he is therefore unwilling to discuss them and expects the engineering geologist to propose the foundation method only. It is usually the owner who loses in this situation.

The second group of information to be provided by the engineering geologist concerns the quantitative values of the properties of foundation soils. In order to be capable of meeting these requirements, he should know how the mechanical behaviour of foundation soils is related to the statical behaviour of the structure, which other factors of geological character might affect the structure, and what are the approximate costs of different foundation methods.

10.1. Demands made on the foundation of structures. In order to illustrate how serious is the influence of the geology of the building site, let us first review what is expected from a good foundation. Several of these requirements are comprehensive and they will be analysed individually in Sections 10.2 to 10.6. Those which can be answered rather briefly will be treated here. Let us discuss the demands in the order of their importance, i. e. according to the damage which would arise if they were neglected.

10.1.1. Stability of the ground. Even if not the first in the order of frequency, the sliding movement of a site is accompanied by considerable property loss. The problems of sliding phenomena were discussed in Chapter 7. From the point of view of foundations of buildings and industrial structures, the building up of unstable ground is not advisable, since the cost connected with the stabilization of a slope is generally much greater than the value of the building itself.

It is not easy to find out whether the ground is susceptible to sliding. The topographical features, however, often provide evidence of this danger and an experienced civil engineer should be sufficiently qualified to notice them. The irregular undulation of contour lines on a sloped area should incite an analysis whether they are associated with a slope movement or not.

In exceptional cases the engineering geologist has to yield to the interests of the owner and analyze the possibility of building up a site even if it is endangered by landslide movements. Two cases may be distinguished:

(a) The sliding movement is relatively deep, in contrast to a sheet slide. Only the building up of the toe can be tolerated provided that the structure is massive enough and is surrounded by a stabilizing fill. This work is very expensive since the risk that excavations could provoke a movement must be eliminated. Therefore, the work should be planned in stages; the foundation beams are to be constructed one after another and the surrounding area of an excavated trench for a beam under construction should be ballasted by temporary dumpings of earth. Heavy foundation beams or reinforced concrete walls of the ground floor must stiffen the foundations into a statical unit. The orientation of the foundation beams should coincide with that of thrust exerted by the slide. Often foundation on piles is to be preferred with the piles rising above the ground and buried in a fill below the base of the structure.

(b) A site threatened by a recent or potential sheet slide. This case is frequent when planning settlements on the periphery of towns since suitable vacant areas are rare. Such a site should be stabilized as soon as possible. Important is the magnitude of the slope angle as steep slopes are difficult to stabilize. Every building, road, sewer should be studied in view of the danger of slide development in the course of excavation. Deep foundation pits should be excluded and all underground lines should be oriented downhill. As previously stated, the stabilization of the site should be accomplished beforehand; this also provides additional experience of the site and leaves time enough to complete the drainage if unsatisfactory.

A delicate case arises when deep excavation is to be performed near existing structures, as occurs in brick factories, where unstable slopes of loam pits can endanger factory buildings. Fig. 10-1 shows an example of the disturbance of a building by the slide of a loam pit. The production in the pit had to stop, the stabilization of the slope was accomplished and the foundation footings of the building were underpinned. The case of buildings damaged by the slide of a quarry slope is shown in Fig. 10-2. Raw material was worked in a high steep slope the

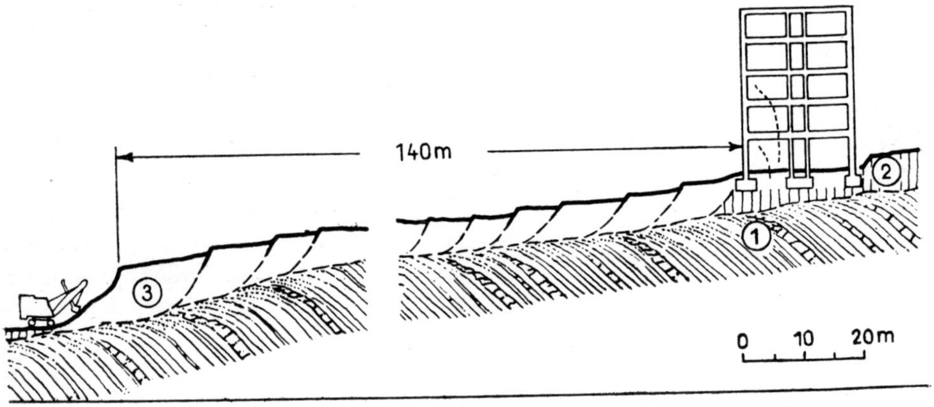

10-1. Building endangered owing to a hazardous loam working. 1 — Clayey shales and sandstones, 2 — slope loam and debris, 3 — debris affected by slide movements.

lower section of which consisted of Cretaceous marl. The sliding movement also involved the bottom of the quarry and affected several buildings. After a stabilizing berm has been filled, the deformation stopped.

10.1.2. Earthquakes affect buildings in a catastrophic manner. The science of seismic effects on structures is very broad and does not include questions of engineering geological character only. From the latter point of view several questions should be considered:

(a) Is the risk of damage greater for buildings that are located near tectonic faults in seismic regions? To answer this it is necessary to distinguish the occurrence of dislocations along the faults from that of the intensity of shocks.

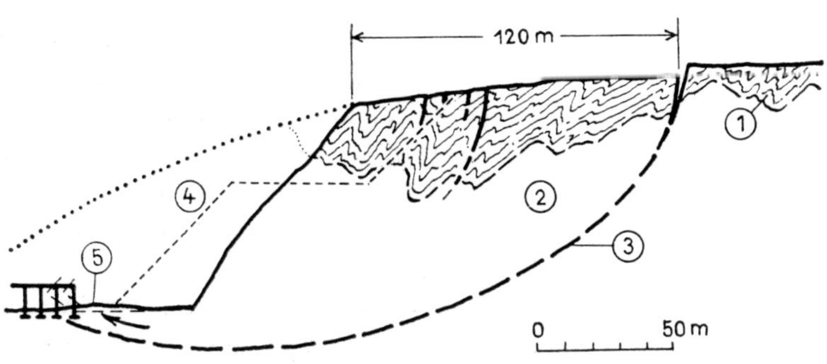

10-2. The building site for an annexion of a cement factory affected by bulging deformation of marl. 1 — Limestone, 2 — Cretaceous marl, 3 — slip surface, 4 — stabilizing fill, 5 — upheaval of the quarry floor.

When active faults are present, the occurrence of slip movements is to be expected especially in areas where the cover is shallow. For example, after the large earthquake in Turkey in July 1967 (Ambraseys and Zátopek, 1968), strike slips of the order of 1.9 m and throw slips of about 1.2 m were found along the Anatolia fault. But relatively small dislocations were observed where the cover was thick.

The greatest seismic intensity and the accompanying damage of structures are not necessarily associated with the presence of faults, since they can occur over a distance of several kilometres from the fault. This was also found by the investigation of the earthquake in Turkey as outlined above as well as of that in Peru in 1970 (Schneider and Mencl, 1970). The explanation can probably be found in that the fault is inclined so that the contact of blocks at depth where seismic shocks originate (hypocentrum), does not correspond to the outcrop of the fault on the surface.

(b) Another question concerns the rate at which the foundation soil affects the seismic behaviour of the structure. In general, a weak soil may almost double the amplitudes, while at the same time it lengthens the time of swing. Therefore normal buildings (not tower-like), possessing short time of natural vibration (about 1/3 s) tolerate the earthquake well, especially when their foundation is stiff enough, as it is with piles. In turn, this type of buildings suffers from resonance when founded on rock.

This general rule, although not quite inapplicable, has been subjected to many discussions, especially since the catastrophic earthquake in Mexico City in 1957. The opinion prevails that each structure will eventually meet an earthquake possessing a frequence identical to its own. The duration of the earthquake in Peru in 1970, 45 seconds, exceeded all durations experienced in the region and this made the earthquake disastrous. The Alaska earthquake in 1964 was also characterized by a long duration of 4 minutes and this resulted in great damage. For this reason the general requirement tends to stiff structures, resistant to shocks of any frequency.

(c) Good earthquake resistance cannot be attained if the foundation is poor. Moreover, when analyzing earthquake effects, the destructions caused directly by seismic shocks have often been considered, while the influence of settlement was neglected. The latter can be caused either by the contraction of the foundation soil due to vibrations or by its liquefaction if the ground water level is high. Therefore, in weak soils the pile foundation type is increasingly preferred, provided that the piles are long enough to penetrate a layer which is not susceptible to contraction or liquefaction. An example of the collapse of an urban area owing to a high ground water table, resulting in a large slide, is the earthquake catastrophe in Anchorage in Alaska in 1967 (Idriss and Seed, 1967).

(d) The question whether the topographical position of a site can increase the degree of damage cannot be answered satisfactorily. However, cases where the structures located on salients or ridges have suffered more than in depressions (if uplift was not present in the latter) can be evidenced. This was observed with the earthquake in Alaska (Mc Culloch and Bonilla, 1967), as well as with that of Peru (Schneider and Mencl, 1970).

10.1.3. Cavities, either of natural origin or man-made, are dangerous if they are present beneath the structure. Their existence can be presumed on the basis of the geological situation and local experience. Cavities are often very large (see also Chapter 14). Their presence is often hidden for a long time by a loam cover, but the ground surface begins to subside after heavy traffic, irrigation of gardens, etc. have been applied. Artificial hollows exist in historical areas of towns; their stabilization is expensive and is of great concern to municipal engineers. Old mining works also left many openings unfilled and even their identification requires a dense network of prospecting holes.

10.1.4. The unsatisfactory bearing value and settlement of foundation soils are other factors which may cause damage to structures. Due to the large extent of this subject it will be treated in Sec. 10.2.

10.1.5. Ground water at the site is another issue of great concern. A series of problems is associated with it and Sec. 10.3 will be devoted to them.

10.1.6. Volume changes of foundation soils and rocks. Damages to structures promoted by these changes are not rare. The engineering geologist should conclude whether their occurrence is to be anticipated at the particular site, and if so, which measures are necessary to avoid them.

Volume changes can originate due to several processes but freezing and desiccation are most common.

(a) The more clay particles the soil or rock contains, the lighter is the structure, the more shallow is the foundation and the greater is the depth of freezing, the greater the danger to the structure. The increase in volume of coarse-grained soils is relatively small as only the water present in them freezes into ice. But clay particles in fine-grained soils have the capacity to absorb water from underlying layers and to transmit it to the thin ice layers. Because of this, the ground surface upheaves several tens of centimetres after a long period of frost. This upheaval affects light structures or less stiff parts of structures, for example, the parapets of windows on the ground floor.

Damage due to freezing has been known to exist for a long time and the necessary measures for preventing it have been incorporated into the building standards. A satisfactory foundation depth is generally required.

However, the fact that rock may also increase in volume is sometimes neglected. Fig. 10-3 presents the example of frost damage to a structure founded on a Cretaceous sandy marl. The structure serves for the storage of clover seed. Its outer walls are brick-built and founded at the depth of 0.8 to 1.2 m, while the foundation depth of pillars is 1.8 to 2 m. After several years of service the structure suffered cracks after a heavy winter in 1928/1929. Due to the increase in volume of the rock the walls but not the pillars upheaved. Cracks appeared and the window panes were crushed by the vertical thrust of the longitudinal bracing beams. At first sight, the cracks suggested damage caused by the settlement of the pillars, but ice inlets between the bedding planes of the rock were found beneath the wall foundations.

Another case, hardly anticipated by the engineering geologist, but often imputed to him, is the upheaval due to frost of a floor in an unheated cellar. The cave paving and insulating layer are often torn from the pillars, which do not upheave owing to large loads and deep foundations, and water penetrates into the cellar in the spring.

Upheavals of clayey subsoil are frequent beneath cooling plants. In spite of thermal insulation frosty conditions develop in the soil after several years of steady cooling. Conversely, the soil beneath the circumference of the structure is subjected to a supply of warmth from the surroundings. Therefore the central part of the structure upheaves by about 10 cm or even more and damage to the structure occurs. In order to prevent this the cooling rooms should be set on cellars or on artificially thermal-treated sheets.

(b) *Desiccation* is another frequent source of damage associated with volume changes. Damage caused by desiccation of natural origin is common in Cretaceous and Tertiary clays. In several regions, e.g. in the London Clay, the volume of damage was so large that the Building Research Institute started an information campaign among the population.

Damage appears on walls exposed to sunlight. The water content of the foundation soil decreases by several per cent beneath the outer boundary of the wall as compared with the inner boundary. Therefore the foundation tilts out from the building. The wall above it either suffers a vertical crack, or — if the ceiling is stiffened by anchors or by collar sets — the crack is oblique. The corners of buildings are often damaged in this way.

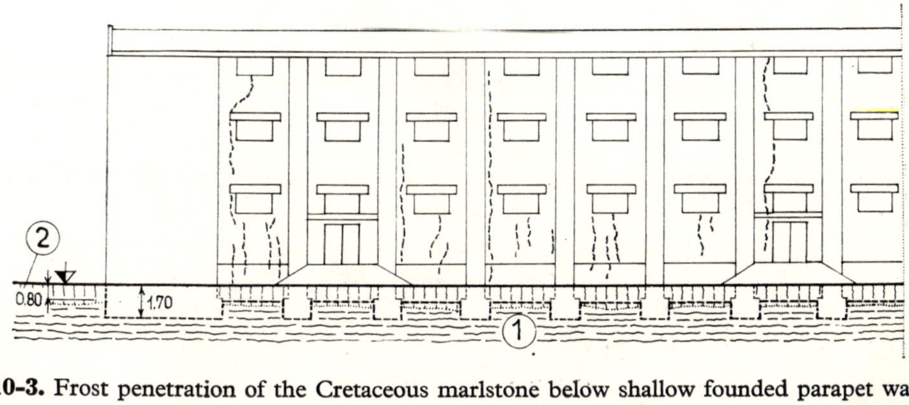

10-3. Frost penetration of the Cretaceous marlstone below shallow founded parapet walls caused their breaking from the deep founded columns during the heavy winter of 1928 to 1929. 1 — Hard sandy marlstone, 2 — sandy loam.

The effect of *tree roots* is known. In the case shown in Fig. 10-4 absorption of water by the roots decreased the water content of the Neogene clay to about 24 % within the reach of the roots, while a value of 29 to 32 % was determined at other places. Moreover, roots reduce the soil substance. Poplars, however, produce such a dense tangle of roots that they actually replace the soil.

When a permeable sand lies beneath a clayey layer the desiccation can affect the total depth of the latter. The ground exhibits distinct vertical cracks, which cross pavements, garden walls and often also light buildings. Cracks in walls of buildings cannot close during wet weather as fragments of brick and mortar become stuck in them.

The best measure is to build higher, rather heavy structures in areas of clayey soils. The ground plan of the building should be close and free of light additions. When this is not possible, minor precautions are necessary. Trees absorbing water avidly from the earth, such as poplars or willows, should be planted far from the walls (the British Building Code of 1948 recommends a minimum distance of 7.5 m). Pavements, instead of a lawn, should be laid around the building. The near lawns should be sprayed during dry seasons and water from the eaves should be allowed to infiltrate partially into the ground. In Czechoslovakia, the building standard orders a foundation depth of at least 1.4 m. If the buildings are equipped with a heating boiler, the foundation depth is measured from the cellar floor. In England the foundation on short reinforced concrete piles has often been preferred.

(c) Artificial desiccation occurs in brick factories, where the shrinkage of clayey subsoil around the kilns causes damage due to differential settlement. Therefore

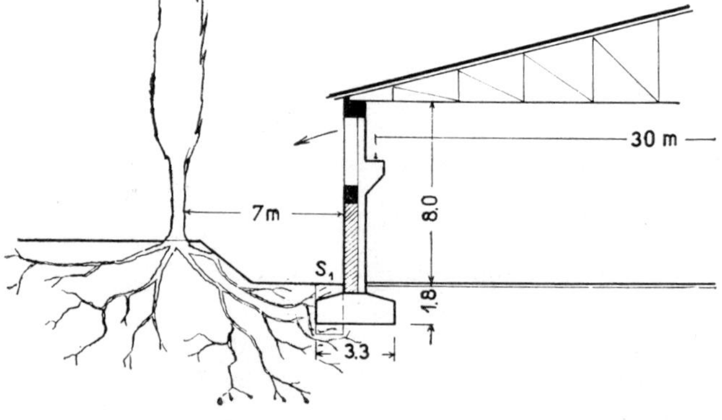

10-4. Columns of the factory hall tilting due to soil desiccation caused by the action of roots of a row of poplars.

gravel layers saturated by water are placed beneath the kilns. Also of interest is the reciprocal process: after the work of the plant has stopped, the ground surface upheaves again. An example from the Ostrava region can be quoted, where the foundation plate of a new coking plant under construction upheaved in the area which had been occupied by the old coking plant. The foundation soil consists of Neogene clay. The upheaval amounted to 12 cm but the reinforced foundation plate was not damaged in spite of the differential upheaval. What matters more is the future development, after work in the new coking plant begins, as coking plants are susceptible to differential settlements. Fig. 10-5 shows the hall of a

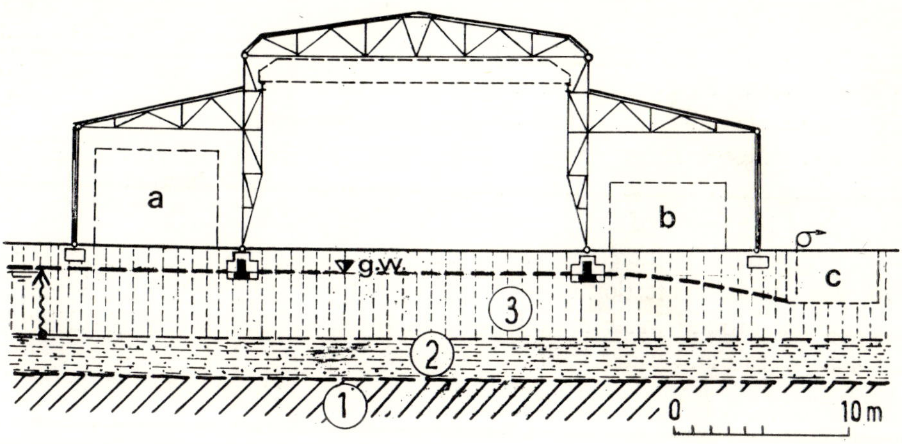

10-5. Casting house suffered additional settlement of foundations due to the desiccation of the clayey soil caused by foundry furnaces. 1 — Neogene clay, 2 — sandy soil with organic content, 3 — clayey soil; a, b — furnaces, c — excavation.

casting house, the columns of which subsided after 30 years of service. This manifested itself by the irregular settlement of the crane runway. The ground-water table was previously at the level of the foundations of the columns, but during the reconstruction of the plant the water table was lowered by pumping and desiccation of the clayey subsoil occurred in the vicinity of the furnaces (Fig. 10-6).

(d) More particular is the case of damage caused by the increase in volume of certain rocks. In this way, alum shales caused damage to the building of the Museum of Natural Sciences in Oslo.

10.1.7. Another group of requirements concerns the effects of vibrations caused by machine operation or traffic. The base of machines should damp their vibrations, the propagation of vibration into the surroundings should be prevented and vibrations coming from the vicinity should not damage the structure.

The engineering geologist can help to solve these problems by determining the elastic properties of the subsoil as well as the level of the ground-water table.

In order to understand the essence of the problems he should be familiar with the principles of vibration effects:

By providing machines with a large base, the contact stress in the foundation is reduced and therefore also the elastic settlements decrease. This foundation is therefore stiff from the point of view of vibrations. If, in addition, the base structure is light, straddled, and well stiffened, the natural frequency will be high and the magnitude of amplitudes small. This is why slowly operating machines (less than approx. 1000 rev/min) were provided with large base foundations, in order to differentiate the natural frequency of the foundation from that of the machine. The drawback to this solution is that the vibrations propagate far into the surroundings.

Contrariwise, weak foundations promote low natural frequencies, smaller than those of the machine revolutions. A rather small base area and higher foundation columns, or special weak elastic dampers, are necessary to lower the natural frequency. The vibration period is long and vibrations do not spread into the surrounding area. On the other hand, great amplitudes of vibration result, and can damage the frame of the machine, fittings, etc. Therefore the weight of the foundation should be increased (but not the base area), in order to decrease the amplitudes. Good protection of the near structures can be obtained in this way, provided the ground-water table is low. In the latter case, the uncompressible complex of soil and water spreads vibrations far from their origin. Therefore, areas with a high ground-water level should be considered as less suitable for the buildings of factories containing large machines. A pile foundation is often necessary for the improvement of the vibration conditions.

The effect of vibrations coming from the neighbourhood, e.g. caused by traffic, can be partially eliminated by digging a trench or any lightly covered hollow around the building. The preservation of old buildings of cultural value often finds a use for this technique.

10.2 Mechanical behaviour of foundation soils. A problem associated with this question has been the subject of theoretical and practical soil mechanics

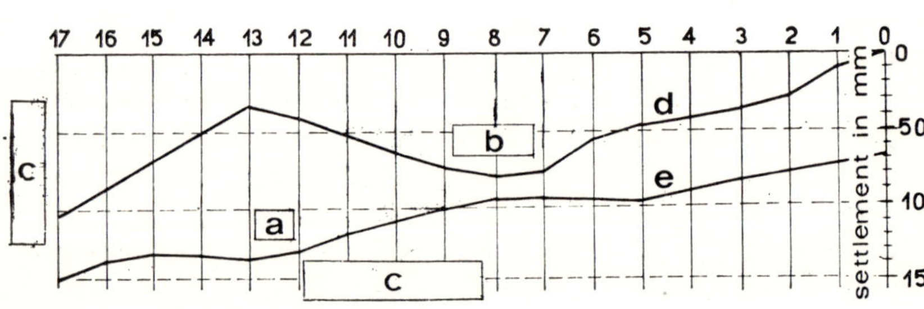

10-6. Settlement of the left-hand (d) and right-hand (e) rows of footings as in Fig. 10-5, caused by the desiccation of the foundation soil. a, b — Furnaces, c — excavation.

studies. Therefore, only a review is presented in the following paragraphs, with an emphasis on the influence of engineering geological conditions. As stated in Sec. 3.8, the development of statical calculations covers the analysis of the behaviour of both the foundation soil and the structure from the onset of loading until the theoretical state of collapse. But such a calculation cannot be done unless a preliminary design of the structure and of its foundations is drafted. For this, much experience is needed, and regional geological conditions should be considered. This can be done by the engineering geologist provided that he is familiar with the fundamentals of soil mechanics and of the statical behaviour of structures.

Damage due to the settlement of foundation soil is common. For practical purposes settlement caused by the deformation of surface layers is distinguished from deep-seated deformations.

10.2.1. Settlement due to deformation of near-surface layers. Several areas provide foundation in a compressible surface layer of less than 5 m thickness. Shallow foundation may also result from the presence of ground water. Loamy Holocene alluvial deposits are representative of this category. Residual soils, slope debris, loess layers, fills, etc. also constitute a relatively shallow compressible layer.

In order to assess the order of differential settlements, the idea that the settlement at any point of the foundation area is in direct relationship to the contact pressure between the foundation base and the soil at that particular point, is to be adopted. Therefore settlements are expected to be uniform below the uniformly loaded foundation area. However, loading below a uniformly designed structure is not uniform; e.g. corners and chimneys are heavier than other structures. Therefore differential settlement and the danger of cracks are to be expected. This is accentuated by the instant response of a given soil to the loading (Sec. 3.3). Moreover, the deformation properties of these soils vary from place to place.

With regard to the complexity of the problem, the assumption is made that the differences in settlement are proportional to the total settlement. Therefore the limitation of the total calculated settlement is standardized in the majority of building codes, its magnitude being e.g. of the order of 4.8 cm for brick buildings founded on sand (ČSN 73 1001). Nevertheless, this approach is not sufficient in many cases. Reinforced foundation beams are usually necessary on the given foundation soils, often laid in both directions so that a foundation raft is formed. Foundation plates are necessary under heavy structures. Nowadays, these structures are statically analyzed by means of computers. Therefore a more exact definition of the deformation properties is expected from the engineering geological investigation today than in the past. While the mean (it is rather low) modulus of deformation used to be applied in the past, the need for the determination of the actual variability in this parameter has prevailed recently. For example, the designers in the U.S.S.R. introduced a mean value of the E modulus and moreover a quantity α indicating the ratio of the maximum to minimum E which exists at the site (Lishak and Vronskii, 1967). Another quantity to be indicated concerns the probable spacing over which α varies from minimum to maximum.

Several interesting findings appear when comparing the values of α of several soils. In Czechoslovakia, Holocene deposits display the greatest values, as their modulus of deformation varies from about 50 to about 300 kp/cm². Not far behind them are the Pliocene deposits having a close alternation of clay, giving a value of E often less than 60 kp/cm², and silty sand giving a value of E of 400 kp/cm². Residual soils of igneous and metamorphic rocks constitute a particular case, as the fresh rock as well as a loose sand may be encountered at the same site.

Within the scope of the discussion of the surface layers of foundation soils the fact that by increasing the foundation area the settlement decreases almost proportionally should be considered.

10.2.2. The effect of deep-seated deformations. The geological situation of the site is often marked by the presence of deep deformable soils. Neogene clays may be quoted as an example, but other deposits may also be typical

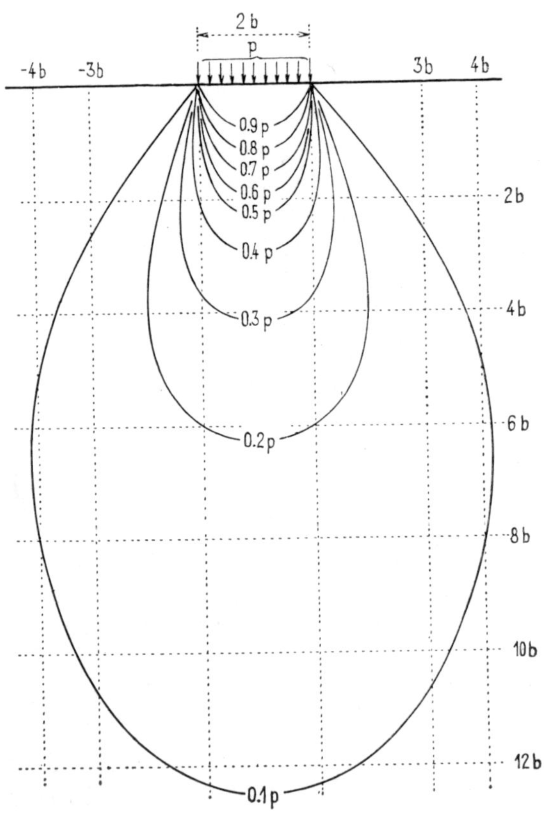

10-7. Isobars (curves of equal vertical stress) in an idealized soil with elastic properties beneath a uniformly loaded strip 2b in width.

representatives (Eoccne London Clay), not forgetting a thick loess cover. The larger the structure, the more the deep layers participate in the settlement.

In contrast to the previous case the settlement conditions may be characterized as follows: when the deformable foundation soil extends to a great depth, the settlement of a uniformly loaded area is not itself uniform. It is greater at the centre and smaller at the periphery. The settlement is produced not only by the loading at one particular point but also by the loading of its surrounding. This is because stressing does not propagate in the vertical direction only, but also spreads sidewards. In Figs. 10-7 and 10-8 isobars are plotted, indicating the

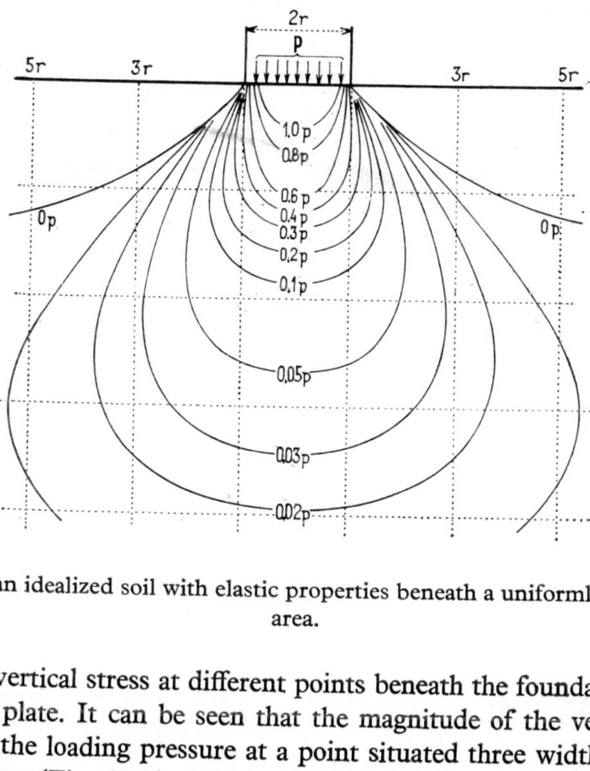

10-8. Isobars in an idealized soil with elastic properties beneath a uniformly loaded circular area.

magnitudes of vertical stress at different points beneath the foundation beam and the foundation plate. It can be seen that the magnitude of the vertical pressure is one-tenth of the loading pressure at a point situated three widths of the beam below the surface (Fig. 10-7) and one-and-a-half widths away from the edge. At depth, the pressure below the axis is greater than that below the edge of the beam.

It is sufficient only to be aware of these relations and several features will be apparent:

(a) As stated, the settlement of a uniformly loaded area is not itself uniform, as it is greater at the centre than at the periphery. For example, the settlement of Neogene clay below the foundations of the International Hotel in Brno (length 80 m, width 20 m, loading pressure 13 Mp/m^2) attained 7.1 cm at the centre and 4.5 cm at the corners. As will be shown in Sec. 10.5, the difference between the two values has been partially compensated by the compression of the alluvial layer left below the foundation at a variable thickness.

Since (Fig. 10-8) a lateral spreading of the pressure also occurs at depth, the areas surrounding the loaded area also exhibit settlements.

(b) However, the settlement state of the structure, as previously described, does not conform entirely to the settlement of a uniformly loaded area. Due to a certain stiffness the structure does not deform as much as a free uniformly loaded area. The greater the stiffness, the less is the differential settlement at the centre and the more transmitted is the contact pressure between the structure and the soil to the periphery of the foundation area. Isobars of vertica pressure below stiff structures are plotted in Fig. 10-9. The settlement of the surface of

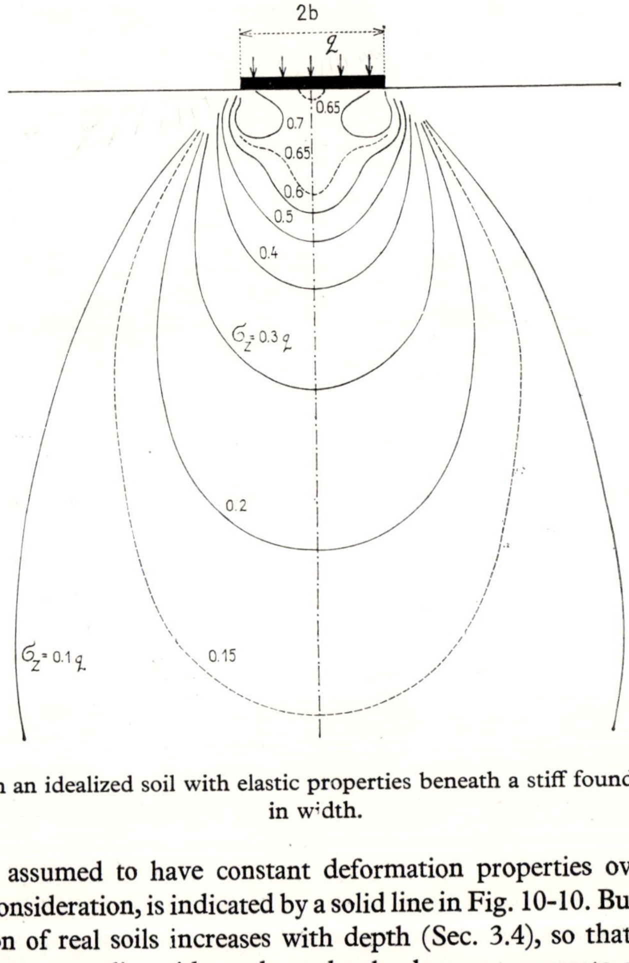

10-9. Isobars in an idealized soil with elastic properties beneath a stiff foundation beam 2b in w'dth.

an ideal soil, assumed to have constant deformation properties over the entire range under consideration, is indicated by a solid line in Fig. 10-10. But the modulus of deformation of real soils increases with depth (Sec. 3.4), so that the pressure of the structure, spreading sidewards at depth, does not generate settlements as large as assumed previously. Therefore, the shape of the real settlement curve (dashed line in Fig. 10-10 for Neogene clay) differs from that of an ideal soil. It is apparent that the analysis of the measured settlements can provide a verification of the actual magnitudes of moduli of deformation[1].

The settlement versus time relations should also be considered if clayey soil

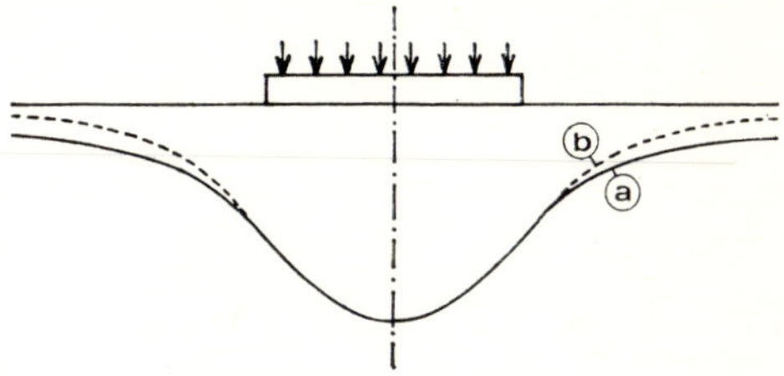

10-10. The settlement curve of the ground surface in the vicinity of a semi-stiff foundation beam. a — For an idealized elastic soil, b — in Neogene clay.

is present in the foundation. The more clay is contained in the soil, the slower is the settlement owing to slow consolidation (Fig. 10-11).

Complex soil mechanical analyses are necessary to obtain the magnitudes of settlement with a satisfactory accuracy. For the sake of the first insight the magnitudes of settlement are plotted diagrammatically in Figs. 10-12 and 10-13 for

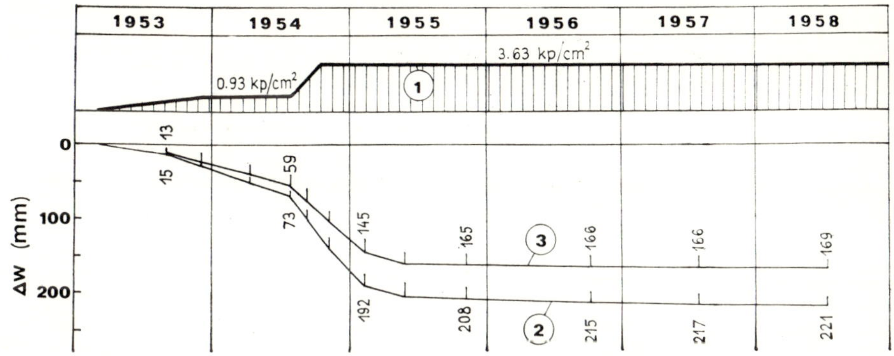

10-11. Time-settlement curve of a bin founded on a Neogene clay. The structure is very stiff and the foundation slab is 2 m in thickness. Dimensions of the foundation are 14 to 39 m. 1 — Time-loading curve during the construction and filling, 2 — settlement of the mid-point of the long side of the foundation, 3 — settlement of the corner of the foundation.

[1] Note: The assumption was tacitly made in the foregoing that the new stress state (after loading) in the soil does not change the values of E and ν (Sec. 3.4), and that a moderate change in deformation properties does not violate the stress field. This was proven by Hruban (1958) in an analytical way, and recently by Morgenstern and Phukan (1969) by the finite element method. Therefore a moderate change in E may be omitted when determining stress magnitudes, but it cannot be neglected when calculating deformations.

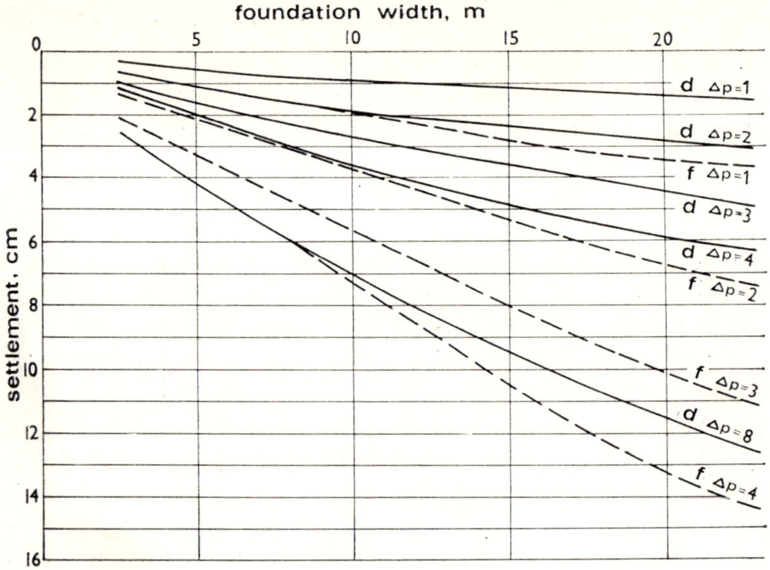

10-12. Approximate magnitudes w (in cm) of settlements of foundation structures of different widths in non-cohesive soils; $\triangle p$ — net loading of the foundation soil in kp/cm², curve *d* corresponds to a dense sand with gravel, and *f* to a fine silty sand.

cohesionless as well as for cohesive soils and for several foundation widths. The net load is that which increases the pressure in the foundation soil. Therefore the weight of the excavated soil should be subtracted from the total load.

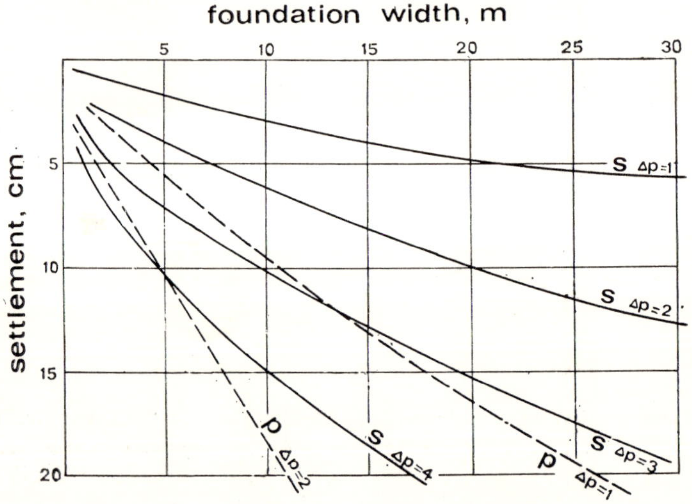

10-13. Approximate magnitudes w (cm) of settlements of foundation structures of different widths in cohesive soils; $\triangle p$ — net loading of the foundation soil in kp/cm², curve *s* corresponds to a very stiff clay, *p* — to a stiff clay.

A comparison of Figs. 10-7 and 10-9 indicates that a significant difference in the pressure magnitudes exists to a depth equal to only about one-and-a-half the considered foundation dimension. This is the depth to which the soil properties should be investigated with great care. Deeper layers produce a rather uniform settlement only.

(c) The case of a uniformly loaded area was studied in Sec. (a) and the prevalence of the settlement at the centre emphasized. The behaviour of structures built close together does not differ from this, as they tilt towards each other. But exceptions to this general rule can be encountered when one of the structures is heavy and the other is light but large and stiff. The latter structure should lean towards the former, but due to its stiffness it does not, and a very thin gap forms under its foundations in the section close to the heavy structure. Of course, the structure in the overhanging state is prone to vibrations. Thus an administration building of a steam power plant, founded on Neogene clay in northern Bohemia began to vibrate due to its proximity to a main block (Fig. 10-14). The work

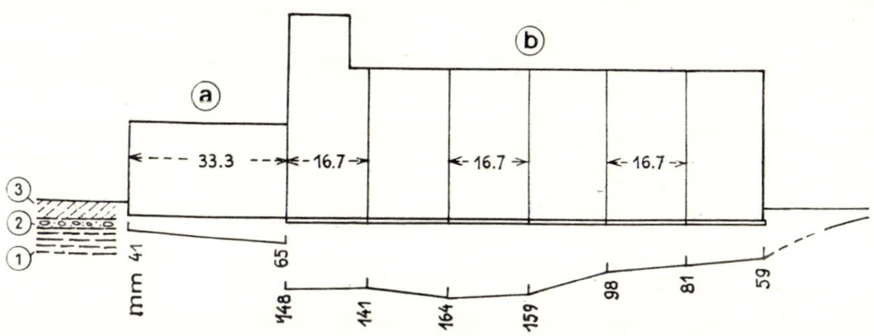

10-14. Settlement of the administration building (a) of a steam power plant, founded on Neogene clay (1), and annexed to the gables of the generating plant (b), 2 — sandy gravel, 3 — Holocene deposits. Building (a) displayed vibrations during the first several years of service. (An analogous case of settlement of a power house annexed to the toe of a high gravity dam is shown in Fig. 13-40.)

of the personnel was seriously affected. However, the remedy was simple, as the filling-up of the gap by grouting was the only necessary measure. The vibration of the structure also diminish when it settles and bears against the ground again; this is mostly accompanied by formation of cracks.

(d) Another case is encountered in towns when building up vacant space or extending the existing plants. A new structure is often built close to the old one. The soil at depth near the old structure has been compressed by the pressure of the old structure. Therefore the new structure tilts away from the old; tilting is conspicuous particularly when the structure is high (e.g. a new elevator tower, church tower, Fig. 10-15, etc.). Provision against this can be made by arranging steps in the foundation base (Sec. 10.5). The old structure often experiences vibrations as discussed previously, but it soon begins to exhibit cracks, which also appear later at a greater distance from the new structure, while vibrations begin to diminish.

Since the process of consolidation is slow at depth, the tilting of structures, vibrations, cracks and increased settlement at the centre appear only many months after the construction period.

10.2.3. Unstable foundation soils. In general, soils which exhibit quick, large settlements belong to this category. From the point of view of the mechanism of deformation the small angle at which the pressure propagates laterally is also characteristic. Loose sand, loose fill and loess are representatives of this group.

(a) Loose sand and silt, as well as poorly compacted fills, exhibit large settlements, especially when exposed to shocks which are omnipresent nowadays. Density is the decisive technical property of these soils (Sec. 3.3) and the measurement of porosity may be employed to quantify it. As stated earlier, a porosity greater than about 39 to 40 % is associated with loose sand but a lower value is to be considered if large shocks are to be expected. The difference in behaviour is of such influence that a different foundation technique is necessary. While simple footings are sufficient in dense sand, deep foundations are required in loose sand and silt.

Fills consisting of fragments are poor foundation soils due to the presence of large voids instead of pores. Considerable settlements occur but intensive shocks are necessary to provoke them. To provide safe foundation, fills should have a poro-

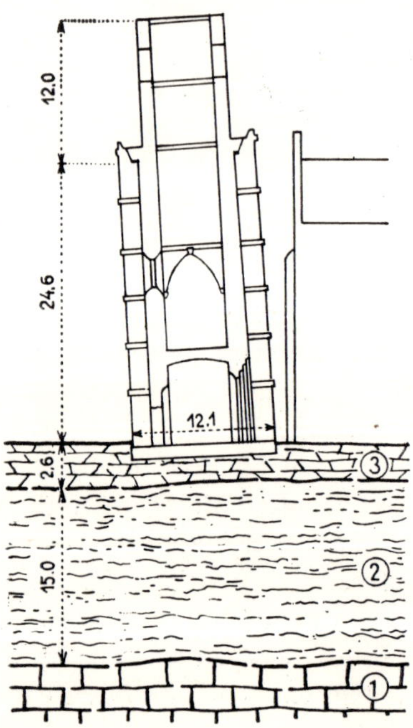

10-15. Tilting of a tower, additionally annexed to an older nave (Estampes, France). 1 — Limestone, 2 — sand with organic material, 3 — weak travertine.

sity smaller than about 32 per cent. Density can be improved by filling the voids, e.g. by cement grouting.

(b) *Loess* is a distrusted foundation soil as it is subject to abrupt settlement when saturated. Because of this, paradoxical situations occur. For example, old, poorly founded buildings do not show fissures, while new structures in the vicinity designed with great care exhibit serious damage. This results from the access of water to the foundations, often made possible by the recent practice of sloped foundation pits. Braced pits as used earlier offered less routes for surface water to penetrate the subsoil.

As discussed in Sec. 3.4, compressibility of loess varies from place to place. For example, loess blown out from Neogene clay, as is common in Brno, has a rather high clay content and therefore acquires the character of a less compressible loess loam. Properties also change with depth. Some loess layers had been exposed to weathering and to the action of water on the fossil ground surface. At the site of the power plant near Trnava in Slovakia, collapsible loess possessing the dangerous porosity of 46 per cent (dry density of 1.46 Mp/m^3) was found to a depth of 8 m. The underlying loess, 8 m thick, was less compressible, its porosity being 35 to 39 per cent (dry density 1.6 to 1.66 Mp/m^3). But opposite cases are not rare. Bally (1965) reported cases in Rumania when loess layers collapsed after saturation but only at depths greater than 8 m.

The importance of knowledge of regional or even local conditions is therefore evident. The experienced engineering geologist may help by indicating the depth of a less permeable clayey layer and by proposing that the foundation level be above it in order to prevent water penetrating the subsoil.

Another question concerns the workability of loess by ramming. Water content and the content of fines are decisive factors. At the site near Trnava as reported above, the 4-m thick layer of compressible loess left after 4 m had been excavated, was compacted 40 cm by special concrete monkeys, 7 Mp in weight.

Soil mechanics analysis has introduced the "number of collapsibility" of loess. Its definition varies. The Czechoslovak Standard No. 731001 defines it as the contraction (ε, in per cent) of an undisturbed sample loaded as by the future structure, after saturation. If this number is greater than 1 %, the loess is to be considered as collapsible. For example, loess with a water content of 10 % and a porosity of 42 % attains this limit under a pressure of about 0.6 kp/cm^2. Up to a pressure of 2 kp/cm^2 the relationship between the pressure and the void ratio (Sec. 3.3) is almost proportional.

10.2.4. Bearing capacity of soils. As noted earlier, the practice of foundation has separated the questions of settlement from those of bearing capacity. When the bearing capacity of a soil is reached, the contact pressure below the foundation is so large that the shear strength of the soil is exceeded and soil is squeezed out from below the foundation. This manifests itself by a steady and accelerating subsidence of the structure. The problem is again more of a soil mechanics character and the engineering geologist can contribute to its solution by defining the geological conditions, the ground-water level and its possible variations. But, as stated above, he should also understand the principles of statical relations.

A compressed wedge develops below the loaded foundation and the surrounding soil is pushed sidewards. This is resisted by the strength of soil. Statical analysis of this process yields the relation:

$$q = \gamma B N_\gamma + \gamma D N_q + c N_c$$

where q is the bearing capacity (in kp/cm² or Mp/m²), γ the unit weight of soil (in kg/cm³ or Mp/m³) and c is the apparent cohesion of the soil (in kp/cm² or Mp/m²), according to Sec. 3.5. The magnitude of the coefficients N depends on the magnitude of the shear strength angle, as shown in Fig. 10-16.

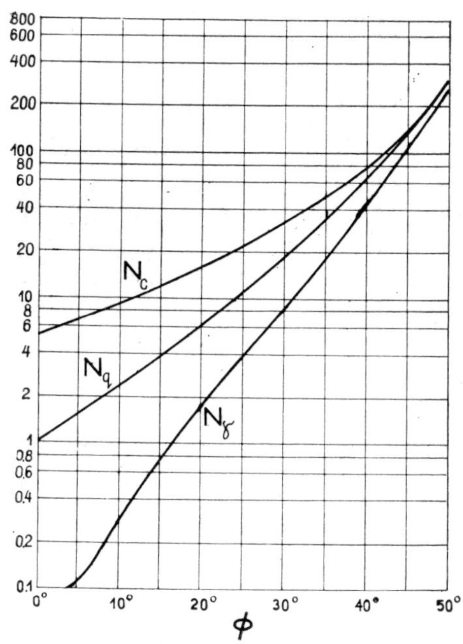

10-16. Diagram showing the values of the coefficients N. Shear strength angles Φ are indicated as abscissae (after Meyerhof 1951).

Three groups may be distinguished from the practical point of view:

(a) *Non-cohesive soils.* Their c is very small, but their increase in strength owing to increasing pressure σ is manifested by a considerable value of Φ. When the foundation width and depth increase, σ also increases, as the slip surface extends to a greater depth. Therefore, the bearing capacity increases considerably with the two dimensions, as is shown in the above equation. The magnitudes of Φ of different soils are decisive for the values of N_γ and N_q. Theoretically, if $\Phi = 0$, then also $N_\gamma = 0$ and $N_q = 1$, while at $\Phi = 35°$ (dense, fine, silty sand), N_γ is about 18 and N_q about 34. It is easy to calculate that with larger values of B and D the bearing capacity is so large that it cannot be employed in practice, as great settlements would result.

In practice a factor of safety of a magnitude of approximately 2 is to be adopted.

The influence of the magnitude of γ should not be neglected. The effective unit weight of soils is about 2 Mp/m^3, but only about 1 Mp/m^3 under water (Sec. 3.3). Therefore the danger of exceeding the bearing capacity of sand occurs with narrow and shallow foundations of walls, especially after the ground water table has risen. The wall sinks into the ground, tilts and collapses.

(b) *Fast consolidating, fine-grained soils.* As shown in Sec. 3.3 and 3.5, loamy soils with a rather small clay content are characterized by a rapid rate of consolidation. Their strength increases with the loading. Consolidation sheets (Sec. 10.5) are set beneath the foundation in order to advance the water drainage from the soil. Consolidation sheets are almost indispensable in fine-grained alluvial deposits. When the effect of consolidation is considered, the essentials of the bearing capacity of fine-grained soil are similar to those discussed in the foregoing paragraph, with the exception of the Φ angle which is smaller (for example, $\Phi = 25$ degrees, yielding $N_\gamma = 3.8$ and $N_q = 12$).

When applying this approach in practice, misunderstandings may occur. After exposure by excavation the soils are so weak that they yield under foot pressure (about 0.35 kp/cm^2). The facts that the width of the sole is extremely small, that its depth is zero, and that the consistency of the soil will improve after consolidation, are easily neglected by critics.

Caution is of course necessary. If the column of weak soil left beneath the foundation is too high compared with the width of the foundation, soil could be pressed sideways. Therefore the depth of the weak soil left under foundations should be restricted, which affects the foundation level (Fig. 10-17).

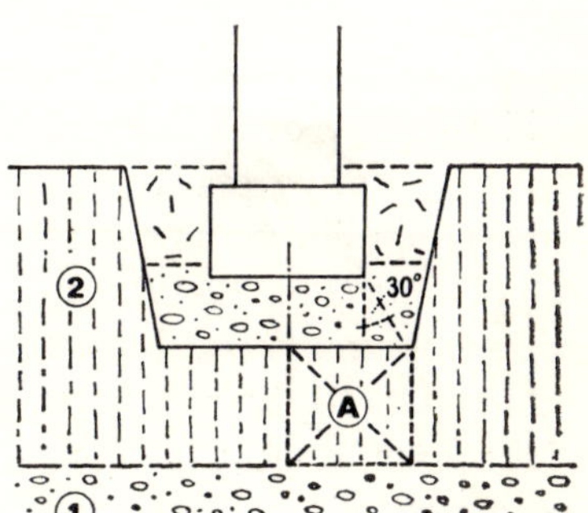

10-17. Approximate estimation of the width and thickness of a consolidation and bearing gravel layer beneath the foundations in Holocene deposits. The estimation is based on the width/height ratio of rectangle A in Holocene soil (2) left beneath the bearing gravel layer, 1 — sandy gravel.

(c) *Slowly consolidating cohesive soils*. In Czechoslovakia, typical representatives of this group of soils are Neogene clay and thick deposits (boulder clay). of glacial origin. The process of adjustment of particles to the increase in pressure is slow. The delay in the growth of strength compared with the construction period is so long that the increase in strength at the end of the construction period is often negligible. This is taken into consideration by adopting $\Phi = 0$. Then $N_\gamma = 0$, $N_q = 1$ and N_c equals about 5.4. If a universal factor of safety is assumed, its magnitude should be about 2.5 to 3. As no effect of consolidation is presumed, many standards relate maximum safe bearing capacity to the consistency of the soil. The most usually applied values of allowable pressure are of the order of 0.5 kp/cm^2 for soft clay, 1 kp/cm^2 for medium stiff clay and 2 kp/cm^2 for stiff clay. The pressure of the excavated soil can be added to these values.

The settlement of the structure indicates that the process of consolidation occurs and that the bearing capacity of the foundation soil increases. That is why failures occur in structures which are loaded by a relatively large service load a short time after being built (storage bins and reservoirs, Fig. 10-18).

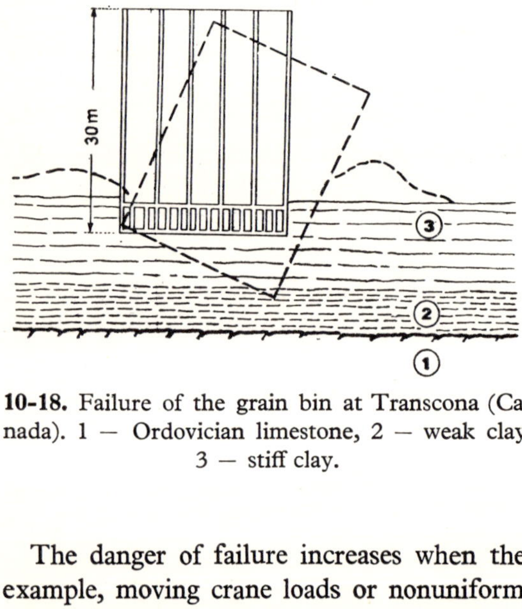

10-18. Failure of the grain bin at Transcona (Canada). 1 — Ordovician limestone, 2 — weak clay, 3 — stiff clay.

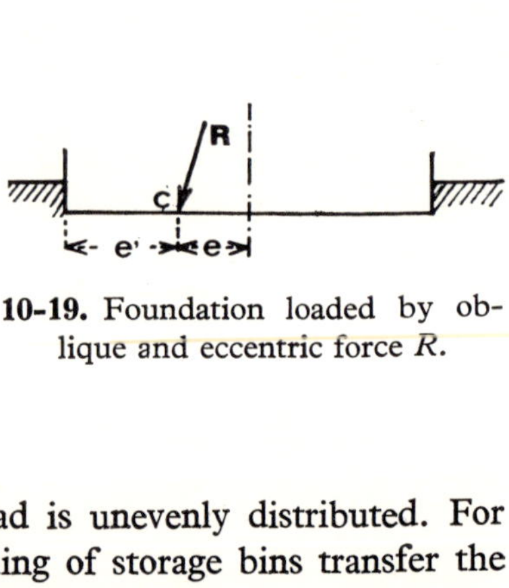

10-19. Foundation loaded by oblique and eccentric force R.

The danger of failure increases when the load is unevenly distributed. For example, moving crane loads or nonuniform filling of storage bins transfer the point of application of the resultant. Often the loading is not only irregular but also oblique (Fig. 10-19), since horizontal forces are also present (wind pressure, earth pressure, lateral impacts of cranes).

An exact calculation is complex and can be accomplished only by the application of the finite element method. As compared with the vertical loading, the vertical component of the oblique load should be smaller since the horizontal component helps to press the soil up from below the foundation. For a crude estimation the criterion may be used that the reduction in the vertical component (as compared with the vertical loading) depends on the angle of inclination of the load with respect to the normal. For example, with the inclination of 10 degrees, the reduction factor is of the order of 0.5 to 0.6 with non-cohesive soils and of the order of 0.8 with cohesive soils; or with the inclination of 20 degrees the magnitudes are 0.2 to 0.4 and 0.6, respectively. The influence of asymmetry can be incorporated into the estimation by substituting 2 e' for B in the above relations (Fig. 10-19). This estimation is based on the assumption, proven by several findings, that at least the portion of the foundation area symmetrical to the point c is capable of resisting the asymmetrical load.

From the practical viewpoint, a noteworthy fact is that the most stable situation does not require the point of application of an oblique force to be at the midpoint of the foundation width. Due to the oblique orientation of the resultant R soil can be pressed up in one direction only (to the left side in Fig. 10-19). The greatest bearing capacity is therefore associated with an asymmetrical position of the point of application of R (to the right of the midpoint in Fig. 10-19).

10.3. Foundation and ground water. The presence of ground water affects the problems of foundations to a considerable extent. A change in the foundation method is often necessary and construction costs may rise without yielding beneficial results.

Cases occur, indeed, when a foundation below the water table is inevitable as, for example, with the tippler hopper of power stations, pumping stations, etc. However, the foundation is often situated below the ground water table due either to the inflexibility of the architect or to a false determination of the water table.

Every soil contains water necessitating the technique of waterproofing. However, damp-proof insulation cannot support the head of a few decimetres of water, so that expensive, but not always safe, heavy water-proofing is necessary, bearing fully against a reinforced concrete plate. The latter should be capable of resisting the hydrostatic pressure which acts on the water-proofing.

For these and other reasons it is reasonable to locate at least the major portion of the foundation above the water table. If the total height of the foundation can be situated above the water table, excessive costs associated with the construction can also be saved, although the higher foundation level often makes pile foundation necessary.

If the water level at the site can be lowered by drainage into the sewage, suitable drainage can be effected by a set of drains spread over the foundation area and a deeper main drain at the periphery. If the inflow is small, the function of the drains may be taken on by a gravel sheet. Horizontal boreholes can also serve as a drainage system if the site is inclined, but their cost is rather high.

Certain care should be taken with clay. Although seemingly dry during construction, the foundation pit may be accessible to water owing either to the presence of fine sand layers, or a net of fissures, or, as is frequent, to a bad compaction of the backfill of the corners at the periphery of the former sloped excavation pit. Therefore even a clay subsoil necessitates drainage of the foundation.

If the gradient of the area does not make a gravitational drainage possible, drains are led into one or several pumping wells. As heavy, extensive water-proofing can be replaced by a light one, this solution is economically effective, provided that the pumped inflow does not exceed 10—20 litres per second. The storage capacity of the wells, including additional storage pits, should be great enough to prevent water rising up to the level of the water-proofing during a possible break in the electric power supply.

The danger of a potential rise of the ground-water table should also be anticipated since the bearing capacity of soils could be reduced because of it (Sec. 10.2.4).

In contrast, a *lowering of the water table* promotes undesirable settlements, and cracks appear in neighbouring buildings. Settlement is caused by an increase in the effective pressure in drained soil. Even the brief pumping of a foundation pit can cause damage to the buildings in the vicinity. Widely known is the case of Venice, where the intensive pumping of artesian water by the industrial plant at Mestre increased the settlement from about 1.5 to about 5 mm a year.

The question of the lowering of the water table is also of great importance for structures supported by wooden piles as the decay of wood can be caused. But this danger should not be overemphasized. After construction of a new section of a thermal power plant in Brno, vibration of the old turbo-generators commenced. This phenomenon was ascribed to the dewatering of the foundation pits in the new section of the plant, which, as alleged, caused the decay of the wooden piles. But investigation of the piles has proven that they were well preserved in alluvial clayey deposits. In reality the vibration was caused by the small subsidence of piles because of a negative friction resulting from the vicinity of new heavy boilers. However, a permanent lowering of the water table can involve the decay of piles.

Corrosive water, present at the site, requires a particular consideration. Investigation of the chemical nature of water and of its effect on the foundation structure are out of the scope of this book, but the engineering geologist has to deal with several associated questions.

Firstly, he has to give advance warning of the probability of the occurrence of corrosive water on the basis of the general knowledge of the geology of the site. For example, the presence of pyrite suggests the occurrence of sulphate in the water. Secondly, the intensity of the source of corrosive capacity should be considered; this cannot be expected from the expert in chemistry. For example, the corrosiveness of the subsurface water of residual soil of igneous rocks will be relatively low, since corrosive concentration may diminish after several measures have been taken (drainage of moorlands, treatment with limestone). In contrast, the corrosive activity of water flowing out from pyrite-bearing rocks will be rather constant. The case of the site of a power and heating plant at Brno can be quoted, where the sulphate concentration was found to be of the order of 306 mg/l in a boring in 1929, during the construction of the first section, but only 161.9 mg/l was found in the same borehole in 1949. The decrease in concentration resulted from the increased circulation of water in the alluvium due to the construction of a weir on the river. While foundation on wooden piles was preferred in 1929, concrete piles were used in 1949.

According to its chemical nature, water can be classified as soft and highly mineralized.

Three groups of soft water occur:

(a) soft water with a low mineral content,

(b) soft water containing dissolved carbon dioxide, and

(c) other acid waters.

Pure *soft water* dissolves calcium bicarbonate and also, gradually, silicates and aluminates of the cement binder. Frequently the effect of agressive carbon dioxide participates, even when its content is relatively small. Water of this kind is present in areas of igneous and metamorphic rocks, where water cannot be enriched by mineral constituents. Soft, acid water flowing from moorland is dangerous due to the high content of humic acid.

The degree of acidity or alkalinity of water is specified by the hydrogen ion concentration. An excess of hydrogen ions causes the water to act as an acid, and an excess in hydroxyl ions as alkaline. Neutral water requires an equal content of both ions, as in pure water, and contains 10^{-7} ions of each of the two groups in one litre at a temperature of 22° C.

The negative logarithm of hydrogen ion concentration indicates the degree of acidity (the sum of the content of both groups is constant). This quantity is designated as *pH* and can attain values of zero to 14. When it is equal to 7, the water is neutral; acid water has a *pH* lower than 7 and alkaline water higher than 7.

During the foundation of machine works at Žďár (Moravia, in gneiss), ground water exhibited the hardness of 1.46 grades and pH = 6.28. With a content of 18.3 mg of HCO_3^- and 9 mg of SO_4^{2-} per litre it was reactive. But drained water flowing after the construction has been finished displayed pH = 6.9.

The concrete pipes of the drainage set around the sanatorium building at Starý Smokovec (High Tatra Mts.) were decomposed after several years and the lower section of the pipes dissolved, while the upper section remained untouched. Water coming from granite debris was essentially pure atmospheric water and its carbon dioxide content was produced by the decay of organic matter of the moorland.

Highly mineralized water is a hard water containing an excessive quantity of compounds. Sulphate- and carbon dioxide containing waters are frequent in Czechoslovakia.

Sulphate-containing water forms gypsum ($CaSO_4 + 2\,H_2O$) when in contact with calcium hydroxide. Gypsum results in expansion of concrete. More serious is the effect of calcium aluminum sulphate (Candlot salt), which crystallizes with 31 water molecules and therefore increases in volume. Crystals of felt-like structure disintegrate the concrete and turn it into a slurry. Sulphate-containing water is usual in rocks containing pyrite and other sulphate compounds. The peripheral parts of coal measures also contain pyrite and marcasite and waters flowing from them are rich in sulphate.

In the area of Prague the highest sulphate content occurs in water coming from the joints in the Ordovician shales, but water of the valley terrace is also mineralized to a high degree. Greater mineralization and higher sulphate content appear in ground water near valley slopes, decreasing towards the river. A profile across the alluvial deposits is plotted in Fig. 10-20. Wells and boreholes, the amount

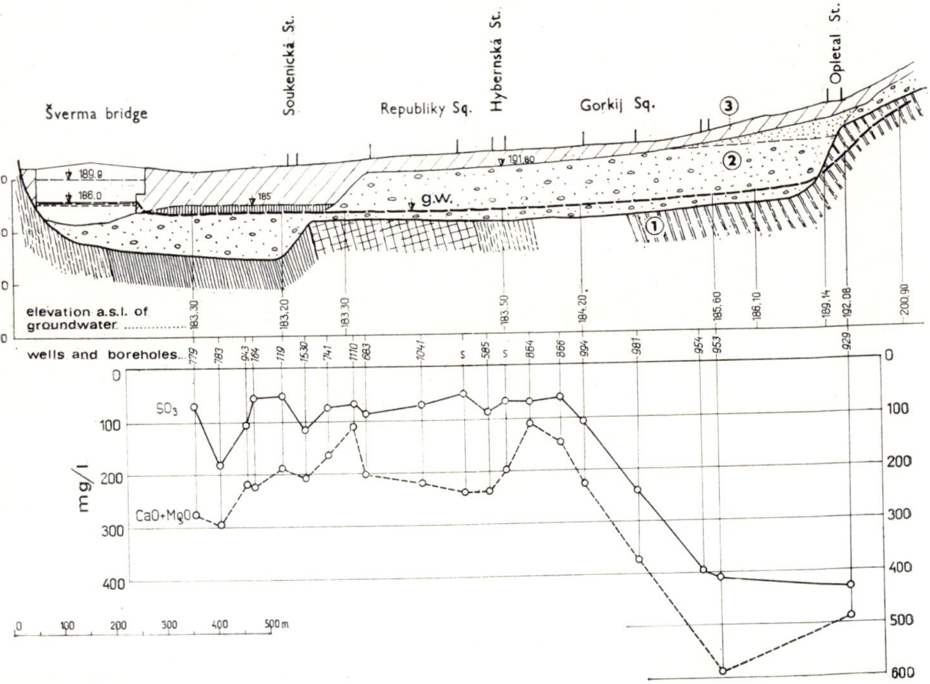

10-20. Diagram showing the dissolved solids content of the ground water in the wells and boreholes in Prague (Záruba, 1948). 1 — Ordovician shales and quartzite, 2 — sand and gravel, 3 — fill.

of SO_3^{2-} and the sum of CaO + MgO in mg per litre, contained in water, are indicated. The diagram illustrates the prevalence of mineralized water in wells near the rock slope which borders the valley terrace.

Water may become corrosive after being in contact with rocks containing harmful ingredients. For example, at the site of a bank construction in the Old Town of Prague, water pumped from terrace gravel showed 154 mg/l of SO_3^{2-}, but after the excavation was deepened into Ordovician shales, the SO_3^{2-} content rose to 271 mg/l, as water flowed along the joint planes coated by gypsum.

An increased amount of sulphate is not rare in water percolating through waste fills of ash and slag. Thus 1603 to 6700 mg/l of SO_4^{2-} was measured near Prague. Old mining dumps in the environment of Kutná Hora enrich water with sulphates due to the presence of pyrite and chalcopyrite.

Cement with a low lime content is preferred in an environment rich in sulphates. The admissible concentration of SO_4^{2-} ions depends on the conditions under which water acting on concrete will rejuvenate. For example, the Czechoslovak Standard No. 731001 allows a concentration of 360 mg/l in permeable gravel, but 600 mg/l in an impermeable soil.

Carbon-dioxide water is corrosive due to the high content of CO_2. Mineral waters belong to this category and, as their corrosive action on concrete and iron is known, their piping requires a special technique.

10.4. Foundation excavation. As also occurs in other fields of civil engineering, a particular construction stage may be more hazardous than the finished structure. This is particularly true for the stability of slopes and bottom of the excavation and for its sealing and dewatering. The engineering geologist has to face several problems of his own.

The first of them concerns the question whether the bottom of the excavation will be free of ground water or not.

Absence of ground water is always favourable for the stability of the pit. Owing to the use of mechanical devices, a sloped excavation has been preferred recently. As a temporary measure only, the slope may be inclined at a relatively higher angle than is usual with slopes of road cuttings. Dry loam is often excavated at an angle of 45 degrees. More care is necessary in Neogene clay or in Palaeogene and Cretaceous claystone. Even here the slope can be maintained over the summer period at an angle of 45 degrees. But this depends on how quickly the strength of soil will decrease both upon unloading due to excavation and as a result of dilatancy. Slope failure may be advanced by several features of geological character:

(a) Fractures give rise to collapse, as the soil resistance has been weakened along them and water is often present in them;

(b) whether the clay is fissured or not;

(c) great horizontal residual stress, present in clay, as well as stress concentration if the pit is to be excavated near the toe of a valley slope, give rise to curved slip surfaces as well as to fissures in the slope.

As a result of the above features, failure occurs several weeks after excavation (Fig. 10-21). Another phenomenon is the upheaval of the clay surface on the periphery of the excavation, so that water is prevented from flowing out from permeable overburden and weakens the clay in the slope. Horizontal displacement of the slopes involves loosening of the loamy cover at the periphery of the excavation and sliding destruction occurs.

The prediction of the behaviour of the slopes of the pit on the basis of laboratory tests on small samples only may be unsatisfactory. On the other hand, large blocks of soil can be obtained only after the excavation has begun.

(d) The stability of a slope is always weakened by the presence of even only a small inflow of water, which promotes the increase in volume of clay and its weakening. Water may be present in joints, in thin sandy layers, etc.

In built-up areas the use of braced foundation excavations are often preferred to sloped pits. Several techniques are available nowadays to protect the pit walls, be they concrete diaphragm walls or various cast-in-situ pile walls. Their execution is of concern to the engineer, but the engineering geologist is expected to describe the properties of soils, especially the presence of boulders, the depth of the rock surface, its strength, etc., in short all properties governing the performance of excavation machines.

Other necessary and important information concerns the preliminary evaluation of the *resistance of anchors* which stabilize the walls of the excavation (Fig. 10-22).

10-21. Slope failure of a foundation pit in Neogene clay. (Photograph by Mencl.)

The safe resistance of anchors is generally of the order of 10 to 15 Mp in clayey shales, but it increases considerably in rocks, in which grouting is effective. Thus even more than 50 Mp can be obtained in sandstones with shale interlayers, over 100 Mp in firm limestone, etc.

Care is necessary in built-up areas when fixing the anchors to the foundation structure of existing buildings. Small displacement of the pit walls due to residual stress in the soils augments tension in the anchors, which are therefore loaded more than expected.

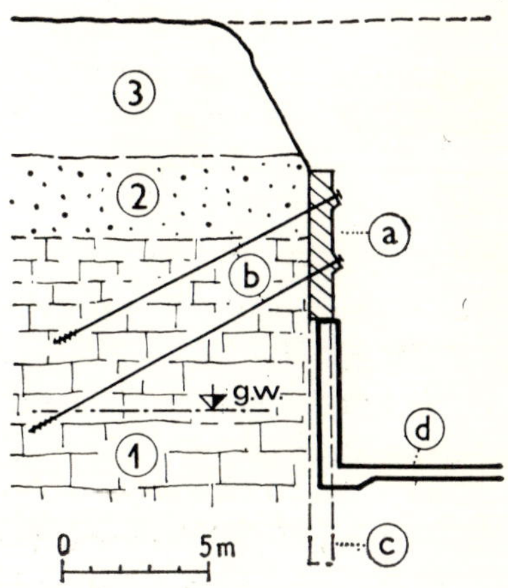

10-22. Example of an anchored wall in a foundation pit (Société Nouvelle de Construction et de Travaux, 1968). 1 — Dolomite disintegrated near the surface, 2 — Pleistocene sand, 3 — fill, a — side wall supported by anchors (b), c — piles spaced 3 m apart, bearing the side wall and the loads of the future building, d — casing of the basement.

Difficulties associated with the *drainage of the foundation excavation* as well as the danger of slope failure are added when the excavation extends beneath the ground-water level.

A great inflow of water into the foundation excavation is to be expected in the valley floor. In order to evaluate the expected quantity of water that will enter the pit in the course of construction, a pumping test is often performed during the investigation period. The inflow depends very much on the presence of thin loamy layers in the gravel, since they prevent water from percolating from the upper into the lower beds. The depression curve in gravel is rather flat, so that the slopes of the foundation pit are not endangered by seepage forces. A drainage set near the toe or a surcharge fill on the lower section of the slope are the usual measures. Nevertheless, the inflow into the pit is often so great that a sheeting is necessary, either by means of sheet-pile walls or by impermeable diaphragms. As far as possible, the sheeting reaches down into an impermeable layer. The engineering geologist is expected to describe the properties of the soils in sufficient detail for the construction of the sheeting. Remnants of trunks or boulders make the construction of the sheeting difficult.

When the bottom of the excavation is below the groundwater table and the surrounding soil is not sufficiently permeable, while a permeable layer exists underneath (e.g. gravel), the problem is more complicated. Even when the pit

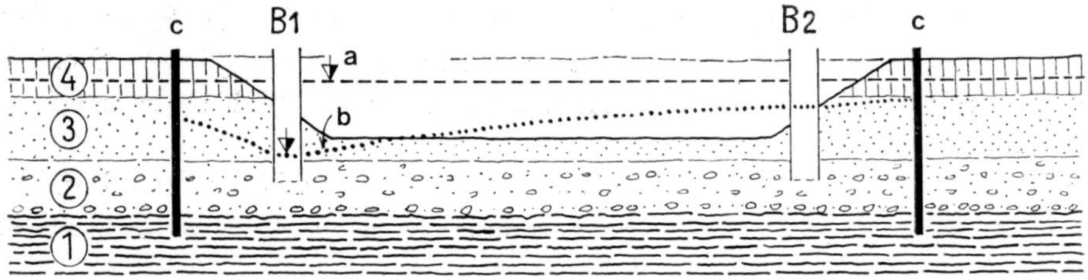

10-23. Unsatisfactory drainage of foundation excavation. Water can penetrate above the bottom of excavation in the vicinity of well B_2 where pumping has been neglected. Seepage forces loosen the fine sand in the foundations. 1 — Neogene clay, 2 — sandy gravel, 3 — fine sand, 4 — Holocene loam, a — original ground-water level, b — drained ground-water level, c — sheet wall, with many openings due to the presence of tree stems in the deposits.

is sheeted (Fig. 10-23), a small inflow of water penetrates the gravel layer and from it passes into the pit. This is sufficient to stir up and loosen the bottom of the pit and to slump its slopes. Therefore the rule is to be observed that the pumping of water should be done from the most permeable layer. Fig. 10-23 presents an example where the permeable layer was drained by means of pumping wells.

The lowering of the water table by means of pumping wells is also necessary when the pit is surrounded by fine sand, silty soils, or by a mixture of both. This arrangement is necessary since the pumping from a sump at the bottom of the pit would involve a steep hydraulic gradient (Sec. 3.5) in soil in the immediate vicinity of the pit. Again a stirring up and loosening of the soi results. In contrast to the case shown in Fig. 10-23 the range of action of a welll is small, as the low permeability of the soil makes the hydraulic gradient around the well very steep.

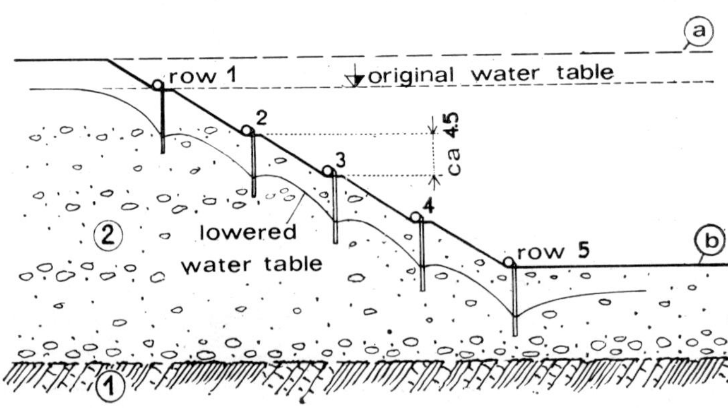

10-24. Lowering of the water table by pumping from several rows of wells shown schematically. 1 — Bedrock, 2 — sandy gravel; a — original ground surface, b — bottom of excavation.

Therefore, either deep pumping wells equipped with borehole pumps, or several rows of closely set shallow wells, are used (Fig. 10-24). A point of interest to the engineering geologist is also the required diameter of the borehole necessary for deep pumping wells.

In order to increase the discharge capacity of pumping wells a vacuum technique is used in soils possessing a coefficient of permeability of the order of 1×10^{-4} to 1×10^{-5} cm/s. In still less permeable soils (silt), the discharge velocity of water towards the wells can be increased by an electric current, a method known as drainage by *electroosmosis*.

10.5. Choice of foundation method. This problem is the object of analysis of the designer, but the engineering geologist cannot neglect it if he wishes to avoid mistakes during the investigation of the site. The most important question to be answered concerns the choice between the spread foundation and deep foundation (e. g. pile, pier and well or caisson foundations).

Spread foundation is generally of advantage when the bearing capacity of the soil is sufficient, when large differential settlement of the structure is not to be expected, and when ground water cannot influence the structure or disturb the construction. As the differential settlement can be reduced by increasing the stiffness of the foundation, the bearing capacity, the presence of ground water and the cost of piles remain the most important items.

In order to consider the necessary stiffness with regard to the differential settlement two cases may be distinguished (see also Sec. 10.2): (a) The shallow, surface layers (e.g. residual soils, Holocene deposits or debris cover) are responsible for the differential settlement; (b) the deformation of deeper layers is also to be considered, as it is a typical case with Tertiary deposits.

(a) When only shallow layers are of influence, simple foundation footings or beams, with a width just sufficient to form a footing needed for the exact location of the walls of the superstructure, are not always enough. Frequently the widths of foundation structures must be determined by calculation with the aim of obtaining equal settlements of walls or columns.

The columns of industrial as well as public structures are often far apart and therefore loads are great in them. Large foundation areas are necessary. In order not to break, footings or beams should be thick enough, which necessitates an increase in foundation depth. As a result, more subtle foundation structures, supported by piles, are often preferred.

In contrast, houses and office buildings possessing closely spaced walls or columns make a foundation on beams easy and economical. When the foundation soil is compressible, while its bearing capacity is sufficient, the foundation beams can contribute to the stiffness of the superstructure. A simple foundation raft is usually an economical solution. When creating a well-compacted, firm gravel layer underneath the beam (Fig. 10-25), the pressure is distributed not only laterally but also in the longitudinal direction of the beam, and the differential settlements decrease. The statical effect of the gravel layer is the larger the better compacted it is. Gravel layers compacted by vibration possess a modulus of deformation smaller than 500 kp/cm² and their capacity to distribute the pressure is relatively low (see the angle of 30 degrees in Fig. 10-25). Therefore, if a more

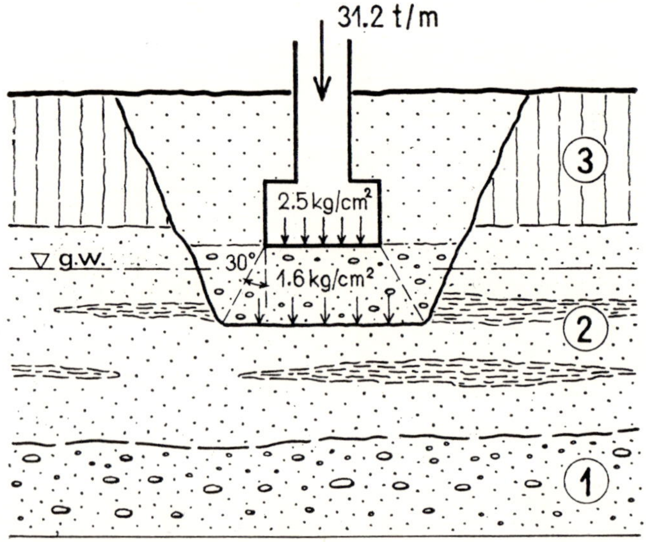

10-25. Gravel layer distributing the foundation pressure and contributing to the stiffness of the foundation structure. 1 — Sandy gravel, 2, 3 — Holocene deposits.

intensive effect is to be attained, tamping is necessary. The modulus of deformation increases to about 800 kp/cm² and the angle in question exceeds 45 degrees. Due to difficulties resulting from the use of heavy tamping devices the less intensively compacted layers are often preferred while the width and thickness of the foundation beams are increased. The task of the engineering geologist is to indicate the modulus of deformation of the gravel layer, for which purpose field loading tests are often used.

The gravel layer loses its raison d'être if it cannot assist in spurring the consolidation of the subsoil, for example, of dry loess, whereby the surface water may percolate through it downward to the loess. However, this does not apply for cases where wet loess loam is present.

When the shallow layer below the foundation is not only compressible, but also exhibits a low bearing capacity (e.g. Holocene deposits) and if the pile foundation is not a useful method, the necessary beam width is so large that a foundation mat results. As shown in Fig. 10-26, uniform settlement cannot be warranted in this case either. A heavy grain storehouse exhibited settlement near a gable

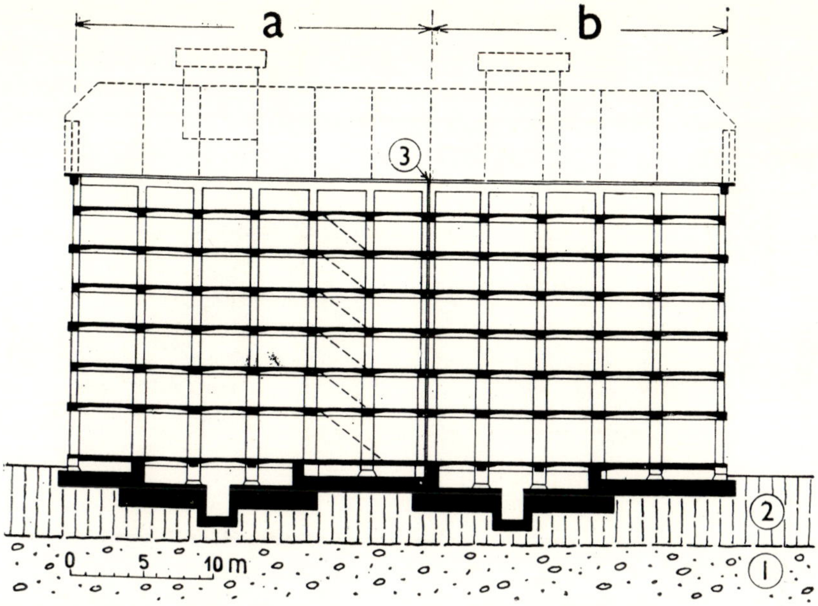

10-26. Tilting of block (b) of the building founded at a shallow depth on loess loam (2). 1 — Sandy gravel, 3 — opened dilatation joint.

(on the right-hand side in Fig. 10-26), which was greater by 7.8 cm than that of the other part of the building. This was due to the non-uniform thickness of the loess loam left under the foundation. When designing the thickness of the mat the stiffness as well as sensitivity of the superstructure towards the differential settlement should be observed. Thus, for example, brick buildings with concrete floors are stiff enough and do not suffer from differential settlement as compared with built-up concrete panel structures which are relatively fragile. Therefore a monolithic basement is often preferred in the latter case, thus forming a foundation box.

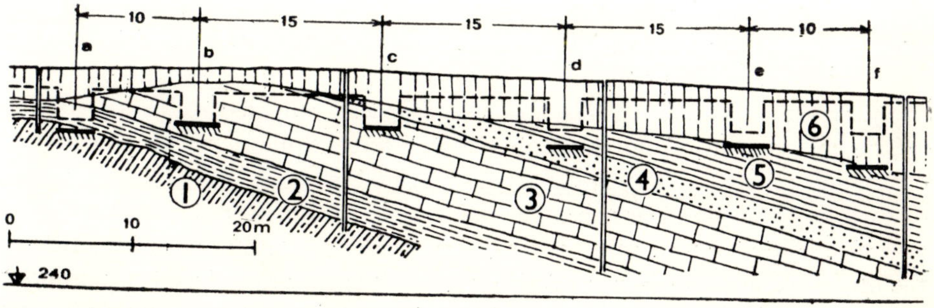

10-27. Building site on a heterogeneous foundation soil. 1 — Algonkian slate, 2 — kaolinitic clay (weathered slate), 3 — Cretaceous sandstone, 4 — clayey sandstone, 5 — Turonian marl, 6 — loess loam.

When the quality of the foundation soil varies from place to place, the differential settlement should be calculated and the lay-out of structures, the dilatation joints and the foundation method must be adapted to the geological conditions. Fig. 10-27 shows the site of an industrial plant with nonuniform foundation

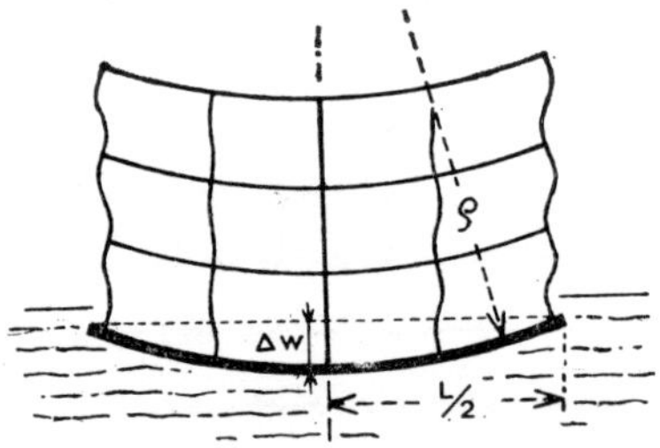

10-28. Schematic plot of parameters important for na approximate evaluation fo the allowable deformation fo the structure.

conditions, in the area of Cretaceous formation, north-east of Prague. Instead of the initially proposed uniform depth of footings, various foundation depths were adopted. In this way weathered layers of Algonkian slate (beneath row *a*) as well as loess loam (rows *e* and *f*) were avoided as foundation soils. Anyhow rocks of different compressibility have still remained as a subsoil. In order to obtain a rather uniform settlement, various widths and foundation depths of individual footings were proposed, and the firm sandstone in the foundation was loosened by drilling and wedging to a depth of 40 cm, in order to equalize the settlement with that experienced on marls.

(b) A somewhat different approach is necessary when the foundation soil is compressible to a large depth, as are Neogene clays, deposits of deltas, etc. The compression of deep layers is to be added to that considered in the previous paragraphs. This appears also with individual footings or continuous beams. In many cases the bearing capacity of Neogene clay makes a pressure of 2 kp/cm^2 admissible, but the width of foundations is governed rather by the calculated settlements. As in case (a), the greater the load, the larger footing is necessary. But the larger the footing, the deeper reaches the effect of loading. Therefore, even under the conditions of an equal contact pressure, the larger footing settles more than the smaller one. In order to obtain uniform settlement the contact pressures under large footings should be reduced as compared with small footings.

The enlargement of individual footings or of continuous beams also prevents occurrence of unequal settlements at sites located on slopes. As a smaller amount of earth is excavated near the outer border of a platform, the settlement of the structure is greater near this border than at the opposite inner border. Although the difference in settlement is relatively small, it is easy to reduce it by proposing larger foundations at the outer boundary of the platform. This solution is a prerequisite when the structures are sensitive to differential settlement (e.g. gas tanks).

Large foundation mats exhibit dish-like settlements (Sec. 10.2) and two close structures tilt towards one another. Since floor structures are incompressible in a horizontal direction, the dish-like settlement causes the outer columns to suffer bending deformations (Fig. 10-28). In the same way tensile cracks originate in the walls of brick buildings. Therefore the allowable radius of curvature (ϱ) of the settlement curve of the building depends on the length (L) of the latter and is of the order of $\varrho = 250\ L/2$ with most types of structures. If necessary, the dish-like settlement of a structure can be reduced by loading its corners, as was done for the building of Moscow University (Fig. 10-29, NIIOPS, 1964), or as has previously been practised by the builders of ancient castles. An artificial gravel layer of varying thickness may also assist in reducing the differential settlement as used e.g. with the foundation of the building of a hotel in Brno (Fig. 10-30).

When the curvature of the structure cannot be tolerated, it can be divided by

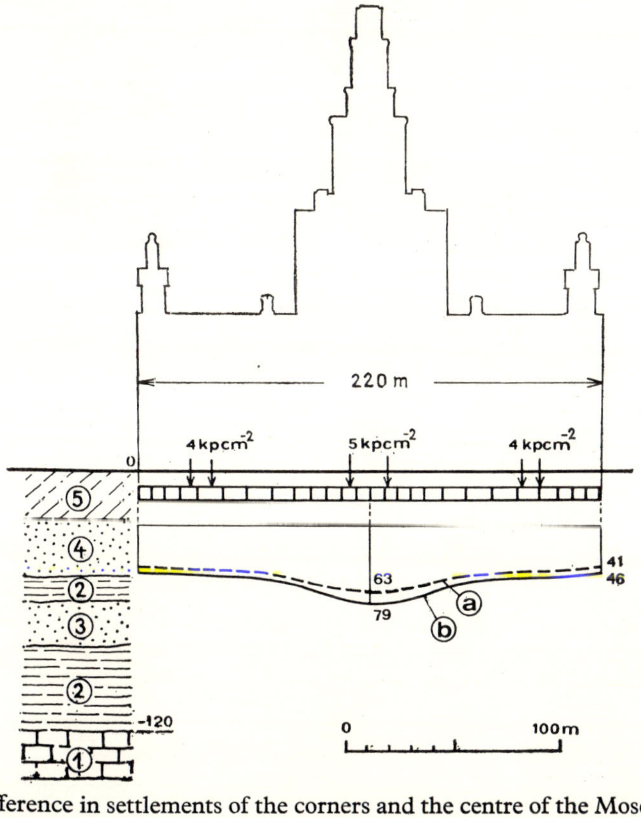

10-29. The difference in settlements of the corners and the centre of the Moscow University building was reduced by about 15 mm due to the loading of the corners. a — Curve of settlement (in mm) after the construction, b — 8 years later; 1 — Carboniferous limestone, 2 — Jurassic clay, 3 — Jurassic sand, 4 — Cretaceous sand, 5 — sandy loam and moraine. (After NIIOPS, 1964.)

dilatation joints into blocks, the length of which obey the requirements of allowable curvature. This is especially important with rigid structures, where the allowable curvature is more restricted. For example, $\varrho > 1000\ L/2$ is required with monolithic reinforced concrete bunkers, and the length of blocks is therefore smaller.

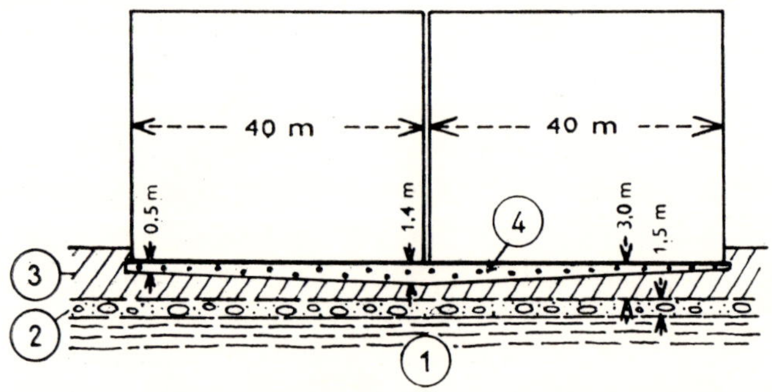

10-30. Foundation of the hotel building at Brno using a gravel layer of uneven thickness. 1 — Neogene clay, 2 — sandy gravel, 3 — Holocene deposits, 4 — compacted gravel layer.

As a result of a dish-like settlement the dilatation joints close near the top of the structure. This should be respected by allowing the joints to gape slightly in order to avoid the crushing of blocks at the contact. The tilting of close structures can also be eliminated by arranging an irregular foundation, as was done with the bunker shown in Fig. 10-31. A varying thickness of loam, left under the foundation may also help to obtain a uniform settlement. This is used when the new structure is expected to tilt away from the neighbouring older one.

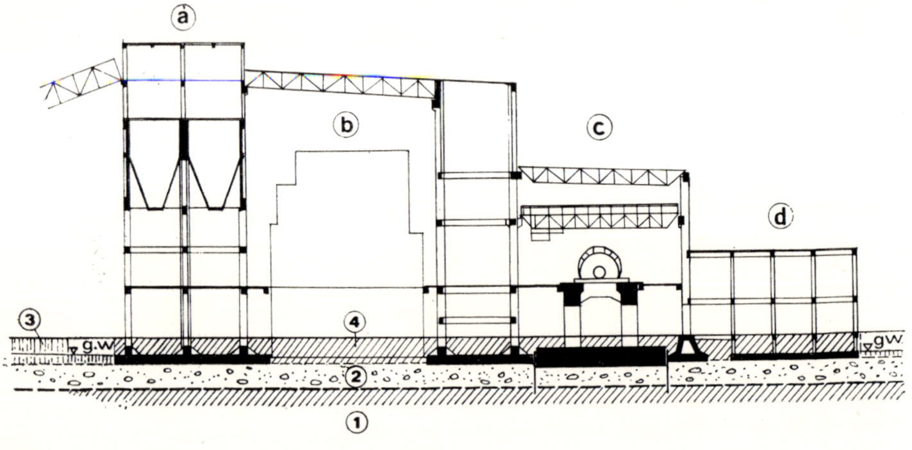

10-31. Foundation of the generating blocks of a thermal power plant on slabs and beams. 1 — Neogene clay, 2 — sandy gravel, 3 — clayey sand, 4 — filling, a — bunkers, b — boiler house, c — engine hall, d — control room.

When the soil is more deformable than the Neogene clay (E equal to about 150 kp/cm^2, see Sec. 3.4.3), the tilting of blocks towards one another in the dilatation joints would be so intensive as to involve damage to the mechanisms of the plant. A deep pile foundation or a foundation box are to be preferred (Fig. 10-32).

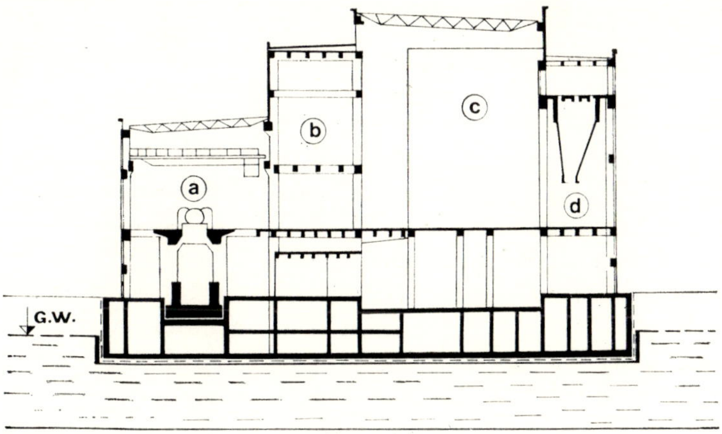

10-32. Foundation of the generating blocks of a thermal power plant on a foundation box. The ground consists of young deposits to the depth of 140 m. a — Engine hall, b — water system, c — boiler house, d — bunkers.

Because some designers are not acquainted with the problems of constructing foundations under the conditions in question, the engineering geologist should also draw their attention to the following: As shown in Fig. 10-9, the contact pressure between a stiff structure and the foundation soil is not uniform; it is of a saddle-like course. In several cases the designing engineer tends to derive benefit from the stiffness of the superstructure with the aim of reducing the thickness of the foundation. In this case, the reactions transmitted by the foundation to the walls or columns at the periphery of the superstructure are greater than if the foundation structure was stiff. Therefore, in order to avoid cracks, the bonding between the peripheral walls or columns and the inner parts of the superstructure should be stronger than is generally assumed.

A pile foundation, although rather expensive, helps in overcoming difficulties associated with foundations on weak soils. Piles transmit the loads to the deeper layers, which generally offer a higher bearing capacity, smaller settlements and especially smaller differential settlements. The base of the foundation structure can be located at a higher level so that problems connected with the ground water can be avoided.

Here and there the opinion may arise that a pile foundation does not require as detailed an engineering geological investigation of the site as does a spread foundation, but this is a mistaken opinion.

When a rather firm layer is present at depth below the base of the structure, the piles can be embedded in it. It may be a layer of Pleistocene gravel, a thick layer of Tertiary sand, or bedrock. The task of the engineering geological investigation is to check that the layer really exists below the entire area in question at the assumed depth and that its thickness is sufficient to support the piles. In fact, the features in question are checked during piling operations, but this is too late. For example, piles have been used on the assumption that a thick gravel layer was present. During ramming the piles penetrated through the gravel layer into the underlying clay, in which their bearing capacity was unexpectedly low. Therefore even simple cases, when the resistance of the pile can be deduced from the ratio of the penetration depth to the number of hammer blows necessary, should be interpreted from the point of view of the geological situation. This deduction is of great value for permeable cohesionless materials, but is entirely misleading for fine-grained soils.

The engineering geological investigation of fine-grained soils with low permeability is even more important. The skin friction presents the important component of the pile resistance and therefore the increase in mechanical properties with

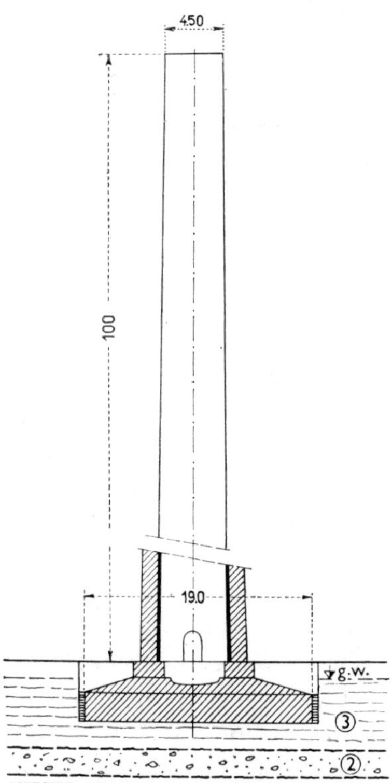

10-33. Foundation of a factory chimney on a slab. 1 – Neogene clay, 2 – sandy gravel, 3 – Holocene deposits.

depth is of great interest. The length of the pile is determined on the basis of geological investigation and should be checked by load tests.

Special attention should be given to the foundations of high structures. Their superstructure is usually very stiff and therefore great settlements should be allowed for. Thus, the U.S.S.R. standard allows a settlement of 30 cm for the foundations of factory chimneys (Fig. 10-33). The centre of gravity of factory chimneys does not lie at great height, but in about the lower third of the shaft. The situation is more delicate with tall buildings (Fig. 10-34). In order

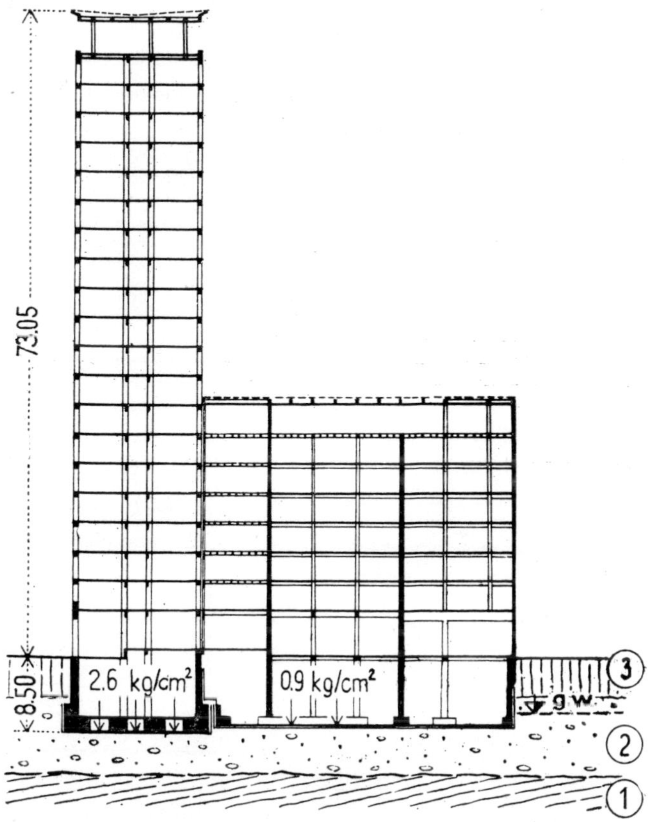

10-34. Foundation of a tall building in Vienna on a box (Pfeffer, 1955). 1 — Neogene clay, 2 — sandy gravel, 3 — loam and filling.

not to tilt, their foundation width should generally be of the order of 1/4.5 of the total height, provided that the foundation soil is firm enough (gravel, dense sand or stiff clay) and the foundation depth is about 6 to 8 m. If rock is present in the foundation, anchoring is possible and a narrower width can be allowed. A deep foundation makes the reduction in foundation width possible. Architects are often opposed to the widening of the basement, the extra width being incorrectly concealed below the ground surface. On the other hand, objections to the cone-shaped widening of shafts of high factory chimneys, television towers etc. do not arise. This straddling of the structure provides a stiff foundation. The 540-m high television tower in Moscow was founded on an annulus 74 m in diameter at a shallow depth on Jurassic clay.

10.6. Investigation of the building site. Graphs in Figs. 10-7 and 10-8 show the depth to which the effective pressure extends below the structure. The same is expressed in Fig. 10—35, in which the volume of subsoil compressed more than

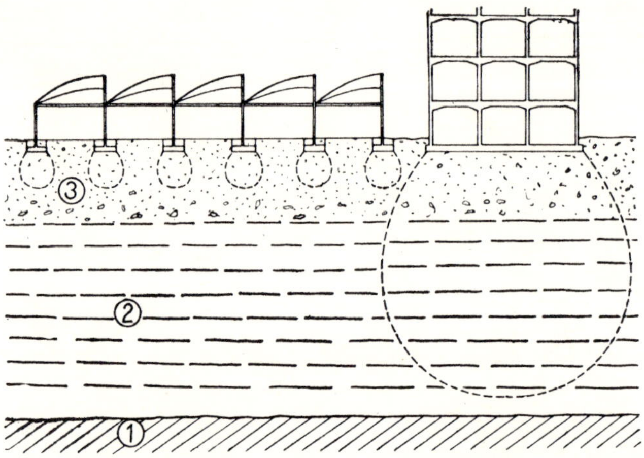

10-35. Schematic diagram of areas to which the statical influence of foundations reaches. 1 — Bedrock, 2 — Tertiary clay, 3 — sandy gravel.

20 per cent of the contact pressure under the footings is shown. The intensity of loading is also of importance and the subsoil should therefore be investigated to a depth of 30 m, or even more where large, heavy structures are concerned. As drilling is expensive, the determination of the number of boreholes and their depth should be given in charge of the engineering geologist who is best competent to estimate the necessary number of holes to be drilled to the bedrock and in the overburden.

As in other spheres of engineering geological practice, boreholes should be located so that the drawing of cross-sections through the given site is possible. Priority should be given to the profiles normal to the contour lines of the ground. There is a danger that the foundation soil might be disturbed by the making of boreholes, especially when large uplift is present, and the engineering geologist should beware of this.

10.7. Evaluation of the site. On the basis of the results of investigation the engineering geologist is expected to supply information indicating the suitability of the site for the particular construction. The foundation costs and the associated increase in costs of the superstructure are also desicive factors, but they cannot be estimated by the engineering geologist, particularly if their presentation is required as a percentage of total construction costs, or even with respect to the technical or economical aspects of the work. A site suitable for the construction of a thermal power station may not be suitable for a housing scheme, since the required foundation depth of a power plant is greater. Therefore, the evaluation presented by the engineering geologist of the suitability of a site cannot be considered as conclusive. Nevertheless, it may provide points for discussion.

Schemes of classification of the suitability of a site, based on geological conditions, may be different. The authors proposed a system which has been incorporated into the Czechoslovak standard for foundation soils. According to this scheme sites can be classified as follows:

(a) A very suitable site, providing the foundation soil does not present any problems, and a small width of footing is possible. Sites on, for example, easily workable rocks of low compressibility, or dense terrace gravel, belong into this category.

(b) A suitable site, differing from (a) in that a problem does arise, such as drainage of the foundation excavation is imperative, or a widening of the footing is necessary. A site in gravel presents an example, provided the simple drainage of the site by pumping can be accomplished and the uplift is so small that it does not require heavy waterproofing.

(c) A conditionally suitable site, if the lay-out and other design particulars of the structure must be adapted to the foundation conditions. This implies, for example, the need for reduction of differential settlements by an ingenious design of the foundations (including pile foundation) and by a modification of the superstructure (e.g. concerning dilatation joints), or the need for reduction of foundation depth due to the ground-water level. Sites on stiff Tertiary clay or on loamy deposits belong to this category.

(d) A site of low suitability, where the foundation presents so many difficulties that the construction costs increase steeply. This occurs, for example, in a valley floor with a high ground-water table, if the Holocene deposits are underlain by a thin gravel layer resting on weak Tertiary clayey silt.

(e) An unsuitable site is that where only exceptionally high costs make the given construction work possible. A site typical of this category is one situated on a landslide or in an area where slope movement due to excavation cannot be prevented.

There is no doubt that the general availability of a site also depends on factors other than those of engineering-geological character, but the former are not to be considered by the engineering geologist. He must confine himself to the suitability from the point of view of his profession.

10.8. Foundation conditions related to the regional geology of the area. The reliability of conclusions concerning the future behaviour of the structure in question may be violated by several unexpected factors. Therefore the value of experience should not be underestimated, especially when it is based on skilled knowledge. As the properties of foundation soils are governed by regional geological conditions, it is advantageous to consider the site in this light. The regional stratigraphic position of the rock should inspire the geologist as well as the engineer to suggest how the structure will behave, which difficulties will appear, and to which factors the investigation of the site should be directed.

Yet not only the pertinence of the rock to a geological unit of certain genesis is of importance, but also the later development may influence its properties. Thus, for example, fine-grained Cambrian deposits are well indurated in Czechoslovakia, their porosity being less than 10 per cent. On the other hand, fine-grained deposits of the same age have the character of over-consolidated clay in the area of Leningrad, exhibiting a porosity of 30 per cent. Therefore the geological approach to the foundation soils should be of regional character.

Let us evaluate the principal types of foundation soils of Czechoslovakia to show the importance of this approach.

Igneous and metamorphic rocks. Rocks in the spacious peneplains of the Bohemian Massif are marked by deep, nonuniform weathering generating residual soils of contrasting thickness. Residual sandy loam with a high mica content is compressible. Because of these factors one part of a structure may be laid on spread foundations on hard rock, while a pile foundation is made for the other part.

The igneous rocks near young depressions are often bordered by faults. The first building site for the International Hotel in Brno, for example, had to be abandoned, since firm igneous rocks were present under one corner, while other parts of the structure were laid on Tertiary clay. Cases do occur when the contact of the two rock types is covered by slope debris and is not reflected in the morphology of the area. An example is the new building of a children's hospital in Bratislava, where the rear wing of the building stands on partly weathered granite, while the two front wings had to be founded on weak Neogene deposits (with a modulus of deformation of only 200 kp/cm^2), which cover the granite sunk steeply along a fault. Attention should also be paid to ground water in granitic rocks, which may be very soft and therefore corrosive.

The oldest non-metamorphosed rocks of the Bohemian Massif are Algonkian slates and greywackes which constitute large areas in central and south-western Bohemia. The Algonkian rocks in central Bohemia provide relatively good foundation soils and stable slopes. In areas of old peneplain they are weathered to a depth of many meters and produce residual soils with a high content of rock fragments. In areas where they had been transgressed by Cretaceous deposits, the products of old, pre-Cretaceous weathering of tropical character have not been denuded and are exposed on the surface as kaolinitic clay. In some areas ground-water demands special attention as it may be corrosive due to sulphates derived from pyrite which is locally disseminated in slates.

Among Lower Palaeozoic rocks those of Ordovician age are of interest as they constitute the largest proportion of the bedrock in Prague. They are represented by thick complexes of clayey shales alternating with coarse-grained quartz sandstones and quartzites. The repeating sedimentation conditions caused rocks of similar lithological and mechanical properties to recur.

Ordovician beds were folded during the Variscan orogeny and disrupted by longitudinal as well as transverse faults, which are accompanied by wide zones of dense fractures. Physical and mechanical properties of the rocks are therefore governed not only by their lithological character but also by tectonic disturbance. In addition the rocks are deeply weathered owing to the deep frost penetration in Pleistocene. Field loading tests yield a modulus of deformation of hard shales with interlayers of quartzite of the order of 30,000 to 50,000 kp/cm^2, while densely fissured clayey shales exhibit 2000 to 8000 kp/cm^2. Shales tectonically strongly disturbed along the fault zones or in the zone of weathering have characteristics of clayey soils with fragments of shales, and their modulus of deformation is smaller than 1000 kp/cm^2.

Several layers of Ordovician rocks contain finely dispersed pyrite. Therefore water in their joints has a high content of sulphate and is corrosive.

The majority of Devonian rocks in Bohemia and Moravia are of limestone type and karst phenomena are frequent in them. If an area is to be built up, the extent of karst phenomena is to be carefully investigated. An example from the Hranice Karst will be outlined in Chapter 14.

Lower Carboniferous rocks cover a wide area in Moravia (the Drahany platform, Odra Hills and Low Jeseník Hills). Shale, greywacke and conglomerate afford good foundation soils with a dense residual cover. The slopes are stable. Close attention should be paid along the borders, which may be of fault origin in the neighbourhood of the Neogene depressions.

Permian rocks, sandstones, arkoses, and argillites afford rather good foundation soils, but landslides may occur on the slopes of deep erosion valleys due to the presence of excessive water.

The Triassic and Jurassic rocks of Slovakia are strongly tectonically affected since they were involved in the Carpathian nappe structure. In addition, the limestones are often karstified and dolomites heavily disintegrated.

The Cretaceous of the Bohemian platform is characterized by thick complexes of sandstones, sandy marlstones, claystones and marls. Sandstone and sandy marlstone afford good foundation soils, but claystone and marl are rather weak near the surface and suffer volume changes (Fig. 10-36). Care is also necessary at sites near valley borders where deep open cracks may exist in the sandstone or sandy marlstone due to the deformation of the underlying claystone (Fig. 10-37); these also promote the occurrence of block sliding.

The Carpathian Flysch Belt is of great concern from the point of view of engineering geology. In Late Cretaceous and Early Tertiary a large foredeep developed at the periphery of the Inner Carpathians and was filled with marine deposits. Thick complexes of alternating sandstones, claystones and conglomerates originated. This alternation of sandstone beds with shale layers is typical of the Carpathian flysch area. Flysch complexes were folded and overthrust even upon the Neogene of the foredeep. The tectonic disturbance of flysch rocks is reflected in their technical properties. Sandstone beds were fractured and disrupted by open transverse cracks. Claystones have been compacted and folded and their strength along bedding planes has been reduced.

10-36. Characteristic cracks in the shallow-founded buildings caused by shrinkage of Cretaceous marl (Turnov district). (Photograph by Záruba.)

Flysch rocks succumb readily to weathering, so that many slopes are mantled with a thick cover of clayey debris susceptible to sliding. The stability of slopes has been often weakened by stream erosion or by human activity. The old tradition that settlements should be located on more stable ridges or in the valley plain deserves consideration. The hydrogeological conditions of the site should be investigated especially with respect to wet periods.

Neogene rocks afford building sites of varying suitability. Miocene sand is usually very dense and presents a good foundation soil along the border of the Carpathian foredeep. In contrast, Miocene clays and marls are feared owing to the danger of landslides, to the possible large settlements of structures, and to the possible destruction of buildings by volume changes. The bearing capacity and compressibility of clays depends on the degree of preconsolidation. An allowable

10-37. Cracks in the foundation of a water reservoir on Cretaceous sandstone near Rájec (Moravia).

loading of 2 kp/cm^2 is usual, but almost twice this value is employed in the Ostrava district, where clay had been preloaded by the continental ice sheet.

Pliocene deposits differ greatly from Miocene deposits — besides clay also lenses of silt and sand are frequent. Owing to the rather silty character and the presence of water in sand lenses, clays break easily and even small foundation cannot be excavated in them without installing drainage prior to excavation. The density of silt and sand is rather low.

Products of weathering and other Quaternary deposits are often encountered as foundation soils and display varying properties.

— Residual soils reflect the character of the parent rocks as well as the mode of weathering, which often extended to varying depths, producing weak deposits (previously discussed in Chapter 6 and earlier in the present Section).

— Rock debris on mountain flanks are often several tens of metres thick, are generally loose and display a creep movement downslope resulting in the disturbance of roads, chalets, etc.

— Detritus on slopes may afford either good or bad foundation soil, depending on their mode of deposition as well as on the bedrock. Detritus often prevents water from flowing out of the bedrock or gravel terraces. As a result uplift is present and sliding can easily develop. Therefore the ground-water level should be carefully investigated, with special attention being paid to wet periods.

— Alluvial fans of torrents usually contain coarse-grained material and afford suitable foundation sites, often the only suitable site in a mountain valley, but they commonly also contain loam or muddy layers, which may cause disturbance of excavated slopes. Another danger may involve the torrent itself when it shifts its channel during a flood, if not prevented by good protection in the upper part of the fan.

— Gravel terraces usually afford the best sites and have long been preferred for settlements. The ground-water table is frequently rather deep below ground surface. Some care is necessary if the sedimentation process had been interrupted by an erosion cycle, leaving meandering channels in the gravel which were later

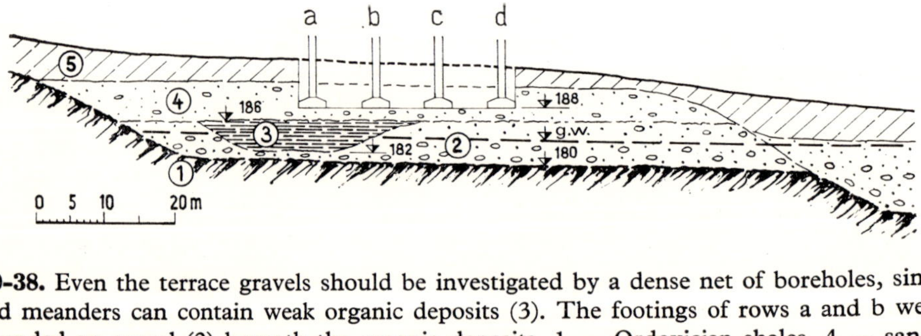

10-38. Even the terrace gravels should be investigated by a dense net of boreholes, since old meanders can contain weak organic deposits (3). The footings of rows a and b were founded on gravel (2) beneath the organic deposits. 1 — Ordovician shales, 4 — sandy gravel, 5 — loam and filling.

filled with clayey deposits possessing a high organic content. An example is shown in Fig. 10-38. The need for deepening the footings in rows *a* and *b* had not been proposed in the design.

The sandy gravel in a valley plain is less uniform. Loam and fine-sand layers, variable thickness and a high water table are unfavourable features. The thickness of the gravel layer should be investigated in detail when the ends of piles are to be embedded in it, and when the underlying beds are rather weak. Layers of clayey deposits, or even organic materials, may be also present in old buried channels below the gravel (Fig. 10-39).

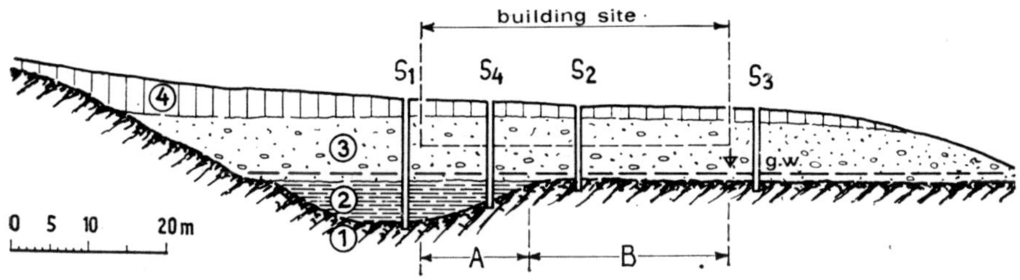

10-39. Bearing capacity of the gravel was reduced, owing to the presence of organic deposits filling a buried channel in the bedrock. 1 — Algonkian slate, 2 — organic deposit, 3 — sandy gravel, 4 — loess loam.

— Deposits of continental glaciers are very nonuniform even when their thickness is great. They require a minute investigation. Moraines generally afford good foundation soils, but clays and sands are usually compressible. The case of the grain elevator at Transcona in Canada, 1913 (Fig. 10-18) has often been recalled; this tilted because the bearing capacity of varved clay was exceeded. Clayey layers in lacustrine sands also present difficulties in excavations because water is withheld in them, making them break and slide.

Moraines of mountain glaciers are usually dense, contain large boulders and offer good foundation soils. When they form dams of lakes care is necessary as liquefaction of soils, piping and similar phenomena due to seepage can occur during excavation.

— Loess as a foundation soil is of problematic character as was discussed in Sec. 10.2.3. Experience of regional nature is of great value, since the properties of loess, especially their susceptibility to collapsing, may vary from place to place.

— Holocene deposits in valley plains are generally weak, compressible and heterogeneous owing to chaotic deposition. As the river branches migrated, deposits varying from gravels to muddy loams originated. In towns they are often covered by fills. Old buildings and industrial structures founded on them on spread foundations have suffered cracks. Due to the high water table a pile foundation is adequate.

Fig. 10-40 shows an example of the foundation conditions on the bank of the Vltava river in Prague. The foundation soil is formed by Holocene deposits 4 to 6 m thick. The majority of old buildings were founded on foundation beams or mats. Each new building caused the opening of fissures in the older neighbouring houses. The large part of an office building, 80 m long, constructed in 1914, was founded on driven piles and does not exhibit any settlement. But the outer parts founded on mats (with the purpose not to disturb neighbouring buildings by the ramming of piles) settled about 11 cm (Záruba, 1948) and also caused settlement in their environments. The granite revetment of the board was disturbed by the differential settlement between the parts founded on the mat and that founded on piles.

10.9. Measurement of deformations of structures. The settlement of important structures is often measured during and after construction. Two aims should be distinguished:

(a) The measurement can prove that the settlement is gradually ceasing and will not exceed the calculated value. The heights are marked on the basement of the building and periodic levelling is carried out.

(b) The measurement is also expected to provide information of a scientific nature, making possible the refinement of calculations of settlement, of the time rate of settlement and of the stiffness of the structure. Settlements in the environs

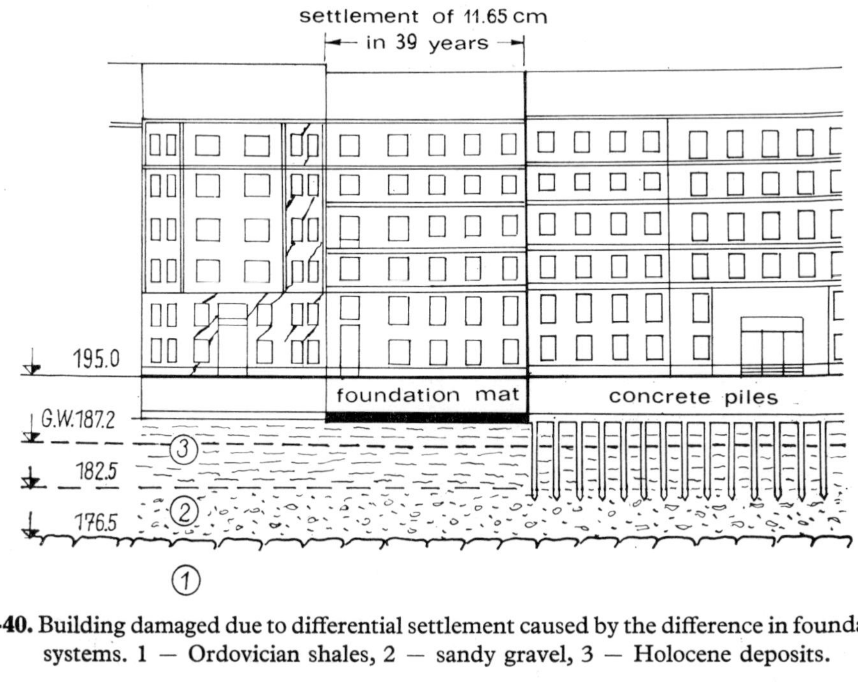

10-40. Building damaged due to differential settlement caused by the difference in foundation systems. 1 — Ordovician shales, 2 — sandy gravel, 3 — Holocene deposits.

of the building should also be measured, as the depth of the zone active in soil deformation can be calculated from it. For the same purpose settlements at different levels in the soil below the structure are measured; deep underground bench marks are therefore set (Fig. 10-41).

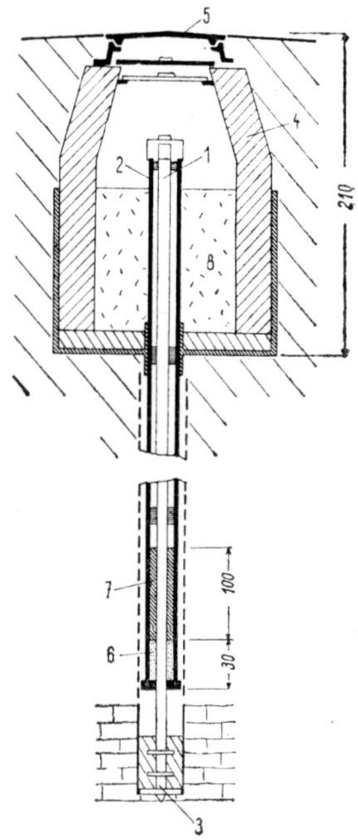

10-41. Deep underground bench mark as used by the Institute NIIOPS Moscow. 1 — Steel tube with bituminous coating and filled with bituminous mixture, 2 — protective casing tube, 168 mm in diameter, 3 — anchoring in Carboniferous limestone, 4 — concrete ring, 5 — cast-iron covering, 6, 7 — bituminous packing, 8 — thermal insulation by sawdust with bitumen.

Special attention is devoted to the buildings which are reveted by stone facings, as these are sensitive to differential settlement. According to the experience of NIIOPS, Moscow, the lining can withstand a deflection equal to 0.001 of the length. The allowable safety value has been estimated at 0.0007 of the length.

The research institute NIIOPS in Moscow deals not only with the calculation of settlements but also with their measurement. The office building at Smolensk square (Fig. 10-42) was monitored by 76 marks set on the foundation mat as well as by others set on neighbouring structures. The precise levelling was related to deep, fixed points set 35 m apart from the building (Fig. 10-43) and anchored into the Carboniferous limestone at a depth of 28 m (point I) and 20 m (point II) below the ground surface. The control measurement indicated that the reference point II, set into the upper beds of the limestone, displayed a differential settlement of 6.7 mm compared with point I. The foundation box of the building is founded on glaciofluvial sand 11 m in thickness, underlain by a complex of Cretaceous limestone containing interlayers of marls. The settlements of the points shown in Fig. 10-43 are plotted in Fig. 10-44 for the period 1949 to 1964. The central portion of the foundation mat settled 77 mm in 1964, while the peripheral part only

55 mm. The expected settlements at the two points in question were 93 and 73 mm, respectively.

The heaving up of the bottom of the excavation was also measured for the given structure. Before the excavation started, depth bench marks were set in the

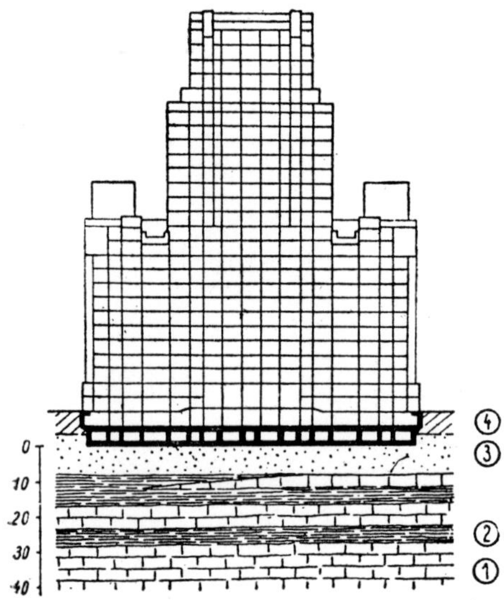

10-42. Schematic section of an office building in Moscow, Smolenski Square. 1 — Carboniferous limestone, 2 — Carboniferous marly clay, 3 — glaciofluvial sand, 4 — filling.

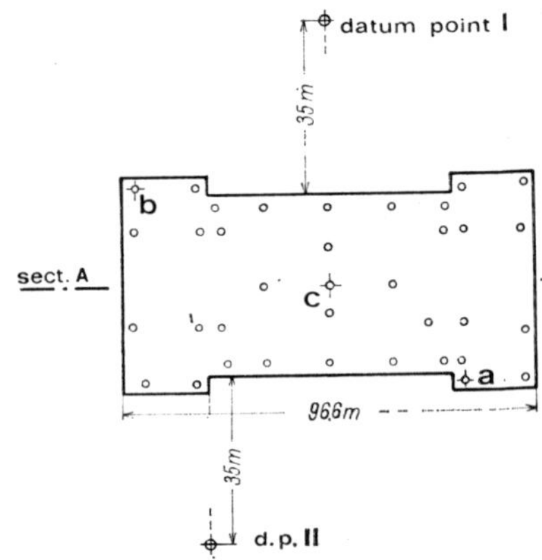

10-43. Schematic ground plan of the foundation mat of the building shown in Fig. 10-42, a, b, c — datum points the settlement of which is plotted on the following figure.

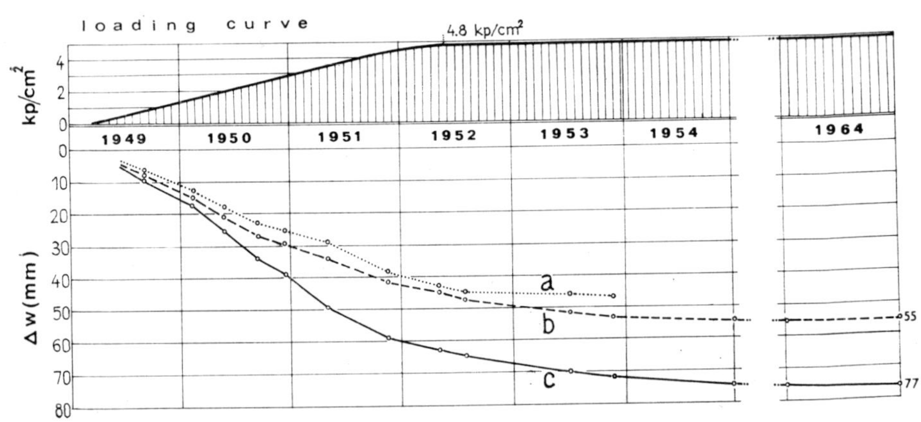

10-44. Increase in loading and course of settlement observed on the building shown in Fig. 10-42 during 1949—1964; a, b, c — datum points. (Egorov, 1954, NIIOPS, 1964.)

boreholes at a depth of 0.5 m below the base level. After levelling of the marks, the boreholes were filled with white sand, the casing was pulled out and the excavation started. Prior to the construction of the foundation mat, the marks were uncovered and levelled. The heaving up of the bottom of the 10.5-m deep and 32-m wide foundation pit measured 52 mm near the centre and 15 to 19 mm at the circumference (Brait and Kozakov, 1950).

Chapter 11

ROADS AND RAILWAYS

Engineers designing and building roads and railway lines have always been aware of the benefit earned from knowledge of the geological situation of the site and from collaboration with geologists. Even during the construction of the early canals and railways several problems of geological character were tackled, such as the question of stability of slopes (Collin, 1846). It is true that the principles of railway location (curvature, ruling grade) were not so challenging with the early lines as they are nowadays, but these new requirements should not decide on the location of the route regardless of the stability of slopes, the suitability of the subgrade, etc. On the other hand, the engineering geologist may initiate unnecessary controversies when requiring intricate measures which can be achieved only by using a large manual labour force.

11.1. Geological requirements on the design of transportation routes. In order to avoid unnecessary construction and maintenance costs, it is necessary that — apart from railway or highway engineering conditions — geological considerations are respected. In a sequence of their importance these may be formulated as follows:

(a) Investigation of the location of the line with regard to the geological setting of the area. This is the most important step of the preliminary investigation. It would be wrong for the designing engineer to define a strictly limited zone for the alignment and ask the engineering geologist to suggest the most favourable location for the line within its boundaries. The engineering geologist should be able to define freely favourable areas for the line and only afterwards should he be asked to compromise his findings with the interests of the designing engineer.

(b) Investigation of the suitability of soils or weak rocks for the construction of embankments or searching for borrow pits when the soils existing in the designed cuttings are unsuitable as embankment material. This is one of the relevant problems which can particularly influence the elevation of the line.

(c) Investigation of the bearing capacity of the subgrade as well as of the need to improve it using materials transported from remote deposits. As the supply of material may considerably rise the cost of construction, this question should be investigated in essentials during the first design stages.

(d) Consideration of the slope angles in deep cuttings and embankments, or of the necessity to construct high retaining walls. This question is usually settled in combination with point (a).

(e) General, standard specification of the necessary work in the subgrade and of the thickness of the subbase.

(f) General investigation of the foundation soils of bridges and of high walls.

(g) General investigation of the excavation conditions (see Chapter 8).

During these investigations the engineering geologist should be in close contact with the designing engineer since the latter is the only person competent of balancing the economic, technical and geological aspects of the design. Nevertheless, the engineering geologist should also seek to understand the principles of the line, its direction and gradient characteristics, the points where the route is determined by environmental conditions, as well as the cost estimate, in order to acquire a general idea of the possibilities of its location.

11.2. Preliminary investigation for the general location of the route. This stage delimits the relevant areas, a difficult task from the point of view of engineering geology. Naturally, when rocks of different properties alternate, the line should be located on rocks of a better quality. Thus, in Cretaceous areas of the Bohemian Massif, sandstones and their weathering products provide better conditions for the location of the line than claystones and marls which exhibit unsuitable properties such as volume changes and water accumulation on their surface resulting in slide movements of the overlying beds.

Rocks of the same kind may possess different properties at different height levels. In the Bohemian Massif, crystalline rocks are less weathered in the lower parts of the valley than higher up. It is therefore advisable to place the road with tunnels near the valley floor.

In contrast, rocks on the high plateaus have been weathered to a considerable depth. Where a large amount of fill material is needed, the deeper cuttings will be economical. But when some cuttings must be excavated in hard rocks, it will be more suitable to elevate the route and procure the material for higher embankments from borrow pits.

The requirement that the volume of excavations and that of embankments should be balanced is therefore questionable, and often even wrong. This holds true, for example, in the Carpathian Flysch zone, where weak clayey shales cannot be used for the embankments. Therefore when laying out routes in West Slovakia, several sections were located in the valley on low embankments constructed of alluvial gravel. Routes can also be located on embankments constructed on the toes of slopes if sufficient drainage of the basement is provided, but when the line has to rise up high above the valley bottom two problems are encountered. The side-hill cuttings are often unstable due to the presence of a thick debris cover. Therefore it is necessary either to cut the line deep into the slope, generally accompanied by the construction of high retaining walls or by anchoring, or to locate the line on a bridge. When reaching higher parts of slopes, their possible stabilization should be considered, for example, by using horizontal drainage boreholes. Stabilization of this kind should be carried out as soon as possible, even before the construction of the line has begun.

In the clayey Neogene deposits of Czechoslovakia almost every cutting is susceptible to sliding. If possible, it is better to avoid excavating cuttings at all. A further reason for this is that the soils excavated cannot be used for embankments.

Other features should also be respected when planning the route. First, wet hill slopes should be avoided. A case is shown in Fig. 11-1 where the presence of springs

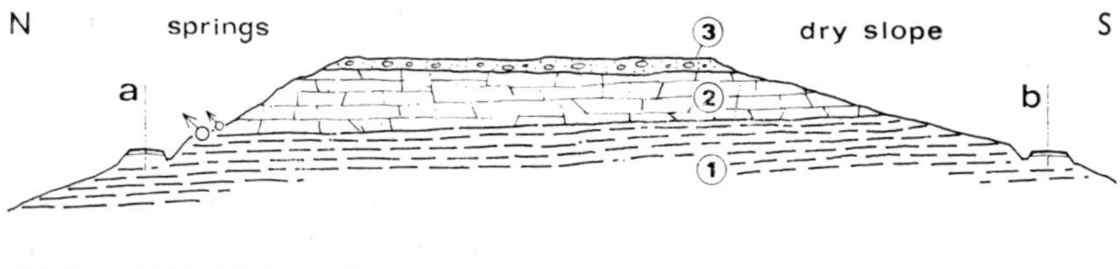

11-1. Wet slope (a) exposed to the north is less suitable for the lay-out of a route than the dry south-facing slope (b). 1 — Cretaceous marl, 2 — sandstone and sandy marl, 3 — sandy gravel.

was a disadvantage. Under similar conditions slopes facing south are to be preferred since they are covered for a shorter time by snow and are generally drier than those facing north.

The preliminary report should contain a description of soils and rocks along the alignment, information concerning their stability, that of the ground- and surface waters, a description of rocks and soils which are expected to exist in the subgrade as well as of the foundation conditions of bridges. On the basis of these data, recommendations for the laying out of the line are presented. The preliminary report therefore contains recommendations as follows:

(a) As a result of the investigation the possibility of the location of the line in cuttings and in side-hill cuttings or on embankments is analysed. The desirable elevation of the line above the valley floor follows. The preliminary estimation of slope angles and sources from which gravel or sand can be provided for the protection of the cutting slope are given. When a tunnel construction is expected, all conditions governing costs should be indicated. An important point is whether slope portals are possible, i.e. whether the line can be directed oblique to the contour lines of the slope in general, or whether a frontal portal is necessary, with the line directed normal to the slope.

(b) The suitability of soils or rocks for the construction of embankments, or the conditions under which they may be used, are given. Rocks unsuitable for this purpose are defined. With regard to conditions found in the basements of future embankments the necessary measures are indicated.

(c) Rocks and soils forming the future subgrade of the route are described and their suitability from the point of view of bearing capacity and frost damage is discussed. The source of materials necessary to stabilize the subgrade and to construct the sub-base is indicated. However, instructions must be given by the designing engineer for the properties required from the material built into the sub-base.

When long routes are proposed, a tabular or diagrammatic presentation is advantageous. The report should inform the reader that the indications presented are of preliminary character and that more detailed investigations are necessary. Recommendations for the detailed investigations are given and the technical methods and procedures are suggested.

11.3. Detailed investigation for route location. This stage of design should comprise a more detailed exploration of the natural conditions, in order that the line may be located definitely in both its horizontal and vertical alignment. When planning the necessary investigation, not only the rocks but also all geological factors which contributed to the present configuration of the surrounding area should be analysed, such as tectonic conditions, water erosion, or slope movements. Since the participation of these factors changed from place to place, each cutting,

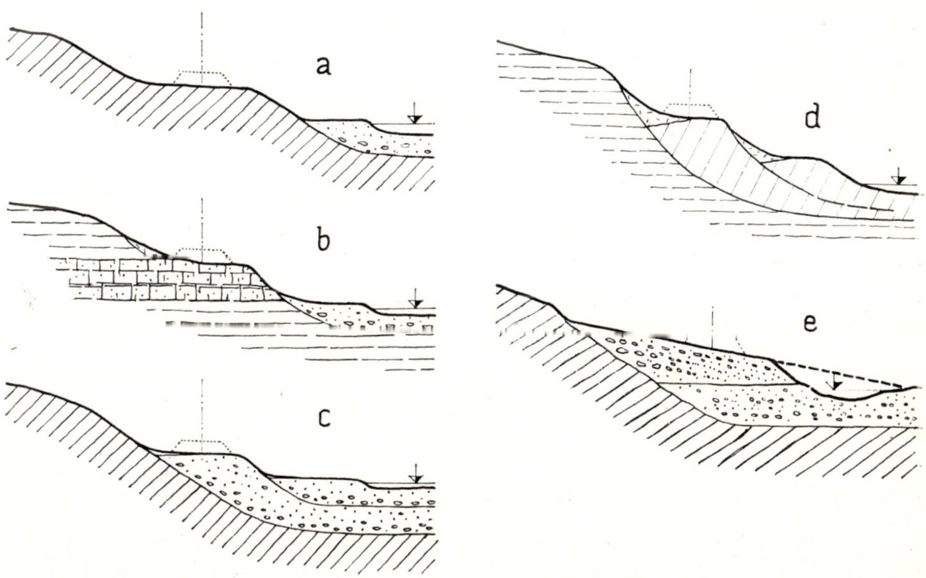

11-2. Different modes of development of a natural bench on a slope. a — Bench formed by lateral erosion, b — by selective weathering, c — river terraces, d — by slope failure, e — alluvial fan cut by lateral erosion.

each embankment and each bridge or wall should be considered as an individual problem from the point of view of engineering geology. That is why the detailed investigation cannot be restricted to the study of surface exposures, but must be based on strategically distributed boreholes. Boreholes and test pits should not be regularly situated along the alignment, but their location must result from an analysis of all morphological features of the area, which are often an excellent indicator of the endogenous and exogenous processes that occurred in the past.

The significance of the analysis of surface features for civil engineering purposes is exemplified by Fig. 11-2. The figure shows different ways in which a natural bench on the valley slope might have originated. Only after analyzing its development can the possibility of the location of the route on it be assessed. A stable base might result in case (a), where the bench is formed by lateral erosion. In case (b), selective weathering shaped the stable platform, consisting of harder rocks. The platform shown in Fig. (c) is a gravel terrace which was deposited during the earlier accumulation periods of the stream. A flat alluvial fan (e), cut by a lateral erosion, also presents a stable base for the route. Nevertheless, care is necessary in this case, as alluvial fans often contain layers of weak, saturated, fine soils. A careful investigation is also necessary when case (d) is found to exist, as the bench shaped by a landslide may be unstable. A borehole located near the toe is instructive, since the slid mass often covered the alluvial deposits of the valley floor. The regularity of the course of the stream, the particularities of its gradient, its curvature and the outcrops of rocks in the river bed can also be good indications of the slope origin in the particular case.

The base of the future embankments, bridges and walls should also be investigated by boreholes. When weak layers occur, it is necessary to bear in mind that stability calculations may result in the necessity to construct a stabilizing berm on the toe of the embankment, and therefore sufficient area should be reserved.

The location of the route is sometimes controlled by the kind of rocks in the future subgrade. Fig. 11-3 shows a railway cutting in eastern Slovakia where the vertical alignment resulted in the location of the subgrade in waterlogged terrace gravels. Drainage of the terrace and sinking of the ground-water table were

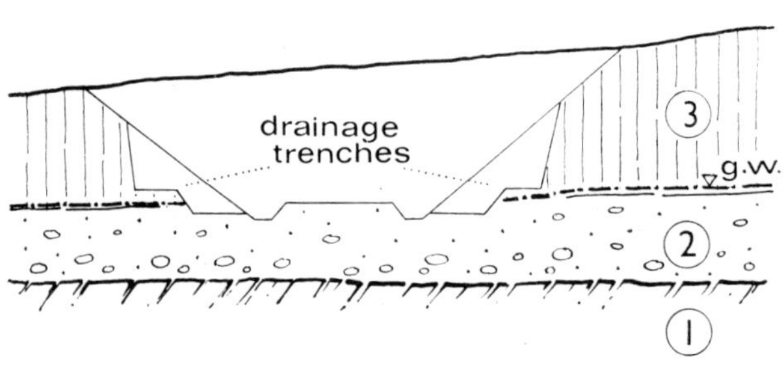

11-3. Deep cutting reaches the saturated terrace gravel; stability of slopes is ensured by rubble drains. 1 — Bedrock, 2 — saturated sandy gravel, 3 — loess loam.

carried out during a gradual excavation of the cutting. If the subgrade were located in the overlying loam layers, uplift water would liquefy the bottom of the excavation and the slopes would subsequently collapse.

Also the denudation remnants of Neogene clays, lying on Cretaceous sandstones in eastern Bohemia, suffer from liquefaction due to the uplift water from the sandstone. Roads should be situated either on an embankment, or on the sandstones but definitely not on the clay, in which even shallow excavations for ditches provoke landslides.

Another example of how the geological situation has caused a change in the vertical alignment is shown in Fig. 11-4a. Ground water flowing from the sandstone

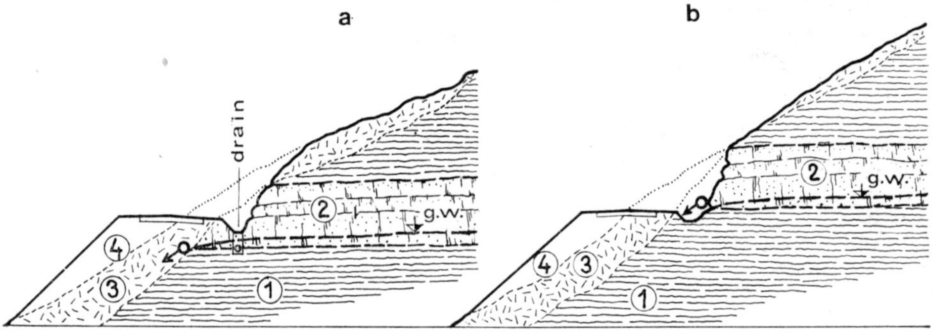

11-4. Ground water flowing out from a sandstone bed weakens the base of the embankment (a) and should be drained. By lowering the vertical alignment of the route (b), water can be drained by a ditch cut in the impermeable marl. 1 — Marl, 2 — sandstone, 3 — slope debris, 4 — fill.

layer seeps into the slope debris in the base of the embankment. In order to avoid a costly drainage system, the arrangement shown in Fig. 11-4b was designed.

A very unsuitable situation arises when the subgrade of the route is situated lower than the surrounding ground surface, so that the necessary gradient for drainage is unavailable. This occurs in flat regions, such as broad depressions or flat watersheds. They should be avoided. Often a roof-like reverse gradient must be created after many years of service in order to obtain sufficient drainage of the cuttings.

When the geological survey is made after the design has been drawn up, the designing engineers are often reluctant in changing the design to suit the geological conditions. The investor may also refuse to spend the necessary money although the costs involved represent only a fraction of the total expenses which will be required later because of necessary reconstruction and because of the remedies necessitated by landslides. In several regions, typically in the Carpathian Flysch zone, stable and unstable areas repeat with such frequency that the route has to be planned in the field. Only after corrections of the route to fit the geological situation have been carried out can the axis be considered as fixed.

During corrections of the axis a continuous analysis of the areas which are to be avoided or stabilized is necessary.

11.4. Cuttings and half-cuttings. The problems concerning the design and construction of cuttings and half-cuttings may be approached differently when considered from a rock-mechanical or engineering-geological point of view.

There is no doubt that rock mechanics will produce effective slope design, when an improved statical calculation method is developed enabling a reliable definition of the stress state to be made for any point along the slope, taking into account the non-homogeneity and anisotropy of the rock. The engineering geological investigation should supply the analysis with all necessary characteristics and parameters. However, it should be realized that methods of investigation are limited. A thorough numerical definition of all necessary parameters can rightfully be demanded in uniformly bedded, tectonically little disturbed rocks, but the system of planes of discontinuity and weakness is so intricate in many rocks that the orientation of these planes must only be assumed. Very detailed, expensive investigation would be necessary to define the orientation with more precision but still the results could remain inaccurate.

For this reason the engineering geological approach based on experience, estimates and simple calculations should not be overlooked. Apart from his findings at the site the engineering geologist should consider:

(a) The experience gained with natural and artificial slopes in the given rock. By evaluating all surfaces of weakness observed at different localities a review of the frequency and other features of these surfaces can be obtained, and a comparison with the actual problem can be made.

(b) The assumed mechanism of a potential failure of the slope and statical consideration emerging from it. When one or more surfaces meet forming a wedge which could slide into the cutting, the mechanism is clear. A prerequisite is that the shear strength along the surfaces is known (Sec. 3.5.4). In contrast, when the surfaces dip into the slope the method proposed in Sec. 3.5.5 can be used, assuming that the general direction and magnitudes of the principal stresses can be estimated. The major principal stress is relatively small near the slope surface and its direction is parallel with the slope (Fig. 3-11). When moving from the surface, the direction of σ_1 tends to a vertical and the magnitude of σ_1 attains the value of the overburden pressure. The magnitude of σ_3 can also be approximated. A further procedure is shown in Fig. 3-31. However, when high slopes are dealt with in weak or loose rocks, the zones near the toe of the slope should be specifically investigated. Great compression can give rise to contractant behaviour at this place, provoking a progressive failure of the slope. In order to avoid this, a reduced shear stress is allowed in this zone, for example, that corresponding to one half of the shear strength. This requirement often leads to a reduction in the slope angle.

(c) The resistance of rocks to weathering should also be considered. Rocks in slopes are not protected against the penetration of surface water and are exposed to weathering, which often proceeds relatively fast. Slopes in the railway cuttings in the biotite and cordierite paragneiss of the Bohemian—Moravian Uplands have shown weathering immediately after the first winter. Similarly several slopes of railway cuttings near Banská Štiavnica in Slovakia, excavated in the young volcanic and pyroclastic rocks, have decomposed after twenty years of service into sand which slips into the cutting, filling ditches and borders of the subgrade. For these reasons the slopes of the half-cuttings of the railway Brno—Královo Pole, situated in tectonically disturbed granodiorite, were excavated at an angle of 45 degrees and covered with vegetation similarly to soil slopes.

On the other hand, stability conditions allow slope angles of up to 65 degrees in igneous rocks containing sporadic irregular and discontinuous closed joints. The stability in the bulk, however, does not exclude the danger of falling out of individual, rather small blocks from the slope surface. Therefore protection using rock bolts, flexible nettings supported by cables, or even revetment walls, is necessary.

(d) Great attention should be given to the occurrence of water in rock slopes. Either local seepage of water along the cracks or a continuous line of springs can appear. Since the stability of the slope is reduced by seepage forces, especially during the frost period when an impermeable crust forms on the surface, drainage boreholes should be installed.

Recent construction methods do not favour the construction of walls in rock cuttings because of the manual work associated with it. *Anchorage* of slopes is preferred. The arrangement of anchors, designed in collaboration with the engineering geologist, should respect not only statical relations, but also technological feasibility. First the position of a potential slip surface is to be assumed. The orientation of anchors should not be normal to the direction of the assumed slip of the rock wedge. Anchors should be slightly deflected towards the horizontal so that they are stretched with the displacement. Another problem is whether the rock will behave dilatantly or not during the slip displacement. If dilatancy can be assumed, prestressing of anchors is not necessary and tight contact with the rock only requires a light tension in the anchor. In contrast, contractant behaviour requires considerable prestressing. Often a combination of anchorage and grouting is necessary. The prerequisites of statical calculations are that the shear strength of the rock and the necessary coefficient of safety are established.

As stated, the arrangement of anchors is also governed by feasibility. When a cutting is relatively short, the drilling of anchorage boreholes would delay the excavation works. Therefore, a small number of rows of anchors is preferred, drilled, if possible, from one working level. Steep slopes such as natural slopes high above the road, raise the problem of how to fix the drilling platforms.

As an example, the landslide stabilization in a Tertiary marl and marly limestone forming the slope above the highway near S. Remo in Italy may be quoted (Fig. 11-5). The sliding movement occurred along a bedding plane, dipping about 45°. The shear strength as measured along the planes, was about 22°. In order to ensure the stability, nine 30 m deep shafts, 24 m centre to centre apart, have been excavated and filled with concrete. By this a reaction to about two thirds of slide thrust (450—550 Mp/m) was provided. The remaining thrust of about 160 Mp/m

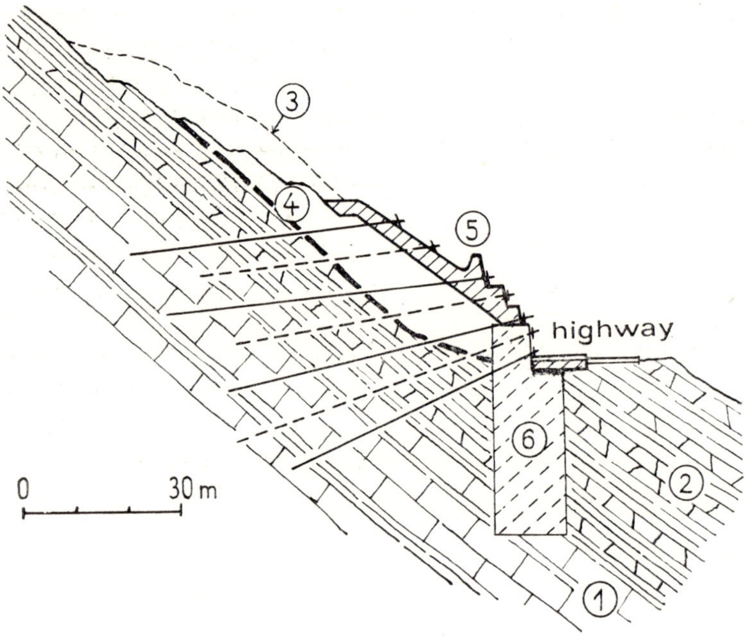

11-5. Stabilization of a slope above a highway near S. Remo in Italy using anchors and concrete piers (Baldovin and Fattore, 1974). 1 — sandstones with marls, 2 — Tertiary marls and limestones, 3 — original slope surface, 4 — slip surface, 5 — reinforced concrete slab, 6 — concrete piers.

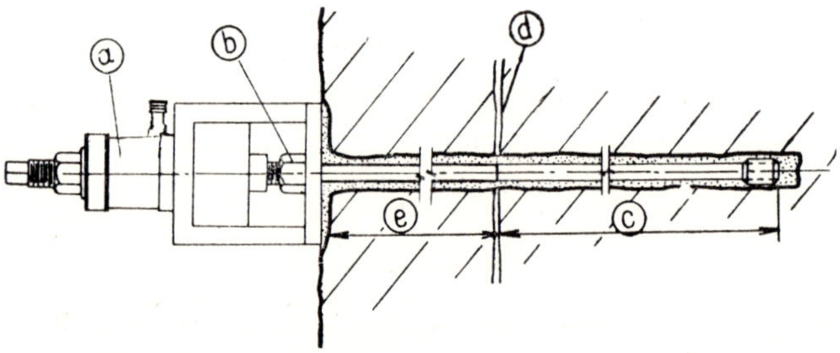

11-6. Prestressed anchors in the canyon wall above the spillway of the Boulder Dam (Simonds, 1958). a — Oil jack, b — turnbuckle, c — grouted end of the anchor, d — fault, e — shank of the anchor wrapped by paper.

11-7. The Innsbruck-Brennerpass mountain motorway is laid on bridges for a considerable length to avoid the disturbance of the stability of slopes. (Photograph by Záruba.)

was resisted by anchors, 3×4.5 m apart, each presenting an operational pull of 120 Mp. More than 300 anchors 45—75 m long were used in order to reach the slope portion in which sandstone beds prevail (Baldovin and Fattore, 1974).

Fig. 11-6 shows details of a prestressed anchor used to stabilize the canyon wall in the vicinity of the Boulder Dam on the Colorado River. The stability of the slope consisting of Tertiary volcanic breccia was threatened by the presence of a fault, dividing a block of rock 62 m in length and 37 m in height. The portions of anchors between the fault and the slope surface are wrapped in paper in order to concentrate the action of anchoring force on the fault surface. Anchors were set in a grid of 2.3 to 2.3 m size and the prestressing force was 28.5 Mp. After the prestressing, the fault was grouted under a pressure of 1.75 kp/cm^2 (Simonds, 1958).

A provision necessary to ensure the permanent action of anchors is the prevention of rust. Therefore special steel, special sealings, or other securings are needed. This is expensive and, therefore, anchoring as a permanent measure is limited to cases which cannot be stabilized in other ways.

Broad half-cuttings for motorways cannot be excavated on steep hillsides because their stability would be undermined and their natural appearance disturbed. The construction of the motorway between Innsbruck and Brennerpass can be quoted as an example. The roadway was supported by circular hollow piers. Fig. 11-7 shows a section of the route on the steep slope of the valley of the river Sill, which consists of phyllite and gneiss, disturbed tectonically and often even mylonitised. In several sections where creep of rock debris could endanger the stability, the piers were founded deep in the rock and empty space was left around them. This makes a displacement of debris of 1.7 m possible without the piers being affected.

Cuttings in *non-cohesive soils* could suffer from the presence of seams and lenses of fine-grained soils, as they can be found, for example, in alluvial fans. In these cases the dip of the cutting slopes is reduced from 1 in 1.5 (34 degrees) to 1 in 2 (27 degrees) or less. The advantage of this is that the slope can be constructed using machines. If water is present, drainage should be accomplished at a right time. As the region is often flat and no gradient is available for the installation of horizontal boreholes, provisory vertical pumping wells should be installed first and only after the excavation has begun are they replaced by horizontal drainage boreholes.

The design of safe slopes in *cohesive soils* has always been a subject of discussions. The engineering geologist can contribute to the analysis by calling attention to the

11-8. Failure of a gentle slope of a cutting in the Neogene clay near Česká Třebová (eastern Bohemia), caused by the presence of subsurface water. (Photograph by Mencl.)

existence and effects of adverse geological features. These are usually associated with the geological character of the particular rock and therefore experience gathered in the given region is of great value.

When designing a safe slope, the mode of a potential failure should be considered. In principle, the slope may be endangered by a shallow slip or by a deep landslide with a curved or planar slide surface.

(a) *Shallow slips* of the slope surface may occur under the following conditions:

With low slopes, which have been excavated at a steep angle (e.g. 1 in 2, 27 degrees) because no ground water was found during the exploration of the site. Loss of cohesion results from the infiltration of rain water into the surface layers. The normal stress σ is small near the surface and thus the section of the strength curve in the Mohr diagram is valid where the curve bends down towards the co-ordinate origin. This section of the Mohr curve displays a very small magnitude of τ_0, but Φ' equals 40 to 45 degrees. Considering the statical loading by penetrating water, the safe angle of the slope should be less than $\Phi'/2$, i.e. less than 20 degrees (1 in 2.75). Slopes steeper than 20 degrees deform, suffer from cracks which are penetrated by rain water, and the surface layer collapses.

Frost penetration makes this situation more serious, as the soil is enriched with water sorbed from the underlying layers, and a break-up follows in the spring. This phenomenon also occurs when the slope gradient is less than 20 degrees, and can be prevented by covering the surface with a thick humus or sand layer (about 50 cm).

Rather gentle slopes in soils suffering from considerable shrinkage are liable to similar failures resulting from the penetration of water into the shrinkage cracks. It is the case of Tertiary clays in the regions of neovolcanic rocks.

Relatively gentle and, therefore, apparently stable slopes are in a critical state when the ground-water table is higher than the cutting bottom. With a gentle slope (in the case shown in Fig. 11-8 the angle of slope in Tertiary illitic clay with organic material is 11 degrees; recently, the angle of slope has been reduced to 7 degrees but after two winters new destructions are observable) not even the deep drainage under its foot was capable of lowering the water table sufficiently. The water table remains near the slope surface in areas to which the action of drainage does not extend. Often the ground water flows out from the slope, which is subsequently marked by wet areas and moisture-loving vegetation. In both cases destruction occurs due to frost action and during a period of several years the state of the slope deteriorates and the ditch, as well as the subgrade, are buried under the slushy soil mass. Often the foot drainage is also destroyed by pressures occurring in the foot.

Therefore, rather steep slopes are preferred, but they should be provided with a protecting sheet in the lower section, for example, using a fill of coarse-grained sand. This sheet is connected with the foundation of the roadway and contains the drainage system (Fig. 11-9). It is recommended that the protecting sheet be arranged immediately after the excavation, long before the foundation sheet of the roadway is constructed. If the dimensions of the cutting need to be spared, a pile wall installed near the slope foot can serve as a support. The protecting sheet need not necessarily consist entirely of sand or gravel but its underside should provide the drainage. In short cuttings (up to a length of 150 to 180 m), the water table can also be lowered by horizontal drainage boreholes situated under the level of the cutting and drilled from the zero points.

(b) In considering the possibility of *deep landslides* the following cases should be distinguished:

In firm, dilatant over-consolidated clays landslides are principally of the character of rotational slides, and in this way they are also analyzed statically. The engineering geologist, who is aware of the development of these landslides in natural conditions, will draw attention to several particularities:

— Owing to the energy necessary to propagate the mechanism of dilatancy, the slip surface is frequently not circular. In sections where a small normal stress exists, i.e. near the root area, the slip surface runs deeper into the slope so that it is almost vertical. On the other hand in the central section it runs higher than that of the corresponding circular slip surface. This decreases the coefficient of safety by about 7 % as compared with the classical solution.

— Dilatancy involves an increase in volume along the slip surface. Therefore, the water content also increases, by about 5 per cent. This is why the characteristics of the shear strength of the soil should be investigated under similar loading conditions. The increase in water content is not immediate, and time will elapse before the soil in the slip surface absorbs water from the surrounding layers. Therefore, the slope can remain at a steeper angle for several days. This is of practical importance when supporting the slope by a protective fill.

At the outcrop of the potential slip surface near the toe of the slope, where normal stresses are small, the relieved clayey soil sorbs more water than in another part. Accordingly, under a normal pressure smaller than about 0.5 kp/cm^2, the

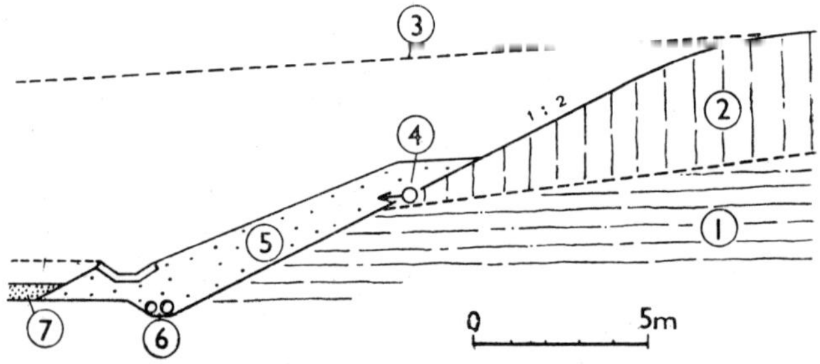

11-9. Arrangement of a supporting fill in a cutting. 1 — Neogene clay, 2 — loess loam, 3 — original ground surface, 4 — water seepage, 5 — gravel fill, 6 — drainage, 7 — consolidated subgrade used as a service road.

strength curve bends down in the Mohr diagram, i.e. it displays a higher Φ' but very small τ_0. This state worsens when water flows out from the slope. It is necessary to load the slope toe with a berm of draining material, with a low retaining wall or with a pile wall combined with a berm. One of the tasks of the engineering geological investigation is to consider where the necessary berm material can be found. The berm material should have a permeability ten times greater than that of the soil, but it should not be so coarse that it allows the soil particles to be washed into the berm. The construction of a wall requires a considerable amount of manual work. A pile wall, when used, often requires anchorage, as the horizontal resistance of a pile is not much greater than between 10 and 20 Mp.

— Natural anisotropy of the soil is another factor which influences the shape of the slip surface. Anisotropy is governed by the fact that even along the little recognisable bedding planes the shear strength is smaller than in other directions. For example, even the apparently isotropic Neogene marl of Brno exhibits a difference of about 20 %. Therefore, the slip surface makes use of this horizontal weakness and turns from the upper curved section into a rather horizontal flat lower section.

— The presence of large horizontal inherent (or residual) stress in soils (see Sec. 3.6) promotes the increase in volume of the soil in the bottom of the excavated cutting. The bottom upheaves and the toes of the slopes approach several centimetres towards one another. This generates tensile stress in the ground surface near the top of the slope, which in turn gives rise to slide movements of the upper parts of the slopes. This can be prevented by cutting a berm into the slope face in the upper part. The development of local slip planes is more hazardous. They start at the toe and disappear into the depth (Mencl, 1965), weakening the stability of the slope until large slides develop, with a root area a large distance beyond the slope top. Reduction of the slope angle could prevent these processes; nevertheless, unfavourable consequences, as reported above, do not make this an attractive solution. Another corrective measure is a very thick, drained fill at the toe of the slope. A good, but expensive procedure is to construct compound, strutted retaining or pile walls.

— In Sec. 3.4.5, attention was directed to the problem that even *over-consolidated clay* behaves in a *contractant* manner when subjected to relatively low stress. Many stiff clays begin to behave in a ductile manner when the normal stress amounts to 2 kp/cm^2. Therefore many slopes higher than only 10 m can show deformations characteristic of ductile shear failure. Since the volume of the soil decreases in the shear zone, pore water must be squeezed out. This generates a temporary pore-water pressure which undermines the stability.

Slide movements of this category develop rapidly, mostly during excavation works. No slip surface can be found in the excavated soil mass, but only a dense system of partial surfaces, and for the most part, only a kneaded material. One or several distinct slip surfaces, however, appear on the periphery of the slide area where the normal stress is small and the material therefore behaves as dilatant (Fig. 11-10). A characteristic feature of this mechanism is the S-shape deformation of the slope surface. The contractant zone undergoes large displacements even under a shear stress smaller than the shear strength. Dilatant parts are forced to participate in these large displacements and therefore exhibit distinct slip surfaces. Water penetrating the cracks aggravates the situation. Therefore it is recommended that those parameters of the shear resistance which correspond to the strains at which the dilatancy (under small σ') and contractancy (under greater σ') begin to appear be introduced into the statical analysis of the slope. For this solution, the coefficient of stability can be low, having a value of about 1.2. When plotting the assumed slip zone, it should be considered that it tends to develop deeper in the slope than the slip surface formed as a result of dilatant behaviour.

— In the geological conditions occurring in Czechoslovakia it is frequently found that the upper part of the cutting is situated in Quaternary non-cohesive soils, while the lower part lies in Cretaceous or Tertiary clay or claystone. In this case the clayey soil also deforms in a ductile way, while the overlying non-cohesive layers suffer rather extension cracks and their contribution to the shear resistance is small.

— The engineering geologist should warn of the presence of fissured clay (Fig. 11-11), as lower parameters of shear strength must be used than those for a compact clay in the statical analysis. It is, however, of advantage that fissured clays can be drained by horizontal boreholes. One of the causes of slope failures in fissured clay is that fissures open because of stress relief and desiccation, so that water can penetrate the soil. Therefore early protection of the slope by a humus cover is desirable, as well as a fixation of the toe by a fill, retaining wall, or pile wall. Again, protection against frost is required. The installation of deep drains

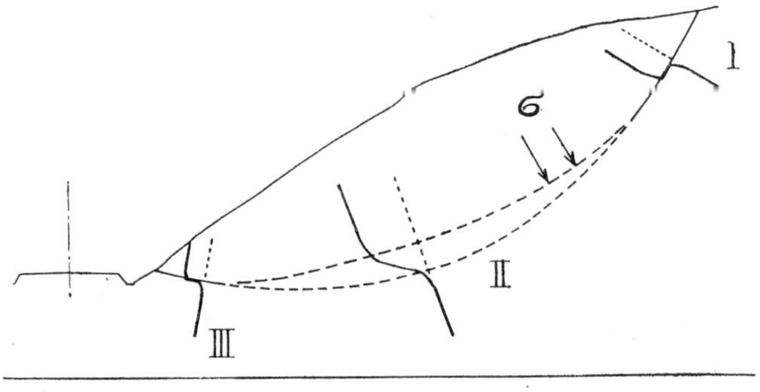

11-10. Progressive failure of a slope developing due to the origin of zone (II) of a ductile slip failure in the area of greater σ. Near the slope surface the zone passes into slip surfaces I and III.

11-11. Fissured clay (Neogene) in the railway cutting near Most. (Photograph by Mencl.)

under the ditches can be abandoned with regard to that the necessary machine work may trigger slope movements. Drainage holes can be drilled through the fill bank or through the wall.

— Ground water present in the slope decreases the effective stress between the clay particles and reduces the shear strength, while tangential shear forces do not diminish. The stability of slopes containing water is therefore much less than that of slopes without a continuous water table. A general rule can be applied (valid more strictly for non-cohesive soils) that the factor of safety of a slope in soils containing ground water equals one half of that when the water table is absent. But even when the slope is excavated with a half-angle of inclination, it will be less stable than a steep slope in the absence of water. This is partly due to the action of frost and partly because deformation increases significantly with increasing pore-water pressure. Therefore, even a gentle slope suffers cracking. Again, deep drainage is desirable; otherwise the slope should be covered with a permeable protection fill so that the escaping water freezes outside the clay surface. The angle of the slope may then be steeper than discussed above.

— Equally hazardous is the presence of water pressure in relatively more permeable intercalations or thin layers, as are beds of sandy detritus at the base of clayey deposits (Fig. 11-12), sand partings between soft Tertiary clays or sand layers in glacial deposits. The water may even rise from the bedrock.

An effective drainage is not easy. The principle that the most permeable materials should be drained, requires that all water-bearing beds should be crossed. But the beds are often not planar but bent or sagged (e.g. because of a fossil landslide) and water may remain in them even after drainage, thus generating water pressure. The drainage of the bedrock is also difficult as permeable and impermeable layers often alternate.

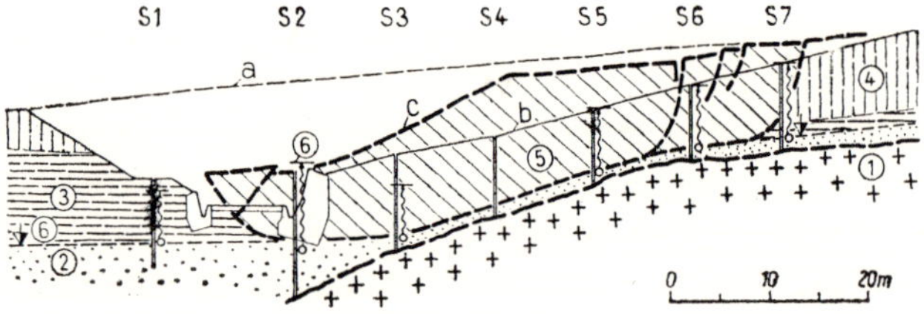

11-12. Effect of the uplift on the stability of a cutting. 1 — Granite, 2 — Neogene sand, 3 — Neogene clay, 4 — loess loam, 5 — slid mass, 6 — ground water under pressure, a — original ground surface, b — slope after stabilization, c — slope after sliding movement.

The curved shape of permeable layers may obey the particularities of the geological structure. For example, gravel terraces deposited on Tertiary clay beds have been upheaved near a valley slope due to the removal of weight after the erosion of the valley. Water can accumulate behind the upheaved border and this gives rise to sliding failures of the excavated slope. Besides draining by boreholes, trenches, etc., also vertical boreholes draining water down into a more permeable layer, should be considered.

(c) The third group of slope movements in cuttings are *sheet slides* along *predestinated geological surfaces*. They may originate when the surface layers are of different permeability and are periodically saturated with water. Sliding movements in flysch areas are most frequent where impervious clayey layer covers more permeable rock debris. The underlying rock debris layer is periodically saturated with water, which penetrates either from above, or more frequently, from the bedrock. Due to the inclined position of the debris layer, great hydrostatic pressure develops near the toe of the slope. A slide occurs along the lower surface of the clayey layer due to the uplift, and affects the slope of the excavated cutting or even the base of the route. In accordance with the intensity of the uplift these movements develop on slopes at an angle of 12 to 14 degrees, but they are known to occur even at angles less than 10 degrees. The best measure is an early drainage of the slope, installed long before the construction of the route has begun. The drainage system is generally dry for a long period of the dry season, but water appears after several rainy days. Along the borders of Tertiary depressions sandy layers are often almost parallel to the existing slope. In this case the presence of water is also only periodical. One of the most important tasks of a geological investigation is to establish where these layers come near to the surface in the lower part of the slope, since the drainage system should be installed at these positions.

11.5. Embankments. Engineering geological investigation must deal with two problems: a study of the embankment material and an investigation of the stability of the embankment foundation.

Nowadays, embankments are generally compacted in order to reduce their settlement. This matter is of primary importance when comparing the suitability of different fill materials. The sequence of soil suitability is as follows: GW, SW, GP, GM, GC, SP, and SC. SM and CL groups can also be used, but the possibility of their insufficient compaction during wet weather should be taken into consideration. In the same way siltstones and shales should be considered. The compaction should be so effective that at least 40 % of fragments are less than 2 mm in size, and the liquid limit of the fines should be smaller than 50 %. In embankments made up of other rocks, especially of rocks readily disintegrating, the voids between blocks should be filled with small fragments. This can be controlled to some degree by the technique of blasting, by drilling the holes close together and by using small charges. A millisecond blast is advisable. Although the compaction of embankments is not an engineering geological affair, it is necessary to keep in mind that one of the important causes of the undesirable deformation of embankments is contractant behaviour (Sec. 3.4.5) due to poor compaction.

The base is often where failures begin to develop in embankments. Its stability analysis follows the rules of soil mechanics, but the engineering geologist should draw attention to several features which might remain unnoticed:

(a) Large pore pressure may be present in the subsoil. This applies, for example, to the margins of Tertiary depressions where uplift is supplied by the lower, more permeable layers or by the bedrock. The slip surface extends to an unexpected depth. (Fig. 11-13).

(b) The weight of the embankment produces a pore-water pressure in the little

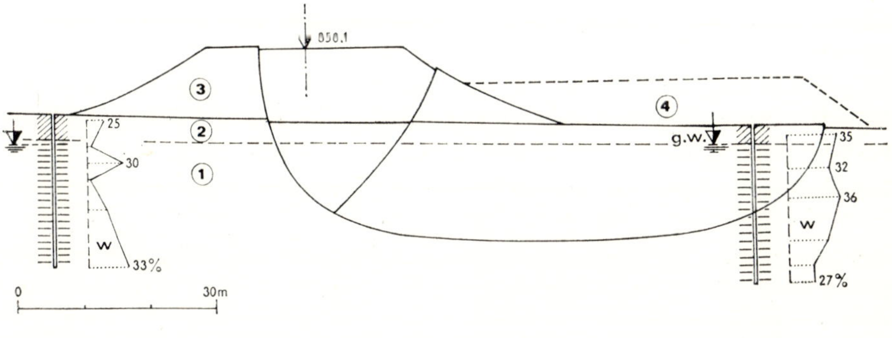

11-13. First stage of a failure of a motorway embankment founded on Neogene clay (1), 2 — loam, 3 — embankment 11 m in height, 4 — stabilizing berm.

permeable subsoil. Its magnitude can be calculated on the basis of laboratory tests on soil samples. However, pore-water pressure also affects areas outside the embankment base, and this should not be neglected. The statical effect is the same as if a subsoil outside the toe of the embankment was partially excavated. Therefore stabilizing berms should be constructed, generally to a greater height than would be presumed.

(c) If the subsoil contains beds of fine sand saturated by water and covered by impermeable soils, hydrostatic pressure caused by the weight of the embankment spreads throughout the sand bed far beyond the embankment toe and the surrounding area deforms slightly due to uplift. The deformations can be determined by means of levelling. When the embankment has been constructed on even a gentle slope, slow downhill displacements appear. Uplift is spread downwards some distance from the embankment and the ground begins to slide. The mass of the embankment is often involved in the slide movements (Fig. 11-14). Therefore, sand layers should be drained before the embankment is constructed.

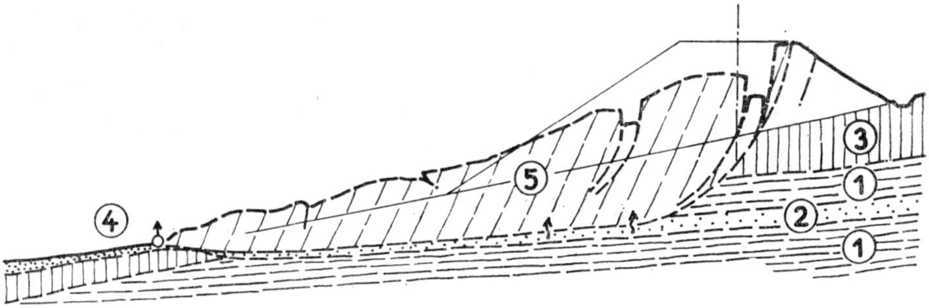

11-14. Embankment failure caused by ground water under pressure present in permeable Neogene sand (2). 1 — Neogene clay, 3 — loess loam, 4 — sand washed out during sliding, 5 — collapsed portion of the embankment.

(d) Thick layers of alluvial deposits also cause failures of embankments. However, in contrast to clay deposits, their consolidation proceeds in a shorter time and, therefore, lower stabilizing berms are sufficient. To advance consolidation, the ground surface under the future embankment should be covered with a sheet of gravel, slag or another rather permeable material.

(e) In gorges or other valley-slope depressions drainage of the bottom is necessary.

Roadways lying half in cuttings and half on embankments are adversely affected by water from drains laid under the ditches of the cutting side. Water from them is often drained by cross-drains to the other side of the roadway, across the embankment. This drainage system is incorrect, as settlement of the embankment cannot be prevented; water penetrates the fill and often also its base, giving rise to slides (Fig. 11-15).

11-15. Failure of highway embankment due to water escaping from a damaged drain constructed along the left, excavated portion of the roadway. (Photograph by Mencl.)

11.6. Roadways and their subgrades. Requirements on the bearing capacity and frost resistance of roadways are increasing, as the axial load of vehicles increases and as snow is nowadays removed from roadways. A thickness of up to 90 cm has been achieved for roadways. It consists of layers having different functions, namely the top course, the bearing course, and the sub-base. The sub-base has to prevent water penetrating from the subgrade to the bearing course. The subgrade is mainly required to resist disturbance by vehicles during the construction of the roadway.

The construction of a roadway is expensive and the knowledge of geological conditions may contribute to the reduction of its cost. First, the question of the sub-base comes into consideration, as permeable sand or gravel are necessary to break the capillary lift. The sub-base can be omitted when the subgrade consists of pervious, unsaturated soil or rock. If not strong enough, the bearing capacity of the subgrade should be improved by stabilization, for example, by rolling gravel into the soil. The engineering geologist should investigate where in the longitudinal profile of the route soils or rocks of a desirable quality such as hard rocks, eluvial sand or terrace sandy gravel can be found. The engineer should

11-16. Unsuitable subsoil of the roadway slaked after spring thaw is squeezed out through the cracks of the deteriorated roadway. (Photograph by Záruba.)

provide such gradient conditions on the line that these layers could be readily exploited.

Another problem is the drainage of the subgrade. A routine design adopting simple drains of 30 to 50 cm deep, situated under the subgrade verges, may not be the most effective solution. In order to avoid their blocking by suffosion, the drainage collectors should comprise at least two pipes, one laid higher, draining the subgrade, and other laid lower, draining the soils and the toe of the cutting slope. This method is rather expensive, but it is better than drains situated under ditches, which undercut the slope. The engineering geologist should also help to decide at which points water should be conducted away from the longitudinal drains, and to select excavation materials which should be stockpiled and subsequently applied as a fill material for the construction of the subgrade.

The bearing capacity of a finished subgrade is determined by several conventional tests. Thus, for example, when designing bitumen top courses, the „Californian Bearing Ratio“ (CBR) test is used. A cylinder 5 cm in diameter is pushed into the base of the designed layer and the pressure needed to push the cylinder 2.5 mm into it is measured. The percentage ratio of this pressure to that of 70.8 kp/cm^2 (the latter is the resistance exhibited by a good, crushed stone base) is taken as a characteristic of the bearing capacity of the base. It is a deciding factor, together with the magnitude of the wheel pressure, for the thickness of the roadway.

Older roadways often upheave in winter, as the underlying clayey soil increases in volume because of frost as water is absorbed from the underlying soil (Fig. 11—16). The top course is destructed during spring thawing. The repair is usually connected with a radical reconstruction of the road. Repair using horizontal boreholes is preferred by railway engineers since it does not disturb traffic and little manual work is needed.

As stated, the sub-base of the roadway should consist of permeable gravel or sand. In contrast, the material for the bearing course may be less permeable, provided its bearing properties are good. Therefore fragment material with a small amount of loamy material is satisfactory. With regard to the large quantity of material necessary, familiarity with the geology of the area may be expedient in finding suitable deposits. The ease of compaction of the material is an important factor to be considered.

11.7. Geological investigation for bridges and other structures. The foundation of bridge piers and abutments requires a serious geological investigation. It should not be forgotten that valleys and other depressions of the earth surface did not generate fortuitously, but especially because of the presence of weak rocks.

Small bridges (culverts, under-line bridges), mostly situated in the floor of small depressions or ravines, require an investigation of the thickness of weak deposits and of the depth of the water level. It is often recommended to relocate the structure near the zero-point of the longitudinal profile of the line where the bedrock may be found nearer the ground surface (Fig. 11—17). This is advantageous when a surplus of fill material is available, as the depression or ravine upstream of the road embankment must be filled to the level of the bottom of the culvert. The brook is then conducted at this level. The bottom of the depression under the fill should be drained in order to minimize uplift in the infiltrating water and therefore also the danger of the embankment failure.

If this is not possible, for example, because of a relatively flat relief of the de-

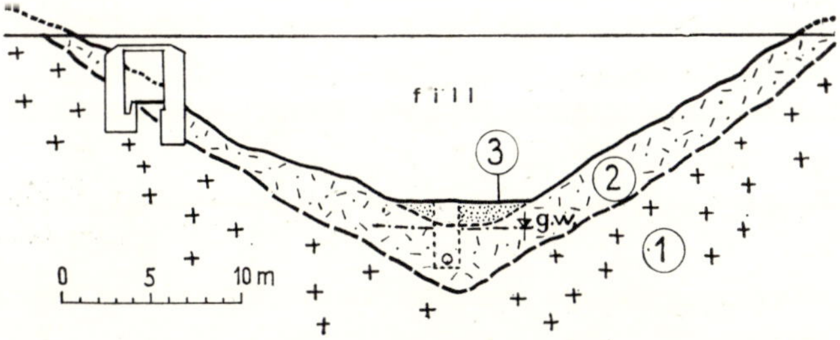

11-17. Lay-out of a culvert on a valley slope may be of advantage since better foundation conditions often exist on the slope. 1 — Bedrock, 2 — slope debris, 3 — muddy deposit consolidated by a drain constructed along the valley floor.

pression, two small bridges can often be constructed advantageously. A small culvert is located at the bottom of the depression, while an under-line bridge is located on the slope of the depression. Because the embankment is founded on weak deposits, the culvert can also be founded in the same way. The culvert settles together with the embankment base and its structure should be designed not to suffer from deformation. Prestressed and prefabricated closed frames, installed in segments, offer a good method. The bottom should be elevated according to the anticipated settlement of the subsoil. Due to small, unexpected differences in settlement, a rather larger passage section than is strictly given by hydrotechnical calculations is recommended. The consolidation gravel layer at least 20 cm in thickness is installed under the foundation of the culvert. A prerequisite of this method is a sufficient bearing capacity of the subsoil under the embankment. When a stabilizing berm is needed to improve the stability of the subsoil, only a partial extension of the culvert is necessary, while for the rest an open ditch is satisfactory. The stabilizing berm must be filled at the same time as the embankment.

The subsidence of an embankment situated on flat sloping ground also produces downslope displacements. Therefore the dilatation blocks of the culvert become separated from one another. On one railroad line in the Carpathian Flysch belt this process has not stopped for forty years since construction. Fig. 11—18 shows an arched culvert deformed by settlement and slope displacement. The structure of the culvert or other under-line bridges should be sufficiently strong not to suffer from local overstress, and the outlet should either be founded deeper or supported by a pile wall so that its displacement is decreased.

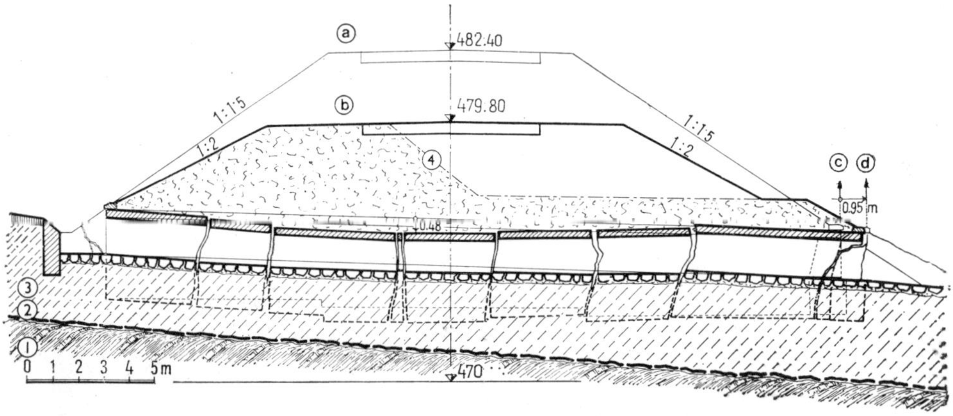

11-18. Failure of a culvert founded on a compressible, weak soil owing to the weight of the embankment. 1 — Marly shale, Palaeogene, 2 — old slip surface, 3 — clayey sandy slope detritus, 4 — state of the embankment at the initial stage of failure, a — embankment height as designed, b — reduced height of embankment, c — original face of the culvert, d — displaced face of the culvert.

Bridges across broad valleys pose other problems. Even in hard rock the foundation conditions of one of the piers may be poor. This situation can be anticipated from the geological investigation of a wide area. Since the origin of the valley is often associated with tectonic weakening of the rocks, one or several fault zones of different width may follow the direction of the valley (Fig. 11—19). The character

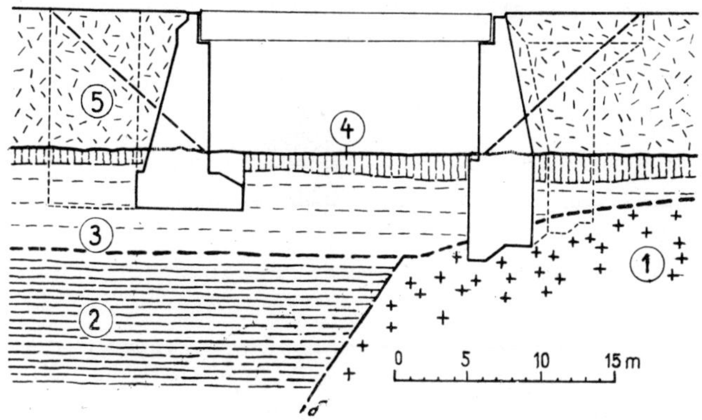

11-19. Statically determined bridge structure allows the use of different methods for the foundation of abutments, since the structure sustains differential settlement between the supports. 1 — Granite, 2 — Neogene clay, 3 — redeposited clay, 4 — Holocene loam, 5 — filling.

of the bedrock may also vary from pier to pier, as it does with steeply bedded formations. Statically determinate or supple beam structures are often necessary when differences in settlement are expected.

Fig. 11—20 presents an example of how the geological conditions influenced the design of a structure. In Teplice (north-western Bohemia) an arched bridge

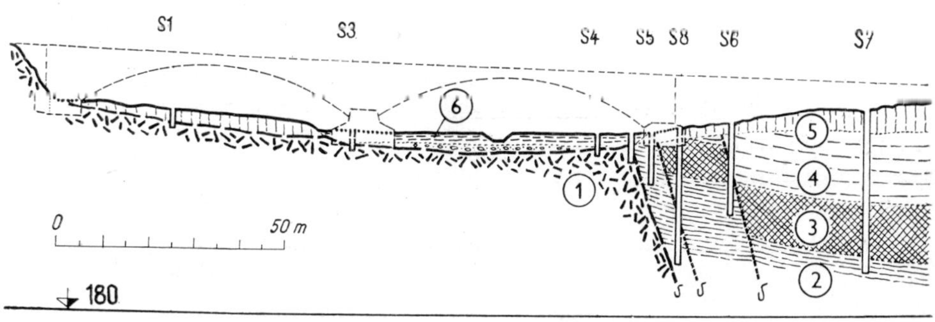

11-20. Foundation of the arch bridge at Teplice (Bohemia). As Tertiary clay and the coal seam below the designed northern abutment are very compressible, the span of the openings has been reduced so as to found all supports on porphyry (1). 2 — Neogene clay, 3 — coal seam, 4 — overlying clay, 5 — loess, 6 — alluvium.

with two 30-m span openings was designed. Exploratory borings revealed a firm quartz porphyry under the future southern abutment and under the intermediate pillar. The foundation rock of the future northern abutment consisted of Tertiary clay and contained a coal seam. The contact between the porphyry and the Tertiary deposits was tectonic. The step-like course of the coal seam indicated that the principal fault was accompanied by several parallel subsidiary faults. Tertiary deposits were relatively compressible as compared with the porphyry, and the surface layers of the coal and of the weathered, overlying clay were both weak. Therefore a relocation of the bridge was recommended in order that all the supports be founded on the porphyry.

Statically indeterminate structures can suffer from foundation displacements and they are more endangered by creep or relaxation processes. On the other hand, continuous beam structures are preferred nowadays if a good rock is available. Again the geology of the site should be analysed in time, as faults may be associated with the origin of the valley. If high piers are situated out of the reach of faults, difficulties connected with the deepening and extension of the foundations can be evaded. Also other, apparently secondary problems, may arise from poor geological conditions. This applies, for example, to the choice of which support under a continuous beam structure should be loaded by the fixed bearings. Such a support is loaded by relatively large horizontal forces. Brake forces of railway bridges may attain two-thirds of the vertical force at a particular pier. Transverse wind forces are more important with road bridges. There is no doubt that great horizontal forces increase the loads acting on the foundation rock and therefore, if possible, fixed bearings should be located on the supports with the best foundation conditions.

How the geological structure affects the foundation of bridges can be exemplified by the Nusle bridge in Prague. The erosive valley was cut in weak Ordovician shales (Fig. 11—21) traversed by several strike faults. Rather hard shale layers form the valley slopes. In order to reach less decomposed rock, the piers in the valley bottom were each founded on 8 cylindrical shafts 3 m in diameter.

Overdeepened river channels under the valley deposits have advanced the idea of large-span bridges. Washington bridge in New York presents an example. The span of its central opening is 1064 m. The eastern abutment (Fig. 11—22) on the left bank is founded on Precambrian gneiss and the western abutment on Triassic shale and sandstone, at a depth of 26 m. The top of the western valley slope is formed by an intrusive body of diabase, in which the anchoring galleries of the back-stays have been situated. Although the valley cross-section and the sites of supports were carefully investigated, unexpected difficulties arose with the foundation of the western abutment. Excavation and sealing of the foundation pit were impeded by the bouldery diabase debris covering the bedrock.

11.8. Activity of the engineering geologist during construction and maintenance. As with other large civil engineering structures, the construction of roads also requires the co-operation of the engineering geologist. Large cuttings extending deep into rocks demand continuous examination and survey during construction. Temporary exposures on the cutting slopes and on the walls of excavation pits provide important data on the geology of the area and therefore they should be thoroughly logged. The accuracy of the results of geological investigations can be verified at the same time.

Collaboration of the geologist with the resident engineer is often necessary, as the sequence of excavation works, transient shapes of excavation slopes, the dimensions of stabilization berms, drainage works, etc., often need to be adjusted to the actual conditions found. The geologist must also see that the excavation is performed in layers, that the transient excavations are drained, and that the stockpiles of humus material are not deposited so that they surcharge the slope edges and that water can accumulate behind them.

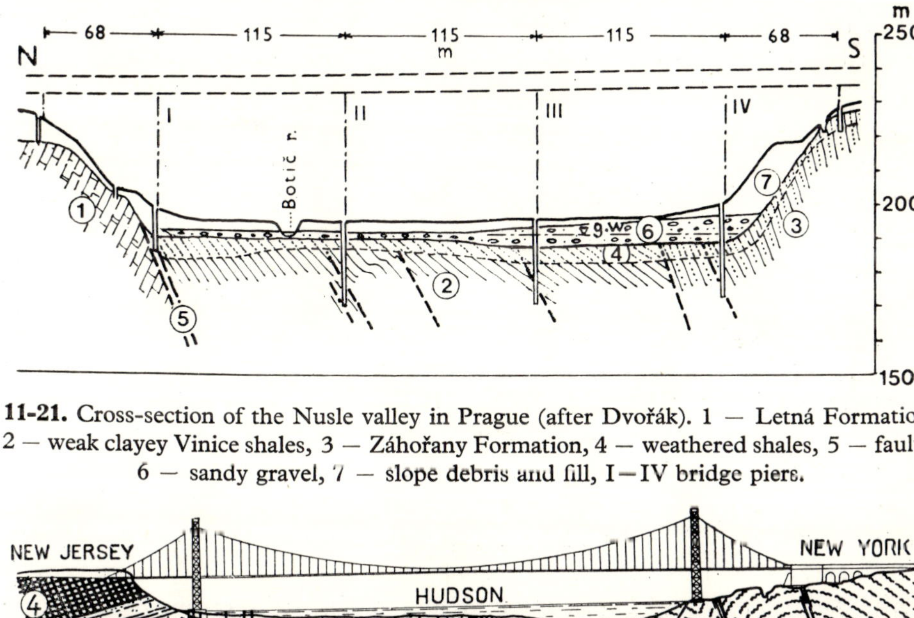

11-21. Cross-section of the Nusle valley in Prague (after Dvořák). 1 — Letná Formation, 2 — weak clayey Vinice shales, 3 — Záhořany Formation, 4 — weathered shales, 5 — faults, 6 — sandy gravel, 7 — slope debris and fill, I—IV bridge piers.

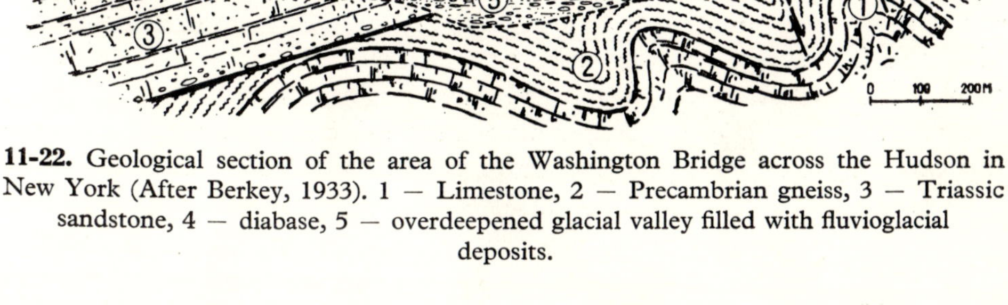

11-22. Geological section of the area of the Washington Bridge across the Hudson in New York (After Berkey, 1933). 1 — Limestone, 2 — Precambrian gneiss, 3 — Triassic sandstone, 4 — diabase, 5 — overdeepened glacial valley filled with fluvioglacial deposits.

Chapter 12

TUNNELS AND UNDERGROUND POWER PLANTS

12.1. Introduction. Geological investigations for tunnel construction are of a very responsible nature, since geological conditions are decisive for the choice of the route and the correct construction methods; they greatly affect the cost and safety of operations.

In tunnel construction, geological studies were first applied to civil engineering works situated at a great depth below the land surface, so that all assumptions of the deep geological structure were immediately checked. Early geological reports did not prove quite satisfactory in this respect, as effective working methods had yet to be developed and practical experience had yet to be gathered. The existing ingenious hypotheses concerning the geology of the earth surface layers came nowhere near to meeting the requirements of tunnel construction. Only detailed field work could afford the necessary information. The reputation of geological investigations suffered from the erroneous assumption that geological conditions could be reconstructed accurately from rare natural outcrops. These, however, can only serve as a basis for the preliminary studies of tunnel alignment and feasibility of the work, but not for the compilation of construction plans and working procedures. Where there is a lack of natural outcrops, a sufficient number of boreholes, galleries and test pits should be carried out, enabling a detailed geological map and reliable geological profiles of the study area to be constructed. The following two examples will show to what extent the profiles constructed ahead of tunnelling can approach the true picture of geological structure.

Several profiles of the Hauenstein Tunnel on the Basel—Olten railway in Switzerland are plotted in Fig. 12—1. The first profile has been compiled by F. Mühlberg in 1910 and the second by A. Buxtorf in 1911, on the basis of geological studies conducted before the beginning of excavation. The third section was constructed by Buxtorf in 1916 after the tunnel had been driven through. The preliminary profiles proved to be very near to the actual conditions; some major differences have been established only along the faults.

Figure 12-2 shows several profiles through the Grenchenberg Tunnel in the Jura Mountains in Switzerland. Intricate geological conditions permitted several conclusions to be drawn from reconnaissance investigation. The tunnel route cuts through two peaks — Grenchenberg (1405 m) in the south and Graitery

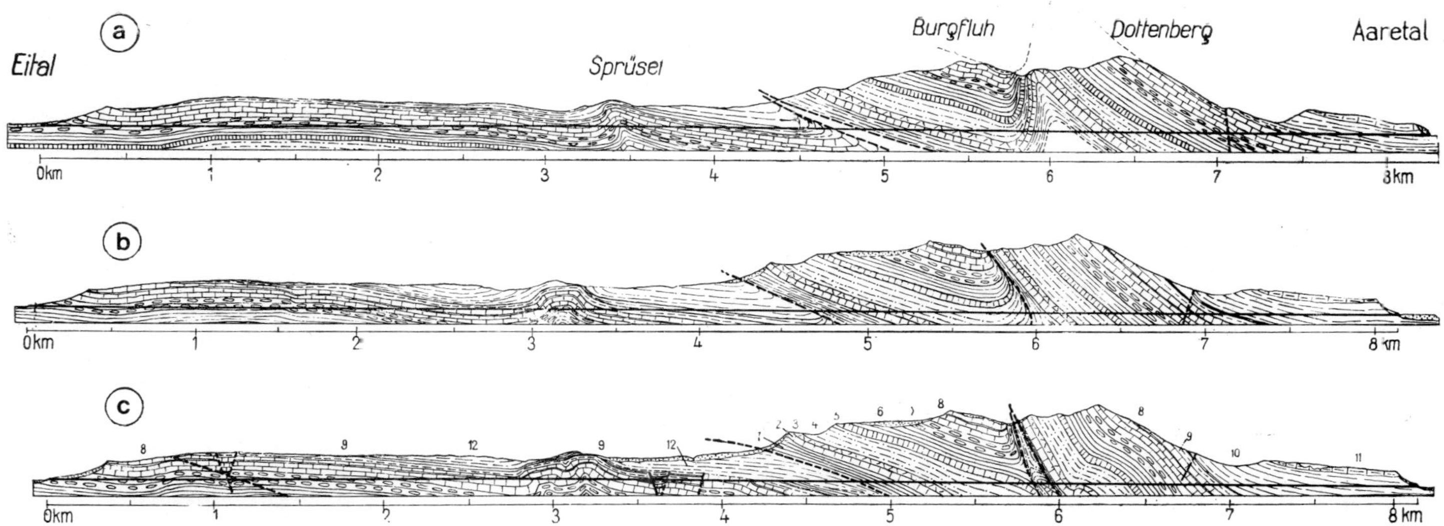

12-1. Geological sections of the Hauenstein Tunnel (Buxtorf, 1916). a — Preliminary section compiled in 1910, b — in 1911, c — in 1916 after tunnel construction; 1—3 Muschelkalk complex, 4 — Keuper, 5 — Lias, 6—9 — Dogger, 10, 11 — Malm, 12 — Tertiary.

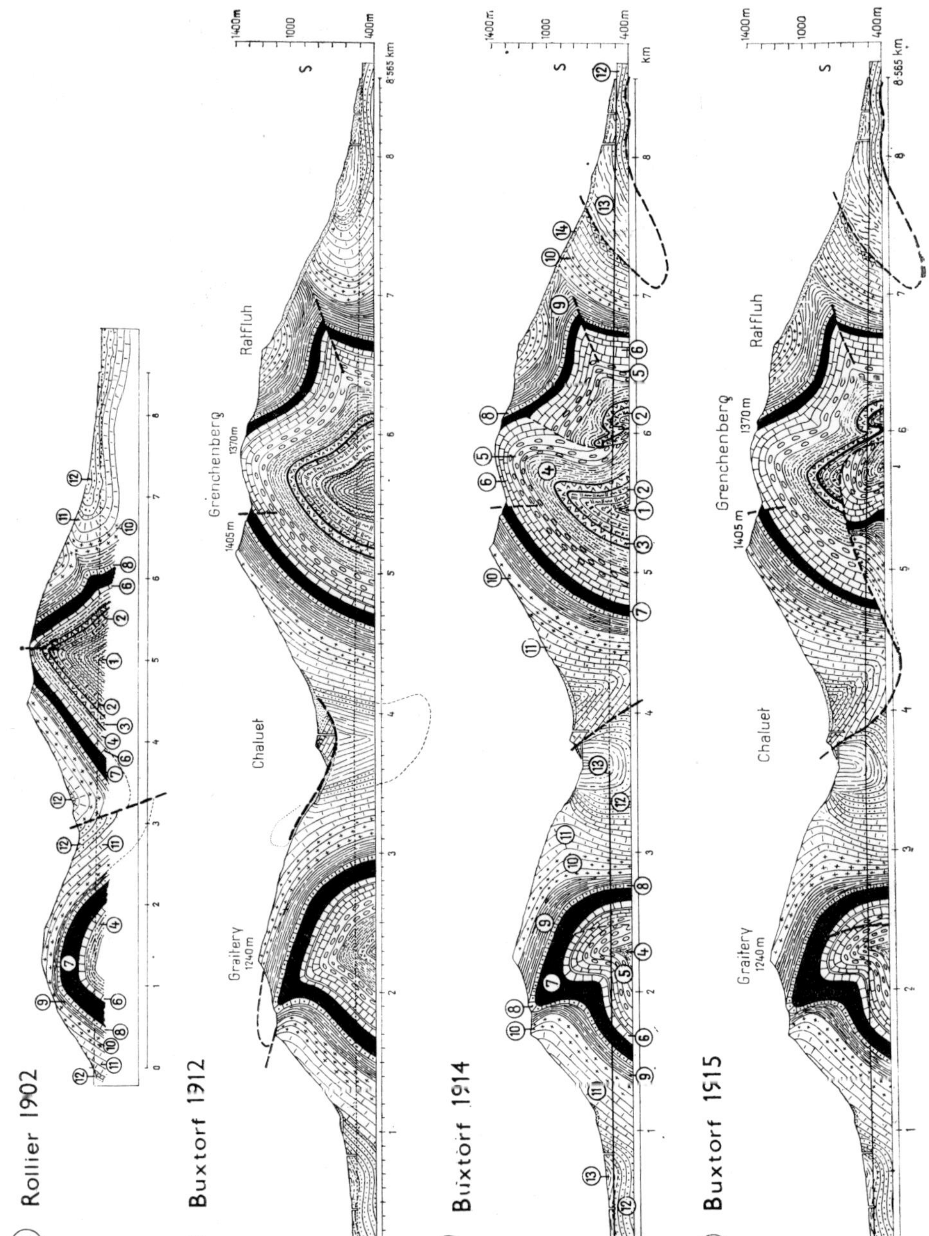

12-2. Geological sections of the Grenchenberg Tunnel (Buxtorf, 1916). 1 — Lower and Middle Triassic, 2 — Keuper, 3 — Lias, 4—6 — Dogger, 7—11 — Malm, 12 — Oligocene Molasse, 13 — Miocene Molasse, 14 — slope debris.

(1240 m) to the north. The tunnel was driven through Middle Triassic to Upper Jurassic rocks and Tertiary deposits. L. Rollier, who prepared the preliminary profile (1902), presumed the tunnel would pass through two major anticlines of nearly symmetric structure, which form the two peaks. He thought the intervening Chaluet ridge to be a double syncline disrupted by a thrust fault, along which the Jurassic overrides the Tertiary Molasse of the southern limb of the Graitery anticline. In this case the preliminary profile showed features very different from those that were to follow. The difference between the first profile and others is also due to the fact that it related to a somewhat different tunnel route.

The detailed geological investigations of A. Buxtorf and others carried out before, during and after construction gradually disentangled the intricate structure of the area. The following three profiles show that the structure of the Molasse Zone at the northern tunnel opening was recognized correctly. The Graitery anticline was found to be more complex than Rollier thought. Its apical part is overturned slightly northwards and faulted; the synclinal bend disturbing its northern limb has been found only in driving the advance heading.

As to the Chaluet zone intervening between the two peaks, Buxtorf (1912) presumed that the Tertiary molasse formed a simple syncline, on which a block of Jurassic rocks slid from the southern slope of Graitery. However, several boreholes sunk before the start of tunnelling showed that the syncline is divided by a thrust fault along which the Jurassic crops out. Rollier's original opinion proved to be nearer reality.

Geological and tectonic conditions of the Grenchenberg anticline were still more complicated. Although Buxtorf's profile of 1912 was very ingeniously conceived, it does not agree with existing conditions. After completion of the advance heading it was found that the Jurassic rocks of the southern limb had been thrust southwards on the Tertiary molasse and that the core was disrupted by the overthrust of the Chaluet zone. The profiles of the tunnel from 1914 and 1915 show how difficult it is to visualize the tectonic details along the tunnel route. Many features that could not be anticipated from the results of reconnaissance investigation were resolved only after the middle part of the tunnel had been driven through.

In the above case co-operation of engineers with geologists was advantageous to both sides. Geologists assisted in recognizing the geological structure of the area with respect to the demands of tunnel builders, but at the same time they obtained valuable information about the structure of mountain ranges. Construction of deep Alpine tunnels, in particular the Simplon Tunnel, yielded many valuable data on the geological structure of the Alps. The construction of the Harmanec Tunnel on the Banská Bystrica—Horná Štubňa railway line (Slovakia) furnished evidence of the long-presumed nappe structure of the Velká Fatra Mts. Overthrust planes, along which Triassic rocks were thrust on younger Cretaceous beds, were exposed at several places by tunnelling (Záruba, Andrusov 1939).

12.2. Tasks of engineering geological research. The extent of engineering geological studies depends, as with other civil-engineering structures, on the stage of the project and on the importance of the work. They include reconnaissance and detailed investigations, and co-operation during construction. The problems that should be studied are the following:

(1) Study of geological conditions of the area and of the properties of the rocks. Results are taken into consideration in choosing the general alignment of the tunnel, with due regard to other requirements. The „suitability" of rocks for tunnelling changes with the development of technological methods. Rocks that exert great pressure will be undesirable for tunnel workings in any case, but since modern high-speed machines react sensitively to changes in the hardness of rocks, tunnels driven in rocks exerting a uniform medium pressure may be less expensive than those driven in strong rocks whose strength and hardness vary greatly over short distances.

(2) Hydrogeological investigation of the area along the tunnel route, from which may be estimated in which sections and in what quantities inflows of water into the tunnel can be expected. Hydrogeological conditions are of particular importance in planning tunnels for water supply, undergrounds, and all tunnels crossing areas where corrosive waters are present. The occurrence of corrosive water is so undesirable that, if possible, such sites should be avoided. In some cases, the presence of gas and the temperature of the environment need to be taken into account when assessing the suitability of the site.

(3) When the general alignment of the tunnel has been selected, features controlling its exact position are investigated. The tunnel should not be located in rocks disturbed by weathering or disrupted by faulting. If the fault zone cannot be avoided, it is desirable to align the tunnel at right angles to it. The major joint and fracture systems are also the controlling factors, particularly in underground openings of large diameter (e. g. for underground power houses).

(4) Location of tunnel entrances. Care must be taken not to disturb the stability of slopes by approach cuttings. Sliding movements and pressures could delay the working progress.

(5) In order to estimate the cost of the tunnel construction and to elaborate the detailed project, pressures on tunnel lining must be determined and, in connection with it, tunnelling methods should be considered.

(6) Assessment of drillability and excavability of rocks is necessary for providing the construction site with adequate equipment.

(7) Co-operation in securing necessary aggregate material for tunnel lining, since its supply must be continuous as the work progresses.

(8) Study of the effect of tunnel working on the surrounding area. The drawdown of the ground-water table may result in the disappearance of water in wells and springs. Subsidence of the ground surface may damage buildings, transportation routes and others.

Reconnaissance investigations are based on geological maps and detailed field study. Usually several test pits and boreholes are sunk to expose the bedrock or fracture zones buried by superficial deposits. The results of reconnaissance studies are plotted on the map and in preliminary geological sections of alternative tunnel sites.

Detailed engineering geological investigation must be carried out, to refine assumptions on the geological structure of the area deduced from preliminary studies. An advance heading is the most useful tool for obtaining the necessary information. It is not always used in modern tunnelling, which is to be regretted from the geological point of view. In excavating the diversion tunnel (13.5 km long) of the river Isère (French Alps) in a complex of crystalline schists, the work had to be interrupted when the tunnel entered talc schists and anhydrite and the tunnelling method had to be changed. At first, horizontal test borings were used for investigation, but these proved unsuccessful because of small core recovery, so then an exploratory drift was drilled ahead to a length of 1 km (Olivier-Martin, Kobilinski 1955).

The use of heavy machines in modern tunnelling, which are difficult to transport over large distances and which need a free, untimbered space, makes it very important to know the geological structure along the tunnel route in the greatest possible detail. To achieve this, the advance headings are replaced by vertical boreholes, if tunnels are to be driven at a small depth, and by horizontal test holes drilled ahead of the face in deep tunnel constructions.

12.3. Investigation of the geology of the area and general alignment of the tunnel. In cases where a geological map of at least 1 : 25,000 scale is not available, reconnaissance investigation must involve detailed mapping of a wider area along the tunnel line. Mapping of a narrow belt would not give a complete picture of prevailing geological conditions. For deep mountain tunnels the whole drainage area of the designed tunnel should be mapped.

During reconnaissance investigation, dips and strikes of beds cropping out in all valleys and gorges transecting the mountain range are measured, particularly in those running parallel to the tunnel route. Deep valleys and gorges in its proximity are the best sources of information concerning the mode of deposition and stratigraphy of rocks to be encountered in the tunnel. Where there is a lack of natural outcrops, subsurface exploration is necessary. If the boundaries of individual strata are covered by superficial deposits, the use of test pits is advisable. On the basis of the geological map, complemented by the results of field and subsurface exploration, one or several longitudinal profiles, depending on the number of alternatives, are constructed. It is also recommended to construct several cross-sections (Fig. 12-3).

The prediction of geological conditions for deep, long tunnels is a difficult task. Because of the lack of well-exposed boundaries between geological formations

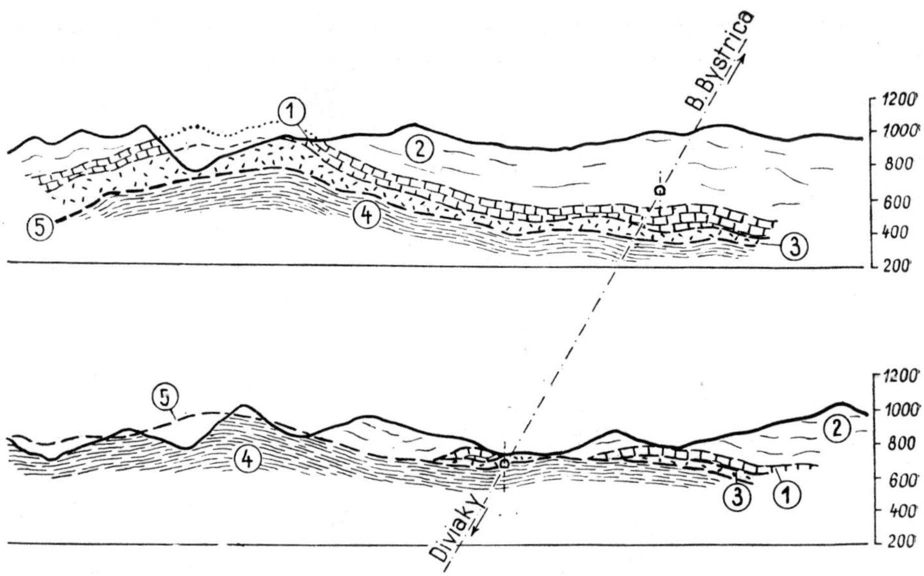

12-3. Geological cross-sections of the area of the Harmanec Tunnel. 1 — Triassic limestone, 2 — dolomite of the Choč nappe, 3 — dolomitic breccia, 4 — Neocomian marly limestone of the Krížna nappe.

and minor inaccuracies of topographical maps, geological methods do not always furnish an unequivocal solution for constructing geological profiles. In areas where stratigraphic and tectonic conditions show a certain regularity, the demands on preliminary geological sections can be stricter than in mountain regions with complicated tectonic structure.

Two geological profiles of the Harmanec Tunnel in Slovakia are shown in Fig. 12-4; the first was compiled in 1937, before the start of construction, from detailed

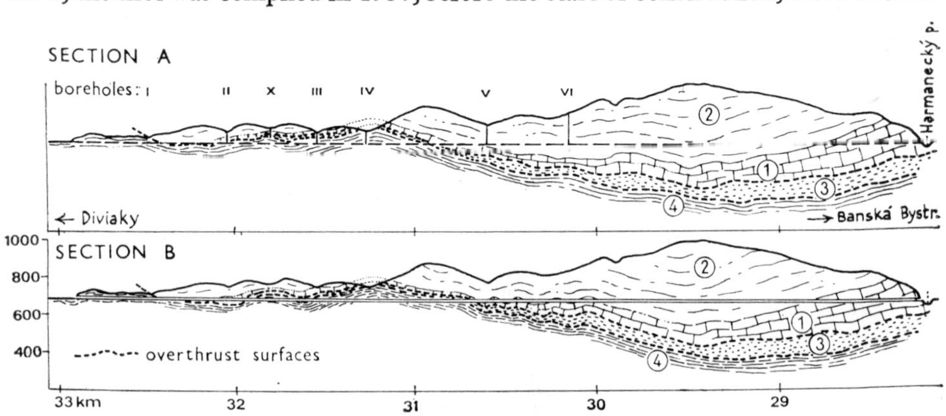

12-4. Geological sections of the Harmanec Tunnel (Záruba, Andrusov, 1939). 1 — Triassic limestone, 2 — dolomite, 3 — dolomitic breccia, 4 — Neocomian marly limestone of the Krížna nappe.

mapping and a few deep boreholes, and the other in 1939 after the advance heading was excavated. From the two profiles it can be seen that the geology of the area corresponds broadly to the predictions of geologists. The major part of the tunnel was driven through the Middle Triassic limestones and dolomites of the Choč nappe, which had been overthrust on the Lower Cretaceous marly limestones of the Lower Subtatric nappe. A layer of shattered dolomite and dolomitic breccia intervenes between the two nappes, probably belonging to the lower one. In the course of excavation thrust surfaces were exposed at several places; these confirmed the assumption of the overthrust of older Triassic rocks on younger Cretaceous complexes.

The two profiles differ most markedly between km 30 and 31 where the thrust surface runs higher than was presumed. As a result, the limestone sequence recurred four times along the tunnel line and the thrust surface was encountered about 100 m nearer to the Harmanec portal. Between km 31.5 and 32.5 the downwarp of the thrust surface was greater than the authors of the geological profiles anticipated. These minor inconsistencies are partly balanced against one another so that the portions of single rock formations encountered in the tunnel do not differ substantially from their predicted lengths.

	Presumed	Actual
In dolomites	50 %	47 %
In Triassic limestones	16 %	19 %
In Cretaceous marly limestones	34 %	34 %

Tunnel designers often regard the length of the tunnel as a major factor in estimating the cost of the work. They do not realize that the cost of excavations in unfavourable rocks can be twice as high as in rocks of suitable properties. In Moravia the shortest possible route for a water-pressure gallery was chosen, although it ran in soft chlorite schists. If the gallery were located in firm gneiss and mica-schist surrounding the belt of these rocks, its length would be longer by 20 %, but excavation and lining would be appreciably cheaper.

Since the geological conditions are not the only factor to be considered in tunnel planning, unfavourable rock complexes cannot always be avoided. This was not the case in planning the northern branch of the Trans-Iranian Railway. There, the geological conditions could and were respected and the alternative alignments of numerous tunnels were appraised according to the length of sectors to be driven in unfavourable Jurassic shales (Rabcewicz 1938).

12.4. The hydrogeological conditions of the tunnel area should be studied carefully as ground water is one of the most serious hazards in tunnelling. It is not the authors' intention to discuss hydrogeological problems in detail but only to point out the importance of a correct estimation of the inflow of water into a tunnel and of its pressure on the lining.

Inflow of water into the tunnel increases construction costs by about 20 %, as water-proofing of the lining, and often also drainage behind the lining, are necessary. With large underground openings, such as the caverns of power stations, drainage of the lining is imperative as it could not resist the ground water pressure. In contrast, the external water pressure is favourable for water-pressure tunnels (Sec. 12.9), and its estimation is therefore desirable. Extremely high inflows below the river valleys make the drainage of tunnels impossible.

In general, the requirement is to estimate the volumes of water inflow at specified points along the proposed tunnel axis. The question of whether concentrated flows or dispersed inflows are to be expected is of great concern since the former may threaten tunnelling operations. The dimensions of the temporary and permanent collectors should also be found prior to tunnelling and the total cost of water-proofing should be estimated in advance in order to avoid difficulties of economic character. The chemistry of the ground water should likewise be determined, as corrosive water can be deleterious to the lining. The estimation of tunnel sections in which an increased water inflow is to be expected is of particular importance when the tunnel is to be driven downward.

Table 12.1

Mont d'Or	3000 to 10 000 l/sec
San Jacinto (California)	to 3000 l/sec
Simplon	to 1200 l/sec
Apennines	to 1200 l/sec
Harmanec (Czechoslovakia)	to 800 l/sec
Grenchenberg	to 800 l/sec
Tecolote (California)	to 680 l/sec
Weissenstein	to 450 l/sec
Gotthardt	to 230 l/sec

The railway tunnel across the Apennines between Bologna and Florence was driven from both portals as well as from the inclined adit in the mid-length of the tunnel. The inflow of water into the downward-driven section was so high that the pumping station had to be enlarged several times; it contained 37 pumps with a discharge of 1200 l/sec before the tunnel has been holed through.

Discharges from several tunnels are given in the table above, which shows the extreme inflows that can occur and the necessity to provide adequate drainage measures.

The increased inflow of water can occur in rock tunnels:

(1) Along *fractures and faults*, especially where rocks are strongly disrupted tectonically. A heavy inflow of water was encountered during the driving of the advance heading of the Bral Tunnel (railway Štubňa-Handlová in Slovakia). Water was stored in and fed by a fault separating Palaeogene shales from an andesite body. Water burst into the heading under high pressure so that the gallery caved in. An auxiliary drainage gallery had to be driven before the driving of the advance heading could be recommenced.

(2) Through *permeable rocks* of high porosity, such as porous sandstones, arkoses and weathered granite. Inflows are greater near the contact of these rocks with less permeable rocks.

(3) By *karst openings* in limestone which can conduct even underground streams into the tunnel in areas lacking all traces of water on the ground surface. During the construction of the Grenchenberg Tunnel a strong underground stream had stopped the driving of the advance gallery. Only after two months, when the inflow sank from 800 to 570 l/sec, was driving recommenced. During the construction of the Harmanec Tunnel (Fig. 12-4), the face of the advance heading was flooded by a concentrated inflow of 400 l/sec in January 1938. As the possibility of inflow had been anticipated and provisions had been made, the work was interrupted for only two days.

To estimate the amount of water that might invade the tunnel is not easy. An approximate calculation is based on the delimitation of the catchment area, from which water could be drawn into the tunnel, in the geological map. The boundary of the hydrogeological catchment area does not generally coincide with the boundary line of the geographical basin. When permeable and impermeable rocks alternate, the extent of the catchment area is defined by the outcrops of the lowest impermeable stratum which will be encountered during the tunnel driving. During the hydrogeological investigation of the Harmanec Tunnel (Fig. 12-4), the catchment area was found to be elongated from north-east to south-west. This direction was parallel to the axial planes of gentle folds, which deformed the tectonic nappes in the tunnel environment (Fig. 12-3). The folds have a significant bearing on the hydrogeological conditions of the tunnel. Only water stored in the syncline of Lower Cretaceous impermeable marl will be drained by the tunnel. After the demarcation of the catchment area has been carried out, the annual rainfall should be determined. A dense net of rain-gauge stations exists in Czechoslovakia which can provide reliable records of the rainfall measured in the last 90 years. The total annual volume of rain can therefore be calculated for the catchment area.

Another question is the determination of the infiltration quotient. The relationship between the quantities of water which evaporate, flow out and infiltrate varies considerably according to local conditions of the catchment area. First it is the climatic character of the area, especially the distribution of rainfall, the duration

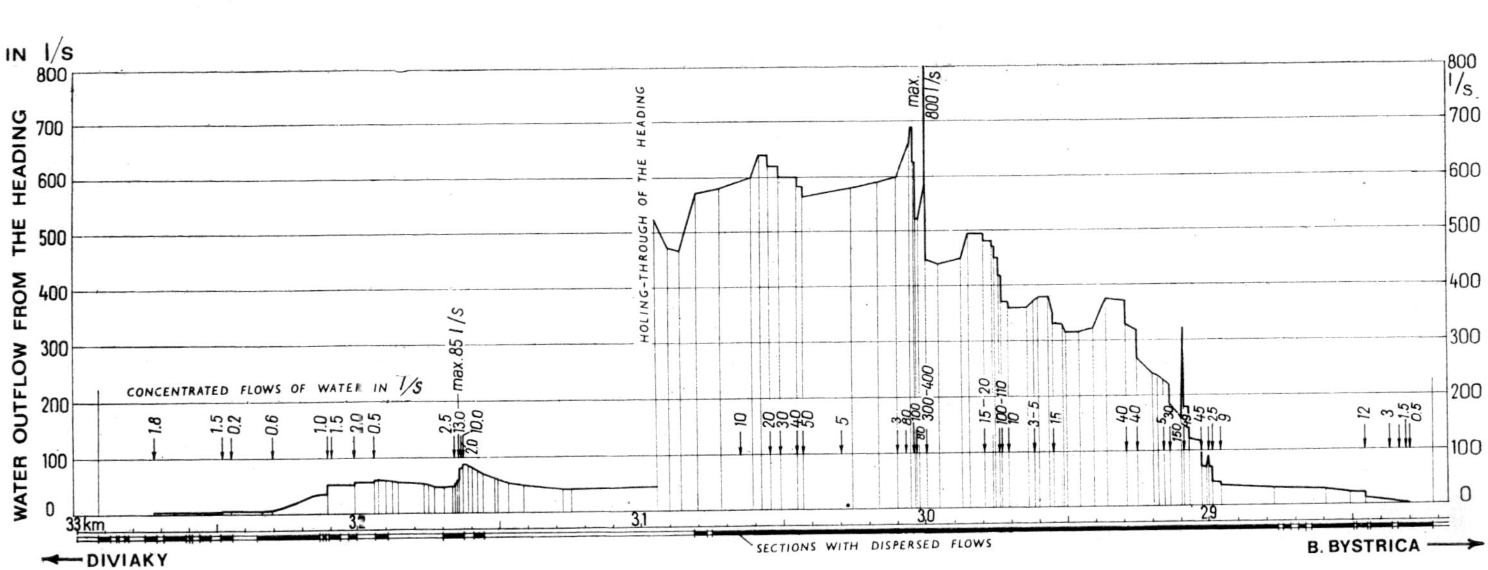

12-5. Time-discharge diagram during the driving of the advance gallery of the Harmanec Tunnel; discharge measured at both entrances. (Záruba, Andrusov, 1939.)

of rains and the ratio of rain to snow. The temperature and humidity of the air also govern evaporation. Other relevant factors are: (a) The shape of the ground surface and its geological character. Steep slopes consisting of impermeable rocks promote runoff while infiltration prevails on gentle slopes covered by permeable products of weathering. (b) The nature of the vegetation, i. e. whether the area is cultivated or whether forests are coniferous or deciduous, etc.

Analysis of these questions for the Harmanec Tunnel in Slovakia showed a catchment area of about 26 km² overgrown with two types of forest, and formed by steep slopes with outcrops of dolomite and limestone. These rocks are permeable, since they contain numerous fractures and underground channels. Therefore an infiltration quotient of 40 % was adopted. The evaporation quotient of 40 % corresponded to a surface covered by deciduous forest. Only 20 % was attributed to the runoff, as the valleys, especially their upper sections, were almost dry. Surface-water flows existed only during downpours. On the basis of the given data an average discharge from the tunnel of 330 l/sec was estimated. The assumption was made that the flow was not uniformly distributed along the tunnel length but was concentrated in several sections, in which the short-term outflows could attain even a multiple of the calculated average discharge.

After the tunnel had been holed through, the given forecast proved true. Short-term discharge from the tunnel was 800 l/sec (on 17th January, 1938, Fig. 12-5) at the Harmanec entrance. After the holing-through it decreased to about 400 l/sec and to an even lower value in dry years. It appears that the infiltration quotient had been rather underestimated and that for the permeable dolomite a value of 45 to 50 per cent should be considered.

When the impermeable strata form a basin and are overlain by permeable rocks, an investigation of the ground-water level is required. If the tunnel was situated above this level, the flow would be small. In contrast, with the level of the tunnel below the ground-water table, concentrated flows much higher than those calculated, are to be expected (Fig. 12-6) until the water table is lowered to the tunnel

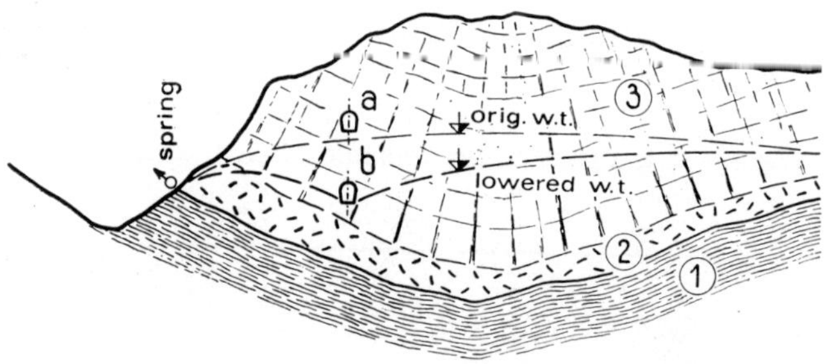

12-6. The alignment level of a tunnel in relation to the ground-water level controls the discharge from the tunnel. 1 — Neocomian marly limestone, 2 — dolomitic breccia, 3 — cavernous Triassic limestone.

level. For example, near the portal of a tunnel near Brno an outflow was measured and found to give a discharge of about 600 l/sec. Water was stored in the Tertiary sands and trapped by marly debris. The discharge diminished gradually to about 1 l/sec corresponding to the yield of a former spring. In order to avoid slaking of marl debris a short drainage gallery was driven. Horizontal drainage boreholes would be preferred nowadays.

During preliminary investigation attention should be given to the consequences of the drainage for *the tunnel environs*. With the exception of tunnels below the valley and also water-pressure tunnels, a tunnel lining usually allows water to be drained out from the rock (Fig. 12-7). In populated areas wells commonly dry up and an alternative water supply should be provided.

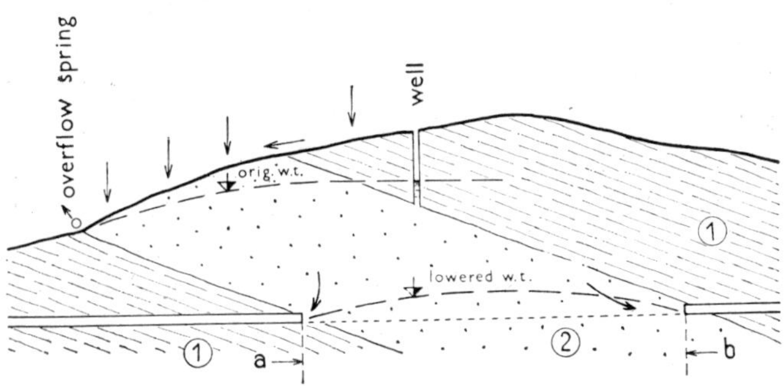

12-7. Lowering of the ground-water table caused by tunnel driving. Water in the well will also disappear. 1 — Impermeable shales, 2 — permeable sandstone, a, b — sector of increased inflow.

A particular task of hydrogeological investigation, which arises especially when driving tunnels below the valley bottom or when constructing a water-pressure tunnel (Sec. 12.8), is to evaluate the *pressure of ground water* to which the lining will be subjected. This question is far from being simple with water-pressure tunnels crossing hills high above the bottoms of adjacent valleys. Dry and saturated sections often alternate along the tunnel. Since the estimation of groundwater pressure is of great importance for the thickness of the lining, the exact measurement of pressure is frequently demanded. There is no need to explain why the results are not reliable in many cases, especially when the tunnel has temporarily performed a drainage function. Therefore, hydrogeological investigation of the mountain range prior to tunnel driving is of good assistance.

12.5. Detailed investigation for the location of the tunnel. After the general location of the tunnel has been proposed, especially with regard to the rocks to be traversed as well as to the hydrogeological situation, the more precise selection of the alignment is the next task. This is made with regard to important exogenous and endogenous factors which have affected the rocks in the past.

Many cases exist where even deeply located tunnels have been afflicted by factors extending from the ground surface. Several of these cases are still remembered by engineers and geologists due to their tragic consequences. Such is the case of the Lötschberg Tunnel (1908) traversing the granite massif of the Berne Alps. The possibility that the glaciofluvial filling of the valley of the Kander river should be encountered at a depth of 170 m under the valley was not anticipated. The seepage from boreholes in the front face of the gallery provided a warning. After the blast, water and sand flooded the advance heading and killed 25 tunnelmen. The alignment has since been relocated by forming a by-pass to avoid the dangerous area.

The builders of the Gotthardt Tunnel (1872—78) also nearly faced a similar catastrophe. Boreholes drilled in the 1940's upwards from the tunnel at km 2.7/8, where a high pressure had been encountered during the tunnel driving, revealed that the thickness of the rock overburden was only 40 m and that the remaining 280 m to the ground surface consisted of the saturated glacial drift of the Andermatt basin. The Wäggital water tunnel was driven under the Schwendibach valley (Fig. 12-8). Because the valley was covered by a moraine, an exploratory shaft

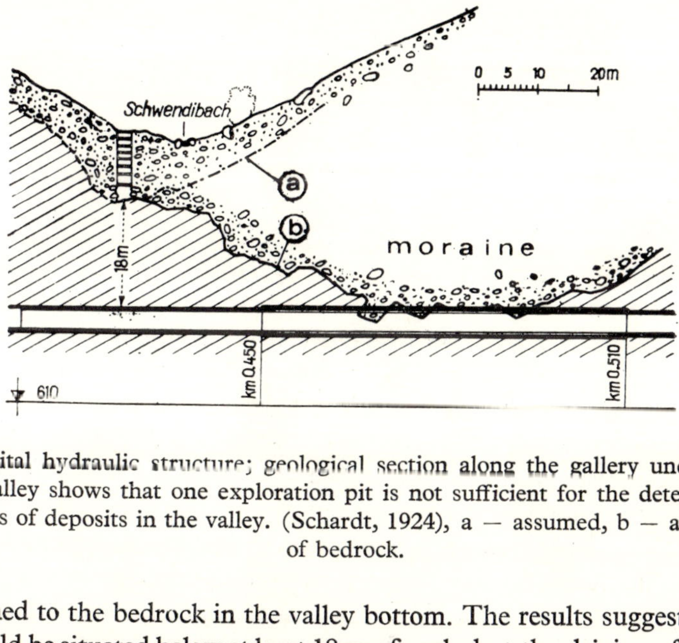

12-8. Wäggital hydraulic structure; geological section along the gallery under an overdeepened valley shows that one exploration pit is not sufficient for the determination of the thickness of deposits in the valley. (Schardt, 1924), a — assumed, b — actual surface of bedrock.

was deepened to the bedrock in the valley bottom. The results suggested that the gallery would be situated below at least 18 m of rock, but the driving of the gallery showed that the lowest point of the bedrock does not lie under the valley axis, but about 50 m farther to the north. Tunnel operations in this section proved to be very difficult owing to high water flows and considerable pressure of the cohesionless moraine material.

The construction of the Weißenstein Tunnel (near Gänsbrunnen, Switzerland) in 1907 also met with great difficulties for which a false estimation of the thickness of alluvial deposits of the Raus brook was responsible. Figure 12-9 shows the longitudinal profile of the tunnel in km 3.4—3.6. There a complex of Tertiary sandstones and marlstones was presumed (a), but instead the tunnel had to be driven in alluvial muddy deposits (6) and for the following 120 m its upper part through Pleistocene gravel with great water inflow. The caving-in at several places resulted in the creation of sinkholes.

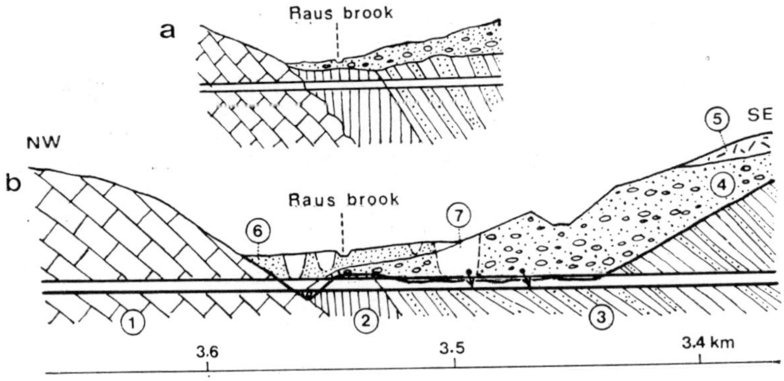

12-9. Longitudinal profile of the Weißenstein Tunnel in Switzerland (Krantz, 1928) a — Profile after preliminary investigation, b — actual profile. 1 — Jurassic limestone, 2 — Tertiary ferruginous sandstone, 3 — Tertiary sandstone and marl, 4 — Pleistocene gravel, 5 — slope debris, 6 — alluvial muddy deposits, 7 — sinkholes produced by caving-in of the tunnel.

Another source of difficulty is unexpectedly deep-weathered rock. Weathered and weakened Silurian shales were met at a depth of 150 m when driving the 1.2 km long highway tunnel below Mt. Big Walker in Virginia. Because the pillar left between the two 10.5-m wide tunnel openings was only 8.4 m thick, its stability was endangered and the driving operations had to be modified.

Tunnels situated in the valley slopes may be endangered by the release of the natural horizontal stress in the rock in the valley slopes (Sec. 3.6), and by the gravitational slope movements. The case of the Unterstein Tunnel in the Salzach river valley near Salzburg is instructive (Figs. 12-10,12-11). Construction of this tunnel started in 1874 in order to cross a rock spur consisting of chlorite schists. Because the beds were dipping into the mountain their stability was not doubted and the axis was situated near the slope surface. The work, in 8-m long sections excavated one by one, continued without difficulty until February 1875, when the advance heading had been holed through for the entire length of 164 m. The lining had been finished over a length of 64 m, 15 m were under construction and 25 m were in excavation. In order to advance the operations, several sections were opened in the 15-m length in the spring of 1875. This resulted

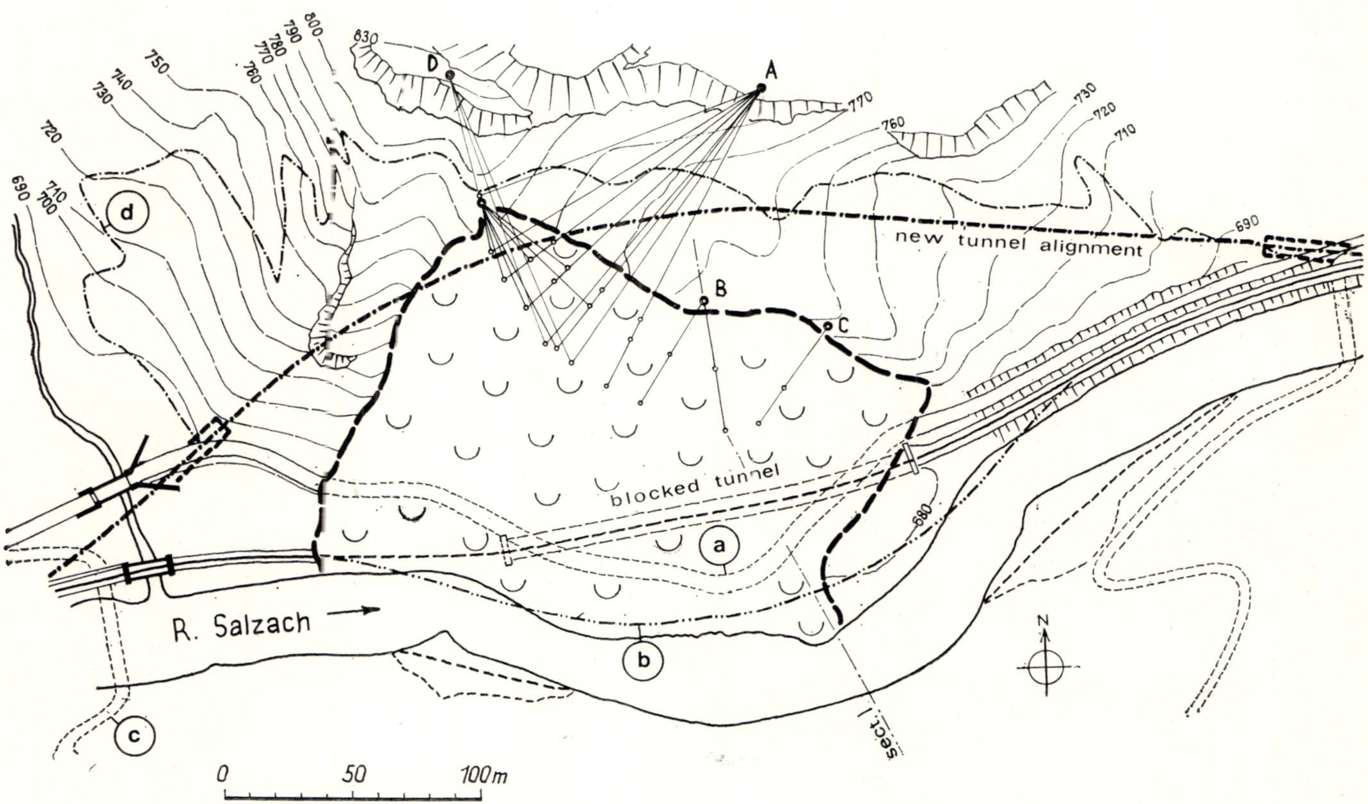

12-10. Caving-in of the Unterstein Tunnel owing to the collapse of the rock slope (Wagner, 1884). a — Blocked highway, b — provisional railway route, c — relocated highway, d — foot path.

in the caving-in of two sections. The rock was found to be transected by fractures perpendicular to foliation, i. e. dipping steeply into the valley. Large pressure began to build up, the dolomite and conglomerate masonry lining of the finished sections was crushed and had to be replaced by granite blocks. However, on 10th June, 1875 the operation had to be stopped and as soon as the crew retired, a total caving-in and collapse of the slope occurred. The highway was blocked and had to be relocated on the other bank of the river. The collapsed slope was investigated by means of galleries and its displacements were measured. Results led to an abandonment of the proposed line and a new tunnel, 400 m long and located deep in the slope was constructed over the years 1877—88.

The unexpected occurrence of weak rocks or fault zones may also present an obstacle to tunnel operations. The existence of weak zones in a rock can often be detected from the morphological features of the ground surface. They promote depressions and saddles in the relief. As a saddle presents the possibility of the shortest tunnel, it is attractive, but the sceptical opinion that the rock in the saddle area affords reduced resistance to external agents is to be encouraged. In Slovakia, a tunnel which was situated below a saddle, which originated along a major fault, had to be driven in saturated, running crushed gneiss.

Underground openings of limited length provide the possibility to situate the axis normal to the major surfaces of weakness. This orientation is possible, for example, with the auxiliary caverns of surge chambers, access tunnels of underground power stations, etc. The wider and higher the opening, the more urgent is this requirement and it is fully respected during the design of caverns of power stations (Sec. 12.9). Major fault zones should be crossed at right angles, even by longer tunnels, and the tunnel axis must be adapted to this requirement by inserting suitable bends.

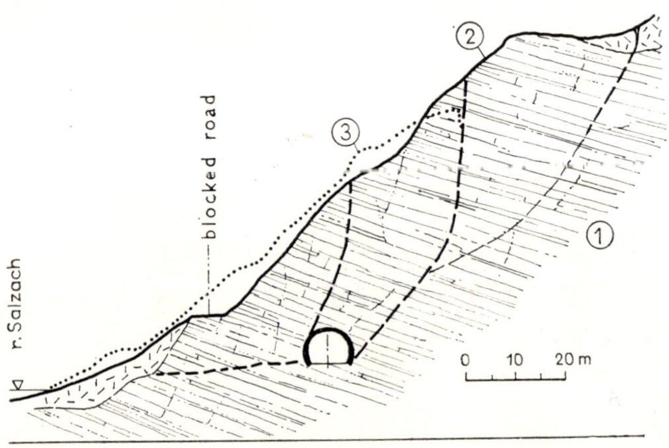

12-11. Cross-section of the collapsed slope above the Unterstein Tunnel (Wagner, 1884). 1 — Chloritic schists, 2 — original slope surface, 3 — slope surface after sliding.

The effect of the bedding orientation on *tunnel driving in stratified rocks* is illustrated in Fig. 12-12. The most favourable case occurs when the tunnel axis runs at right angles to the strike of beds dipping in the direction of the driving. Somewhat less favourable is the case of crossing the beds which dip into the ex-

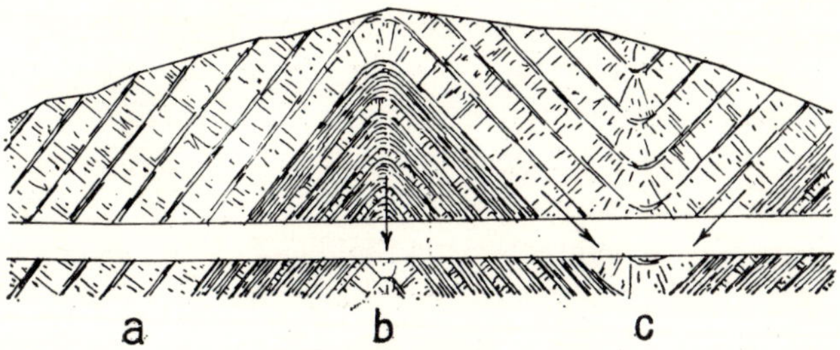

12-12. Tunnel driven through a folded rock complex perpendicular to the strike of beds. a — Beds dipping into the excavation make the blasting easy but the danger of blocks sliding from the face arises, b — when crossing an anticline the driving meets a wedged and fractured rock, c — when driving in a syncline water flows can be expected.

cavation (Fig. 12-12a). Blasting is easy but the danger of blocks sliding from the face arises; the tunnel walls are stable in both cases. When crossing an anticline in the direction perpendicular to the strike of beds, the driving meets a wedged rock and this obstructs the excavation. The danger that the blocks may fall increases owing to a dense fracturing of the rock (Fig. 12-12b). When driving in a syncline (Fig. 12-12c), water flows can be expected, especially when permeable and impermeable strata alternate.

A less favourable location is where the tunnel is driven parallel to the bedding, especially when the rock is not uniform (Fig. 12-13a). Oblique crossing of beds may involve asymmetrical pressures on the supports and large overbreaks. When the tunnel axis happens to coincide with the anticlinal axis (Fig. 12-13c), the local

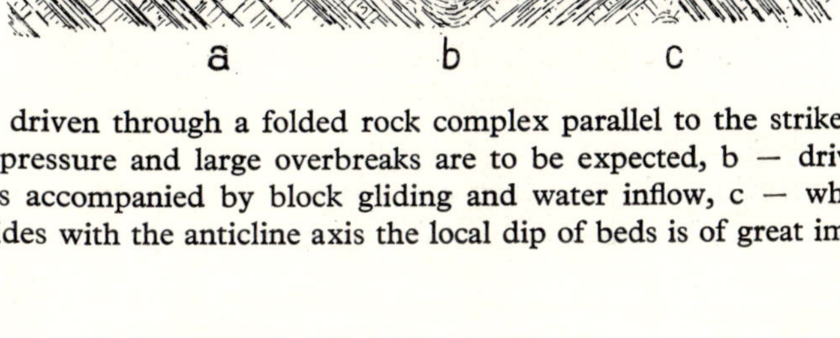

12-13. Tunnel driven through a folded rock complex parallel to the strike of beds. a — Asymmetrical pressure and large overbreaks are to be expected, b — driving along the syncline axis is accompanied by block gliding and water inflow, c — when the tunnel axis coincides with the anticline axis the local dip of beds is of great importance.

dip of beds is of great importance. A gentle dip prevents rock pressure as the roof is arched, but the excavation is not easy. On the other hand, steeply dipping beds do not find support in the tunnel walls and the roof is loaded by the total weight of loosened beds. The location of the tunnel along the syncline axis (Fig. 12-13b) is also to be avoided. Blocks tend to glide into the tunnel from both walls and permeable strata conduct water. Vertical or steeply dipping beds parallel to the tunnel axis are very adverse factors for tunnel driving. The overlying rock mass of great thickness is to be supported and rockfalls high from the roof (chimneys) occur, even when care is taken.

Therefore a thorough investigation of the strike and dip of strata is necessary. The other surfaces of weakness, especially the major joints, also deserve attention.

The investigation of these features from the surface is not an easy task. Prokopovich (1972) performed a detailed comparison of the densities of joints and faults as found in the tunnel and as mapped on the surface. About 1775 joints and 565 faults were mapped over a tunnel length of 3.5 km, whereas only 86 joints and 96 faults were mapped on the surface, although the diorite above the tunnel was only several tens of meter thick. The author recommends opening the surface by excavations or minor blastings in order to obtain better information on the joint pattern.

12.6. Mechanical behaviour of rock and pressure on tunnel lining. The approach of the engineering geologist to statical problems in general has been discussed in Sec. 3.7. In tunnel construction the engineering geologist is expected to master questions associated with the pressure which rocks exert on the tunnel supports.

First, let us realize that no statical analysis can help to prevent the collapse of a roof when two planes of weakness, almost parallel to the tunnel axis, meet roof-like above the crown of a wide tunnel. This can occur particularly in igneous and metamorphic rocks due to the character of their joint sets and fault systems. As all dangerous surfaces cannot be evidenced during the preliminary investigation, the occurrence of their roof-like geometry should be assumed along the total tunnel length. But this assumption would involve a slow and expensive tunnel driving. Therefore, the case must be accepted as a calculated risk, and the risk is taken into consideration by adopting such driving and supporting methods to give advance warning without exposing the workers to great danger. The anchoring of rock and steel supports has been proven to be a good procedure, but in cases of emergency classical methods are of great help. The engineering geologist is expected to evaluate the measurements of surfaces of weakness by statistical methods and to give advance notice when the risk of a meeting of the surfaces above the tunnel is great.

With the exception of the rather abnormal case discussed above, a reliable preliminary picture of the behaviour of rocks around the tunnel can be gained from the analysis of stress and displacement conditions around a circular opening driven into an elastic body. First, let us observe the symbolical model plotted in Fig. 12-14. It relates to a hypothetical assumption that only vertical forces exist in a rock mass. After the opening has been excavated, the vertical forces of the weight of the overlying rock mass have to pass through the abutments. A rock arch develops close to the opening in the walls, but high above the tunnel crown in the roof. Owing to the compressibility of rock symbolized by springs in Fig. 12-14, the spring lines of the arch are displaced apart and a zone of horizontal tensile stress develops in the roof.

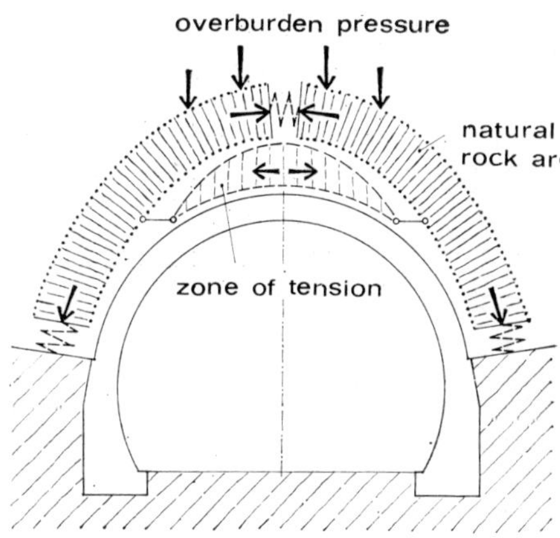

12-14. Symbolical model showing the development of a tensile zone below the induced rock arch above the excavated opening.

Let us observe how the foregoing principal features appear in the results of an elastic analysis. Again, the prevalence of vertical forces is assumed to exist in the rock. The intrinsic vertical stress at a depth $D = 200$ m (Fig. 12-15) equals about $p_z = 200 \times 2.5 = 500$ Mp/m^2. After the tunnel has been excavated, a concentration of stress develops in the vertical section of the tunnel walls, its value being of the order of $3\,p_z$. When plotting the diagram of σ_z we must bear in mind the fact that the two hatched areas in Fig. 12-15 are equal. The elastic analysis (Kirsch 1929, in Timoshenko, 1934) yields:

$$\sigma_z = p_z(1 + B^2/2r^2 + 3B^4/2r^4)\,,$$

in which r is the distance from the centre and B is the radius of the opening. The value $\sigma_z = 1.26\,p_z$ is obtained at a distance $r = 2B$.

The pressure normal to the vertical cross-section is defined by

$$\sigma_x = p_z(1 - 3B^2/r^2)\,B^2/2r^2\,.$$

This yields the tensile stress $\sigma_x = -p_z$ in the roof ($r = B$) and maximum compression stress $\sigma_x = 0.07\,p_z$ at a distance of $1.5\,B$ above the roof ($r = 2.5\,B$).

Let us consider the following facts:

(a) The rock arch in the abutments develops close to the tunnel wall.

(b) In contrast to (a) the rock arch in the roof is widespread and extends very high. A high cover of rock is necessary to provide room for the rock arch to develop. This factor is of importance when designing a tunnel under a shallow cover. The state of stress changes considerably and great shear straining develops.

(c) The horizontal tension in the roof is high.

The settlement in the roof was given for an elastic solution by Borecki (1966):

$$\Delta B = 2p_z B(1 + \nu)(1 - 2\nu)/E .$$

Thus, for example, in a tunnel $2B = 10$ m in width, in a rock with $E = 30\,000$ kp/cm² and $\nu = 0.2$ and for $p_z = 500$ Mp/m², the settlement of the roof equals 1.2 cm, but the actual settlement is rather greater, as the modulus of deformation at tension is small. A slight extension of the opening develops along the horizontal diameter when no horizontal forces are present.

Attention should be given to the tension zone in the roof. Particular danger is associated with tunnel crossings, where accidents often occur due to rockfall from the roof. Therefore, at least a temporary safety support should be errected in tunnel crossings, even in hard competent rocks. The wider the opening, the

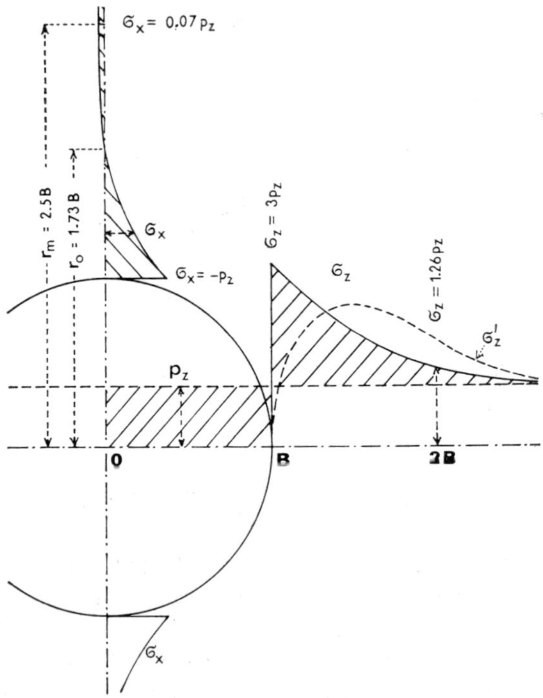

12-15. The influence of the excavation of a tunnel opening on the redistribution of stress in the rock under prevailing vertical loading and small horizontal stress. The hatched areas must equal one another. From this the height up to which (high above the figure rim) the diagram of the compression x reaches can be observed.

higher is the zone of tension stress. For example, the height of the zone would be of the order of 9 m above a cavern 25 m wide; its weight could not be resisted by temporary supports. Even with a rather elliptical opening a small tension zone remains in the roof.

No wonder particular attention is directed to the presence of the inherent horizontal compressive stress in the rock mass (Sec. 3.6). The tension in a roof diminishes as a result of horizontal compression, as can be easily realized when turning Fig. 12-15 ninety degrees. If the magnitude of the horizontal inherent stress was of the order of one third of the vertical stress, no tension zone would originate in the roof. This does not guarantee that the roof will not collapse as the weight of the tensile zone has not been incorporated into the calculation.

Some horizontal stress is always present in rocks in nature, although it is often smaller than the above value. However, near valley slopes, in river bends or along the margins of depressions this pressure is rather small and the danger of caving-in of the roof is therefore great. The caving reaches even several tens of metre high, often up to the surface (Fig. 12-16), due to the presence of vertical joints, the origin

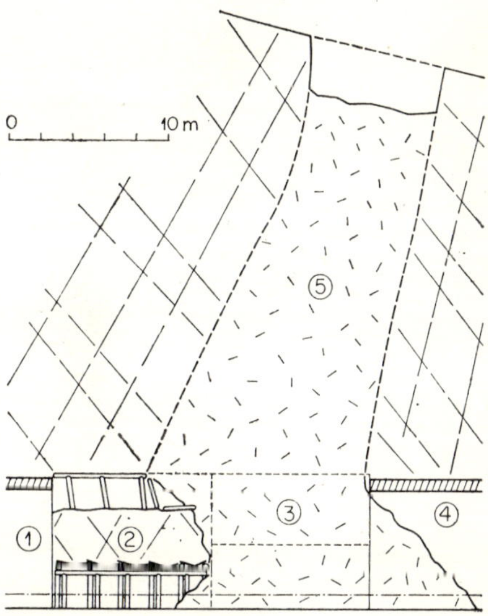

12-16. Caving-in of a slope tunnel. Construction of a double-track railway in the biotite orthogneiss. 1, 4 — Finished tunnel sections, 2 — excavated section, 3 — section under construction, 5 — chimney filled with debris.

of which was also connected with the stress relief upon the erosion of the valley.

Let us observe the stress state in the tunnel walls. Fig. 12-15 shows that $\sigma_z = 2p_z$ (p_x and p_z assumed to be equal). Therefore $\sigma_z = 100$ kp/cm^2 and $\sigma_x = 0$ exist on the surface of the tunnel wall, at a depth of 200 m underground. The $\sigma_z = 100$ kp/cm^2 is very high compared with a yield stress of rocks (the latter assumed to be a limit above which plastic deformations begin to increase considerably). For example, the value of $\tau_y = 52 + \sigma \tan 53°$ (Mp/m^2) has been found for thin foliated gneiss, displaying planes of foliation normal to the tunnel axis. If so, the Mohr circle yields a value of $\sigma_3 = 80$ Mp/m² for $\sigma_1 = 1000$ Mp/m², for a factor of safety equal to one. Therefore the lining is expected to contribute by its resistance of $\sigma_3 = 80$ Mp/m² to the rock resistance in order to retain a pseudo-elastic state in the rock. This pressure is often called a „true rock pressure" or "pressure of displacement", as displacements accompanied the development of the stress state in question. Yet the resistance of 80 Mp/m², required from the lining, is too great. A reasonable value is, for example, 20 Mp/m², which is high enough as it requires a thickness of the concrete lining of the order of one eighth of the tunnel width.

In order to decrease this pressure, a higher straining of the rock is allowed, but it is associated with larger plastic deformations. The recent development of tunnelling methods is linked to the question of how great a plastic deformation may be allowed. By tolerating an increased plastic displacement in abutments the rock pressure decreases. On the other hand, the rock arch displaces deeper into the rock (the diagram of σ_z' in Fig. 12—15), which is not without consequences.

Three zones develop in the rock surrounding the tunnel. The zone of loosened and partially less stressed rock near the walls, the rock arch or the zone of increased stressing behind the latter, and the zone of rather unchanged stress state farther in the rock. In order to avoid further plastic deformation of the rock the lining must develop a certain resistance. The lining should be relatively thin so that it can adjust its shape to the irregularities of the rock pressure. The lining and the loosened zone form a statical unit capable of resisting the pressure of the rock arch.

The fact that the rock pressure decreases with increasing rock deformation may tempt an exaggerated misuse, resulting in at least two situations:

(a) The loosening of rocks with a high clay content would promote their weakening, i. e. the decrease in strength. Therefore, the rock pressure would increase instead of decrease.

(b) The displacement of the rock arch farther into the rock would be dangerous as the span of the rock arch would increase and therefore not only its stability would become more delicate but also the volume of the tension zone in the roof would increase. Because of this the rock pressure on supports would also increase to the magnitude of the "pressure due to loosening". An old saying originating from the tunnelling in the Flysch area was: „If you allow the walls to deform you obtain a great pressure in the roof pretty soon". A schematic diagram of the contact pressure between the lining and the rock is plotted in Fig. 12-17. The decreasing curve represents the true rock pressure while that going up represents the pressure due to loosening. As seen, the optimum state can be achieved with a horizontal displacement in the walls of the order of 10 mm and a vertical displacement in the roof of about 30 mm, under the conditions as plotted. In this way an almost minimal vertical pressure is obtained in the roof while the horizontal pressure in the walls has principally the same value as the vertical pressure. Therefore, the loads acting on the lining are almost uniform. The attaining of this state is one of the aims of modern tunnelling. As far as the engineering geologist is concerned, the relations as explained should incite him to study the state of rock in the walls of the excavated tunnel. Softening of rock, demonstrating the development of the loosened zone, often proceeds unobserved since it is slow.

The problem as analysed in the foregoing is, however, far from being only two-dimensional. An important additional factor is the longitudinal arching of rock above the section in excavation, between the face and the finished lining. Arching parallel to the tunnel axis considerably improves statical conditions of the section under excavation, but great loads are transmitted from this back upon the supports or lining as well as ahead to the rock face. Hence, the next important and complex questions of modern tunnelling are how to activate the resistance of supports or of the finished lining, how long the unsupported excavated section may be, how great are the deformations in the rock face or how may they be prevented.

It is clear that methods of statical analysis were incapable of answering these questions in their full extent until now. Therefore *in situ measurements* are of particular importance, even at the cost that the statical calculation based on them can be completed only after a part of the tunnel has been driven.

One of the important goals of these measurements is to indicate the thickness of the zone of loosened rock, and the depth in the rock at which the arch develops under the conditions of the applied tunnelling method. Direct as well as indirect

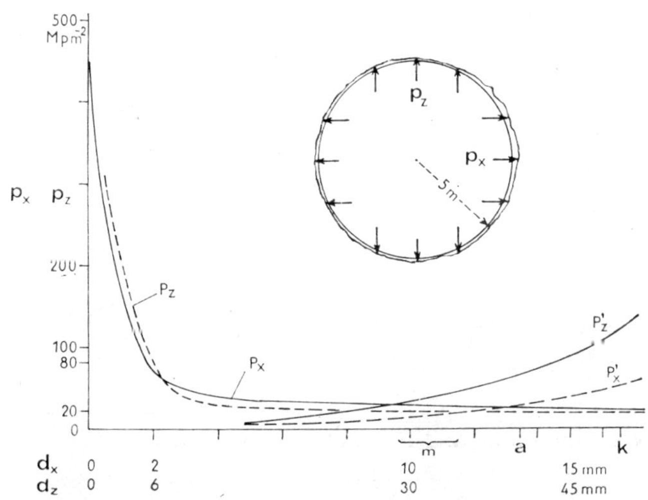

12-17. Yielding of the supports by *d* (mm) in horizontal and vertical directions causes a reduction of true rock pressure *p*, but also an increase in the pressure due to loosening *p*'. Minimum pressure occurs at the yielding of *m*. The common use of steel ribs provides state *a*, old classic tunnelling methods correspond to state *k*. The opening diameter is about 10 m, modulus of deformation of the rock is about 30,000 kp/cm², and the tunnel is at a depth of 200 m.

measurements are used for this evaluation. Measurement techniques continue to develop and therefore only some of them will be described here. Direct methods entail determining the displacements of measuring points fixed in boreholes (the borehole extensometer). The measuring points are fastened at several points of the borehole and by wires connected with its mouth. An abrupt increase in displacements between two points shows the boundary of a displaced block (Fig. 12-18). Boreholes are mostly bored from the tunnel, but sometimes also

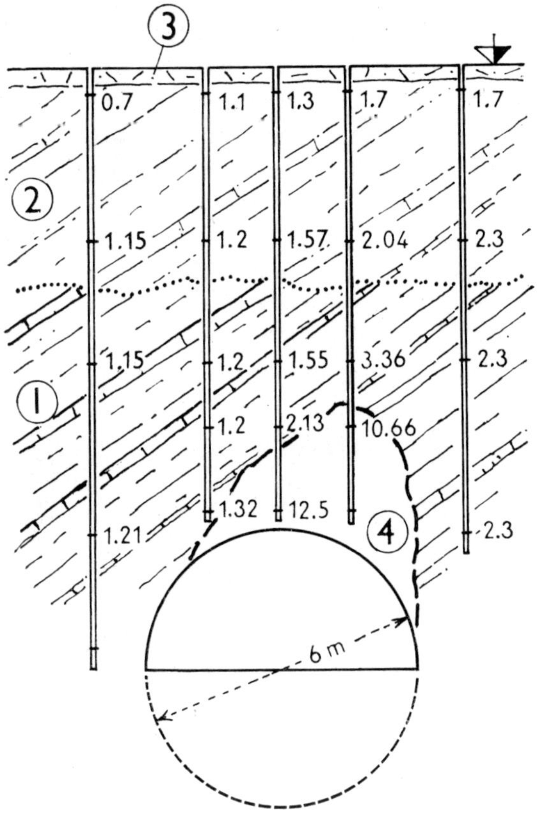

12-18. Magnitudes of subsidence (in mm) of points set in Silurian clayey shales above the excavation for an underground. Supports are steel ribs and gunite. 1 — Weakly weathered rock, 2 — strongly weathered rock, 3 — fill, 4 — loosened block.

from the advance gallery in order to obtain results soon after the full section has been excavated. Shallow located tunnels, as, for example, subway tunnels, make the boring of the holes from the ground surface possible. The displacements of points, measured during the course of driving the tunnel below them, are of interest. The results plotted in Fig. 12-18 were obtained in clayey Silurian shales in the underground construction in Prague. The first distinct settlement of points was of the order of 0.5 mm; this appeared after the tunnel face approached to a distance of 5 m.

Among the indirect methods, *microseismic measuring* is described here. As the pressure in a rock mass increases, the modulus of deformation, and therefore also the velocity of propagation of seismic waves, increase. The results of measurements of velocities of seismic waves around a tunnel are plotted in Fig. 12-19. The velocities were measured between the points of two boreholes sunk 2 m apart in the direction of the tunnel axis. The waves were initiated by a blast. The measured velocities, plotted in Fig. 12-19, indicate the area of increased stress (zone of the rock arch). The measured cross-section is not very different from that of Fig. 12-18, but the results do not coincide. The height of the loosened zone probably varies along the tunnel; nevertheless the finding that the rock arch develops at all is of importance.

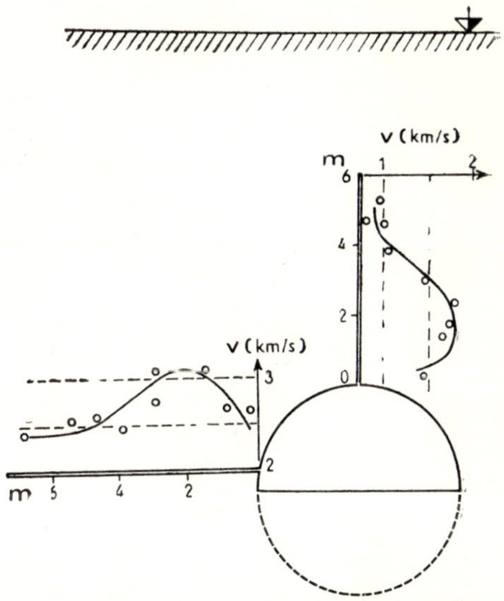

12-19. Seismic velocities measured in the rock above the excavation shown in Fig. 12-18.

Detailed measurements in the course of the widening of an advance gallery into a full tunnel section make possible the observation of how the rock arch, as well as the zone of loosening, develop around the tunnel. This development does not proceed continuously but in stages after each blast, even when they are remote.

The complexity of the nature of rock pressure encouraged the elaboration of different *approximate procedures*. They concern the pressure due to loosening and in principle define the thickness of the loosened zone above the tunnel roof and in abutment on the basis of experience. As early as in 1906 and 1907, M. M. Protodyakonov submitted his proposition that the rock volume ready to fall down is given by the scheme as plotted in Fig. 12-20. The height of the loosened zone loading the lining or supports equals $h_1 = b/f$. The wedges in the abutments cause lateral pressures. According to the values of f, Protodyakonov classified rocks into 10 groups (Dandurov, 1952). Basalt and quartzite are classified as the strongest with $f = 20$, hard granite with $f = 15$, while f of hard sandstone is 6 to 8, and that of stiff clay about 1.

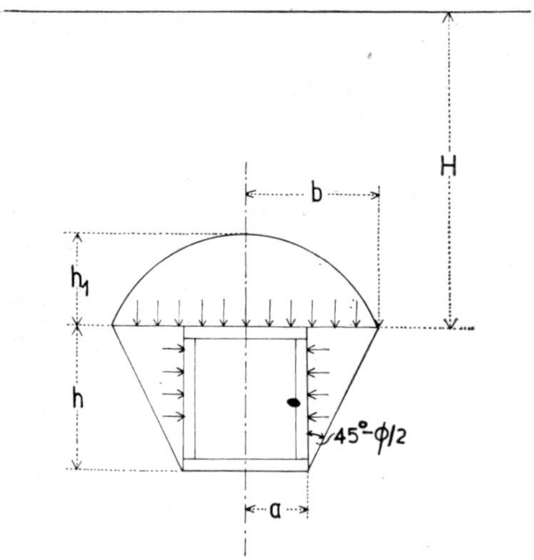

12-20. Evaluation of the loosened zone which loads the tunnel supports, according to Protodyakonov.

Several empirical data defining the height of the loosened zone were also presented by Terzaghi (1946). For example, dense stratified rocks free of clay content exhibit a loosened zone whose height equals one quarter of the tunnel width (Fig. 12-21).

Not without importance are experiences with relatively hard rocks of different brittleness. For example, limestone slabs are pressed into one another and therefore

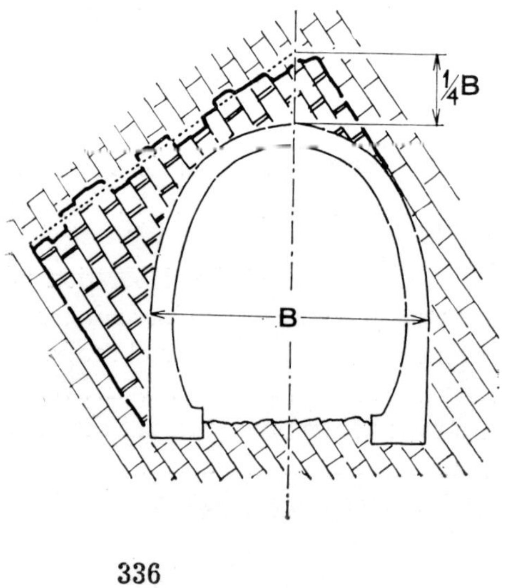

12-21. Loosening of the rock above the tunnel driven along the strike of oblique beds. (After Terzaghi, 1946.)

cave-ins occur rarely, while asperities of hard granite blocks fracture in a brittle way and a cave-in can occur at a moment's noticc and from a considerable height.

Because the rock arch above the tunnel is stressed in compression and shearing, volume changes due to shear straining appear (Sec. 3.4.4 and 3.4.5). As far as the rock behaves in a dilatant way, the compression in the rock arch increases. This can be symbolized by a vault consisting of vault stones and expanded by wedges driven into the bed joints. The opposite behaviour is shown by the contractant behaviour of the rock, symbolized by wedges drawn out from the bed joints of the vault. As a result a thick zone of rock displaces towards the tunnel opening. The loose rock is incapable of initiating the arching effect and the rock pressure upon the lining reaches a magnitude equal to the weight of the overlying rock mass. Only after a considerable narrowing of the opening, of the order of several decimetres, does the rock arch begin to develop. The inability of rock to span any opening is also shown between the individual points of support, for example, between steel ribs. Point supports are therefore unsatisfactory and flat supports should be used.

The *ductile deformation* of the rock promotes displacements ranging over a considerable distance from the tunnel. Mining under the conditions in question is called working at critical depth in coal mining. Rock behaviour of this kind can be encountered from a depth of about 50—100 m in stiff claystone, and from a depth of several hundred metres in weak shales of the Flysch Belt. The critical depth is of the order of several thousands of metres in stronger rocks, such as Carboniferous sandstones and slates. When the critical depth at about a thousand metres is reached in some bituminous coal mines, another phenomena called pseudocontractant behaviour, known from rocks of flyschoid nature, are associated with it. Due to the concentrated stress the slates are squeezed out from the tunnel walls. The rock volume decreases and therefore ductile behaviour appears. The fundamental prerequisite is therefore the immediate erection of a flat supporting system (reinforced gunite or reinforced concrete panels, grouted at the back immediately after errection, etc.).

Hard compact rocks can resist well the increased stress in the tunnel walls associated with the stress relief normal to the walls. Abnormal features appear when the concentrated stress is extremely high owing either to the great depth of the tunnel or the large inherent stress in the rock mass. The rock is then incapable of resisting the great stressing and buckles. Since the shape of the opening is generally concave, spalling plates are shaped so that they are elongated in the direction of the tunnel axis. *Rock bursts* are often abrupt and accompanied by a pop noise (the so-called *Bergschlag* in the Alpine tunnels). Rock bursts appeared in the Simplon Tunnel even after it was opened to railway traffic and caused upheaval of the ballast, bending of the rails, and several derailments. Rock bursts in mines are often of such dimensions that the openings cave in and the shock waves spread through the surrounding area. The early erection of supports in tunnels is important. Other measures are used in mining, e. g. the weakening of rock by drilling holes or water injections.

In *non-cohesive grounds* attention should be paid to the maintenance of the original density of the rock during the tunnel driving. Special care is necessary with sand which escapes through gaps between the liner plates, causing development of cavities and a progressive loosening of the soil. Weak cement bonds in the sand are also affected. The natural arch collapses so abruptly that no escape of the crew is possible. Therefore the gap between the temporary supports and the ground face must be grouted as soon as possible, or the ground face must be gunited, etc.

The loads on tunnel supports increase as a result of *seepage forces* if the sand is saturated. Tunnel driving becomes difficult and special procedures are necessary (shield tunnelling under compressed air, ground freezing, grouting or lowering of the water table). As long as drainage by the tunnel itself is sufficient, deep drainage holes should be bored in order to transmit the effect of the seepage forces deep into the ground, far from the opening.

Swelling phenomena appear in rocks with a large clay content, as in overconsolidated clay. The rock surrounding the tunnel adsorbs water from the environment after the opening has been excavated and the walls stand temporarily because of the negative pore pressure. If unsupported, spalling in the walls and caving-in of the roof occur in a short time. Therefore the method of tunnel driving must enable the setting of the lining shortly after excavation. Pressure on the lining increases due to swelling and after about one year almost reaches the magnitude of the original natural stress. The measure for the swelling pressure to be reduced as a result of gaps left between the lining and the rock rather impairs the state of affairs. The decompressed rock weakens, and moreover it weakens nonuniformly around the opening, so that the pressure on the lining is nonuniform, and finally attains the magnitude of the original natural stress.

12.7. Tunnelling methods. The need for a purposeful approach to the questions of tunnelling from the point of view of rock mechanics has already been discussed. Several recent developments in tunnelling methods reflect this requirement since they tend to attain the state denoted by *m* in Fig. 12-17. This is reached by decreasing the unsupported length between the lining and the tunnel face. Important load-bearing elements, e. g. bolts, are erected close behind the excavation. In weak rocks, the lining consists of steel ribs stiffened by a gunite layer, the latter often being reinforced by a wire mesh, etc. In a bad rock, the face or the excavation benches are also provided with a gunite layer early after being excavated. Considerable savings in the thickness of the concrete lining can be made in this way, as the thickness is often only half of that necessary with old, classic tunnelling methods (the state with the latter denoted by *k* in Fig. 12-17). The thickness of the concrete lining is not the only economy measure. The velocity of tunnel driving is also important, but is not satisfied with several recent methods.

The less elaborate full-face driving method with steel rib supports (pressure state denoted by *a* in Fig. 12-17), combined with the construction of a concrete lining far behind the excavation, is more apt for obtaining great driving velocities. Therefore many engineers, especially when driving long tunnels, adhere to this method. The statical relations, shown in Fig. 12-17, have influenced this method in that the gap between the back of the steel ribs and the rock is carefully blocked, the rock is anchored, etc. A more cautious method, as symbolized by the pressure state *m*, is considered fully in inhabited areas.

However, in several cases return to *classical methods* is necessitated when relatively short lengths of rock behaving in a ductile manner must be crossed by a tunnel situated in hard rocks. This danger is to be considered especially when the use of heavy drilling and hauling equipment is expected, as the changing over of driving operations to classical methods involves considerable expense and loss of time.

The questions pertaining to engineering geological classification of rocks were discussed in Sec. 3.8. There is no doubt that these classifications are too general for the purpose of tunnel driving. Therefore, the practice of tunnelling led to the development of a proper, applied classification.

If such a classification is expected to provide a basis for building contracts, it should take into account two important cost factors, namely rock excavation and temporary supports. Since hard rock generally transmits only a low pressure to the supports, the classification need not be too detailed. During the construction of railway tunnels in Czechoslovakia, the grouping of rocks into three categories according to the driving facility proved useful:

A driving facility of degree I was given to rocks which allowed excavations either without any temporary support at all, or with only a very light one to prevent loosening of individual small blocks. The lengths of excavated unlined sections were unlimited and full-face driving was possible; light lining was sufficient. In order to appreciate also the cost of excavation of very hard rocks, a supplementary category was given to excavation in so-called tough rock.

A driving facility of degree II was attributed to driving in less competent rocks requiring heavy supports and a heavy lining, with only short sections opened and with multiple-stage face driving, etc.

A driving facility of degree III was given to weak saturated rocks requiring elaborate operations (shield tunnelling, complicated multiple-stage advance with forepoling).

The thickness of the lining was usually not estimated in accordance with the degree of the driving facility but from experience gained during the driving of the advance heading as well as during the engineering geological evaluation of the rock.

Nowadays more elaborate classification systems prevail, serving both for the estimation of driving difficulties as well as for the design of the lining. Terzaghi's classification (1946), modified by several authors, has been widely used. The particular rock is classified as: 1 — hard, compact rock, 2 — hard, stratified, or schistose rock, 3 — massive and only moderately jointed, well-interlocked rock, 4 — hard, but jointed, well-interlocked, less competent rock, 5 — strongly fractured or densely jointed, incompetent rock, 6 — crushed rock, non-cohesive sand, 7 — ductile rock at moderate depth, 8 — ductile rock at large depth, 9 — swelling ground.

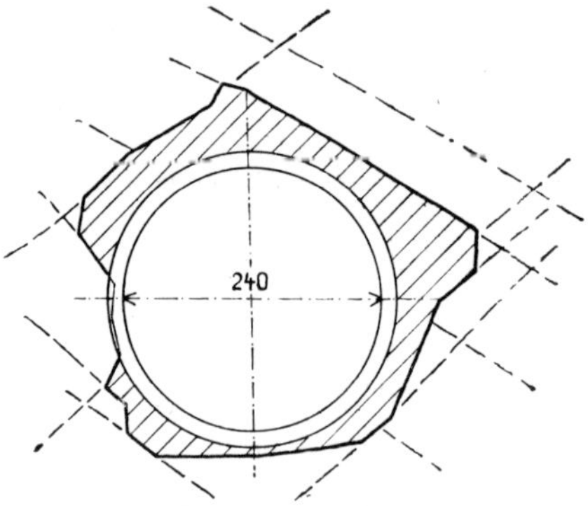

12-22. Overbreak in a gallery in stressed and jointed granite.

There is no doubt that recent methods of statical calculation require a quantitative presentation of rock properties (Chapter 3), but the practice of tunnelling still rather tends to an easier approach by way of categorising rocks. The engineering geologist can help the engineers most effectively if he is capable of showing them the characteristic outcrops of the rock and of comparing the future rock behaviour in the proposed tunnel with that encountered in a tunnel built earlier.

An estimate of the *quantity of explosives* to be used is also demanded from the engineering geologist. The quantity used is governed not only by the rock itself, but also by the applied method of driving. In order to illustrate this let us compare two cases: Classical tunnelling methods included drilling with air hammers; the holes were individually located in order to obtain the greatest effect from the minimum charge. The explosive consumption was of the order of 0.5 kg per m^3 of rock and up to about 0.8 kg with extremely hard "tough" rock. Modern full-face driving, using long drill holes and a rigid drilling pattern, requires about 2 kg per m^3 of excavation.

Another question connected with the method of driving is that of *overbreak*. The tendency to overbreak is frequent with stratified rocks when they are so hard that they require more explosives. Modern blasting technique tends to reduce the overbreak by applying a dense pattern of holes, a set of uncharged rim holes, and a dense and exact sequence of blasts (so-called millisecond blasting). Often a perfect cross-section is obtained ("*smooth blasting*"), which is also favourable for the stability of rock walls. An example of the overbreak as occurred in the stressed, fractured granite of eastern Bohemia is shown in Fig. 12-22.

Because the excessive volume must be backfilled in order to prevent softening of the rock, the question of overbreak should be studied during the preliminary investigation as it influences construction costs. It should be mentioned that not only geological conditions but also the care and ability of the crew are an important factor and cannot be overlooked. The most undesirable are the economic consequences of excessive overbreak with water pressure tunnels, since the overbreak must be filled with dense concrete.

Geological investigation during tunnel construction also includes the mapping of the tunnel roof and walls. The results are plotted at a scale of 1 : 200. The importance of detailed mapping is often only appreciated after many years, when repairs are needed; if information concerning the state of rock is unavailable, neither the cause of defects nor the extent of repairs needed can be estimated. In recent tunnelling which allows only a short time for the inspection of the working face, the nearly permanent presence of a geologist and a continuous photographical documentation are desirable.

12.8. Rock temperature and ventilation. The choice of the ventilation equipment should be made on the basis of the preliminary studies; therefore the estimation of the rock temperature and of the possibility of gas escapes must be submitted. The rock temperature is important with deeply located tunnels. As is known, rock temperature increases by about 1.8 — 3 °C per 100 m of depth, i. e. the *geothermal gradient* is about 30 to 33 m, although much more under mountains. Geoisotherms (surfaces of equal temperature) run almost parallel to the ground surface in flat regions. The geothermal gradient decreases under valleys and increases under mountains, so that the differences equalize at depth.

Temperatures measured in several tunnels are shown in Table 12-2. A comparison of the highest temperatures shows that depth is not the only deciding factor. The difference between the rock temperatures in the Gotthardt and Simplon tunnels is particularly instructive. Rock temperature is also influenced by the dip of strata, as rocks conduct heat better along the strata. Therefore mountains composed of steeply dipping complexes exhibit a lower temperature than those where strata run parallel to the surface. Steeply dipping rocks occurred in the Gotthardt Tunnel, while strata rather parallel with the surface prevailed in the nappes above the Simplon Tunnel.

Water is another influential factor. Cold water percolating through the rock reduces its temperature. In contrast, thermal water increases the temperature. The geothermal profile along the Grenchenberg Tunnel (Fig. 12-23) shows well the reducing influence of cold water flows on temperature in the tunnel. The temperature conditions can be explained only after considering the geology of the area. The tunnel traverses jointed Jurassic limestone in the chainage from 4.2 to 4.5 km and from 6.7 to 7.2 km, where cold water flows appeared.

Chemical processes may also influence the rock temperature such as the oxidation of coal seams, the decomposition of pyrite, the transformation of anhydrite into gypsum; these are all processes which must be anticipated on the basis of the preliminary investigation.

The danger of possible *gas escapes* must also be evaluated before calculating the necessary ventilation output. Methane is more frequent than carbon dioxide and hydrogen sulphide. Emissions of methane often occur during the driving of tunnels in the Carpathian Flysch Belt and even cause breaks in the driving. Large explosions and fires were caused by methane emissions during the construction of the Great Apennine Tunnel on the Bologna-Firenze railway line.

Explosive gases are encountered in rock series containing coal seams as well as in areas of oil fields and salt deposits, bituminous rocks, or their remnants. Care should be given to the detection of gas appearance during the drilling of exploratory holes; gas emissions should be registered and samples taken. The best approach is to invite the collaboration of a mining expert so that the necessary ventilation and other provisions may be made.

Table 12.2

	Gotthardt	Arlberg	Simplon	Lötschberg
Tunnel length, m	19944	10240	19803	14563
Altitude of the portal, m	1109	1302	687	1200
Altitude of the ridge, m	2861	2030	2840	2917
Maximum height of overburden, m	1706	720	2135	1673
Highest rock temperature, °C	30.8	18.5	56	34

12.9. Hydraulic tunnels. One of the important types of hydraulic structures are hydro-electric power plants, which derive water from a reservoir and conduct it to a place where a higher head can be obtained. It is easy to realize that the geological structure is a controlling factor even when drafting the first, general scheme. However, the problem is not only the situation of the construction in suitable rock, but also often the choice of a suitable type of the work. The construction shown in Fig. 12-24 is characterized by its expensive, long *water-pressure tunnel*. Due to fluctuation of the reservoir water the conduct tunnel must be situated below the minimum operational water level. Therefore with the reservoir full, the water head is several tens of metres and sometimes more than

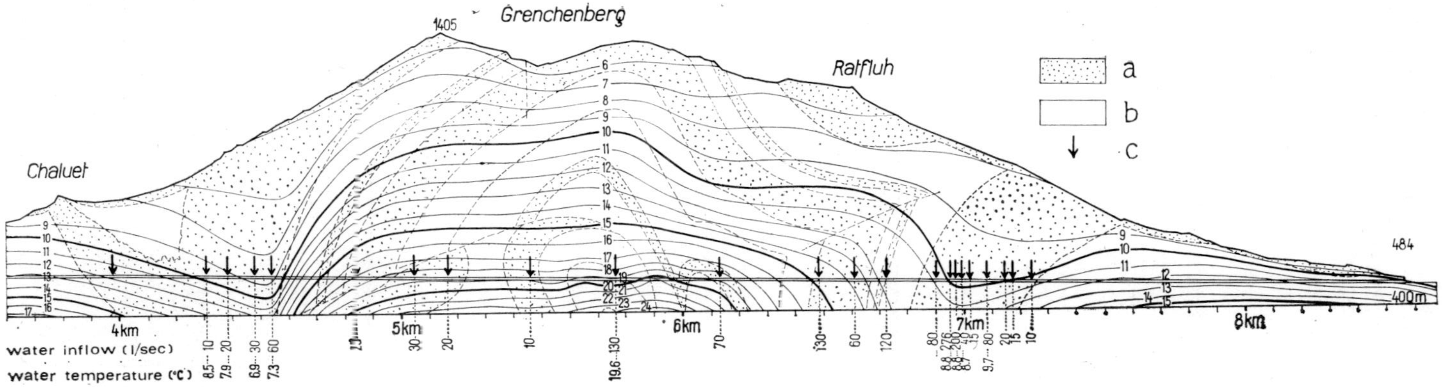

12-23. Geothermal gradient in part of the Grenchenberg Tunnel; the temperature was decreased by inflow of cold water (Buxtorf, 1916). a — Permeable rocks, b — impermeable rocks, c — concentrated water inflow.

50 m above the tunnel. The lining of the tunnel does not only support the rock, but is also stressed by the internal water pressure and transmits it partially to the rock. Therefore the design of the lining requires attention to be concentrated not only on the rock pressure, but also on the capacity of the rock to interact with

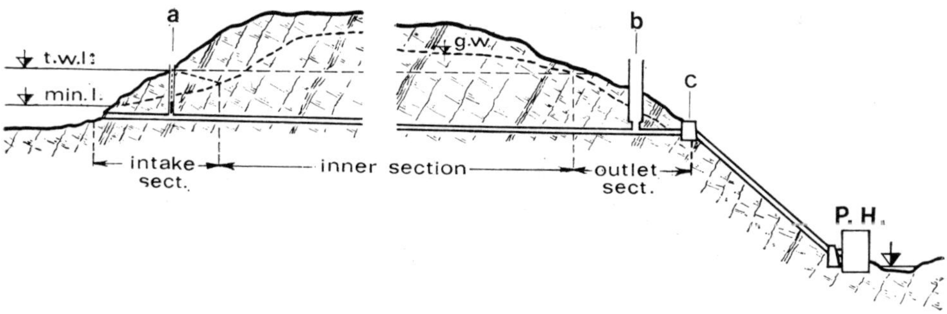

12-24. Schematic longitudinal profile of a water pressure tunnel. a — Valve shaft, b — compensating chamber, c — quick operating valves.

the lining in order to resist the internal water pressure. The effect of the lining can be reduced when ground water is present in the rock keeping a high head. This is why the hydrogeological conditions are of great importance for the design.

An underground power station associated with a *long tailrace tunnel* (Fig. 12-25) is characterised by two expensive structures, the cavern and the tailrace tunnel. When geological conditions are favourable for the construction of a cavern, the choice of an underground power station may be encouraged when the rock is free from ground water. The tailrace tunnel lining, if necessary at all, can be thin, and is often limited to the roof. Considerable saving arises from this as compared with the expensive water-pressure tunnel described above. The difference in costs is the greater the longer the tunnels are. Before dealing with the engineering-geological conditions influencing the type of lining of water-pressure tunnels, let us examine the hazards threatening the individual structures of such a tunnel. Leakage of water from the tunnel, even when small, is of increased importance when the head is great, as the loss in energy production cannot be neglected. Other accidents can also occur. The intake area of the tunnel is under the influence of water in the reservoir, therefore there is no danger of water leakage from a full tunnel. On the other hand, the tunnel lining must resist the rock pressure as well as the full water head when the tunnel is emptied. Another problem is the location of the intake structure, as the stability of the slope decreases during drawdown of water in the reservoir and the possibility of slide movements arises. Therefore much more consideration must be given to the location of the intake structure of a water-pressure tunnel than to that of portals of other tunnels.

The central section of a water-pressure tunnel can suffer water losses only when the ground-water level is below the water head in the tunnel. In this case water can escape towards the outlet of the tunnel. Therefore the end of the central section, where it meets the steel lining of the outlet section, is provided with a ring of grouted drill holes. The longitudinal drains, conducting water during the construction of the concrete lining, should also be poured by concrete.

The ground-water level in the outlet area always stands below the water head in the tunnel. Water escaping from the tunnel could endanger the stability of the slope and with it the safety of the outlet structures. Therefore, the rock mass surrounding the outlet section of the tunnel must be treated as an earth dam. A considerable section of the tunnel outlet should be lined with steel. But with the empty tunnel the steel lining would be loaded by the pressure of water from the rock. This water must therefore be drained by boreholes, galleries (often using former adit or exploration galleries), etc. The drainage system should not be installed near the tunnel, as cavities caused by erosion would be very dangerous for the stability of the steel lining.

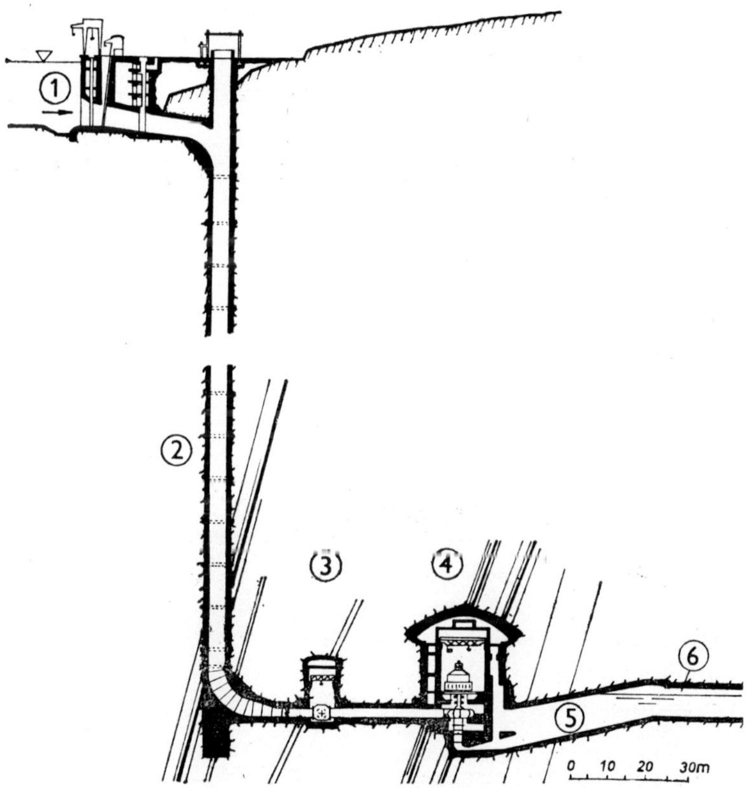

12-25. Schematic longitudinal profile of a water scheme with underground power station and a long tailrace tunnel. 1 — Intake, 2 — penstock shaft, 3 — valve chamber, 4 — main cavern, 5 — compensating chamber, 6 — tailrace tunnel. (Záruba, Zajíc et al. 1967.)

The participation of the rock in the resistance of the lining against the internal water pressure is the more effective the greater is the modulus of rock deformation. The world's first standard on the design of water-pressure tunnels introduced in the U.S.S.R., established the "coefficient of rock resistance", k, which equals the ratio of the internal water pressure p (kp/cm^2) to the increase in tunnel radius u (cm). The elastic analysis yields:

$$k = p/u = E\nu/b(1 + \nu)\,,$$

in which b is the radius. Since the error in the estimation of ν is of little influence, only the magnitude of the modulus of rock deformation E must be carefully investigated. Rocks with E greater than 50 000 kp/cm^2 present good conditions for the construction of water-pressure galleries up to 3 m in diameter, and to a head of several tens of metres. Of course, the demands on the rock quality increase as the tunnel diameter and the water head increase. However, rocks of poor quality also cannot be dismissed when other conditions are favourable for the project.

The displacements of the tunnel walls due to water pressure are subjected to investigation in order to obtain the magnitude of E. As the operation in the tunnel does not allow measurements to be made, a lateral adit is often driven for this purpose. The method of exerting pressure on the rock surface by means of flat jacks, instead of water, has recently become widespread. Flat jacks are set around the circumference of the gallery and they bear against a stiff steel structure (system TIWAG, Fig. 12-26). The magnitudes of the modulus of deformation obtained

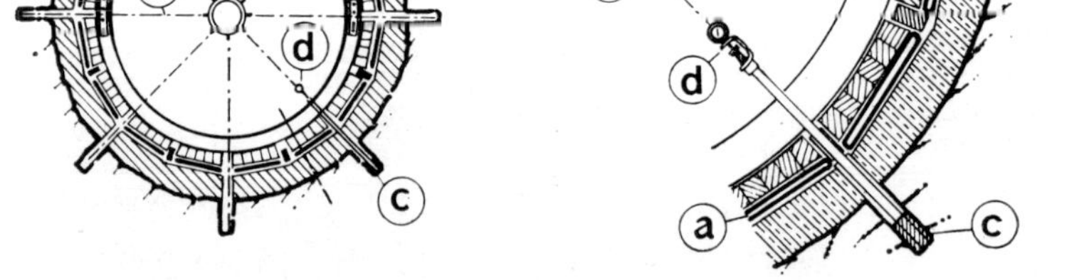

12-26. Determination of rock deformability in a trial gallery by TIWAG method (Laufer, 1966). The concrete-lined circumference of the gallery is loaded by flat jacks (a) which support on wooden and steel construction (b). Shifts of points (c) anchored in the rock are measured by indicators (d) attached by wires (e) to a firm central tube (f). Deformation of the circumference of the gallery is measured additionally with the use of wedges (g).

in this way are often by about one-third greater than those obtained by plate-loading tests.

In order to attain a good interaction of the rock and the lining, the gap between them as well as the loosened rock zone should be grouted. The modulus of deformation is thus also often increased. But the ease with which the rock can be grouted must be considered previously. Most satisfactory evidence can be obtained from TIWAG tests, which indicate the moduli of deformation before and after grouting.

Other problems of engineering geological character concern the location of several openings of the outlet area. The most important is the *surge tank* (Fig. 12-24). The dynamic impact caused by the halting of water flow due to the quick closing of valves must be balanced in this shaft. Because of this, the surge tanks are located as close to the outlet as possible. Since rock cannot exert a large resistance near a valley slope, its interaction with the tank lining is poor. This is why compromise location is often preferred, rather remote from the slope, with a steel-lined shaft. This in turn poses problems associated with the pressure of the ground water on the steel lining when the tank is empty, similarly as discussed earlier.

The construction of the deep shaft of a surge tank of large diameter is not easy particularly where the connection with the tunnel is concerned, as this involves weakening of the rock. For example, the excavation for a surge tank caused rock movements at Malnitz in the Austrian Alps, which were the more dangerous due to the presence of a railway above. The slope was formed of gneiss and mica schist. The construction of thick retaining walls was necessary before the reinforced concrete structure of the tank could be realized (Hammer, 1927).

The *inclined pressure conduit* from the tunnel to the power station is an expensive and important unit of the scheme. The slope on which it will be founded should consist of stable rock with a thin cover of debris, since this could endanger the anchoring blocks. In contrast, the location of the conduit underground, into an inclined tunnel, under the assumption that the rock should support the lining against the internal water pressure, is reasonable only when the rock is of very good quality and the overburden is thick enough. An interesting example of the disturbance of a conduit by the movement of rock debris, which occurred in New Zealand, was described by Benson (1946). The conduit, consisting of four pipes each 1.5 m in diameter, with a head of 200 m, was located on a slope formed of chloritic gneiss and covered with slope detritus. Displacements of the supporting blocks were registered soon after the operation of the plant started. They attained a value of 10 cm in three months. Due to the thick cover and to the deep weathering of the rock on the slope the conduit has been relaid into the tunnel. This was driven deep in the slope, as the geophysical investigation indicated that fresh rock occurred at a depth of 60 m. The longitudinal profile of the conduit was therefore not straight (Fig. 12-27).

In a hydraulic work consisting of an underground power station and a long tailrace tunnel, the construction of a *large cavern* is of major concern. As its influence on the statical conditions of the surrounding rock mass is considerable, the engineering geologist should try to economize as far as possible with the width and height of the cavern. Sometimes a large cavern is excavated but it is divided into small compartments which could be located elsewhere, for example, in the elongated part of the cavern. Conduit tunnels, surge chambers and adit tunnels should be located with respect to the minimum disturbance of the integrity of the rock mass.

Several questions concerning the orientation of the cavern with respect to the joint pattern were discussed in Secs. 12.5 and 12.6. As boring is not sufficient to provide adequate information of the joint sets, an exploratory gallery is driven into the future cavern area in order to investigate the joint pattern, their density and filling. Sometimes situations arose where the rock began to slide from the chamber walls, and expensive corrective measures were necesaary. A block 22 m high and more than 40 m thick of biotite gneiss began to move from the wall of the cavern of the underground power station at Morrow Point in Colorado (Dodd, 1967). The sliding movement occurred along a fault dipping 17 degrees

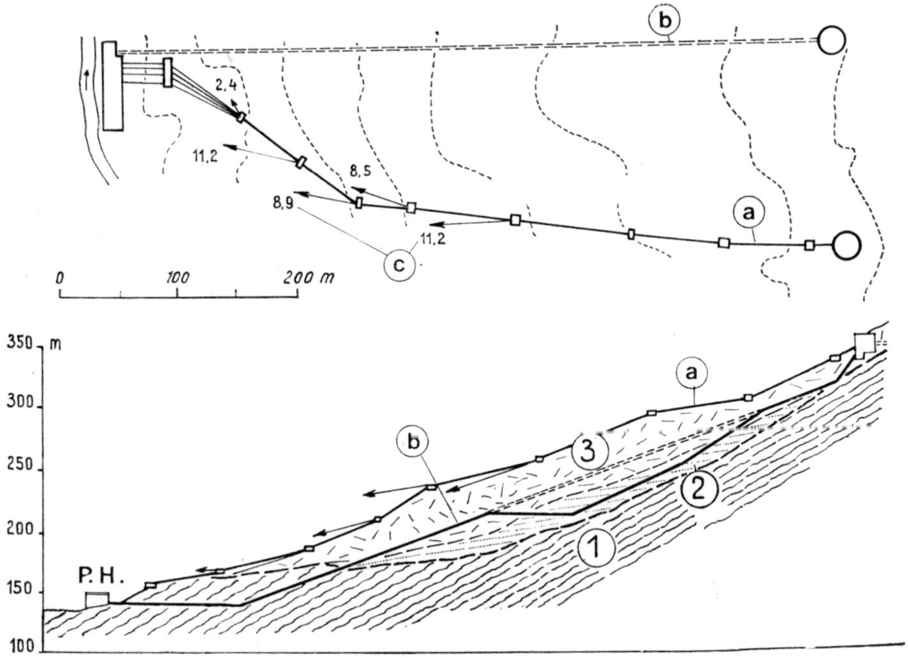

12-27. Deformation of penstock due to the foundation on unstable slope; pipeline had to be relocated underground (Benson, 1964). 1 — Chlorite gneiss, 2 — weathered gneiss, 3 — slope debris; a — original position of the penstock, b — penstock relocated into a gallery, c — movements of blocks in cm.

into the chamber. After about 6 cm of displacement the movement was halted by the use of an elaborate system of anchors and dowels, and by draining the back of the block by drill holes into a drainage gallery. An example of an advantageous location of the main chamber based on the analysis of the tectonic disturbance of rocks is presented in Figs. 12-28 and 12-29 (Karl, 1954).

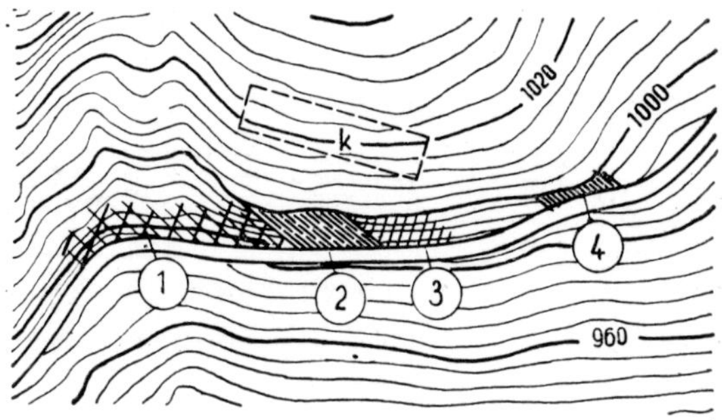

12-28. Determination of tectonic zones in the surroundings of a designed underground cavern (Karl, 1954). 1—4 — Zones with different degrees of tectonic disturbance shown in Fig. 12-29, k — cavern.

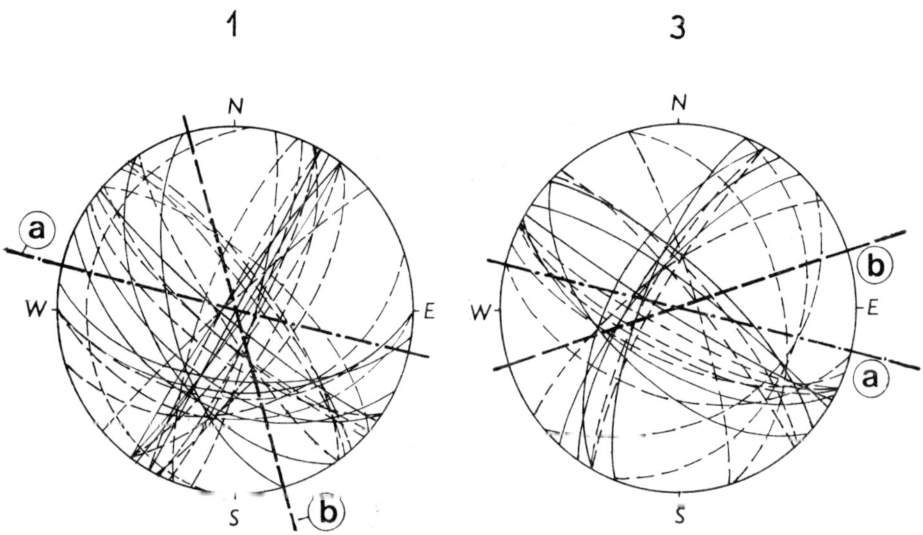

12-29. Diagram of the tectonic disturbance of zones 1 and 3 shown in Fig. 12-28. Joints are represented by their intersections with the mantle of the lower hemisphere (Karl, 1954). a — Cavern axis as originally designed, b — recommended axis of the cavern.

Another question is where the cavern should be situated in order that it does not lie either in rock free of horizontal residual stress or in the area of a large stress. Therefore considerable attention is paid to the measurement of inherent stress when designing underground power stations. This requirement leads to its siting rather below the level of the valley floor.

The ingenious location of caverns (from the point of view of geology and hydrogeology) at the Braz and Prutz power stations in the Austrian Alps deserves attention. Fig. 12-30 shows the longitudinal profile of the power station at Prutz on the Inn river, in the Tyrol. The main chamber was excavated in thick-bedded dolomite and limestone, normal to the strike of beds. The rock complex dips at an angle of 50 degrees into the mountain and contains a stratum of marly slate (Partnach slate). The latter is impervious and divides the slope area into a lower part, which is dry and in which the cavern was located, and an upper part where the dolomite and limestone are strongly water-bearing. During the construction, the water-pressure conduit was drained by an adit gallery with a discharge of 80 to 100 l/s. The cavern also presents an example of economical approach, since its width is only 18 m, maximum height 25 m and length about 55 m (Fig. 12-31), as compared with the installed capacity of 3 × 40 MW. About 220 m^3 of excavated rock equals one installed MW. Economy in excavation of the surge chamber also is to be noted.

The Braz power station displays similar geological conditions. It is situated in the complex of the Liassic limestone, and the marly Kössen slates protect it from water percolating from the overlying dolomite.

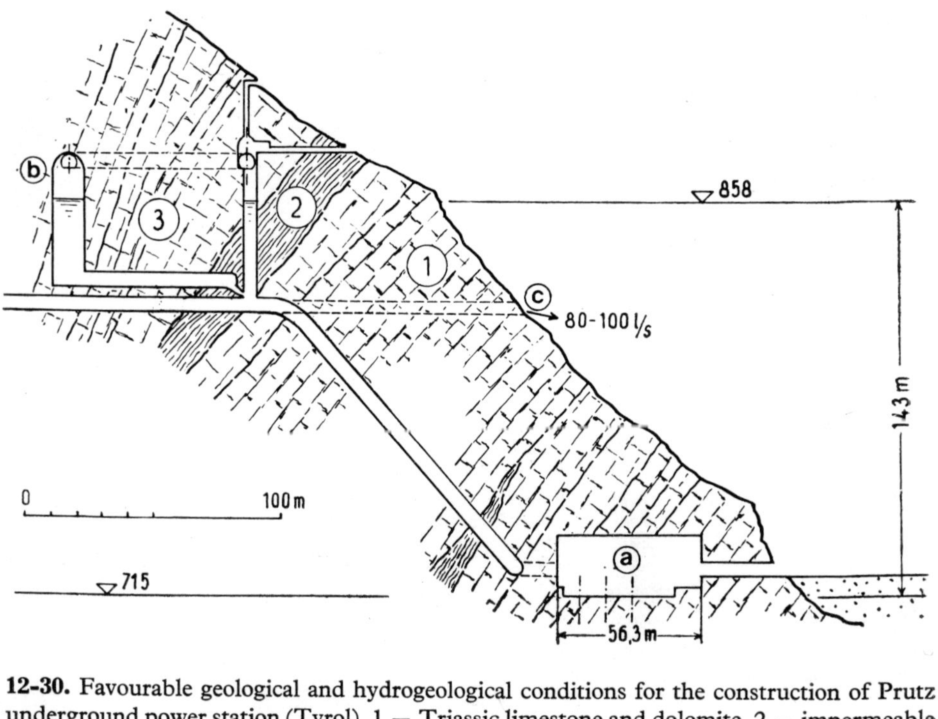

12-30. Favourable geological and hydrogeological conditions for the construction of Prutz underground power station (Tyrol). 1 — Triassic limestone and dolomite, 2 — impermeable marly slates, 3 — water-bearing limestone and dolomite; a — underground power station, b — compensation shaft, c — adit gallery.

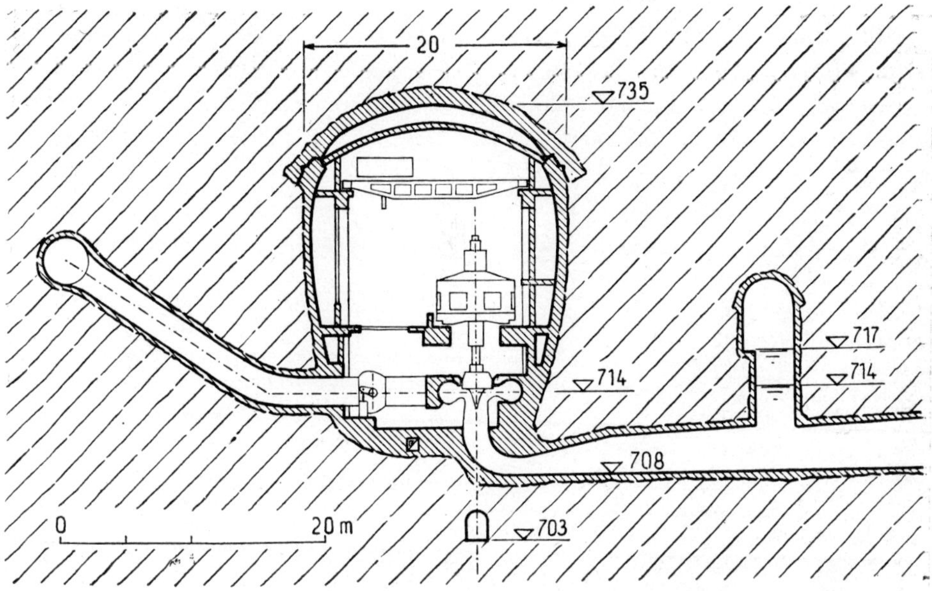

12-31. Cross-section of the Prutz underground power station (Austria) in Triassic dolomite An example of economical dimensions of a cavern (Tiroler Wasserkraftwerke, Innsbruck)

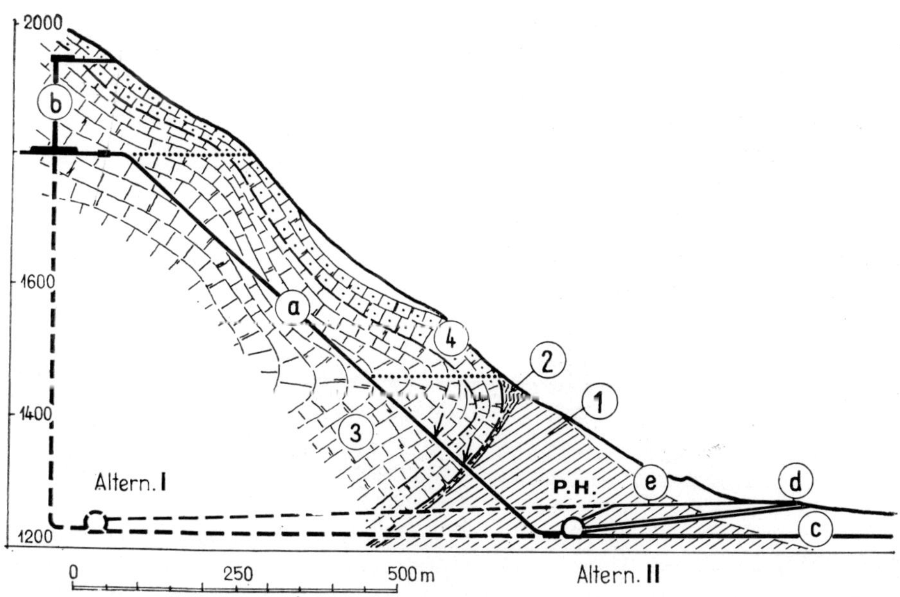

12-32. Geological section of the Premadio underground power station, Italy (Carati, 1958). 1 — Phyllite, 2 — sericite schists, 3 — hard dolomite, 4 — jointed dolomite, a — inclined penstock tunnel, b — compensating chamber, c — tailrace tunnel, d — access tunnel, e — ventilation gallery.

12-33. Machine hall of Hojum underground power station excavated in granite (Trollhättan, Sweden).

Another example of the influence of the geology on the design of an underground power station is presented by the Premadio scheme in the Italian Alps (Carati, 1958). Two alternatives were studied (Fig. 12-32). The first consisted of a vertical shaft situated in the Triassic dolomite and a 1400-m long tailrace tunnel. The second consisted of a steel-lined conduit located in an 850-m long inclined tunnel, a cavern situated near the slope toe in the phyllite, and a tailrace tunnel 730 m long. This second alternative was used. Driving of the inclined tunnel could be advanced with this alternative by providing a supply adit at mid-height. Most important was the isolation of the phyllite area by a layer of impervious sericite slate from the saturated, jointed dolomite. Also the hydrogeological conditions of the environment, near the Bormio spa, have been less affected by using this method.

The excavation of the chamber and the disposition of the lining are other problems to be studied. The little disturbed granite in Scandinavia makes the full-face excavations of the chamber possible without any support. The roof is anchored and provided with a thin reinforced concrete shell against rock fall and trinkling water. The walls of the chamber are supported only by a file of piers, which also support the crane runway. The rock face is left uncovered where smooth walls are not required (Fig. 12-33).

However, the rock is frequently not sufficiently stable and besides the roof the walls also require supports. Competent rock makes possible the excavation of the lower part of the chamber at the same time as the excavation of the arch, but *wall anchoring* is necessary and should be accomplished as soon as the wall excavation proceeds. The application of anchors is now widespread and two cases can be distinguished:

(a) The function of the anchor forces is not only to increase friction in the bedding surfaces or joints, but also to introduce new forces into the rock mass. The stress state in the rock surrounding the cavern is improved in this way. L. Müller (1953) applied this arrangement in a Liassic jointed limestone at the Braz power station, as mentioned earlier. Fig. 12-34 shows that relatively long anchors were required to be effective in the manner as indicated. The area, which would be affected by the anchor forces was determined by means of a photo-elastic model. Fig. 12-34 shows that the anchor direction is almost normal to the major joint set.

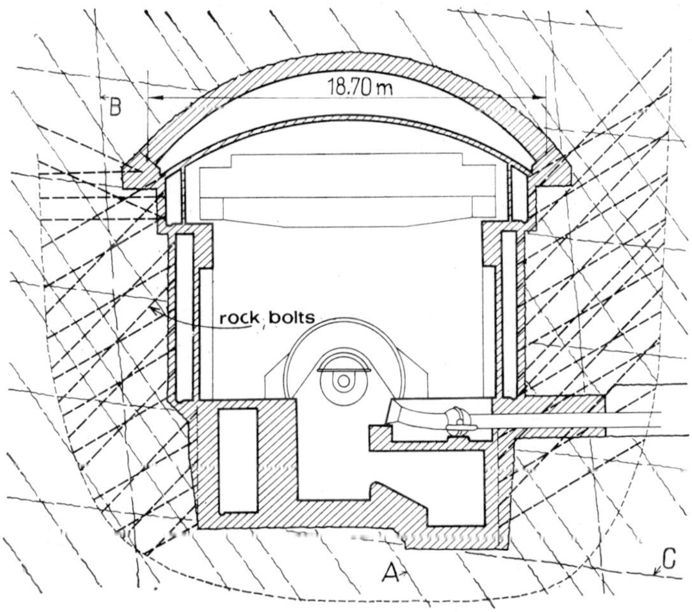

12-34. Cross-section of the Bratz power station (Austria) with anchoring of walls in limestone dissected by joint systems A, B, C. (Müller, 1953.)

(b) In several cases the effect of anchors is restricted only to the closing of joints, thus reducing the danger of block sliding. Anchors may be shorter than those in the preceding case, and the angle of their direction with the planes of weakness should not only bring about an increased friction but also provide beneficial tangential components of the anchoring forces (Fig. 12-35).

In rocks of rather poor quality tunnelling by use of the German method is necessary. One of the walls of the cavern at Prutz (Fig. 12-30) was constructed in this way.

With recent developments in power systems, the engineering geologist must face problems associated with *pumped-storage power plants.* Sometimes a river reservoir is used for the storage, but in mountains the storage reservoir is constructed on a flat mountain plateau. Although the individual elements of the scheme do not differ in principle from those as discussed earlier, still the problem as a whole is more complicated. This is due to the fact that the storage reservoir, the pressure conduits, the power station, and the lower dam must be constructed in one valley profile alone. There is no need to explain that such an ideal profile is not easily found. Therefore concessions must be made: for example, the design with an underground power station must be abandoned. Generators sets with horizontal axes are provided in power station, since those with vertical axes require a great chamber height; this is at variance with the mechanical properties of the rock as well as with the hydrogeological conditions. The storage reservoir is usually sealed, e.g. by a bituminous membrane. Difficulties arise rather due to the

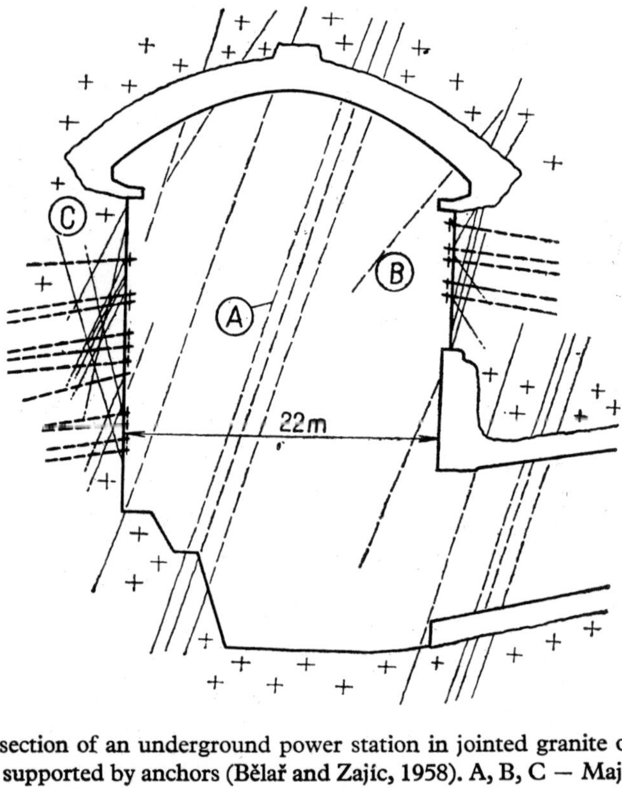

12-35. Cross-section of an underground power station in jointed granite of the Bohemian Massif. Walls supported by anchors (Bělař and Zajíc, 1958). A, B, C — Major joint systems.

position of enclosure dams of the reservoir near the hill slopes and to uncertain density of rock detritus covering the plateau and hill slopes.

As an example, the Cruachan pumping storage power plant can be described (Fig. 12-36). The situation of the chamber was determined by geological condi-

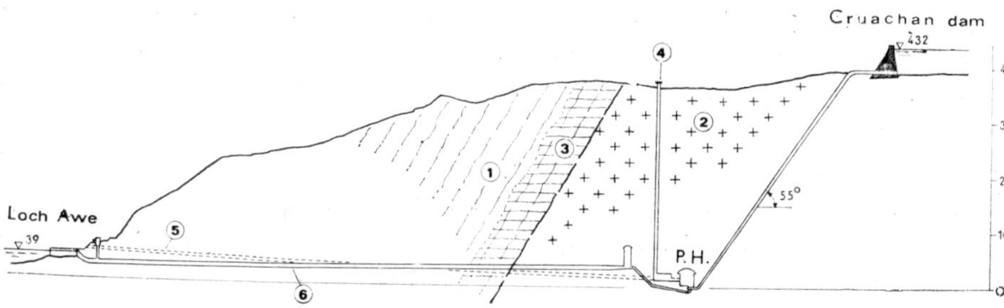

12-36. Cruachan pumped-storage power plant in Scotland. The cavern is located in hard granodiorite. 1 — Phyllite, 2 — granite, 3 — contact hornfels, 4 — ventilation shaft, 5 — adit tunnel, 6 — tailrace tunnel.

tions. The tailrace and the adit tunnels cross phyllite and hornfels rocks, which envelop the granite massif, and the latter supports the storage reservoir. The adit tunnel is 1200 m in length and was driven as an exploratory gallery in order to investigate the area of the cavern. Granodiorite in which the cavern was located had excellent properties which made it possible to leave the walls unlined and protected only by the piers of the crane runway. The roof of the chamber spanning 26 m is protected by concrete vaulted strips 2.3 m in width, and the free rock surface between them, 20 cm in width, is provided with polyvinol chloride slabs to protect the cavern against trickling water.

12.10. Underground railway tunnels. The construction of undergrounds is a special branch of tunnel engineering, since saturated valley deposits must occasionally be traversed. Temporary draining of deposits is out of the question, because of the large quantity of water as well as the danger of subsidence of buildings in the environment. As a result, tunnels are shield-driven under compressed air.

Difficulties associated with tube driving under the valley floor as well as with the securing of adjacent buildings and with the relocation of city services, lead to the lay-out of the underground deep in the bedrock. One of the important tasks of the engineering geological investigation is to estimate the minimum necessary thickness of overburden above the tunnel, which depends on the character and permeability of rock. In claystone, which is practically impermeable, only a few metres above the tunnel crown are necessary to separate the ground water of the valley deposits from the excavation, provided that a reasonable driving method has been used (shield tunnelling).

Tunnels of the Budapest underground were driven about 20 m beneath the surface of Tertiary deposits. This rather great depth made driving without a shield possible on the right bank of the Danube, where stiff marl of Lower Tertiary age was traversed. Air pressure was used, which helped to support the excavated roof. Shield driving was necessary only to construct the wide tubes of the stations, but air pressure could be spared. In the central area of Pest, in Tortonian marl, shield driving without air pressure was used. In the eastern area of Pest, in weak silty Sarmatian marl with thin sandy interlayers, shield tunnelling under compressed air was necessary and the driving was difficult.

A great water inflow should be suspected in hard, jointed rocks, even when the overburden is several tens of metres thick. Permeability of stratified rocks will increase because of the opening of joints and of bedding planes associated with the stress relief resulting from excavation. Therefore, a higher location of the tunnel (denoted by *a* in Fig. 12-37), beneath the valley floor is hazardous, while

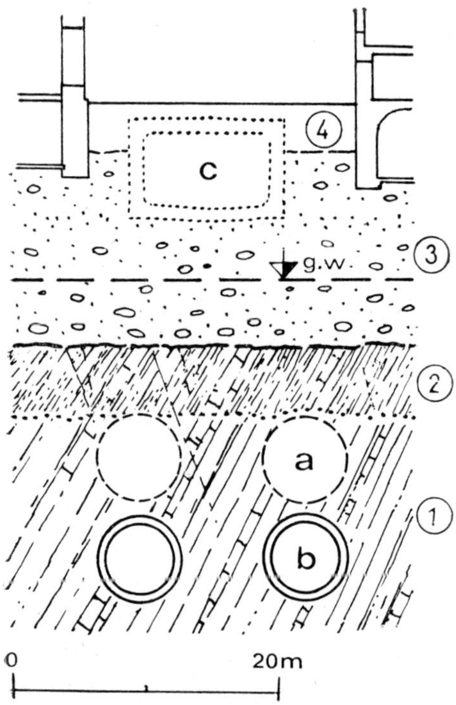

12-37. Effect of geological and hydrogeological conditions on the location of the underground railway. Geologically, location (c) above the ground water level is the most suitable. Route (a) is less suitable than (b) because of a small thickness of the bedrock above the tube. Considerable discharge and rock pressure are to be expected. 1 — Ordovician shales with quartzite interlayers, 2 — weathered shale, 3 — sandy gravel, 4 — fill.

a lower location *b* is safer. The location as denoted by *c* is most satisfactory from the point of view of geology since the structure is not at all influenced by the ground water. Nevertheless, the passing from a near-surface location to the deeply located section (Fig. 12-38), if necessary, is associated with difficulties.

Important experience concerning the influence of geology on the driving of underground lines was gathered during the construction of the new East-West express line in Paris. The earlier lines of the Metropolitain were mainly located at a shallow depth in Quaternary deposits. Only in several sections, as below Montmartre, below the Seine, and at several line crossings were the lines located in Tertiary sand and marl.

The new line required the location of the route at a depth of 25 to 35 m, due both to the underground service lines in dense inhabited city districts, and to the direction parameters of the line. A running speed of 100 km/h required a curvature radius of 500 m, and the line was situated irrespective of the surface street system. The line was driven in Palaeogene beds, only at the crossing of the Seine and at both end sections, where the line has to meet surface railways, it is located in Quaternary deposits. The Palaeogene, in which the tunnels are situated, consists mostly of limestone and marl; only in short sections were either the overlying sand or the underlying clayey sand with lignite encountered.

Tunnelling methods were chosen with respect to the geological and hydrogeological conditions. The system of excavation of the Auber station, as shown in Fig. 12-39, required the excavation of an underground opening 228 m in length, 39 m in width and 19 m in height, below the existing line of the Metropolitain and below the buildings. The lower portion of the station was situated in the limestone and marlstone, and the arch in Tertiary sand. Rocks were hardened and sealed by intensive grouting. Sand and gravel above the arch were grouted first by means of a fan of holes drilled from the top heading. Grouting of the limestone and marlstone followed in order to seal the areas of the future abutments (stage I in Fig. 12-39). Only after this were the side galleries (c) driven; then the grouting of sand above the roof as well as of the limestone below the sole could be accomplished. Grouting of limestone and marlstone was achieved using a

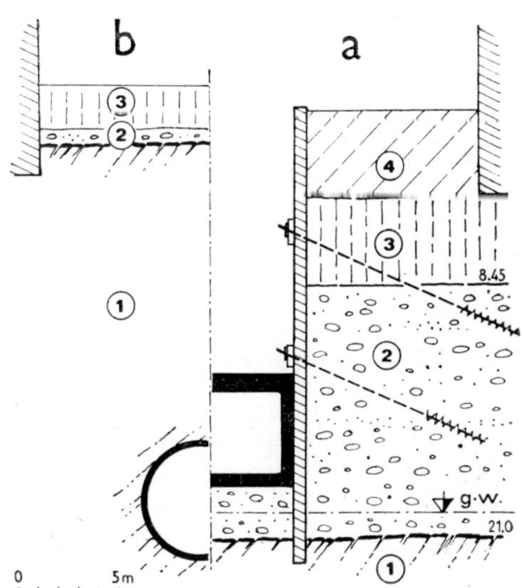

12-38. Cross-section of a deep construction pit (a) with walls supported by anchored concrete diaphragms; transition to the section driven by tunnelling (b). 1 — Ordovician shales, 2 — sandy gravel, 3 — loess-loam, 4 — fill.

cement-clay mixture, Tertiary sand was grouted by resin, and the Quaternary sandy gravel was hardened by silicic acid gel.

Fig. 12-40 shows a schematic longitudinal profile of the section in which the line traverses the Seine valley. From the Etoile station to the La Défense station, only one doubletrack tube 10 m in diameter was driven, by means of a mechanized shield under compressed air. Nevertheless, the experience gained does not show this method to be a good one. Rock pressure and stability of the excavation face cause less difficulties with a single-track tunnel. Another advantage of two single-track tunnels is that the tunnel driven first may serve as an exploratory work for the other.

The portion between the Etoile station and the Seine was very difficult, since the tunnel traversed Palaeogene sand underlying the limestone. Great pressure and gas emanations appeared, the latter originating from the lignite interlayers present in the sand. Previous experience that even little inflammable material ignites under compressed air was confirmed. Large ignitions of the lignite layers

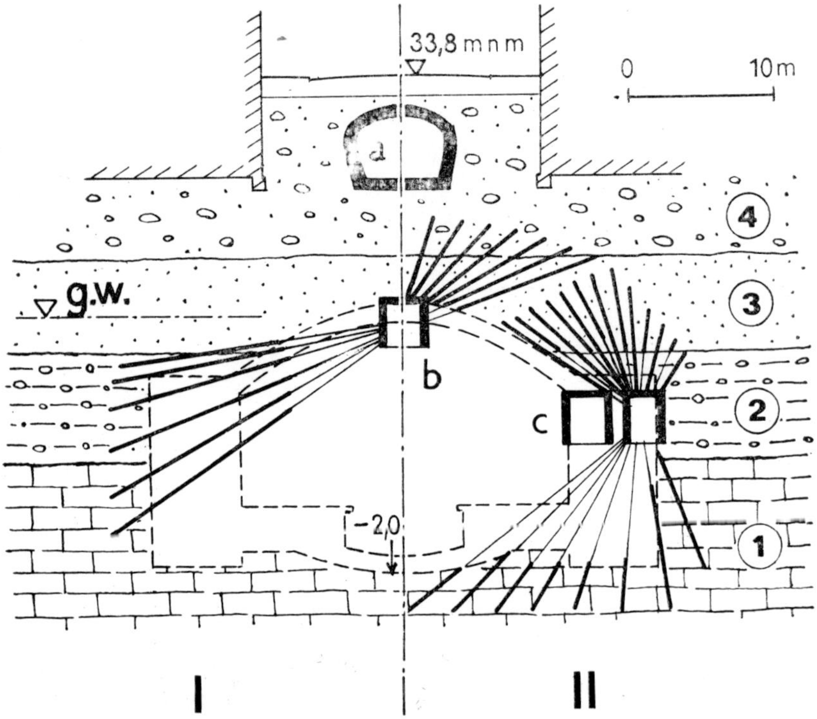

12-39. Schematic section of the Auber station of the new Metro line in Paris. 1 — Limestones, 2 — marlstones with limestone interbeds, 3 — sand with sandstone interbeds (Eocene), 4 — Pleistocene sandy gravel; a — existing Metro line, b — top heading, from which rock above the tunnel and rock in which gallery (c) was driven had been grouted. (Haffen, 1967.)

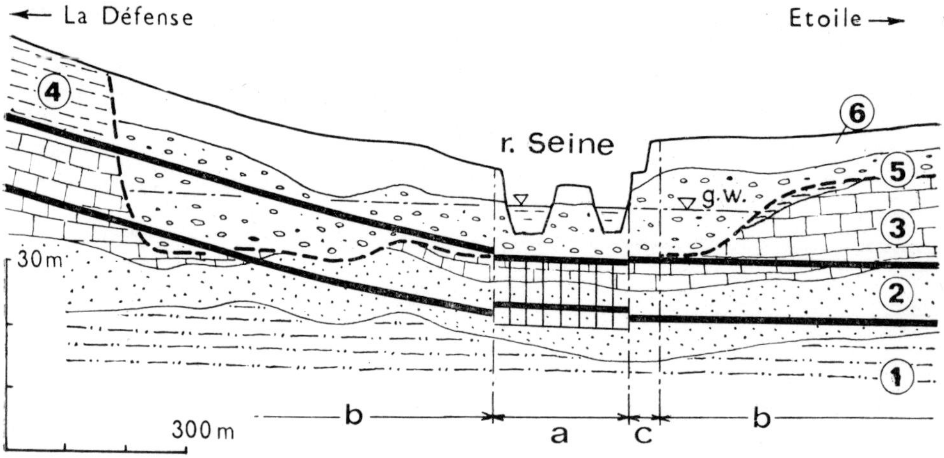

12-40. Schematic longitudinal profile of the under-river section of the Metro in Paris. 1 — Sandy clay with lignite seams, 2 — clayey sand with lignite seams (Palaeocene), 3 — limestone, 4 — marlstone (Eocene), 5 — terrace gravel, 6 — alluvium and fill; a — section of submerged caissons, b — sections driven with the aid of an automatic shield, c — section consolidated by freezing. (Haffen, 1967.)

occurred blocking the driving and necessitating extensive grouting to separate the dangerous area by a grout curtain.

In order not to make the level any deeper than is necessary, floating caissons were used to cross the river bed of the Seine. West of the La Défense station, urban conditions made the excavation of shallow covered cuttings possible and diaphragm walls were applied as a protection.

Chapter 13

ENGINEERING GEOLOGICAL INVESTIGATIONS FOR HYDRAULIC STRUCTURES

The importance of geology in engineering practice was first appreciated at large hydraulic structures, particularly during the construction of dams where many relevant problems are closely associated with the geological conditions of the area. Investigation is not limited to the damsite only, but it must also involve the reservoir area and its environs in order to gather enough reliable information for the preparation of plans for the project. As in other branches of engineering, reconnaissance and detailed (feasibility) investigations are carried out, as well as investigation during construction.

13.1. Reconnaissance investigation provides the basis for the preliminary studies of the project. It should be conducted by an experienced engineering geologist who will compile his report from an inspection of the area, of the outcrops and exposures, and after studying geological maps, relevant publications and records deposited in the archives of scientific institutions. The reconnaissance investigation should cover a greater area than the proposed damsite, since in the course of the design work the project may be altered and other damsites or changes in the height of water to be impounded may be proven necessary.

In case the area has not yet been mapped geologically, basic geological investigation and mapping must be carried out. The investigation should chiefly include the dip and strike of beds, tectonic disturbance of the rocks, folding, zones of faulting and shattering, distribution and character of joints and fractures, as well as the character and depth of weathering.

An engineering geological map of the study area makes it possible to assess all topographically suitable damsites, the watertightness of the reservoir area, stability of the reservoir banks and possible siltation of the reservoir. It is a necessary tool in the search for building materials and in prospecting for possible valuable mineral resources, which should not be flooded by the impounded water.

For some drainage systems special-purpose maps have been constructed which serve for the preparation of water-management plans. In these detailed engineering geological maps river valleys are divided into reaches with similar topographical and geological character with respect to dam construction. These maps serve to provide general information for a designer about the suitability of the respective sector of the stream for the construction of dams.

On the basis of the preliminary studies the engineering geologist, together with the designer, prepares a programme of necessary investigations which may include:

(a) A *survey of the damsite area*, the site of the power plant and the appurtenant parts of the dam, and of the area intended for procuring building materials;

(b) Preliminary *subsurface exploration* of the damsite and the reservoir area, to determine the depth of weathering and the character and course of fault zones which could endanger the feasibility of the work, and to assess deposits of building materials, which are often decisive in the choice of the type of dam.

It is not economic to perform such extensive and costly subsurface exploration for the water-development studies and initial project as for the detailed project, since many projects cannot be carried out for reasons other than geological, and expensive boring would thus be in vain. The driving of galleries and drilling is justifiable only when there is danger of a serious geological obstacle.

(c) For reconnaissance investigation *geophysical methods* are advantageous since the necessary instruments are readily transportable and field measurements are comparatively speedy and cheap. It is particularly the resistivity and seismic methods which enable the thickness of superficial deposits, the depth of weathering and the presence of fault zones to be established.

(d) The *chemical composition* of surface and ground water is also examined to ascertain whether they will damage the structure or not.

The engineering geologist submits a report on the results of the reconnaissance investigation. This should present an analysis of geological conditions of the study area, the outline of stratigraphy and tectonics and of the morphological history of the valley. On the basis of these analyses, the damsite and the reservoir area should be assessed and potential sources of construction material suggested. The report should be accompanied by a detailed engineering geological map, geological sections, detailed boring logs and records of water pressure tests. It should also include a list of the literature used and reports of experts of other scientific branches, such as hydrogeology, soil mechanics, geophysics and possibly mining engineering.

The programme of preliminary exploration should be rigidly scheduled since all investigation works should be completed and evaluated in time so that their results can be applied in the final elaboration of the project.

13.2. Detailed geological investigation is conducted after the preliminary project has been approved. In essentials it concerns the investigation of the damsite (Sec. 13.4), of the reservoir area (Sec. 13.2), of the area used for water discharge from the reservoir (13.17), of the power-plant site (13.18) and of the deposits of construction materials.

The results of detailed investigation of the damsite are plotted on a contour plan, usually to a 1 : 1000 scale. For the investigation of the reservoir area, a base map to a scale of 1 : 5000 or 1 : 10,000 is generally used. The schedule of detailed engineering geological investigation and the location and type of exploratory holes are proposed on the basis of the results of reconnaissance investigation and according to the lay-out of the hydraulic structures. Subsurface exploration is carried out with the use of test pits, trenches, exploratory drifts, core and percussion boring and, if necessary, the valley floor and flanks are laid bare to the bedrock surface. The exploration programme may be modified or complemented depending on the initial results. In some cases test pits, trenches and drifts prove most suitable, since deep mechanical weathering (especially of Pleistocene date) can be most readily evaluated in open excavations (Figs. 13-1, 13-2).

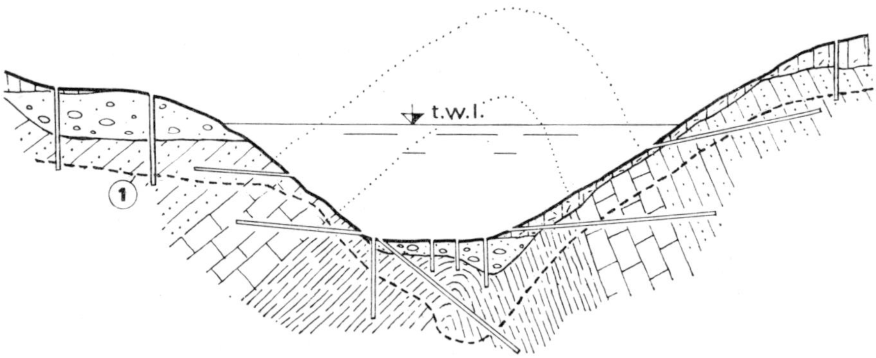

13-1. Plan of subsurface exploration of a damsite. 1 — Depth of weathered zone; the depth of weathering is greatest in the anticlinal bend, where the beds are fractured.

Exploration in solid rock is generally performed by core boring, which provides the possibility of drilling in various directions. A working method that allows for a high core recovery is desirable (Sec. 4.4). Good recovery is possible even in soft and disturbed rocks with the use of large-dimension boreholes and a double-tube barrel. It is, however, necessary to endeavour to obtain as perfect cores as possible and not merely to achieve rapid progress in boring.

The permeability of rocks is determined by water-pressure test and by trial grouting. Grouting boreholes can sometimes be located so that some of the grouted rock is removed by foundation excavation. In this way the sealing of the rock can be best assessed. Trial grouting is a very useful method for investigation since it shows the extent and interconnection of joints and the permeability of rocks at various depths below the ground surface.

Subsurface exploration should be supervised continuously by an expert worker in boring techniques who has a working knowledge of engineering geology and soil mechanics. He will systematically follow the progress of boring and will record all information important for the evaluation of investigations. In accordance with the geologist's instructions he will take samples of rocks and will attend to their correct labelling and deposition in special boxes.

All samples from the boreholes and cores should be carefully preserved, usually at the construction site, in a dry and well-lit room. If the samples are numerous, the boxes should be deposited so as to be easily accessible for further study. After the completion of the structure, typical rock samples with all pertinent data should be deposited with the institution responsible for the subsurface exploration.

The position and altitude of all boreholes should be surveyed. Large-scale surveying is generally carried out by special organizations.

13.3. Co-operation of engineering geologist during construction. The engineering geologist performs the preliminary investigation using separate exposures and boreholes. In constructing geological sections and maps he must supplement and interpolate the results for intervening sections consistently with his conception of the geological structure of the area. As a result, his report will contain, besides well-established facts, some assumptions which may be, to some degree, at variance with actual conditions. It is therefore necessary, especially when large hydraulic structures are constructed in areas with complicated geology, to continue engineering geological research even during construction operations. Excavations and boring work may supply much supplementary information on the geological structure of the area and check the results of previous investigations. If some assumptions do not prove correct during construction, there is still the

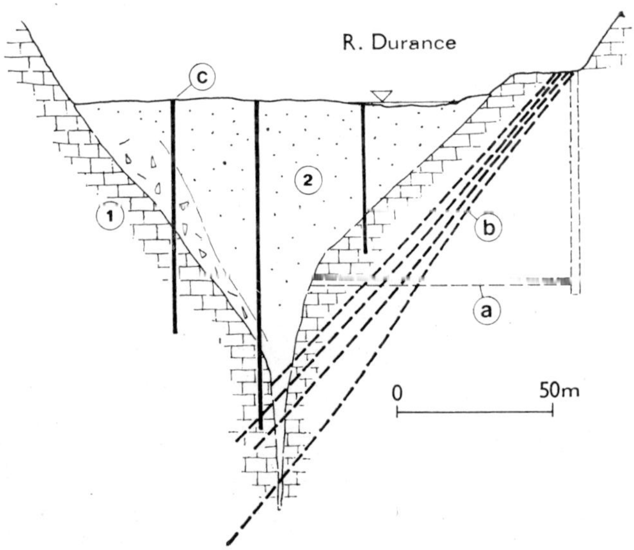

13-2. Investigation of the overdeepened valley of the river Durance for Serre-Ponçon Dam. a — Shaft and gallery from 1913, b, c, — boreholes from 1946; 1 — marly limestones, 2 — river alluvium (sand and gravel) with layers of detritus (Ischy, 1955).

possibility of modifying the construction procedure. Although this may be difficult and cause time and financial losses it can eventually obviate greater difficulties.

The engineering geologist should co-operate with the engineer in planning the method and depth of foundation excavation. This concerns, for example, the depth to which explosives can be used and the thickness of the layer which should be removed without blasting. The exposed surface of the dam foundation bed should be surveyed, geologically mapped and if possible photographed systematically. Of importance is the collaboration of the engineering geologist in grouting; he inspects the grout holes and from the results of water-pressure tests he assesses whether the depth of the grout curtain is satisfactory.

The engineering geologist should be present continuously at the building site in order to examine and assess all temporary exposures and boreholes and to take samples for laboratory tests and for preserving them as documentary material. After the completion of the construction, he should compile a final report containing all information concerning foundation and grouting procedures, the results of technological tests of natural building materials and a detailed geological map of the foundations.

The great importance of continuing collaboration of the engineering geologist with engineers during construction is apparent from many detailed and instructive published reports. When studying some of these reports one can realize what a richness of invaluable experience and both theoretical and practical information can thus be obtained and made available for instruction and the progress of engineering geology. This is true of many papers given at the International Congresses for Large Dams which furnish ample evidence of the fruitfulness of this close co-operation.

13.4. Investigation of the damsite. In view of the enormous growth of the demand for electric power, dams have recently been erected mainly for energy supply. More and more dams are being built, bolder and bolder constructions are designed at sites less and less suitable from the geological point of view. This implies that the demands on the extent and thoroughness of engineering geological investigation are steadily increasing. The choice of the damsite is most important in every hydraulic work, as it is decisive for the effectiveness and economy of the project, as well as for reliability of operation and the safety of the downstream region. The following principles should be observed in assessing the suitability of a damsite from the geological point of view:

(a) Damsites are generally chosen in constricted parts of valleys with steep rocky slopes. The configuration of the damsite controls not only the lay-out of the dam but also the scheme of water diversion during construction and other aspects of the building project.

(b) The geological setting of the damsite area must be studied very thoroughly, particularly the stratigraphical and tectonic conditions, in order to recognize reliably the geological structure of the site. For this purpose the most modern methods and techniques and all available sources of information should be used.

(c) At the selected site the depth to a suitable foundation bed must be determined, as well as the technical properties of rocks and their permeability with special reference to leakage around the dam and uplift under the dam.

(d) The stability of slopes above the crest of the dam should be assessed since excavation for the abutments may disturb their equilibrium.

(e) In an earthquake-prone region, it is necessary to assess the degree that the earth's tremors may possibly attain.

(f) The chemical composition of surface and ground water should be analysed to determine whether the dam and its foundation will suffer from their effects.

(g) The rocks in the valley floor and its flanks should be examined with respect to their workability (see Chapter 8).

(h) Very important is the supply of building materials. The engineering geologist must examine their properties and consider whether the excavated material at the site is usable or whether other deposits must be found. With regard to sand and gravel aggregate for concrete, its purity and strength and the petrographical composition of individual fractions must be examined, as well as the form and surface of grains which are decisive for the workability of concrete. It may be necessary to find suitable impermeable material for the sealing of cofferdams and/or for the impermeable core of the dam, and deposits of building stone usable for masonry, revetment, pavement, riprap, etc.

13.5. Geomorphological and geological setting of the damsite. Suitable sites for dam construction generally occur in the constricted parts of a valley with steep rocky slopes, which indicate that they are formed of hard rocks, more resistant to erosion than the rocks in widened reaches with moderately inclined slopes.

Even such a morphologically ideal site may, however, conceal unpleasant surprises. In several cases the constriction of a valley has proved to have been caused by a rockfall. In studying a damsite for the Orlík Dam on the river Vltava (Bohemia) V. Zoubek (1953) identified a large rockfall on the left flank of the valley (Fig. 13-3). The porphyroids in the holes showed incongruent deposition and strong loosening. The existence of an old slope failure was then confirmed by test pits sunk from the exploratory drift at the foot of the slope; gravel of the river terrace was found below the downslipped porphyroid.

For the same reasons several more damsites on the river Vltava had to be abandoned. Frequent rockfalls in the deep river valleys of Central Europe can be accounted for by climatic conditions during the Pleistocene. Because of freeze-thaw action the joints were gradually widened and rocks of the valley flanks were loosened. Where they did not fall down, their stability can now be easily disturbed by excavation for keying the dam.

A similar case from the French Alps is shown in Fig. 13-4 (Gignoux, Barbier

1955). A damsite, apparently morphologically suitable, was studied in the upper course of the river Romanche. The left valley flank is made up of granite and Triassic sandstones and limestones. The more moderately sloping right flank also appeared to be stable; the Liassic shales encountered in boreholes were mostly bedded and impermeable. Since the geological study of the wider area indicated that the valley had been constricted by an old landslide, a deep borehole was sunk from the level of the crest of dam. This revealed water-bearing gravel at a depth of 63 m. The ancient river valley had been dammed by a landslide to a height of 40 m, which shifted the stream channel 70 m towards the left bank. The thick sequence of Liassic shales was thus not in place, but slid onto permeable alluvium of the old valley. This damsite had to be abandoned.

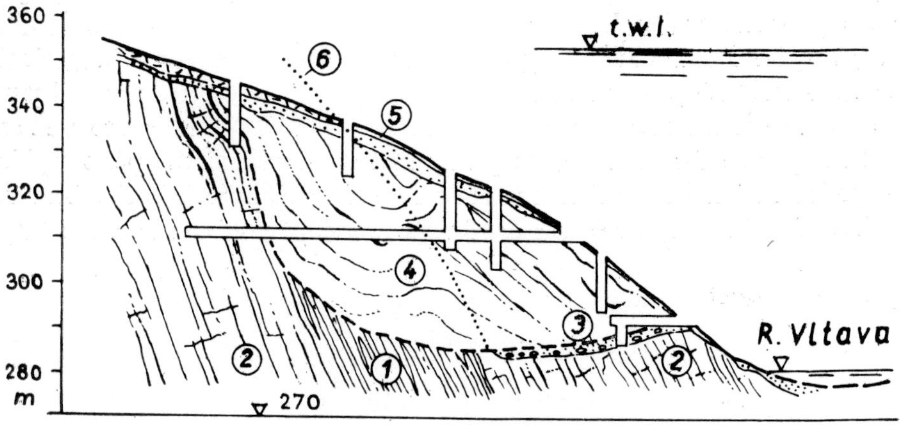

13-3. Geological section of the left bank of the Vltava at the original damsite near Zlákovice. 1 — Porphyroids, 2 — epidiorites, 3 — Pleistocene terrace gravel, 4 — porphyroid material of a rockfall, 5 — slope debris, 6 — slope surface before rockfall. (After Zoubek.)

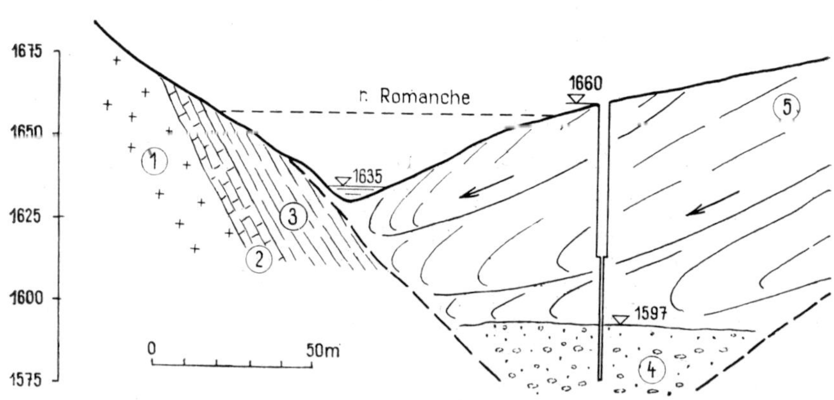

13-4. A damsite depreciated by landslide formed in Lias shales (5). 1 — Granite, 2 — sandstones, 3 — limestones (Triassic), 4 — sand and gravel of ancient valley. (Gignoux, 1955.)

13-5. Cancano arch dam in the Italian Alps. (Photograph by Záruba.)

Morphological configuration of the valley often affects the choice of the type of dam to be constructed. In canyon-like valleys with stable rocky slopes the arch dams are usually most economical (Figs. 13-5, 13-6). In some cases it is difficult to decide which type would be most suitable and several factors must be taken into consideration.

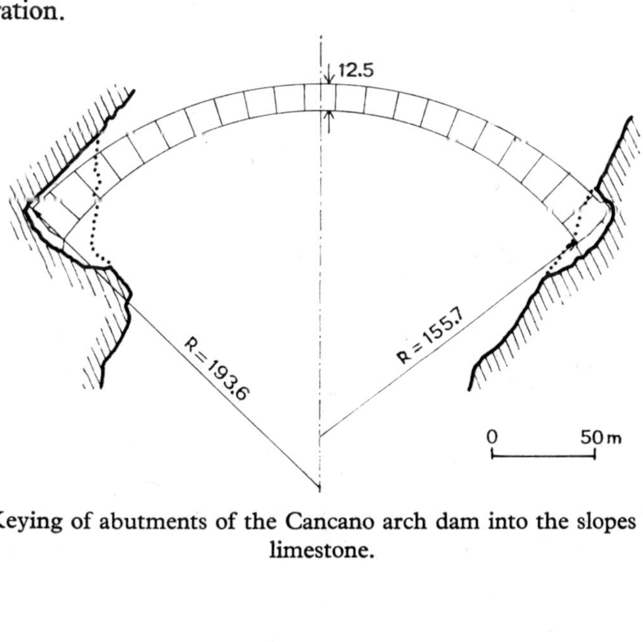

13-6. Keying of abutments of the Cancano arch dam into the slopes of firm limestone.

Several detailed studies were necessary, for example, to select the most adequate type for the Pieve di Cadore Dam (Italy) because of the complicated morphological form of the river valley. The damsite is located in Triassic dolomitic limestones. The cross-section of the valley has the shape of a trapezium 50 m high and 300 m broad. At the right bank the river Pieve cut a 55-m deep erosion gorge. Considering the existing conditions, it proved to be most economical to construct two arch dams separated by an expansion joint (Fig. 13-7).

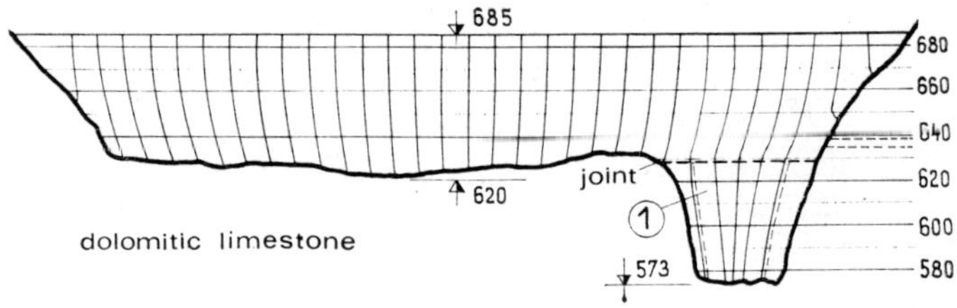

13-7. The type of dam was chosen with respect to the morphology of the site. Erosive gorge vaulted by a separate arch (1).

In the reconnaissance study of a damsite it must be kept in mind that the valleys often developed in rocks of weaker resistance to stream erosion. Many a time an overdeepened erosion gorge occurs in a valley floor, due either to tectonic disturbance of rocks or to the primary character of sediments. It is therefore necessary to understand the development of a valley and its geological structure.

Older *overdeepened valleys* concealed by young sediments may exist in the neighbourhood of present-day valleys. This phenomenon is particularly common in once glaciated areas. Figure 13-8 shows the Pitlochry Dam in Scotland, about

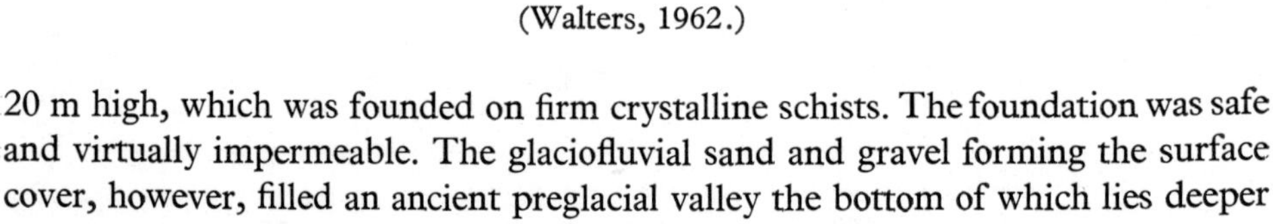

13-8. Pitlochry Dam in Scotland. Ancient overdeepened valley filled with glaciofluvial gravel had to be sealed by a concrete screen. 1 — Firm gneiss, 2 — sand and gravel (Walters, 1962.)

20 m high, which was founded on firm crystalline schists. The foundation was safe and virtually impermeable. The glaciofluvial sand and gravel forming the surface cover, however, filled an ancient preglacial valley the bottom of which lies deeper than the floor of the present river bed. It was therefore necessary to extend the dam by a concrete wall to a length of 280 m.

The examples mentioned above indicate that detailed plans for a dam should be developed only after the geological character of a damsite has been reliably established. Geological investigations for higher dams must be carried out as thoroughly as possible since they determine not only the economy of the design but also the safety of the structure. Geological deficiencies such as fault zones, cavernous layers, buried channels and rock slides, may prove a serious if not unsurmountable obstacle during construction.

Another factor which is decisive in the choice of a damsite is the *stability of the valley slopes*. The founding of a dam demands relatively deep excavations in the floor and flanks of the valley and these can disturb stability conditions. In general, damsites where there are no steep and high slopes above the crest of the dam are regarded as more suitable. In the opposite case, the safety of the structure may be endangered, especially during excavation, which may involve costly measures or even the abandonment of the site. Figure 13-9 shows the section of a rocky

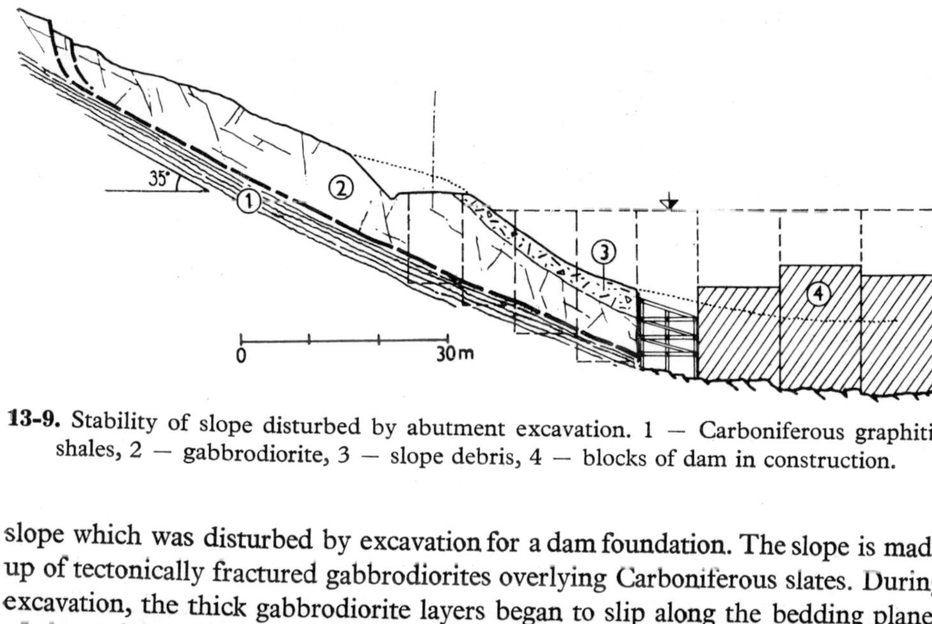

13-9. Stability of slope disturbed by abutment excavation. 1 — Carboniferous graphitic shales, 2 — gabbrodiorite, 3 — slope debris, 4 — blocks of dam in construction.

slope which was disturbed by excavation for a dam foundation. The slope is made up of tectonically fractured gabbrodiorites overlying Carboniferous slates. During excavation, the thick gabbrodiorite layers began to slip along the bedding planes of these slates. To avoid caving-in at the working area, excavation had to be carried out carefully, block by block. The excavated space was supported by heavy struts against the neighbouring blocks which were concreted rapidly to the necessary height. In this case, the original plan for the dam did not have to be changed since the gravity dam was able to resist the pressure from the loosened slope along its longitudinal axis. Excavation for the foundation of a dam may be difficult even in firm gneiss if its beds dip downslope. A partial burying of the dam foundation bed by gneiss, which loosened during the winter period, is seen in Fig. 13-10.

Similar phenomena are fairly frequent in valleys deepened in schistose rocks with the schistosity planes dipping at roughly the same angle as the slope. The

13-10. Excavation may be difficult even in firm gneiss when the dip of strata is unfavourable. (Photograph by Pašek.)

schists are necessarily undermined by excavation and the stability of the slopes is impaired. F. Reuter (1958) reported on such difficulties at the construction of the Rappbode Dam (East Germany); only careful excavation work prevented the occurrence of major landslides.

In connection with the stability of slopes, some remarks should be added concerning the foundation of blocks on slopes prior to the blocks in the valley floor. This is necessary in some cases for easy working. It is always a hazardous procedure, however, possible only in solid rocks with a favourable dip of beds and joints, which are not liable to deep-seated sliding. If exceptional conditions require the use of such a working process, a side block can be concreted only after excavation for its two neighbouring blocks has been completed. This arrangement is necessary in order to ensure the continuity of the foundation surface and to avoid the use of explosives in the proximity of partly concreted blocks.

Geological-morphological conditions of the construction site often decide whether a *diversion tunnel* for temporary passage of water can be constructed or whether the water must be conducted over the building site. A diversion tunnel is more suitable as it facilitates continuous operation. A diversion tunnel can be constructed if the flanks are of suitable form and consist of sufficiently firm rock, since it would not be economical to drive a tunnel in shattered rocks.

When the dam is built in a wide valley, the foundation pit must generally be excavated in sections and the blocks are erected progressively, water being conducted over partly concreted blocks.

The engineering geologist should co-operate with the engineer in planning the *cofferdam.* For this purpose he examines the character and thickness of valley sediments and slope detritus at the foot of slopes. For the design of the cofferdam the probable inflow of ground water to the foundation pit must be established by pumping tests.

In planning hydraulic structures on large streams, the geomorphological conditions (e.g. meander bends) may be used ingeniously to advantage by erecting part or the whole structure on the dry land. As an example, the Wallsee Dam on the Danube (Austria), built in 1968, may be cited. The site was situated in the core of a broad meander, beyond the present river bed, and the location of the power plant, weir and navigation locks was chosen on the basis of detailed geological investigation. After the completion of these structures, the river was diverted into a newly excavated channel. The investigation of construction site II (Fig. 13-11)

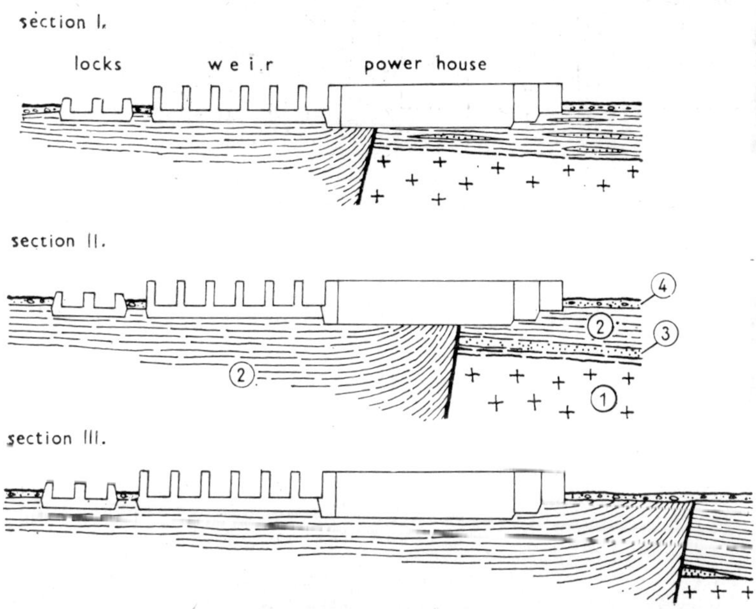

13-11. Geological conditions of investigated damsites I—III for the Wallsee Hydraulic Work on the Danube. 1 — Granite, 2 — Oligocene shaly claystone, 3 — friable sandstone, 4 — sand and gravel.

showed that below the centre of the power plant runs a fault, along which the northern block sank. Oligocene claystones underlying a 9-m bed of sandy gravel were fractured and erected. South of the fault, the thickness of claystones was appreciably smaller and they rested on granite. Because of these non-uniform foundation conditions, a site 200 m downstream was studied. The course of the fault was examined by a number of core borings and the geophysical resistivity method (Fig. 13-12). It was found that the fault would run along the northern periphery of the power plant. Only at the third building site, about 300 m upstream of the site II was it possible to locate the principal structures outside the fault. It was verified by deep boreholes (up to 130 m deep) that the horizontally bedded Oligocene claystones continue down to a depth that could affect the settlement of the structure. Technical properties of the claystones were examined on undisturbed samples in the laboratory and by field tests.

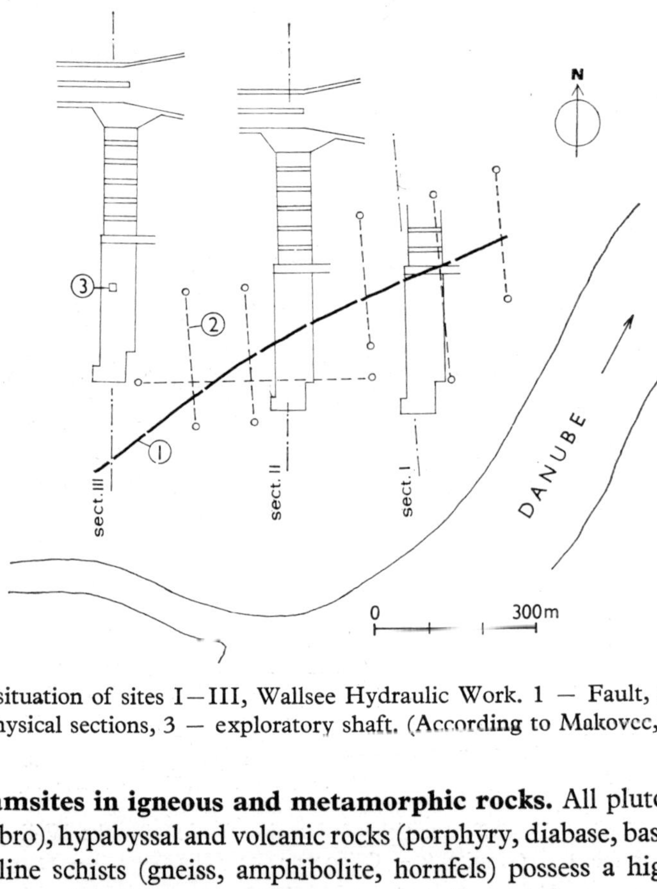

13-12. The situation of sites I—III, Wallsee Hydraulic Work. 1 — Fault, 2 — surveyed geophysical sections, 3 — exploratory shaft. (According to Makovec, 1960.)

13.6. Damsites in igneous and metamorphic rocks. All plutonic (granit), diorite, gabbro), hypabyssal and volcanic rocks (porphyry, diabase, basalt, andesitee and crystalline schists (gneiss, amphibolite, hornfels) possess a higher strength than concrete and their bearing capacity is generally satisfactory unless they are disturbed tectonically or by weathering.

Damsites in *granitic rocks* demand special attention because shattered zones or extensive zones of thermal decomposition may occur in them (see Chapter 6, p. 157). In assessing the character of granites, their geological age and regional tectonic history should be taken into consideration. In general, the older granites are tectonically more disturbed than the younger since they were, as a rule, affected by more folding phases.

Granitic rocks are often strongly altered along major tectonic lines. In investigating the damsites on the river Hornád (Slovakia), for example, the granite near a significant thrust plane was found to be densely fractured and shattered, whereas the same rock farther away was only weakly disturbed.

In Central Europe granites and diorites are also deeply weathered beneath the Tertiary or Cretaceous sediments. The weathering is intensive and of tropical character. The depth of the weathered zone is irregular, in places reaching deep into sound rock, particularly along joints and fracture zones.

Many engineers are of the opinion that damsites in granite provide safe foundation conditions so that they do not require detailed investigation. This is, however, not the case. M. Lugeon drew attention to this error as early as 1933 and in his book demonstrated several case histories. An instructive example is the Fork

13-13. Rockfall in apparently firm granite which caused the Fork Site locality in California to be abandoned. (Photograph by Záruba.)

Site Dam on the San Gabriel River in California. On the assumption that the geological setting of the damsite in granite would be simple, the geological investigation was inadequate. Only during construction was a wide shatter zone revealed on the right bank, within which the granite had been densely fractured and partly decomposed by hydrothermal processes. Excavation caused a large landslide along this zone. The downslipped mass filled the foundation area and since detailed research did not recommend a continuation of the work, the site was abandoned (Fig. 13-13).

In *crystalline schists*, the depth of the weathered zone depends, among other things, on the dip of beds. At a damsite in the Krušné hory Mts. in Bohemia, biotite-paragneiss beds strike parallel to the valley and dip uniformly 30—40° W (Fig. 13-14). On the left flank the dip was roughly coincident with the inclination

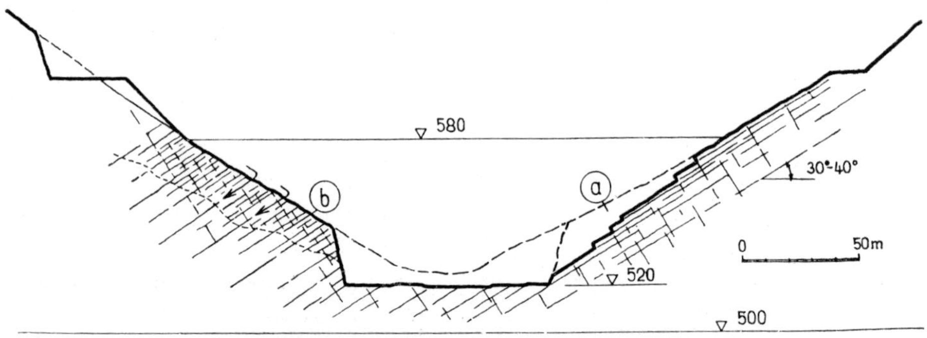

13-14. Damsite in paragneiss in the Krušné hory (Ore Mountains, Bohemia). On the left bank a slide along bedding planes occurred during excavation (a) and on the right bank weathering extended to a great depth (b).

of the slope, whereas on the opposite flank the beds dip into the slope. On the left bank the depth of the weathered zone was relatively small, but during excavation for the foundation, loosening and sliding along bedding planes took place, particularly during the spring thaw. On the right bank, the weathered zone reached to a depth of 12—15 m, particularly because water could penetrate deeper along bedding planes.

The construction of the Prettyboy Dam for the water supply of Baltimore provides an example of difficult foundation work in micaceous gneiss (Fig. 13-15). It was planned as a gravity dam 45 m high. Excavation on the right bank showed that geological investigation was not carried out to a sufficient extent. The rock above the fault plane was decomposed to a depth of 50 m. The foundation had to be deepened appreciably and the dam extended by a concrete wall on the right bank. In relation to the original project, the volume of excavation increased six times, the volume of concrete twice, and as a result completion of the construction work required a much longer time than estimated.

Since even sound rocks are more or less compressible, they will yield under the great load produced by a gravity dam at the bottom of the foundation and that caused by an arch dam at its abutments. It is therefore necessary for the designer to know the *deformational properties* of the rocks as precisely as possible. For gravity and buttress dams the deformation characteristics of rocks in the valley floor must be determined and for arch dams, which transmit water pressure mainly into the valley sides, the deformation characteristics of rocks in the slopes must also be known.

Tests on small rock samples are not really applicable to the determination of the modulus of deformation. Values obtained from rock specimens may be too high since they assume a homogeneous massive substance. The modulus of deformation is substantially reduced by the presence of bedding planes and joints in the rock. The modulus of elasticity of the strongly schistose gneiss on the Lucendro hydraulic structure, for example, was determined on samples to be 390,000 kp/cm^2, but the actual rock gave only 120,000 kp/cm^2. The diabase at the site of the Zitterbach Dam (Stiny 1953) gave 800,000 kp/cm^2 on a rock sample but only 92,000 kp/cm^2 at the site. Moduli of deformation are sometimes greatly different in firm igneous rocks and in crystalline schists. It is unfavourable for the construction of a dam if these rocks meet in the foundations although each of them alone could provide a good foundation bed. Figure 13-16 shows a damsite for which the narrowest section of the valley was chosen, irrespective of the different properties of rocks forming the foundation beds. As a result, the founding of blocks on the left flank met with considerable difficulties.

The *modulus of deformation* can be determined on the basis of either field loading tests on a sufficiently large surface of loading blocks, of water pressure tests performed in galleries, or by the use of seismic methods. Since the rocks near the ground surface and in the valley floor will be more or less loosened, it can be assumed that at a greater depth the bedrock will show better properties than were established near the surface. Seismic tests are preferred since they give

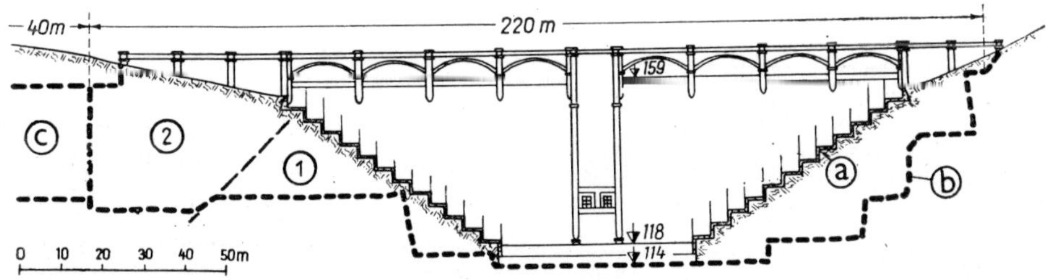

13-15. Unforeseen increase in excavation and volume of the dam in weathered gneiss. 1 — Firm gneiss, 2 — decomposed gneiss, a — planned foundation, b — true depth of foundation, c — concrete wall for sealing the weathered rock. (According to Eng. News Record 1933.)

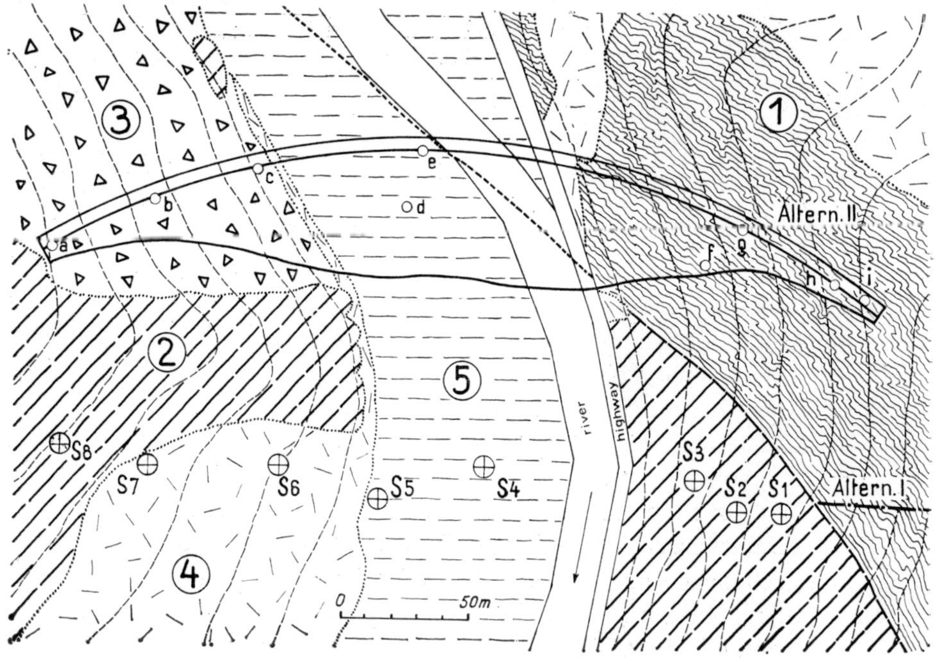

13-16. A damsite with inhomogeneous foundation ground. The left part of the dam was founded on schistose gneiss, the remaining part on firm orthogneiss. Alternative I was more suitable geologically. 1 — Schistose gneiss, 2 — orthogneiss, 3 — bouldery debris, 4 — debris, 5 — Holocene deposits, a—i — localities of surface exploration.

information about the properties of rocks even at great depths. Settlements of dams computed from the results of surface tests are commonly higher than the actual measured values.

Shear resistance. In hard igneous rocks the contact between concrete and the rock is practically never even. The excavation is bounded by irregularly indented surfaces depending on the master joints of the rock. Consequently, there is no possibility of concrete slipping along the surface of the bedrock, unless the surface has been smoothed by a glacier. In this case, the rock surface must be roughened and excavated into steps. A dam failure due to slipping, however, may occur within the bedrock itself. One of the tasks of engineering geological investigation is therefore to examine thoroughly the bedrock forming the foundation beds under a dam to see whether or not they are cut by joints oriented so that slipping under the action of the water pressure could occur along them.

Watertightness of the damsite. In firm igneous rocks and crystalline schists water can seep through open joints and along fractures and faults in the floor and flanks of the valley. Open fissures, particularly in densely disrupted zones formed during tectonic movements, can be sealed by grouting. Crushed or decomposed rocks of shatter zones must be removed to a certain depth and replaced by concrete.

If possible, shatter zones and major faults should be avoided. Figure 13-17 shows the keying of the Shoshone arch dam into granite with the main system of joints running across the valley.

Particular attention should be paid to damsites in basalts, andesites and other young volcanics. Extrusive rocks are generally very permeable since numerous open fissures and beds of cinder originate in them as a result of rapid cooling. They occasionally contain beds of volcanic tuff which may cause uneven compressibility of the foundations.

Unusually complicated, for example, was the founding of some dams in young volcanics in Ethiopia. The Koka gravity dam was founded on alkalic rhyolite. Both the damsite and the flanks of the reservoir had to be sealed owing to dense jointing of the lava flow. Similarly difficult was the sealing of the foundation beds of the relatively low Zula Dam (25 m) on a Recent basalt flow. The open fissures

13-17. Shoshone arch dam (U.S.) on jointed granite. (Photograph by Záruba.)

13-18. Cleaning of the foundation surface by pressure water and compressed air for the Orlík Dam in Bohemia. (Photograph by Záruba.)

were first filled with sand and afterwards with cement-clay grout. The grouting holes were inclined 25° to the vertical so as to fill more and predominantly vertical fissures (Jovanović D. 1970).

The treatment of foundations. Gravity and arch dams can be founded only on sound rock unaffected by surface weathering. In order to plan the correct depth of the foundation, the thickness of the weathered layer must be determined. Excavation in firm rocks should be carried out only with small charges of explosives to prevent unnecessary disturbance of the rock and the opening of fissures. In general, excavation is carried out by blasting to about 2 m above the bottom level of the foundation; for the 1.5-m layer only small charges should be used. The remaining half metre is excavated just before concreting and, if possible, without the use of explosives. In these solid rocks the treatment of the foundation surface is limited to perfect cleaning and washing of the rock by water under pressure (Fig. 13-18). Guniting of the foundations is not necessary. In case the rock is jointed, it is recommended that the whole area be grouted by holes 5—15 m deep depending on the character of the rock and the height of the dam. Contact grouting should be performed after the first layer of concrete has been placed so that a higher grouting pressure can be used.

13.7. Damsites on sedimentary rocks. Investigation of damsites on sedimentary rocks should be conducted with great care since some sediments provide excellent foundation beds while others are quite unsuitable. In considering the foundation of dams, sedimentary rocks can be divided into three groups:

(a) Compact psammitic rocks such as sandstones, arkoses, greywackes;

(b) limestones and dolomites;

(c) clayey shales.

(a) Damsites on sandstones and related rocks. All these sediments were formed essentially by cementing and compaction of sandy beds. Their strength depends on the degree of cementing and the nature of the cement which may be calcareous, ferrous, siliceous, clayey, kaolinitic, etc. The strength of sandstones, arkoses and greywackes is generally higher than the strength of concrete, except for some crumbly sandstones. Sandstones and allied rocks possess relatively great bearing capacity, even if they are poorly cemented. For the construction of high dams exerting considerable foundation pressure the modulus of deformation should be determined, so that the settlement of individual blocks of the dam can be computed. In the foundation beds of the Rossens Dam (Switzerland) the Tertiary coarse-grained sandstones had a modulus of deformation equal to 10,000—20,000 kp/cm^2 parallel to the beds and 8,000—10,000 kp/cm^2 at right angles to them. The modulus of elasticity was established on rock samples at 80,000—150,000 kp/cm^2 parallel to the beds and 30,000—50,000 kp/cm^2 perpendicular to them (A. Moos, F. Quervain 1948).

Sandstones show a relatively high shear strength and when excavated they break along joint or bedding planes. The thinner and more fractured the sandstone beds, the smaller is their cohesion, so that in computations the shear strength must be considered as equal to friction. The shear strength may be reduced appreciably by interbedded clay or clayey shales. The strike and dip of these interbeds must be carefully determined since they are potential sliding surfaces.

Permeability. Water can percolate in sandstones not only through open joints and along bedding and fault planes but also through the pores of the rock. The permeability of sandstones given by interstitial spaces depends on the size of grains and the degree of cementing.

If permeable rocks form interbeds in impermeable rocks (e.g. beds of jointed sandstone in clayey shales) the watertightness of the damsite will depend on the thickness of the permeable beds, on their strike and dip with respect to the position of the dam, on the intensity of their folding and the strike of longitudinal axes of folds towards the dam.

When the beds parallel the course of the valley, the leakage of water downstream of the dam will be more hazardous than when the beds run at right angles to the valley. In the former case a permeable bed will contact impounded water over a considerable length of the reservoir, and rather extensive seepage must be expected, whether the beds are horizontal, steeply inclined (Fig. 13-19) or anticlinally bent (Fig. 13-20a). It is difficult and expensive to seal permeable rocks over a large part of the area to be flooded. Such a damsite could be used if the outcrops of permeable rocks are covered by impermeable surficial deposits, e.g. consolidated clayey debris or clayey slope loams (Fig. 13-20b). The sealing would be necessary only close to the damsite and in the dam foundations. Not a few damsites have been thus improved.

When the beds form a synclinal bend parallel to the valley (Fig. 13-21), conditions will be more favourable. If the dip of beds is smaller than the inclination

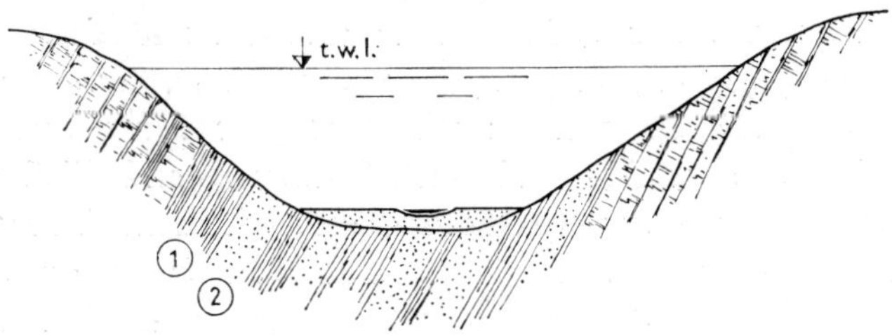

13-19. Beds striking parallel to the valley have a steep dip. Impounded water can percolate through permeable rocks to a long distance. 1 — Impermeable shales, 2 — fractured pervious sandstones.

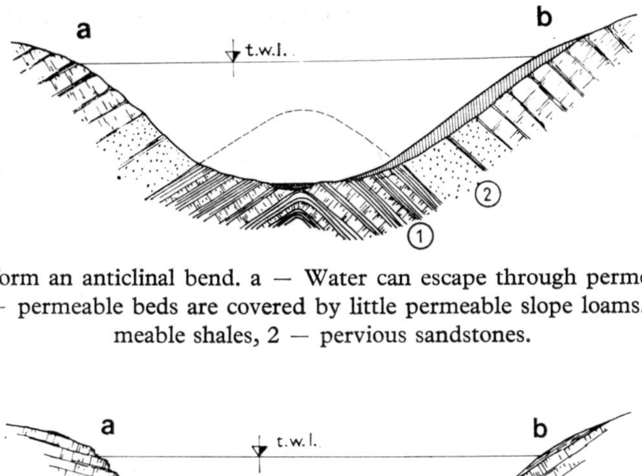

13-20. Beds form an anticlinal bend. a — Water can escape through permeable rocks in the slope, b — permeable beds are covered by little permeable slope loams. 1 — Impermeable shales, 2 — pervious sandstones.

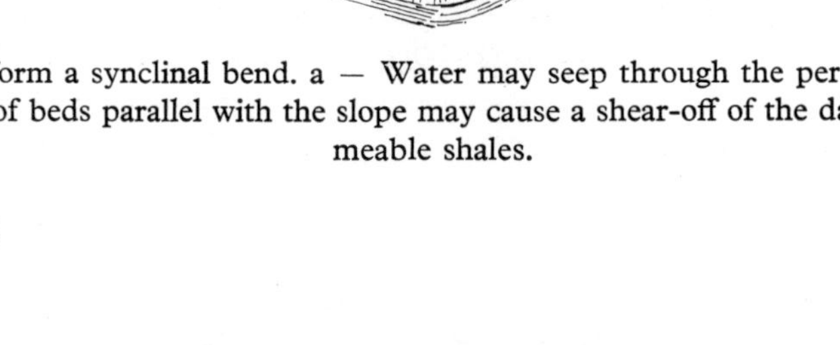

13-21. Beds form a synclinal bend. a — Water may seep through the permeable bed (2), b — the dip of beds parallel with the slope may cause a shear-off of the dam, 1 — impermeable shales.

13-22. Excavation for the right abutment of a dam on the river Arda in the Apennines. Beds of sandstone and shale dip steeply downstream. (Photograph by Záruba.)

of the slope, water can percolate through the permeable rock but to a far smaller extent than in the foregoing case. With regard to watertightness, dip of beds identical with or larger than the angle of the slope is more convenient. Inclined beds are less favourable when they are smooth and persistent since they will possess low shearing resistance. Excavations for the dam abutments may also be hazardous for the stability of the slope.

If beds strike roughly perpendicular to the valley, even a complex with permeable beds may furnish a watertight damsite (Fig. 13-22). The bearing capacity of such a site is also usually satisfactory since the dam will stand on homogeneous rocks in the longitudinal direction. Most favourable for the watertightness of the damsite is a complex of beds dipping steeply upstream. In a complex of permeable sandstones, for example, thin interbeds of clayey rocks are sufficient for the damsite to be almost watertight (Fig. 13-23). It is of advantage in a folded complex to locate a dam in the anticlinal limb plunging upstream (Fig. 13-24). From the point of view of stability, a sequence of clayey shales and sandstone dipping moderately upstream is less suitable since the dam may slip along the bedding planes (Fig. 13-25).

When beds dip at a large angle downstream and are not tectonically disturbed, they provide a comparatively suitable damsite. If, however, the dip of beds changes under a dam, water may escape, mainly at a synclinal bend of beds as shown in Fig. 13-26. In a complex formed of permeable and impermeable beds which dip downstream and are tectonically disturbed, seepage of water along the water-

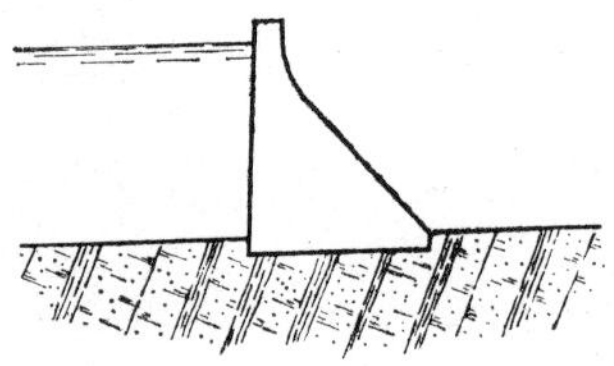

13-23. Steep dip of beds upstream is favourable for founding a dam.

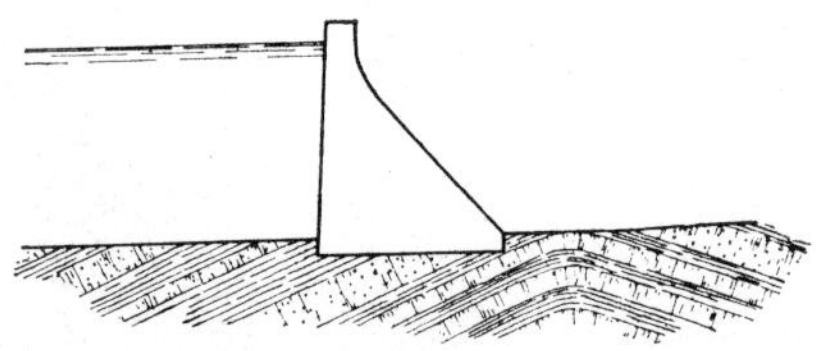

13-24. Dam founded correctly on the limb of anticline dipping upstream.

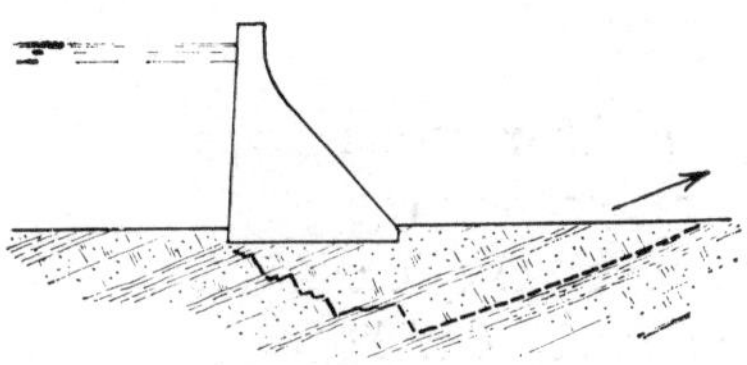

13-25. When the beds dip moderately upstream, the dam may be sheared off along a bedding plane.

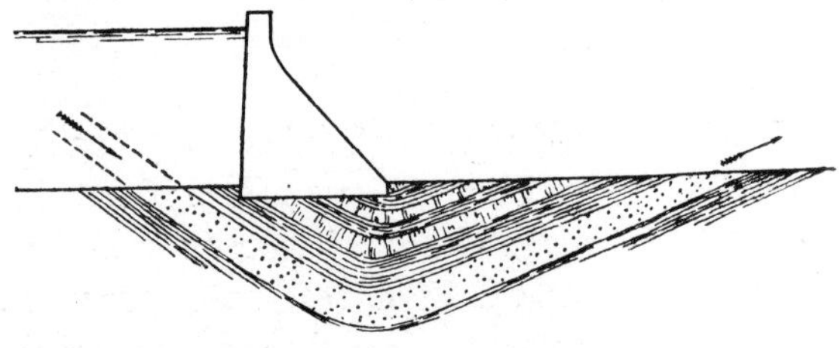

13-26. Unsuitable position of dam on the limb of a syncline with a downstream plunge.

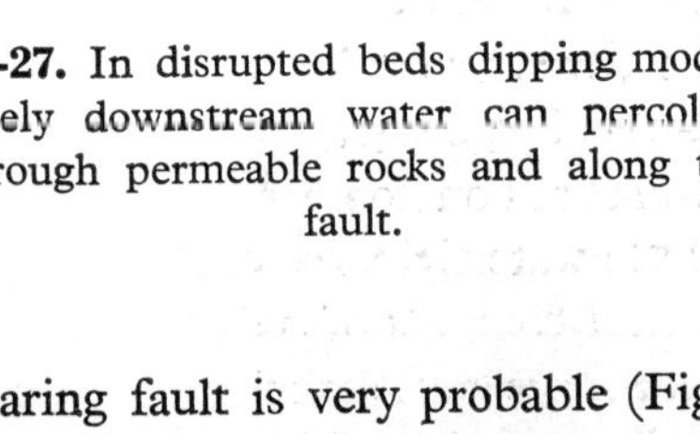

13-27. In disrupted beds dipping moderately downstream water can percolate through permeable rocks and along the fault.

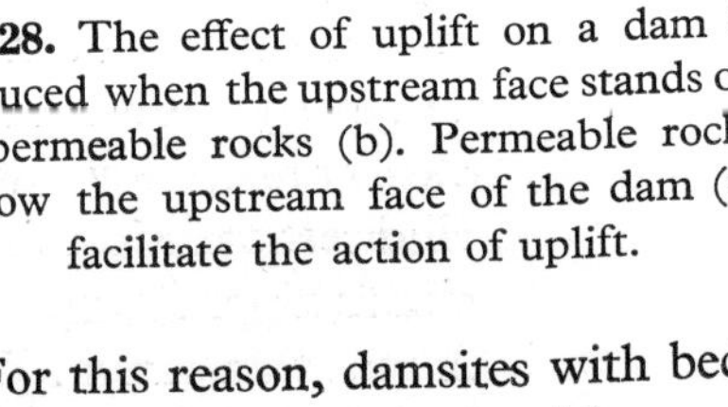

13-28. The effect of uplift on a dam is reduced when the upstream face stands on impermeable rocks (b). Permeable rocks below the upstream face of the dam (a) facilitate the action of uplift.

bearing fault is very probable (Fig. 13-27). For this reason, damsites with beds dipping moderately downstream are generally thought to be questionable.

In a complex with alternating permeable and impermeable beds a dam should be located, if at all possible, so that the upstream face is on impermeable rocks. It would be incorrect to place the dam so that the upstream face would stand on permeable rocks (Fig. 13-28) and the downstream face on impermeable rocks because of the possible effects of uplift.

Treatment of the foundations. No special precautions are necessary to ensure a good bond of concrete with rock in sandstones and rocks that do not weather readily. In principle, all weathered rock and loose blocks are removed. Strongly jointed and schistose sandstones may also be excavated and replaced by concrete. The depth of this filling should be carefully considered. If the joints are open, the filling should extend over the breadth of the disturbed zone and the remaining part should be densely grouted. Deep excavation of a fractured zone may cause loosening of the surrounding rock and impair foundation-bed conditions. If the joints are filled with clay or detritus, the disturbed zone must be excavated to a greater depth in shafts which are opened up and concreted gradually, or another type of dam must be chosen.

Sandstones are relatively brittle and can be greatly disturbed by blasting (Fig. 13-29). The last layer, at least 0.5 m thick, should therefore be excavated by hand and not with the use of explosives.

13-29. Firm sandstone in dam foundation broken by blasting with excessive charges. (Photograph by Záruba.)

If sandstones have marly or clayey cement so that they weather readily, it is recommended that the last layer be removed just before concreting. When a foundation area remains open during winter months, it should be excavated deeper in spring because rock is commonly disturbed to the depth of freezing.

(b) Damsites on limestones and allied rocks. From the technical point of view, crystalline limestones (marbles), massive thick-bedded limestones and flaggy marly limestones, dolomites and dolomitic limestones are considered in this group.

The bearing capacity of limestones is usually satisfactory because they commonly possess a greater strength than concrete. Their modulus of deformation varies depending on the massiveness and jointing of the complex. In jointed dolomitic limestones at the site of the Pieve de Cadore Dam, values of 20,000 to 30,000 kp/cm^2 were obtained. After grouting they rose to 50,000—60,000 kp/cm^2. Solid Triassic limestones on the left flank of the Cancano Dam gave a modulus of deformation up to 200,000 kp/cm^2 at a loading of 20 kp/cm^2.

Limestones may provide a good foundation bed unless their bearing capacity and permeability have been impaired by hidden subsurface cavities. Limestone is a soluble rock and pure jointed limestone in particular is often karstified. Joints may be widened into galleries and caverns, features which have greatly impeded the construction of too many dams. It is therefore an important task of engineering geological investigations to establish whether the rocks at a damsite are disturbed by karstification.

The solution of limestones is a relatively rapid process in a geological sense but very slow with respect to the durability of a structure. A damsite is not endangered by solution of limestones in a valley floor or flanks since concrete is generally more soluble than most limestones.

Shear resistance. Limestones are essentially solid rocks which break irregularly so that there is little danger of shearing between the rock and concrete. The shear strength of limestones is important. In thin-bedded limestones shear may occur along bedding planes. The shear strength may be impaired by interbeds of clayey shales and clay. An engineering geological investigation should establish the disposition of these potential sliding surfaces which must be considered in static analysis.

Permeability. Whenever limestones occur in dam foundation beds the permeability of the site must be carefully studied because of the solubility of this rock. The position, extent, shape, size and depth of underground cavities depend on the development stage of karstification and this is controlled by a number of factors. One of them is, for example, the chemical composition of limestone: the purer the limestone the easier is the leaching, so that karst cavities are invariably larger in pure limestones than in marly limestones under the same conditions. The course of underground cavities can be inferred from the structure of the rock complex and the strike of joints; in anticlinal bends the joints are commonly widened and the solution of the rock is more intensive along them. Karst galleries and caves are generally developed at several horizontal levels following the development of a valley. These levels correspond roughly to the altitudes of Pleistocene river terraces, which are thus a good indicator of the position of the main cavities. In limestone areas which have undergone subsidence, karst cavities formed before the subsidence may occur deep below the level of the valley floor (Fig. 13-30).

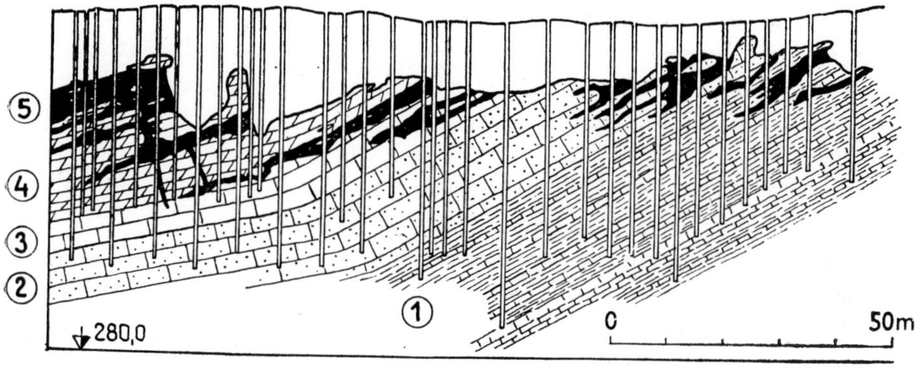

13-30. Karst cavities may extend deep below the valley floor. Cherokee Dam on the Holston River. (After TVA, 1949.) 1—3 — Limestones and calcareous shales of Middle Cambrian age, 4 — Upper Cambrian dolomites and dolomitic limestones, 5 — karstified cavities.

To establish a water-tight curtain in cavernous limestones is very difficult. It is almost impossible to pinpoint all open joints and cavities through which the water can escape in the flanks and below the floor of the valley.

Construction methods used for building dams in limestone terrain are varied. Since the stability of a dam will not be endangered directly by water seepage, some dams are constructed without a grout curtain and only when great losses of water are observed after the filling of the reservoir would measures for sealing be taken. The reservoir can be emptied and cavities then ascertained can be gradually grouted. In most cases, however, grout curtains are constructed before the filling of the reservoir, at least in those parts where seepage is expected.

Cavities and widened joints in limestones may be open or partly filled with washed-in clay and clayey loam. Cavities filled with clayey material may let through only a little water, but its amount may increase progressively as filling is washed out from the cavities. In some cases this phenomenon has been explained erroneously as being due to the rapid enlargement of caverns by the solution of limestones. Detailed investigations, however, have shown that increased water loss was caused by outwashing of the secondary filling from the karst cavities.

In cavernous limestones the sealing of karst cavities by cement and cement mortar has not always proved successful. Cavities are often as much as several metres wide so that it is virtually hopeless to try to seal them with cement grout since this will be washed away by flowing water. At several damsites grouting with hot asphalt has been used with success but this treatment needs a complicated device for heating asphalt in the grouting holes in order that it may be fluid at the cavities to be filled. Asphalt sealing is not permanent; it usually serves as a temporary barrier to facilitate cement grouting. The washing of clay filling out of karst cavities and their sealing with a suitable grout mixture is very laborious and demands a long-term grouting programme.

The degree of karstification is also critical in relation to uplift. In designing gravity dams in limestone traversed by open joints, full value of uplift must be reckoned on and this implies an appreciable increase in the amount of concrete.

Treatment of foundations. All weathered layers and thin beds of firm limestones covering large underground cavities must be removed. The cleaned surface of limestones does not need any special treatment for ensuring the bond between concrete and rock. When a limestone complex contains beds of clayey shales, the shales should be removed to a depth at least equal to the thickness of the bed and replaced by concrete of good quality. The surface of limestones liable to karstification is usually ragged with sharp ridges and deep "pockets" filled with clayey residual material. The surficial layer must therefore be excavated to a greater depth than usual so that the structure will stand wholly within firm unweathered rock.

Since limestones provide hazardous damsites, dams have rarely been constructed on karstified limestones. One which was so founded is the Camarasa Dam in Catalonia, which stands on Jurassic limestones and dolomites. After its reservoir had been filled, numerous springs appeared downstream of the dam with yields of up to 11 m^3/sec. Sealing of the reservoir was difficult since the seepage was distributed over a large area. A grout curtain had to be constructed on both flanks for a length of 1029 m, using 224 grout holes up to 394 m deep (M. Lugeon 1933).

Very unfavourable geological conditions had to be overcome in constructing the De Castillon Dam on the river Verdon in the French Alps (E. Ischy 1948). Morphologically the site was suitable for an arch dam, but the geological setting caused much difficulty. The gorge is cut in Jurassic jointed limestones; on the left flank they dip moderately (about 20°) into the slope and are fairly massive, but on the left flank they dip valleywards and are disturbed by an anticlinal bend. The limestone beds are strongly fractured in three directions as can be seen on the diagram (Fig. 13-31). The fissures are open, some of them as much as several metres wide. They are of tectonic origin, having been formed during anticlinal bending. The arch dam could not be keyed into the fractured limestones without the certainty of reliable support. Very detailed investigations were carried out, using galleries and boreholes to trace accurately the course of the fissures.

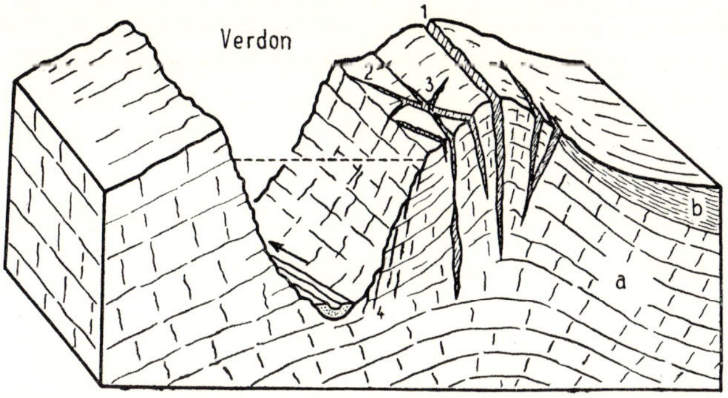

13-31. Block diagram showing the major fractures in limestones in the right valley slope, De Castillon damsite, France. (Gignoux, Barbier, 1955.)

In essentials two problems had to be solved: consolidation of fractured limestones so as to ensure a sufficient support for the arch dam and at the same time the watertightness of the damsite. The compaction was achieved by concreting all open fissures after they had been cleaned out. The rock massif behind the toe of the dam, having a volume of about 200,000 m^3, was then grouted and all loosened limestone blocks at the foot of the slope were reinforced by concrete supports secured by prestressed bolts (Fig. 13-32).

The main problem in grouting was to limit the mass that should be grouted so that the cement would not escape by open fissures to any greater distance. For this purpose peripheral holes, 2—5 m apart, were drilled in several directions from three galleries. The open fissures made the drilling very difficult. Grouting was executed with the use of a special mixture because thin cement slurry would

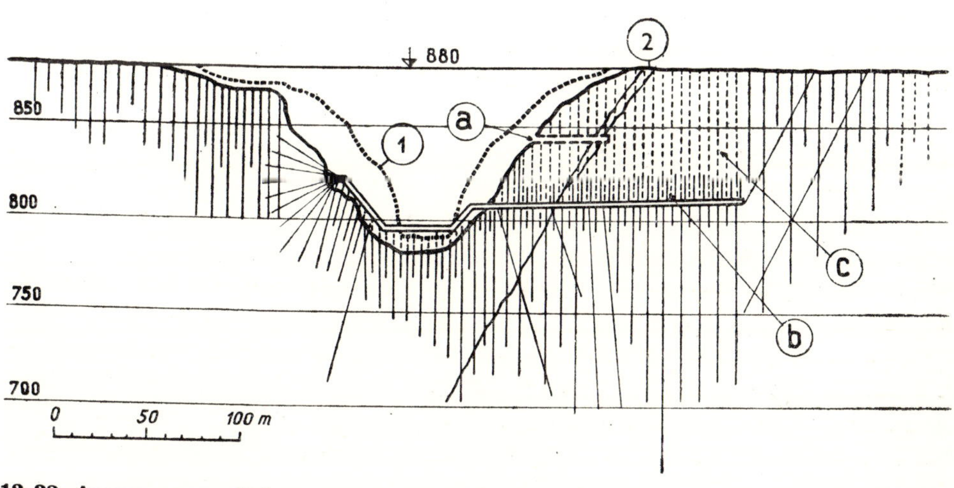

13-32. Arrangement of the grout curtain at De Castillon damsite. a, b — Grout galleries, c — oblique boreholes, 1 — original surface, 2 — open fracture.

tend to flow away through open fissures and would not form a continuous curtain. On the basis of laboratory tests a mixture of cement, bentonite and crushed stone was used. This mixture displayed the consistency of dry mortar but could still be injected by a normal grouting machine. Thanks to the thixotropic character

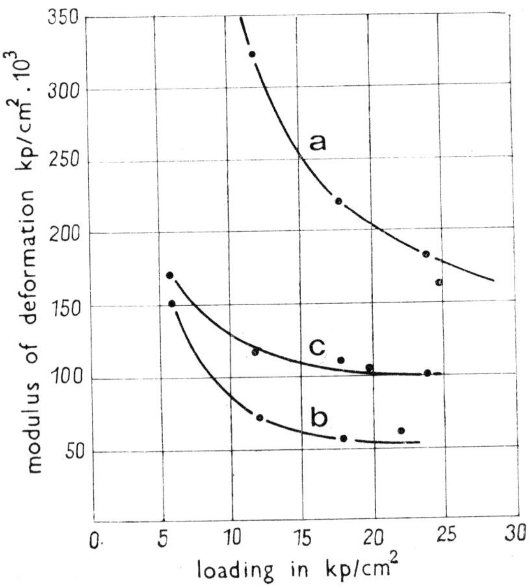

13-33. Deformation moduli measured in marly limestones at Cancano dam-site. a — Left bank, b — right bank before grouting, c — right bank after grouting.

of bentonite, the mixture consolidated soon after it had been injected. The mixture reached a strength of 30—40 kp/cm² after seven days and 80 kp/cm² in 28 days. It was suitable for grouting fissures even several tens of centimetres wide.

The layout of boreholes was arranged in several directions through the massif in such a manner that they could be carried out gradually in accordance with the progress of construction. The estimated length of boreholes was 15,000 m. About 6000 m³ of grouting mixture was used for consolidating a rock mass of 200,000 m³ or approximately 3 % by volume.

In contrast to the foregoing cases, the limestone foundation beds for the Cancano Dam in the Italian Alps, built in 1957 (Figs. 13-5, 13-6) proved favourable. The dam, at first planned to be 137 m high, was founded on Triassic marly limestones, the beds of which dip upstream and are separated by marly interlayers. As a result, the limestones were not karstified. Because of the great load they would bear and also because the rock on the right flank was tectonically disturbed, the deformational properties of the beds were studied in galleries. From the results of measurement shown in Fig. 13-33 it followed that at a loading of 20 kp/cm² the limestone on the left flank had a modulus of deformation 200,000 kp/cm², whereas that on the right bank had one of only 50,000 kp/cm² and this after the grouting of 100,000 kp/cm². The limestones below and around the dam were sealed by a two-row grout curtain up to 90 m deep. The holes were arranged diagonally in plan in order to cut bedding planes and joints. The average consumption of cement was low, being only 11.4 kg/1 m.

(c) Damsites in clayey shales. In technical practice the term clayey shales comprises all consolidated pelitic sediments deposited originally as clayey or clayey-silty alluvium and compacted by compression or cementation and/or recrystallization. Clayey shales exist in nature in various degrees of consolidation. The rate of consolidation is controlled by the following factors:

Size and character of clay particles,
presence of sand seams through which water can escape easily,
thickness of deposits and the load of overlying beds,
period of loading,
strength of orogenic pressures affecting the rock complex.

On the basis of the degree of consolidation shales are divided into two groups:

(1) Shales disintegrating and slaking under the effect of subaerial agencies (Fig. 13-34), particularly on coming into contact with water after partial drying;

(2) shales resistant to weathering which do not slake at alternating drying and saturation with water.

There are also transitional types of shales which break and disintegrate in water but do not slake. To assess this basic property of the shaly pelitic rocks slaking value tests are performed. On the basis of such test results the measurements necessary for the protection of the foundations can be determined.

13-34. Palaeogene clayey shales disintegrating and slaking on freezing. (Photograph by Záruba.)

In Central Europe Palaeozoic and older clayey shales belong universally to group (2), partly because they were exposed to high stresses during the Variscan orogeny. Most Cretaceous and Tertiary pelitic sediments weather readily and slake when wetted after drying. This group also includes many Tertiary volcanic tuffs, tuffites and some sandy shales with clayey or marly cement.

One of the greatest disasters caused by a dam failure, the collapse of the St. Francis Dam in southern California in 1928, was due to the presence of slaking shales in the valley flank.

The bearing capacity of shales of the first group depends on their water content. In general it rises with decreasing water content. These shales possess a relatively low modulus of elasticity so that they show plastic behaviour. Under great loads they could be squeezed from beneath the foundations of the structure. In the floors of deep valleys they are occasionally disturbed by the load of the overlying beds (see Chapter 7). A complex of shales which was disturbed in the Quaternary and has been deformed by the load of the overlying rocks is not suitable for the foundations of a gravity dam. The squeezed out shale beds would also be affected by surface weathering.

Shales of the second group are of greater strength and bearing capacity. Their strength corresponds roughly to that of concrete of average quality. They are not liable to plastic deformation under the load that may be imposed from a dam. The modulus of deformation of these shales is generally smaller than that of concrete so that they show considerable compression under a great load. In choosing the dam type to be built on compressible shales the designer must therefore bear in mind that the difference in loading on the foundations between the full and empty reservoir should be as small as possible. Large deformations of the foundations may damage the grout curtain, loosen the bond of concrete with the bedrock and cause uneven settlement of individual blocks.

Shear resistance. The contact between concrete and excavated shales is usually uneven, so that the shear resistance between the rock and concrete is almost invariably greater than the shear strength of shales. It is thus necessary to determine their strength along surfaces on which shear might occur.

The shear strength along bedding or schistosity planes is appreciably smaller than when the shear plane cuts across the beds. How important is the petrographical study of shales is best demonstrated by the great difference between shear strength parameters along the bedding or schistosity planes. In seams of bentonitic clay, Φ is 12° with a cohesion of about 0.2 kp/cm². The increased content of less active clay, with a simultaneous decrease of the clay component in favour of carbonates, increases Φ to 28° and above. In contrast, the strength across clay beds shows Φ above 20° with a cohesion of 0.5 to 1 kp/cm². In addition, it is generally accepted as more correct not to take cohesion into account in the shear along bedding or schistosity planes, as even at small normal stresses the minute protrusions can be broken off. S. A. Roza (1952) was the first to mention it, but he gave a rather high limit of 10 kp/cm^2 for the normal stress. In connection with the great differences between shear strength along and across the bedding surfaces, the shear plane is often crooked. It cuts across the bed (sometimes breaking-off takes place instead of shearing) and hence it follows the planes of weakness. Statical analysis must take this circumstance into account. In some cases the shear resistance and bearing capacity of clayey shales is so small that the construction of an earth dam will prove to be more suitable.

Watertightness. Clayey shales provide satisfactorily impermeable foundation ground. On account of the plasticity of the shales, loading does not cause the formation of joints and open fissures; in contrast, all existing fissures may be closed as a result of plastic deformation. The rock is virtually impermeable unless the shales are intercalated with beds of permeable sandstone, in which case the shales immediately beneath the sandstone interbed must be thoroughly examined, as their surface can be weakened by water percolating through the porous and jointed sandstone. This disturbance of shales may occur at a far greater depth than the depth of surface weathering.

Compact shales of the second type may be penetrated by open fissures, joints and shattered zones which should be sealed by grouting just as in other solid rocks.

Treatment of foundations. Since shales, particularly those of the first type may change with each alteration of water content, they have some unfavourable properties as foundation beds. In dry weather they may lose water by evaporation and desiccation cracks may start. If concrete is placed on the dried-out shales, the surface layer of the shale may slake by water from the concrete and a thin seam of soft clay may form at the concrete/rock contact. To prevent the formation of this it is desirable to shorten as much as possible the time interval between the finishing of the foundation surface and placing of the first concrete layer. In some cases protection of the rock surface by gunite or asphalt has proved convenient.

In clayey shales, just as in other rocks in their natural state, there may by residual stresses which can be relieved by deep excavation for a dam foundation. Rock may move into the foundation pit and as a result the bedding planes and fissures may open in the bedrock. The opening of fissures is sometimes explained erroneously as being caused by the drying out of shales. This unfavourable behaviour of shales may be prevented only by shortening the interval between the finishing of the foundation surface and placing the first concrete layer and by opening up the foundation pit in small working sections.

The character of clayey shales is decisive for the geotechnical properties of rock complexes containing a large proportion of shales. Such are, for example, the flysch sequences of the Alps and Carpathians, in which shale beds alternate with thin sandstone interlayers. For the assessment of geotechnical properties of these complexes the character of the shales and their tectonic disturbance should be examined. In some regions the flysch sequence is relatively intact, the sandstone

13-35. Impermeable Palaeogene shales with sandstone interbeds, at the Krpelany damsite on the river Váh (Slovakia). (Photograph by Pašek.)

and shale beds extending continuously over considerable distances. They lie horizontally or are moderately folded. Such complexes are generally not very permeable, particularly across bedding, because the undisturbed clayey shales sufficiently seal the permeable sandstone beds (Fig. 13-35).

Most of the flysch complexes, however, were affected by strong orogenic stresses so that they are intricately folded, thrust over one another and traversed by minor and major faults. Special attention should be paid to beds inclined at a steep angle. They are often apparently little disturbed but high pressures which acted perpendicularly to the beds caused the compression, molding, reduction and tight folding of shale interbeds. These provide typical examples of disharmonic folding. The hard sandstone beds separated by plicated shales are frequently densely fractured; the fractures are perpendicular to the bedding planes and open so that the soft shales may be partly pushed into them. Complexes thus disturbed are usually very permeable. The disturbed shales give low values of shear strength, since the bedding planes were smoothed by differential movements which occurred on them. The fractured zones in the flyschoid rocks are not very suitable as foundation beds for dams, particularly for gravity dams. The degree of tectonic disturbance should therefore be correctly established in the reconnaissance investigations.

The Ben Metir Dam, constructed for the water supply of the City of Tunis, is described here as an example of a dam founded on a disturbed shale-sandstone flysch complex of Oligocene age (Stucky 1956). Thin sandstone beds, densely fractured and permeable, are separated by clayey shales which, when unweathered, are firm, but when close to the sandstone interbeds are deeply weathered and plastic. On the left bank the flysch rocks are covered by younger (Pontian) lacustrine clayey and sandy sediments with subhorizontal strike. At the damsite the rocks are heterogeneous and of non-uniform compressibility. In view of these geological conditions, the dam was designed as a rockfill dam at the left abutment (Fig. 13-36) and as a buttress dam on the Oligocene flysch rocks. The buttresses were of such a length that the pressure on the foundation would not surpass 8.5 kp/cm². The sealing of joints between individual buttresses was made so as to sustain a total settlement of 10 cm and up to 5 cm differences in settlement between buttresses. A few low blocks on the right flank, which stands on solid sandstones, were made as gravity blocks (a).

After the reservoir had been partly filled, a settlement up to 58 mm was measured. The differences in the settlements between individual buttresses were less than 1 mm. The watertightness of the damsite was ensured by a grout curtain, 20—30 m deep. Two drainage galleries were driven under the dam to speed up the consolidation of clayey beds and reduce the uplift. Drainage borings, fanwise arranged, were drilled upwards under the dam foundation from these drainage galleries (Fig. 13-37).

13.8. Foundation of gravity dams. The principal requirement for a suitable building site for a gravity dam is the presence of solid rocks with sufficient bearing capacity at an attainable depth in the floor and flanks of the valley. Engineering geological investigations should ascertain the depth of suitable foundation beds below the ground surface to determine the admissible loading and the watertightness of the damsite, especially with regard to the effects of uplift.

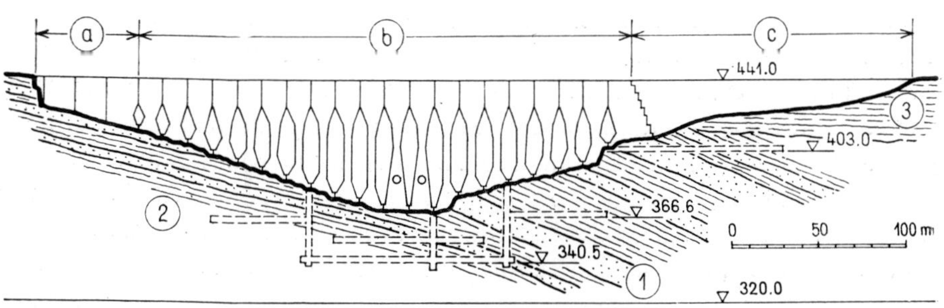

13-36. Longitudinal section of the Ben-Metir Dam (Marty, 1955). 1 — Sandstones, 2 — Oligocene clayey shales, 3 — Pontian clayey-sandy lacustrine sediments, a — gravity blocks, b — buttress dam, c — rockfill dam.

In principle, gravity dams should be constructed on the firm bedrock. Gignoux, Barbier (1955) record an exceptional case when the dam on the river Isère (Savoy) was erected on a vault of 26 m span bridging a canyon 50 m deep. Figure 13-38 shows the layout of borings through which the shape and depth of the canyon

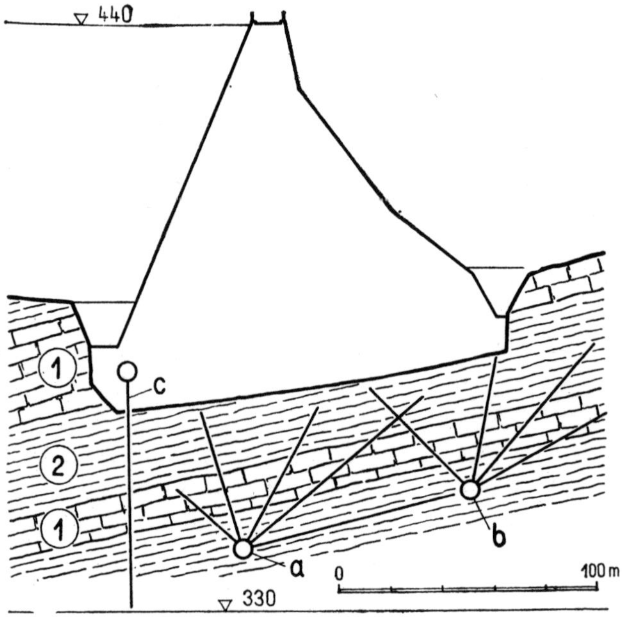

13-37. Cross section of the Ben Metir Dam (Stucky, 1956). 1 — Sandstones, 2 — Oligocene clayey shales, a, b — drainage galleries with drainage borings, c — grout curtain.

were determined. On the left bank deep vertical boreholes were put down, whereas the right flank was examined by oblique core borings driven from a widened recess in the exploration gallery. With the use of oblique boreholes the shape and depth of the canyon were accurately determined. The filling of the canyon (stream and glacial sediments) was removed during construction and replaced by concrete.

Results of detailed investigations are used as a basis for the statical computation of a dam. We therefore think it necessary to discuss concisely the basic presumptions of the computation, at least for gravity dams. For a full reservoir, the external forces acting on the dam are the pressures H_1 and H_2 of water, the load of the dam Q and uplift W (Fig. 13-39). The dam and its foundation must be able to transmit safely the resultant of these forces R. If we resolve the resultant R into normal component N, perpendicular to the foundation surface and tangential component T, parallel to it, the former exerts compression and the latter shear stress on the foundation ground. In statical analysis it is generally assumed that the stress diagram in the foundation area is bounded by a straight line. The maximum pressure on the rock at full reservoir is then under the toe of the dam and under the heel of the dam with an empty reservoir. The rock must safely transmit these stresses; the $T:N$ ratio is decisive for the resistance of the dam to shear displacement.

As seen from Fig. 3-1 the $T:N$ ratio $= \tan \Phi$ is indicative of the stress which the gravity dam will exert on the foundation rock. Figure 3-7 shows that the static interaction between the body of dam and the rock is complex. In view of the great differences in the state of stress at individual points of the foundation area, a gradual weakening may occur when the foundation beds are not homogeneous and very stiff. For this reason the cohesion of the rock is not considered to any great extent and its value is reduced by a high factor of safety. In gravity dams with the usual height to width ratio of about 0.8 the $\tan \Phi$ should be at least 0.75, which corresponds to $\Phi = 37°$. From the values of rock strength parameters in Table 3-1 it is apparent that only a few rocks comply with this requirement. It is surpassed in rocks of small strength or when the strength is reduced along planes of discontinuity. In these cases several solutions are available:

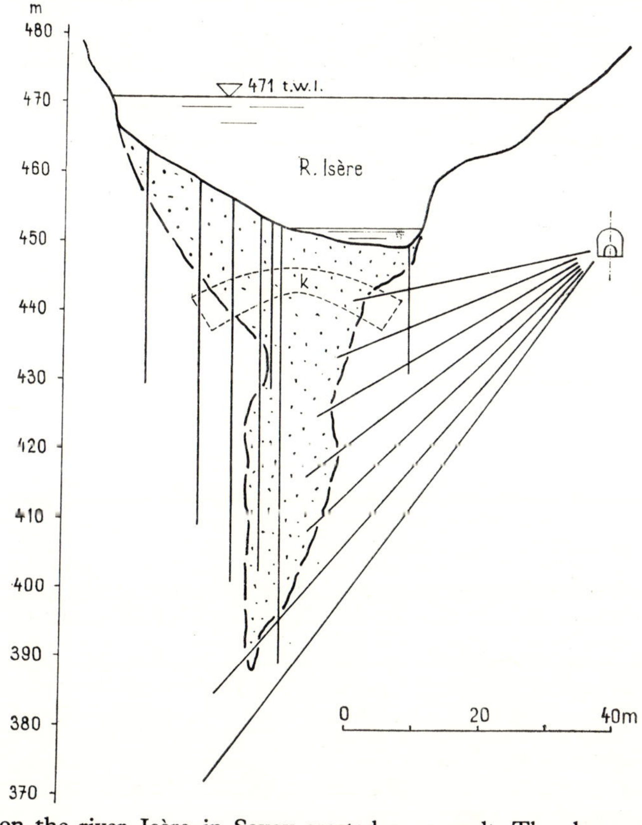

13-38. A dam on the river Isère in Savoy erected on a vault. The shape of the canyon was determined by oblique boreholes sunk from a gallery. (Gignoux, 1955.)

(a) To construct an earth dam instead of a gravity dam. With favourable foundation conditions a gravity dam may be as much as 15 % cheaper than an earth dam but this difference decreases with impairment of rock properties.

(b) To increase the weight of the dam. This procedure, however, is not recommendable in soft rocks, since plastic deformation may occur in them, as was explained in par. 3.4.5. This could cause movement of the dam and possibly also long-time deformation of the rock. Displacements in the foundation ground would be harmful to a grout curtain and other delicate elements of the dam construction. The increase in the weight of a dam must therefore be coupled with an appreciable reduction of its width to height ratio. It is one of the responsible tasks of the engineering geologist to determine the safe limits of normal stress at the boundary of plastic behaviour of the rock.

(c) To deepen the foundation. This solution is directed to achieving an increase in the minimum principal stress at the toe of the dam. It does not exclude, however, the possibility of plastic behaviour of the rock at the toe of the dam.

(d) To change the normal stress diagram in the foundation area and reduce the shearing stress by anchoring the dam to the bedrock. Modern anchoring methods allow for the development of anchoring forces of up to several hundred tons. If the anchoring forces are transmitted to the rock near the heel of the dam, the sum of the normal forces is increased without raising the possibility of plastic behaviour in the rocks at the toe of the dam. The component of the anchoring force which originates through inclination of anchors towards the upstream face

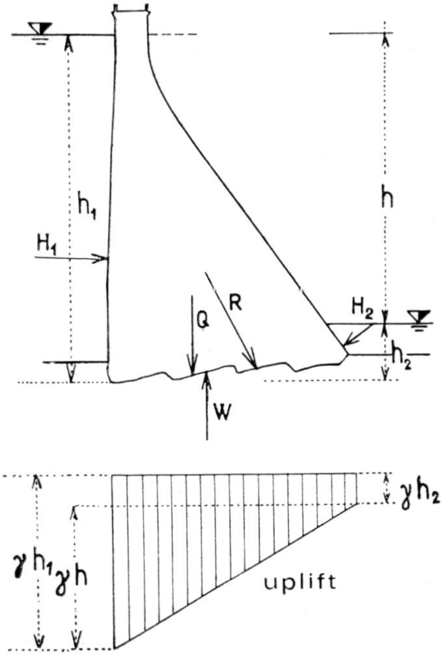

13-39. Principal external forces acting on the body of a gravity dam (see the text).

reduces the shearing stress in the foundation ground. Care must be taken to choose the angle between the anchors and the normal line to the weakness planes in rock smaller than is the angle of strength; otherwise the blocks of the dams could shift towards the upstream face when the reservoir is empty, and the rock could be disturbed. In using the anchoring method a larger proportion of rock cohesion can be taken into account for statical analysis, as the danger of loosening of rock is reduced. The engineering geologist must examine thoroughly the properties of the bedrock so that it can be ascertained by computation if the increase in stability by anchoring will be sufficient. The effect of anchoring must reach below the shear-stressed zone. Since it is always uncertain whether the anchoring will be sufficient and whether the anchors will not fail in the course of time, the statical solution should take into account only seven tenths of the anchor effectiveness in addition to the normal safety factor.

(e) To increase the strength of the rock under the dam foundations by grouting — a procedure which is very expensive on a large scale. Grouting is usually carried out locally, at the toe of the dam, or in disturbed zones, often in association with partial excavation of unsound rock.

(f) Various methods can be used to increase the shear resistance of rocks on their planes of weakness. They are more commonly applied in the construction of arch dams but have even been used for gravity dams. One or more galleries are driven in the foundation beds of the dam along a shattered zone, and part of its material as well as of the over- and underlying sound rock is excavated to be replaced by concrete. The sealed indented zone is strengthened additionally by cement grouting. This method was used, for example, for the foundation of the Green Peter Dam (Corns, Nesbitt 1967). This is a gravity dam 104 m high in the Cascade Mts. on the western coast of the United States. The angle of shear resistance of the subhorizontal tuff interbeds, tectonically affected and strongly weathered, was only 20°. For some blocks, rock was excavated below these hazardous interbeds. The shearing resistance of the concrete was taken as 56 kp/cm^2.

From the foregoing it follows that in statical analyses of foundation conditions under dams in rocks traversed by planes of discontinuity the values of residual shear resistance are generally taken into account (Sec. 3.5). The advantage of this procedure lies in the fact that the scatter of the measured values is usually small so that a relatively low factor of safety may be applied (about 1.5 for masonry dams). On the other hand, the design of foundations in weaker rocks is based rather on the critical loading (Sec. 3.7), since it allows for small displacements and enables modern computing methods to be used.

The movements and settlements of even high gravity dams on the solid bedrock are relatively small, generally only a few millimetres, but they can be harmful to adjacent power houses (Mencl 1970). The compression of the rock by the weight of the dam may extend under the power plant (Fig. 13-40) and the stiff construction of its base may vibrate as a cantilever, similarly to buildings and industrial structures (see Sec. 10.2.2). Vibration will die away after the final settlement, which may show itself by cracks in the concrete construction, or after grouting the thin gap between the foundation and the bedrock on the upstream face of the power house.

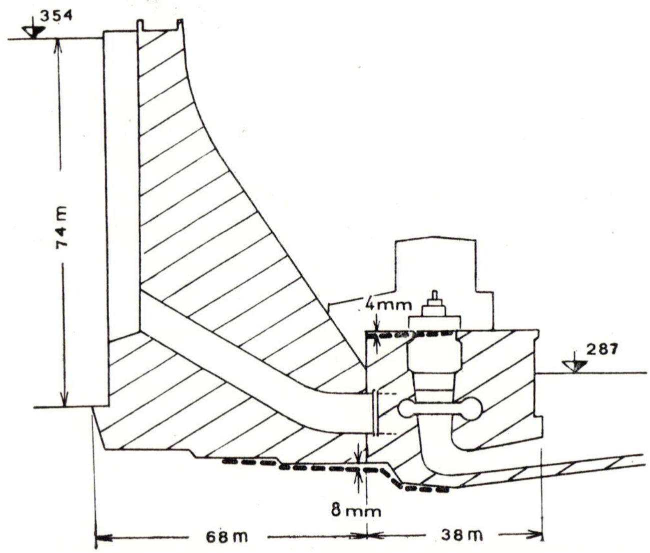

13-40. After filling the reservoir, the compression of rock below the downstream face of a high gravity dam will extend beneath the power plant.

The permeability of rocks affects unfavourably leakage around a dam and its stability, since the *uplift* under the base of the dam is thus increased. Uplift caused by the hydraulic head of water reduces the resistance of the dam to movement, since it decreases the vertical component of the resultant *R*. Determination of the amount of uplift is therefore an important, but to date not satisfactorily solved problem. Uplift is developed by water penetrating below the dam foundation under hydraulic head h_1 (Fig. 13-42), and is controlled by the following factors:

(a) *The properties of the bedrock.* Beds of the more permeable rocks, shattered zones and planes of discontinuity that permit seepage of water must be located by water-pressure tests. It is also necessary to study the effects of impounded water on the filling of the fissures, whether it might be washed out or chemically decomposed. The character of the ground water in the foundation beds should be examined, particularly to determine whether some aquifers may contain confined water.

(b) In most cases, mainly in dams of great height, the reduction of the uplift is achieved by constructing a *grout curtain* near the heel of the dam, which will prevent percolation of water. Detailed measurements carried out on many dams have shown that uplift can be really substantially decreased (Fig. 13-41) since even if percolation of water is not fully stopped, the route of seepage is lengthened. In earlier times a deep cut-off trench would be excavated at the upstream face of a dam, but the excavations would disturb the surrounding rock. Today the grout curtain is preferred, constructed with the use of core borings without disturbing the rock. The depth of the grout curtain depends on the character of the rock and its jointing (Sec. 13.11). In cases where an impermeable bed occurs below the foundation, the grout curtain should be extended to it if possible. The spacing of grouting holes and the grouting technique will be chosen from the test results.

The geologist co-operates in determining the depth of the grout curtain and the direction, spacing and arrangement of the holes of grouting and in the selection of suitable grouting mixture. Water-pressure tests and the results of trial grouting furnish important information in this connection.

(c) *Uplift* can also be reduced by draining the water under pressure from a part of the foundations when the uplift will be decreased to the value given by the water level in the drains. To be on the safe side only part of the reduction in uplift will be taken into account (Fig. 13-42). In firm undisturbed rocks boreholes put down from the same gallery as the grout curtain can be used as drains. Their depth should be at the most 80 per cent of the depth of the grout curtain and their

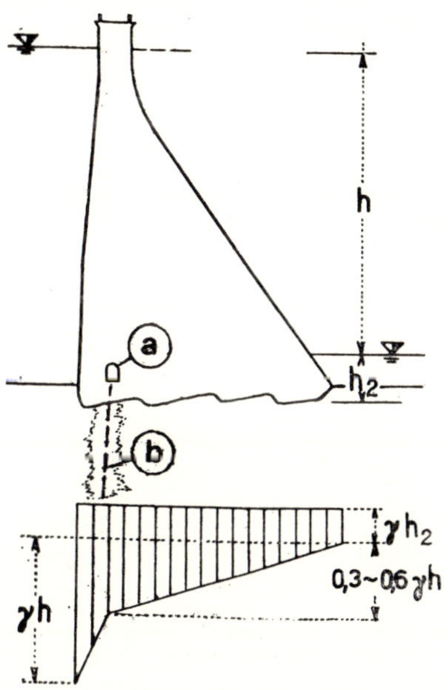

13-41. Reduction of uplift on the foundation surface. a — Grout gallery, b — grout curtain.

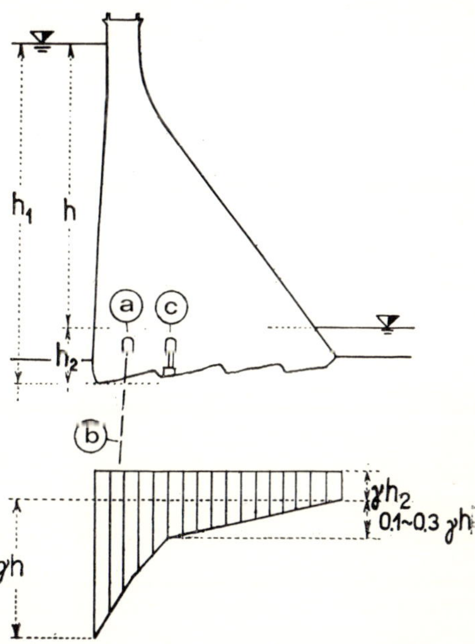

13-42. Reduction of uplift at the foundation surface by the combined effect of a grout curtain and drainage. a — Grout gallery, b — grout curtain, c — drainage gallery.

spacing about 3 m. This procedure can be adopted under the assumption that the rock will not be disturbed or washed out by water flow in the drainage holes and in their neighbourhood.

A special approach to the reduction of uplift below foundations was taken at a damsite on the river Orava (Slovakia), which had very unfavourable geological conditions (Fig. 13-43). The flysch complex, consisting of sandstones and shales, was disrupted by several fractured zones. The designer of the dam proposed to place the foreset blocks on the upstream side of the dam. The uplift curve shows the effectiveness of this arrangement, and of the additional drains.

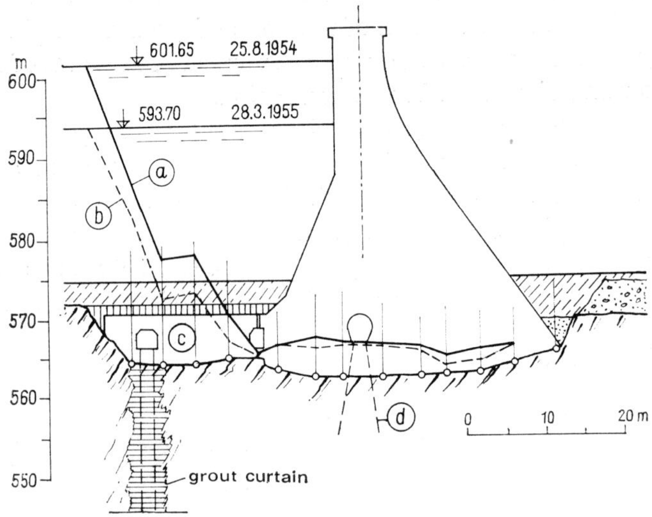

13-43. In the Orava Dam (Slovakia), uplift was reduced by a grout curtain carried out from a gallery in a concrete block (c) in front of the upstream face. a — Uplift measured in 1954, b — uplift measured in March 1955, c — concrete block, d — drainage boreholes. (Čermák, 1955.)

Buttress dams. In mountain valleys difficult of access for the transport of cement and concrete materials, buttress dams are frequently projected since the necessary quantity of concrete is substantially smaller than for gravity dams. The dimensions of buttresses, and particularly their length, can moreover be increased so that the loading on the foundation beds does not surpass the permissible limit.

The Fusino Dam (Fig. 13-45) in the Val Grosino in the Italian Alps is given here as an example. This buttress dam with a straight axis and 78 m high was constructed between 1958 and 1960. It is founded in firm biotite gneiss with amphibolite and quartzite interlayers. A narrow erosion furrow in the valley floor was filled with a concrete block. The relatively broad valley floor, widened by glacial corrasion, is covered on both sides with glaciofluvial and morainic sediments, which are up to 30 m thick at the level of the crest of the dam. In order to seal

13-44. Construction of the Orlík gravity dam on the river Vltava. (Courtesy Hydroprojekt.)

13-45. Buttress dam Fusino in Val Grosino in the Italian Alps. (Photograph by Záruba.)

these permeable sediments the dam was extended into both flanks by a concrete cut-off wall, 2.5 m thick: on the right flank for a length of 58 m and for 106 m on the left bank (Fig. 13-46). The wall was constructed in shafts, in parts 8 m long, from the galleries driven at the level of the crest of the dam.

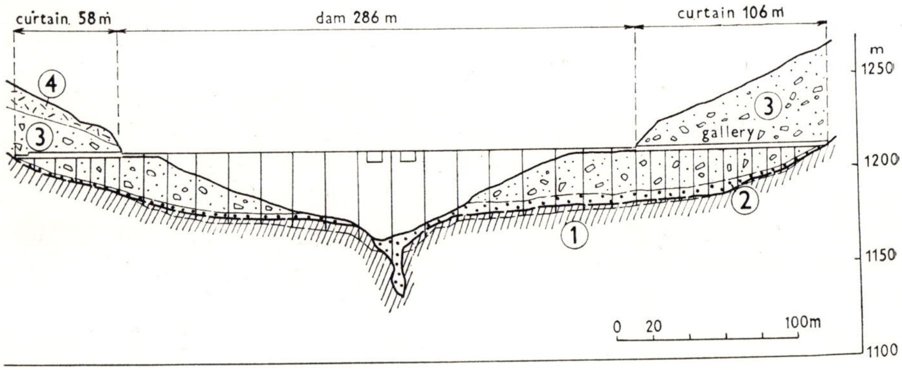

13-46. Longitudinal section of the Fusino damsite. 1 — Biotite gneiss, 2 — glaciofluvial gravel, 3 — moraine, 4 — slope debris.

13.9. Foundation of arch dams. Characteristic of arch dams is the transmission of great forces to the rock abutments by the arching action and the relatively small thickness of the dam compared with that of a gravity dam. The fact that an arch dam is keyed into the abutments obliquely makes high demands on the stability of the slopes which may be endangered by excavation.

Figure 13-47 shows schematically how the water pressure is transmitted through the dam into the rock abutment in order to give a general idea of this feature to the engineering geologist.

(1) *Cantilever action*: In the foundation the dam has a certain thickness, equal,

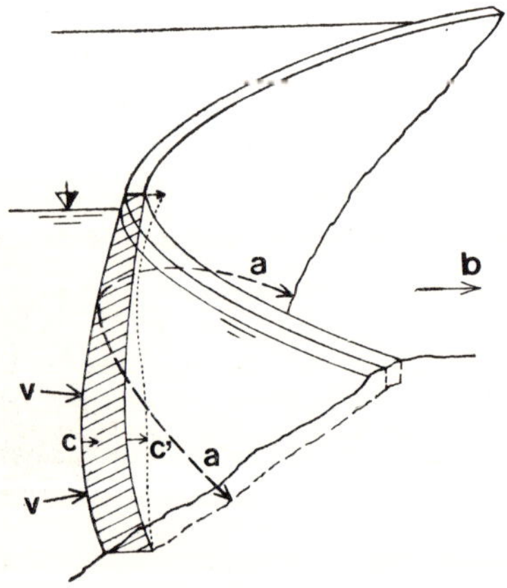

13-47. Simplified scheme showing the principal elements of an arch dam. a — Arch action of the dam, b — downstream direction of the valley, c — cantilever action in the highest vertical section of the dam, c′ — displacement of c section after filling of the reservoir, v — water pressure.

for example, to one seventh of its height. Just as in a gravity dam, this disposition offers a certain resistance to movement, which will be greatest in the lower part of the dam. In the upper part, the deformation of the dam is impeded mainly by the arch action.

As a result of cantilever action the pressures at the toe of the dam are greater than in a gravity dam. In a 100 m high gravity dam the stresses at the toe of the dam will be about 25 kp/cm^2, whereas they may even exceed 35 kp/cm^2 in an arch dam of the same height. At the same time, the stresses at the heel are small in both types of dams. From this it follows that the rock in the lower part of the valley flanks should be strong and not very compressible. Erosive valleys generally comply with this requirement as the lower parts of the valley slopes have been exposed to weathering for a relatively short time. Conditions are more hazardous where the rocks are tectonically disturbed, particularly where a wide fracture zone extends under a valley bottom.

(2) *Arch action* causes considerable compression stresses in the rock abutments. They are greatest between one third and one half of the height of the dam; e. g. in a 100 m high dam they attain about 50 kp/cm^2 at the toe and about 15 kp/cm^2 at the heel. Compressive stresses decrease upwards at a relatively slow rate. From this situation some conclusions as to the geological character of the damsite can be drawn. The upper part of slopes with deeply weathered rocks should not extend beyond mid-height of the dam. The weaker the rocks (e. g. more compressible), the more the cantilever action must be increased, i. e. the foundation must be widened and the dam will approach the gravity type. Alternatively, it will be necessary to construct massive supporting blocks in the abutments and thus transmit pressures over a larger area.

The compressibility of rocks is not the only unfavourable factor. One must take into account the fact that the resultant of all the forces exerted on merely a height of one metre of the rock abutment, when the reservoir is full, will be several thousand tons and that in plan this force is roughly parallel to the valley. Any thought that this force acts in the direction of the arch would be erroneous. The vertical component is of course considerable and the total force thus extends deep into the slope. The direction of the force can be unfavourable if a plane of weakness dips towards the valley and its strike is convergent with the direction of the valley below the dam. The surface may even be irregular as it follows several systems of joints. Modern computation methods make it possible to determine the amounts and directions of the resulting reactions in the individual zones of the abutments. From the diagram of fractures and joints (Sec. 3.3) on which the direction of the resultant can also be plotted, the angle made by the normal of the surface and the direction of the force can be determined. It is necessary to know with certainty the parameters of strength on the surfaces of weakness, their continuity and filling, so as to decide which of them may be dangerous. When these surfaces are not numerous, their strengthening can be carried out by excavation, indentation, filling with concrete and grouting (see e. g. Campo, Peironcely, Maroto 1967). Washing-out by water under pressure (100 atm.) and filling of fractures by concrete was used to a great extent in the Nagawado Dam in Japan (Fujii 1970).

The failure of the Malpasset Dam (France) is suspected to have been caused by a shift of blocks due to the disturbance of the rock in the right abutment at about the mid-height of the dam (Cambefort 1963). The failure of this dam has drawn the attention of designers to the forces in valley flanks which are transmitted under the dam foundation by the water pressure. Since the thickness of an arch dam is small, the block of rock in the abutment which takes up the pressure of the impounded water is also relatively small. It is recommended (Minassian, Sabarly, Londe 1967) to turn the grout curtain and the line of underdrains, as far as possible, upstream of the dam and thus divert the forces into the slope.

The stability of the rock mass in the dam foundation may be impaired by natural weakening of rocks downstream of the damsite, for example, by faulting which caused depressions and furrows in the slope morphology. A ridge limiting a depression would appear to be suitable for dam location as it forms the narrowest cross-section of the valley. It is therefore better to locate a dam so that the ridge supports its downstream abutment. Artificial weakening of rock may be produced by excavations for access roads and for the structures at the toe of the dam.

It is obvious that the engineering geological investigations for arch dams must be particularly conscientious and detailed, but in many cases their cost is compensated by savings in the volume of the dam, which for an arch dam is about one fifth of that for a gravity dam.

The smaller consumption of concrete makes the arch dams suitable for high-mountain areas which are difficult to access. The Curnera Dam built during 1960—1967 as part of the Vorderrhein water-power scheme in Switzerland is shown here as an example (Fig. 13-48). It is 152 m high and dams a narrow valley at an altitude of 1956 m a. s. l. Phyllitic gneiss of the Gotthard massif, forming the foundation ground, strikes favourably at right angles to the valley and dips steeply downstream. The most serious problem that should be solved during construction was the location of site equipments and access roads on steep mountain slopes.

13.10. Geological investigations for earth dams. The engineering geologist is charged today with investigations for earth dams far more frequently than formerly. This is because sites that are geologically and hydrologically suitable for the construction of gravity dams are steadily decreasing in number and because the mechanization of construction can be introduced to a greater extent than in masonry dams. The statistics published by ICOLD (International Commission for Large Dams) show that of 925 dams constructed during 1965—1968, 700

13-48. Construction of the Curnera arch dam in Switzerland in high mountains with difficult access. (Photograph 1965, Courtesy Kraftwerke Vorderrhein A.G.)

were earth dams and 225 were concrete dams (Mermel 1970). The advanced knowledge of soil mechanics makes it possible to construct earth dams of great heights that are both safe and economical. In 1962 construction of the highest earth dam up to this time at Nurek was started in Tadzhikistan; it will be 312 m high.

The engineering geological investigation made for earth dams should deal with similar questions as in the case of gravity dams. The order of their importance, however, is different because of the different character of the structures. The main tasks of the investigation are the following:

(1) To find adequate soil material for the construction of the dam within a reasonable distance of the site; the rocks should ensure the stability and watertightness of the dam and not be liable to volume changes. The deposits of construction material should allow economical working.

(2) To examine the strength of the foundation soil so that the stability of the dam can be assured.

(3) To assess the impermeability of rocks at the damsite, and if necessary to suggest adequate measures for reducing their permeability or for removing the harmful effects of seepage water.

(4) To gather geological data on the basis of which the settlement of the dam and the compression of its foundation can be determined and their effect on the dam and other structures can be inferred.

(5) To find a suitable site for outlets and spillways.

(6) To determine the position of the dam axis so that the dam can be keyed safely into the valley flanks and all other requirements fulfilled.

(7) To assess whether the stability of flanks will not be endangered by excavations for the dam foundation in the floor and sides of the valley.

(8) To assist in the solution of questions concerning the drainage of excavation trenches.

(9) To co-operate in the search for suitable stone and soil materials for the dam itself, drains, backfill, pavements, etc. This co-operation demands continuous contact of the engineering geologist with the designer, so that the requirements following from the geological investigation may be integrated with the general plan of the project.

(1) Available deposits of construction materials generally determine whether an earth dam or a rock-fill dam will be constructed; they also control the layout of the dam. Under similar conditions, an earth dam is usually cheaper than a rock-fill dam. Earth dams may be homogeneous or zoned.

Homogeneous dams (Fig. 13-49) made exclusively of one type of material are only rarely constructed. At least two types of material are usually available, and this makes construction more economical. When a dam is built of one material only, it is not so easy to ensure adequately both its watertightness and stability. To control the effect of water seepage on the downstream face, drainage layers are inserted into the body of the dam, either horizontal or subvertical (Fig. 13-50).

The suitability of soil can be appreciated in general from its grain-size distribution curve. More detailed examination of its properties should be carried out in soil-mechanics laboratories. In Fig. 13-51 the solid lines delimit the area of grain-size curves for soils suitable for the impervious parts of earth dams based on the experiences of the U. S. Bureau of Reclamation. Soils, the curves of which are fairly continuous and extend into the field of gravel, will comply with the requirements of stability and watertightness in homogeneous dams (Fig. 13-49). The earth dam near Luhačovice in Moravia constructed from the slope loams derived from flysch rocks is given here as an example. The grain-size distribution curve of these soils (1) is plotted in Fig. 13-51. To increase the strength at the upstream face, crushed andesite was added to the soil and this part of the dam was more intensively compacted. The slope of the upstream face could therefore be relatively steep 1 : 1.5 and 1 : 2 up to 1 : 2.5 in its lower part. The upstream face is protected against erosion by concrete slabs.

In the Soviet Union, where soils of the same composition form extensive areas, engineers have to use even relatively fine-grained soils for this type of dam. Figure 13-51 shows the grain-size distribution curve (3) for the loess used for

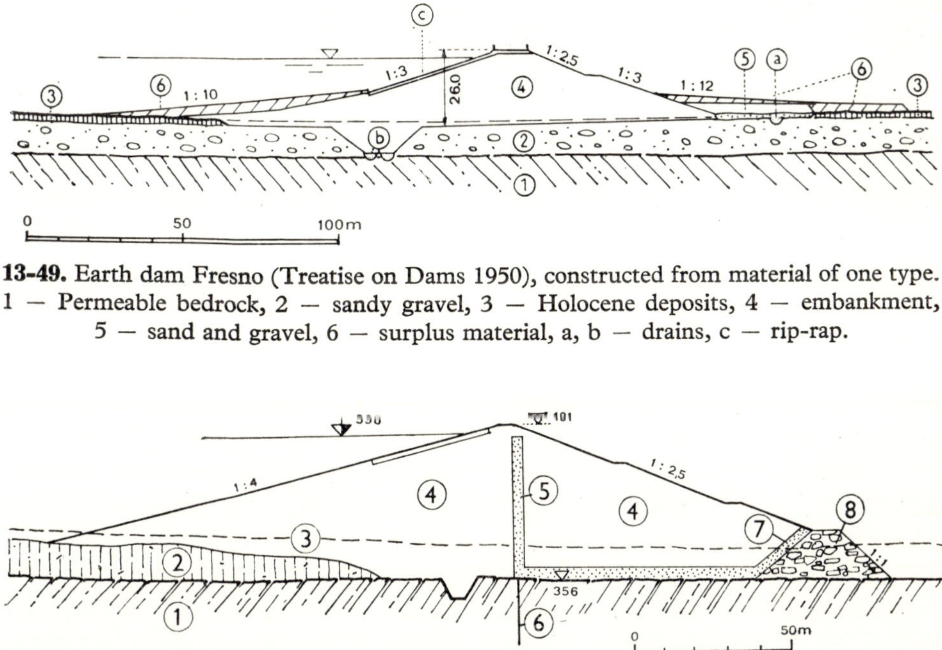

13-49. Earth dam Fresno (Treatise on Dams 1950), constructed from material of one type. 1 — Permeable bedrock, 2 — sandy gravel, 3 — Holocene deposits, 4 — embankment, 5 — sand and gravel, 6 — surplus material, a, b — drains, c — rip-rap.

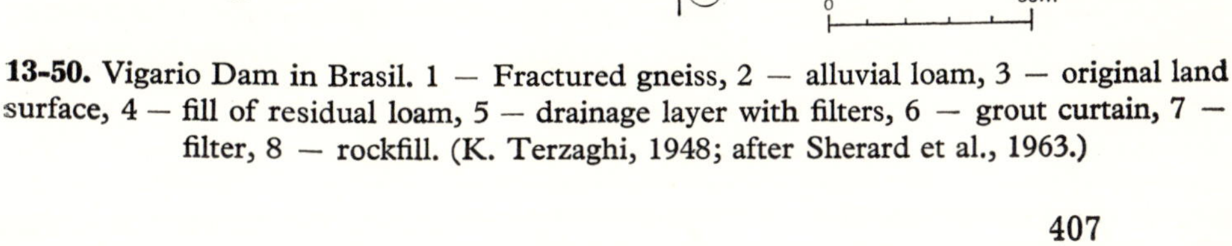

13-50. Vigario Dam in Brasil. 1 — Fractured gneiss, 2 — alluvial loam, 3 — original land surface, 4 — fill of residual loam, 5 — drainage layer with filters, 6 — grout curtain, 7 — filter, 8 — rockfill. (K. Terzaghi, 1948; after Sherard et al., 1963.)

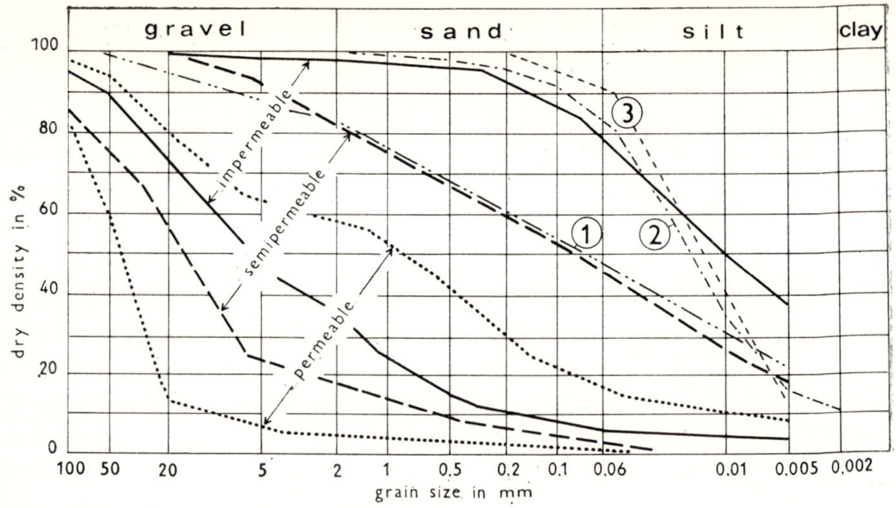

13-51. Grain-size distribution curves of permeable, semi-permeable and impermeable materials for earth dams.

the Katt-Kurgan Dam (Grishin 1954), 31 m high. The slope of the upstream face is 1 : 3, 1 : 4 and 1 : 5 and that of the downstream face is 1 : 3, 1 : 3.5 and 1 : 4. The upstream face is protected by a two-layer pavement.

Zoned earth dams are most commonly planned. They are composed of a supporting fill on the downstream face, an impervious core and a protective layer on the upstream face (Fig. 13-52).

From the point of view of soil mechanics, soils designated GC, SC and CL in the Soil classification (Sec. 3.8) are suitable for the impervious parts of the dam. Soils with silt particles predominating are less suitable as they can be readily washed out if the core is disturbed. They require thick filter layers, carefully placed, particularly on the downstream side. In choosing material, its origin must also be taken into account.

Loess loams are commonly used for the cores of earth dams. Relative to primary loess, they contain less calcium carbonate. Even if its content is larger, there is no danger of its leaching when the soil has enough clay particles. The angle of shear strength of the loess loams is 20°—26° as given by a drained test. A typical curve of grain-size distribution for loess loam (2) for the core of the Těrlicko Dam near Ostrava (Moravia) is plotted in Fig. 13-51.

The loess loams of some areas possess a higher moisture content than is admissible for earth-dam material. Because of climatic conditions the loess loams in the northern parts of Czechoslovakia show 20—26 % water content, which is 4—6 % greater than the optimum moisture content. These soils should not be too dry since this makes them difficult to compact and bind together, nor too wet since an excess of water prevents compaction and reduces the strength (cohesion si smaller and unfavourable pore water pressure develops with settlement of the dam). It has been shown by the use of soil-mechanics methods that a 2—3 % increase in moisture above the optimum value appreciably decreases the stability of the dam. In this case the designer must locate the impervious core inside the dam body (Fig. 13-52) so that the weight of the gravel zone on the upstream side contributes to the stability.

The moisture of soils also affects their workability. Soils with a higher water content cannot be compacted by taper foot roller so that rubber-tired rollers or tractors must be used. The natural moisture of soils and all factors that can cause it to change must be carefully examined.

Slope loams originated by redeposition of weathering material from various soft rocks such as shales, sandstones, granite and gneiss eluvium, etc. Mineralogical study of these soils is directed to the determination of the degree of weathering of individual particles and of their stability. Also of importance is the shape of larger fragments, since the flat, platy fragments, which are frequent in slope and residual deposits, make compaction difficult. The material must be placed in thin layers and compacted by heavy grooved rollers in order to crush these fragments.

Alluvial soils with sandy admixture may be more suitable than loess loams since their shear resistance is greater. It is of advantage to use soils with a larger admixture of sand and gravel, as their pore volume is small and the shear resistance high. Pebbles in soils should not normally exceed 12 cm in size; the method of compaction will determine whether the designer can admit pebbles of larger size. Alluvial soils from a valley floor may have excessive water content and contain organic substances which makes them less suitable. The presence of silty and fine sandy beds, particularly when these contain a large amount of micaceous particles, is also unfavourable.

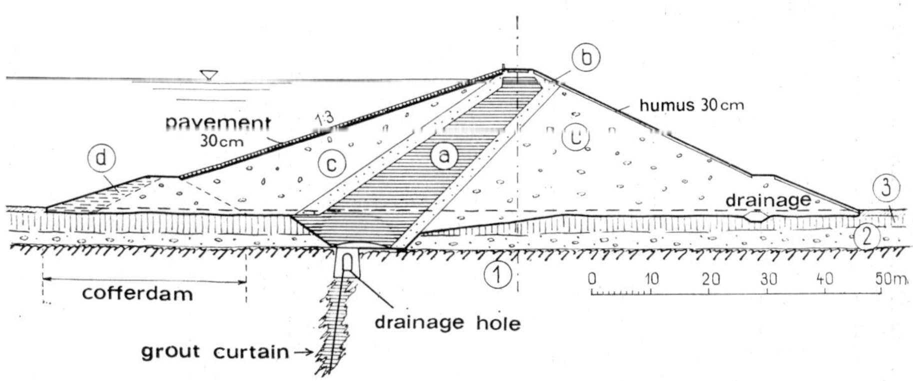

13-52. Zoned earth dam founded on impermeable bedrock. 1 — Marly shales with sandstone, 2 — sand and gravel, 3 — alluvial loam, a — impervious core, b — sand filter, c — sand and gravel, d — sealing of cofferdam.

In some crystalline areas, *residual soils* are the only source of material for the impervious part of an earth dam. Investigation of these deposits must be carefully carried out. The upper parts are usually suitable soil of CL type. The clay particles, however, decrease in amount downwards and the soil takes on the character of silty sand, which cannot be used for the impervious core.

Soils of glacial origin furnish material for the cores of earth dams in wide areas of Pleistocene glaciation, where there is no other earth material available. They are of various character, generally with minor clay content. For this reason several fairly thick layers with progressively increasing permeability must be placed on the downstream side of the dam. When they contain large pebbles, these must be removed.

Soils with a high content of clay particles (more than 40 % grains below 0.005 mm) are unsuitable for earth dams, since they are very difficult to handle. The liquid limit should therefore not exceed 50 %. Tendency to swelling is another unfavourable property of this material.

Shaly claystones, siltstones, and clays of Tertiary age, are used only exceptionally. Shaly rocks must be crushed completely into minute fragments. For example, the London Clays which are worked in a solid state are excavated in thin beds by bucket excavators and sprinkled with water on conveyer belts so that they can be moulded by rolling. Carefully selected filters must then be used but the procedure is economical in that thin cores are sufficient and their small volumes are less affected by frequent rainfall.

The exploration of the deposits should determine their volume, natural moisture content, mode of deposition and the thickness of the overburden. The mode of deposition controls the working method as only a homogeneous deposit can be worked in horizontal layers, e. g. by scrapers. Soil composed of beds of differing properties, which is the usual case, is better worked by excavators since it then becomes more thoroughly mixed. Large pebbles and rock fragments must be picked out. The admissible size of occasional fragments (not cumulative) is about 15 cm, equal to about half the thickness of the compacted layer.

The material used for the supporting part and the cover of the dam on its upstream face should be at least ten times more permeable than the material of the core.

Sandy gravel has a sufficient shear strength and such permeability that, at the drawdown of the reservoir level, water will flow out rapidly from them. They should, however, offer some resistance to water flow to protect the sand filter along the impervious core from washing out. The maximum admissible size of particles depends on the mode of placing the material. Since an embankment is often compacted today by heavy lorries, the limit of grain size is being steadily raised much above 15 cm (one half to two thirds of the thickness of the compacted layer). Rock material with a coefficient of permeability of 0.1—1 cm/sec is the most suitable. Gravel from river terraces and flood plains is used most frequently.

The working method for the deposit is chosen in relation to the homogeneity of the material excavated. Loamy seams in gravel-sandy river alluvium are undesirable since they depreciate the whole deposit, especially during rainy weather. The material should not contain more than 4 % particles smaller than fine-grained sand. In earth dams the height of which exceeds 50 m, contractant behaviour of the material can appear if the rock grains are not sufficiently hard. The petrographical character of grains, the degree of their weathering, etc. must therefore be examined during the investigation of the deposit.

Sandy gravel is not everywhere available in sufficient quantity so that eluvial soils or weathered rocks have to be used. Since they are liable to further disintegration and washing-out and are highly heterogeneous, they should be protected by interior drains from water seeping through the core (Fig. 13-50). In this case the arrangement of filters must be very carefully designed.

Suitable deposits of material for filter layers protecting the impervious core or surrounding the drainage layers are not easy to find. Sands of suitable grain-size distribution are rarely discovered, for example, in flood-plain alluvium; very often they must be sorted from the sand-gravel mixture or obtained by crushing.

The objective of engineering geological investigations for *rock-fill dams* is to find impermeable soil for the core, unless it is to be constructed of other materials, and a quarry where satisfactory stone for the dam body can be worked economically. The main requirement for the stone for a rock-fill dam relates to its strength. For a high dam the stone must have at least the strength of firm sandstones. Material of poorer quality would break and crush under pressure, possibly giving rise to undesirable settlements. Softer stone (e. g. shale beds in sandstone) may be present only in such amount that it does not fill all the spaces between larger stones. The stones must be tightly compacted and not be separated by soft filling of voids. Otherwise the friction would be reduced and excessive settlement of the dam might occur. When the presence of softer fragments cannot be avoided, the material must be sorted during transport and the smaller and softer fragments must be hauled into the zone between the core and the stony part where it can be compacted by tamping. Material for the rockfill itself should not contain more than 10 % of poorer rock. The shape of rock material is also important; elongated flat fragments are undesirable in high dams since cavities may remain under them.

The quarrying of the rock must be thoroughly planned so that transportable blocks are obtained by blasting and the amount of little fragments is small. The amount of fragments below 10 cm in the supporting fill of the dam should not exceed 5 % if possible. The most suitable mode of exploitation should be tried out at the opening of the quarry.

(2) The *strength of bedrock* greatly influences the design of the dam. If the strength is small, the following possibilities of failure must be considered in the analysis of stability:

(a) Construction of the embankment will proceed at a greater rate than the consolidation of the clayey foundation beds.

(b) If temporary openings for the passage of the river are left in the dam, the shear stresses that will be developed at various places in the bedrock must be computed and correlated with the strength of rocks.

(c) The stability of the completed dam is commonly most endangered by a sudden drawdown of the reservoir level. For the statical analysis of this case we must know the strength of foundation rock since in zoned dams, or when the strength of the bedrock is small, the shearing plane will extend from the core into the rock foundation. It is therefore necessary to study thoroughly the strength of all clayey rocks in the foundation and that of the weathered surface layer of the bedrock. Whereas the strength of clayey rocks can be determined readily from undisturbed samples taken from boreholes, that of the bedrock should generally be established by field tests in special excavations.

When the bearing capacity of the bedrock is insufficient, the dam must be designed as an embankment with lateral surcharging fills. Examples are provided by the Fresno Dam in Montana (U. S., Fig. 13-49), the Nechranice Dam (Bohemia, Fig. 13-54) and the Jesenice Dam (Fig. 13-53).

(3) *The permeability* of the rock foundation must be taken into account for the design of sealing and drainage works. In permeable rocks, leakage will result in an increase in uplift and hence in a decrease of the stability of the dam. The water flow, in addition, may wash out fine rock particles so that cavities in the

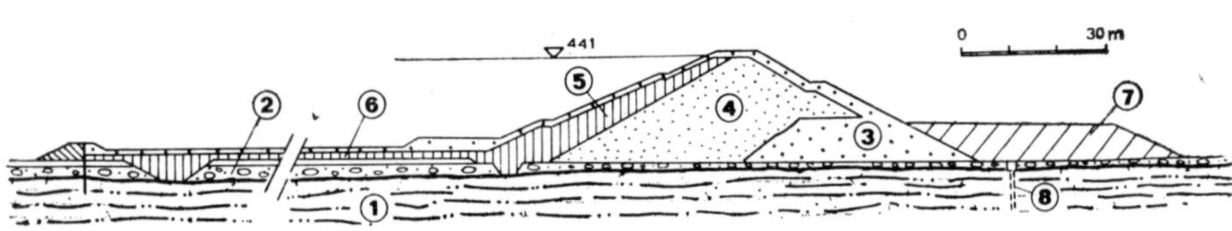

13-53. Section through the Jesenice Dam near Cheb (Bohemia). 1 — Neogene schistose clays, 2 — flood-plain gravel, 3 — gravel fill, 4 — fill of Neogene sand, 5 — impervious core, 6 — horizontal blanket, 7 — surcharging fill, 8 — drainage wells and trench. (Přehradní výbor, 1967.)

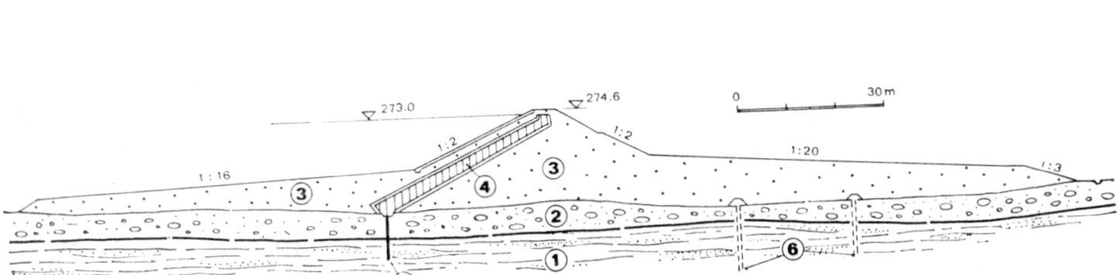

13-54. Section of the Nechranice Dam in north-western Bohemia. 1 — Neogene claystones and sand, 2 — sand and gravel, 3 — gravel fill, 4 — core of loess loam, 5 — impervious clay-concrete wall, 6 — drainage wells. (Přehradní výbor, 1967.)

bedrock may gradually increase and the concentrated flow may eventually rupture the foundation beds.

If impermeable rock is at shallow depth, the impervious core may be extended into it (Fig. 13-52). All joints and fissures in the bedrock, however, must be examined, as clayey soil in the core might be washed out by stronger seepage concentrated in these cavities. Bedrock under the core must therefore be covered by a concrete layer and fully sealed by grouting, or the clay core must be combined with a concrete core.

The requirements for watertightness to be attained by grouting differ from those for the impermeability of concrete dams. The length of leakage paths is decisive to a considerable degree: In dams with a wide core, the depth of the grout curtain can be smaller and the water loss under water-pressure tests greater than for concrete dams (e. g. one and a half times greater than is suggested by Lugeon). In contrast, in dams with a narrow impervious core the leakage path is short so that the grout curtain must extend to a greater depth and the water loss should be smaller than for concrete gravity dams.

If the permeable layer under the dam is so thick that the trench for keying the core would be very deep, a vertical cutoff wall can be constructed under the core (Fig. 13-54) or a horizontal blanket in front of the heel of the dam (Fig. 13-53). For this purpose the natural soil cover can be utilized but its thickness and character must be established by a considerable network of boreholes. If necessary, this soil layer should be complemented by a blanket of thoroughly compacted soils of suitable properties. Care should be taken not to disturb this protective layer by further work.

The permeable layer in the foundation ground may be divided by several horizontal less permeable interbeds, as is often the case in overdeepened glacial valleys. It is not always easy to decide into which of these interlayers the grout curtain should be keyed. Their distribution in plan sometimes makes it necessary to construct separate grout curtains in individual layers.

The permeability of the foundation beds should also be examined on the downstream side of the dam. Depending on the results of water-pressure tests the designer can decide whether it would be more suitable to discharge the water seeping through the foundation through drains or by relief wells (Figs. 13-53, 13-55). The latter prove effective especially where the foundation beds are of different permeability. A correct arrangement of filters in drainage facilities requires a good knowledge of the grain-size distribution of the foundation soils.

(4) *Compression of the foundation beds* can cause a settlement which may affect the outlet works, grout galleries and other construction features in and close to the dam. The settlement can generally be computed beforehand with satisfactory accuracy. Contraction joints can protect the structures from damage but the core of the dam may be fractured. This happens, for example, when compressible alluvium of a flood plain is left on a damsite otherwise formed of firm bedrock.

Fissures may originate in the core and increase if the filters are not well arranged. Similar disturbances can also be caused by settlement of the dam itself especially in valleys with steep rocky flanks dissected by ridges and projections. Evdokimov (1970) mentions that smoothing of the bedrock surface is today required in the Soviet Union.

(5) The *location of outlet and spillway* works necessitates detailed study since improper siting of the individual structures can result in serious disturbance and great difficulties during construction. Diversion tunnels driven in the valley sides are preferred for the passage of water. When this solution is impracticable, either because of the flat form of the valley or the poor qualities of the rocks, the structures are located where settlement of the dam will be uniform. If greater settlement is to be expected, outlets are placed in an accessible tunnel. The tunnel should be divided into rigid blocks with the least possible number of contraction joints, which must be well sealed.

Spillways are placed, if possible, on firm rocks of valley flanks. Where slopes are not adequately stable or where heavy retaining walls would have to be constructed, spillways can be located in the embankment itself.

An example of how to use geological conditions reasonably in the layout of hydraulic works is the Gyumush Dam on the river Zanga in Armenia. The river flows out from Sevan Lake which due to its large size had to equalize the discharge as the equivalent of artificial water storage for several years. The Gyumush Dam (Meier 1936) is the largest structure of the water-development scheme and is located about 65 km from Sevan Lake. The cross-section of the damsite (Fig. 13-56) shows that the weir and the inlet structure are founded on a flow of basaltoid andesite, which crops out on the surface in the central part of the valley. On the right side a rock-fill dam was founded on river gravel whereas an earth dam was constructed on a thick layer of loess loam on the left. It had been assumed that the earth dam would adapt itself better to the deformation of the foundation beds.

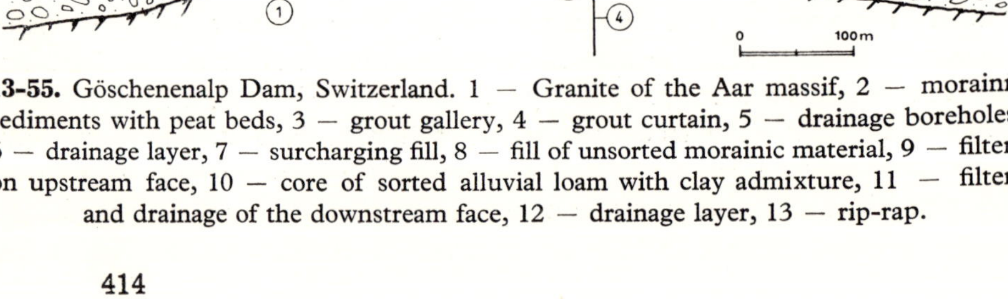

13-55. Göschenenalp Dam, Switzerland. 1 — Granite of the Aar massif, 2 — morainic sediments with peat beds, 3 — grout gallery, 4 — grout curtain, 5 — drainage boreholes, 6 — drainage layer, 7 — surcharging fill, 8 — fill of unsorted morainic material, 9 — filters on upstream face, 10 — core of sorted alluvial loam with clay admixture, 11 — filters and drainage of the downstream face, 12 — drainage layer, 13 — rip-rap.

(6) Since an earth dam is not affected so unfavourably by poorer foundation soil as is a gravity dam, the requirements for its location are not so severe. An earth dam can be constructed even on the site of an ancient landslide provided that it complies with requirements for stability and watertightness. In some cases an earth dam has been so located as to increase the stability of a valley flank.

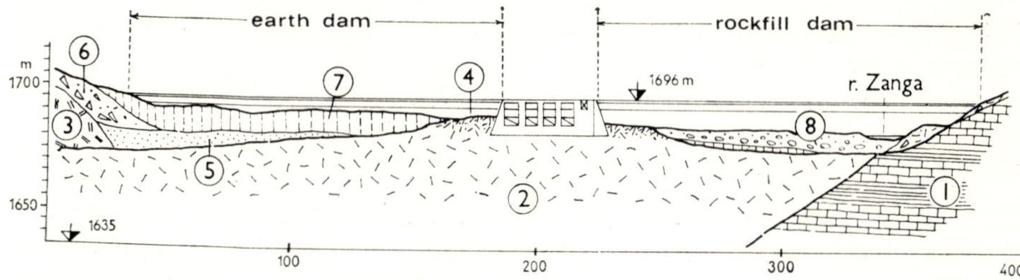

13-56. Gyumush Dam on the river Zanga, U.S.S.R.; several dam types have been used depending on the compressibility of foundation rocks. 1 — Eocene clay and sand, 2 — basaltoid andesite, 3 — younger andesite flow, 4 — fractured andesite, 5 — Pleistocene sand, 6 — slope loam and debris, 7 — loess, 8 — sand and gravel.

It is not always reasonable to key the core of the dam into a protruding ridge in a valley slope. The inclination of a valley slope on the downstream side of such a ridge is greater and the considerable difference in the heights of the embankment on the ridge and behind it may lead to differential settlement and thus to the formation of cracks in the core of the dam.

(7) When the core of a dam is inclined (Fig. 13-52), the trench for the grouting gallery is excavated obliquely upwards the valley slopes. As in this case slope movements may be imminent, the form of the core or of the grout curtain close to the flanks sometimes needs to be modified.

(8) In excavations for dam foundations springs of ground water often appear. The water might have deleterious effects on the concrete or soil during concreting or the filling of the trenches.

In order to prevent this action the water should be conducted by drains laid at the bottom of the excavation to a suitable place where it can be pumped. Since after the filling of the reservoir water under pressure could penetrate into the drains and wash away the core, all drains should be grouted. After the reservoir is filled, the whole dam foundation upstream of the grout curtain stands within the regimen of the impounded water, as will be shown by manometers in the piezometric wells in the grouting gallery. In contrast, the drainage boreholes downstream of the grout curtain should show only a small pressure. They may discharge water leaking through the grout curtain as well as spring water. If the embankment is made of sandy gravel, springs flowing from the foot of the valley slope into the supporting part of the dam are not objectionable. If, however, the embankment is constructed from residual material or partly weathered rock fragments, then springs must be discharged into drains inside the body of the dam or outside it.

A special kind of earth dam are embankments constructed for *waste materials* discharged in liquid form such as fly-ash from thermal power-plants. Many of these materials, after the removal of excess water, show a relatively high density and shear resistance. The main problem is therefore to drain the lower layers of the waste to ensure its co-ordinate action with the primary dam. This can then be appreciably thinner than normal earth dams. The drainage of the waste material is not difficult when the foundation beds are more permeable owing either to jointing, karstification or the presence of a gravel terrace. From this it follows that the engineering geologist must obtain information about the composition of the refuse to be discharged, since it differs within a fairly wide range depending on the plant from which it is derived. Before the selection of the damsite it must be determined whether the seepage of water through the bottom of the reservoir is not hazardous either from the hygienic point of view or because of excessive wetting of the area downstream of the reservoir.

In most cases, however, rocks in the bottom of the reservoir will not be permeable and some installation must be established for the drainage of the sluiced waste such as a drain along the heel of the dam on the upstream face. The construction material for this drain must be approximately ten times more permeable than the sluiced waste but not so coarse-grained that this would be washed in. The spreading should proceed so that the coarser and more permeable fractions settle down near the dam, thus protecting the drain and increasing its capacity.

13.11. Water-pressure tests and test grouting. In recent years valuable experience with the establishment of grout curtains has been gained during the construction of high dams. The refinement of methods enabling permeable rocks to be sealed and consolidated by grouting now makes it possible to erect dams at sites with such adverse geological conditions that two decades ago they would have been unanimously rejected.

The first requirement for successful sealing of the rock foundation is a sound knowledge of the mechanical properties of the rocks and of the geological structure of the site. Experience has shown that each rock type demands its own method of grouting. The permeability of plutonic igneous rocks such as granite differs in character from that of sandstones and shales. Other factors that must be taken into account are the dip of beds and joints, their filling, etc.

The main objective of grouting is to reduce the permeability of rocks and thus the uplift on the dam foundation. In some cases systematic grouting is used to strengthen jointed rocks and thus improve their deformational properties. The engineering geologist should study the following questions:

(1) Determination of the permeability of rocks and its cause and deciding whether a grout curtain would be necessary or not;

(2) Establishment of the admissible pressure for grouting at different depths below the ground surface;

(3) Assembling the facts on the basis of which the depth of grout holes and the general layout of the grout curtain can be proposed;

(4) Choosing a suitable grout mixture with respect to the character of the medium to be grouted and estimating its probable consumption.

These tasks are today generally carried out using water-pressure tests and test grouting. If carefully carried out, these methods provide a reliable basis not only for the appraisal of the permeability of rocks at different depths below the land surface but also valuable information about joints and their filling.

(a) *Water-pressure tests.* Water-pressure tests are usually carried out using the core borings which also serve for grouting the permeable layers of the bedrock. The procedure consists of injecting water into the borehole and measuring the amount of water that can be forced into the tested section of the hole within a given time (e. g. 10 min.) and at a given pressure (e. g. 2—10 atm.).

Water-pressure tests and test grouting are carried out either simultaneously with boring by the "descending" method, or by an "ascending" method after the hole has been drilled to its full depth. In the first method, one stage (e.g. 3—5 m) of the hole is drilled and closed by a packer at the upper end. Water is injected under pressure into tested section, which is grouted up after the end of the test. The grouted section is then bored through, the permeability is re-examined by pressure water and the hole is deepened to a further stage. This procedure is continued to the necessary depth.

Using the second method, the hole is drilled to the full depth and the tests are performed in successive stages upwards. The stages must be sealed by a double packer above and below the tested zone or else the hole is grouted successively from the bottom upwards.

The permeability of the rock within a certain zone can be tested only after the watertightness of the hole has been ensured. The packers used for this purpose are of different types. A simple packer consists of a rubber ring 20—30 cm long, which expands by being compressed by one third and is thus sealed against the wall. The double packer is used mainly for water pressure tests when the hole is drilled to a full length, since it closes the tested zone both at the top and bottom. Figure 13-57 shows a double packer and illustrates how the water or grout may escape through the joint system around the packer above (a, c) or below (b) the tested zone or into the alluvium beyond the casing. In densely jointed rocks longer packers must therefore be used if the closing of the zone is to be effective.

In Czechoslovakia a circulation packer constructed by J. Verfel has proved convenient (Fig. 13-58). The advantage of this packer is the possibility of washing out the hole before the test. The water forced into the borehole through external pipes may discharge through internal pipes which extend to the bottom of the hole. During grouting the mixture is maintained under pressure at constant circulation which prevents the cement grout from settling at the bottom of the drill hole. The circulation packer also enables the testing pressure to be measured directly at the bottom of the drill hole and the grouted zone to be washed after the termination of grouting. The grouted zone thus does not need to be pre-bored.

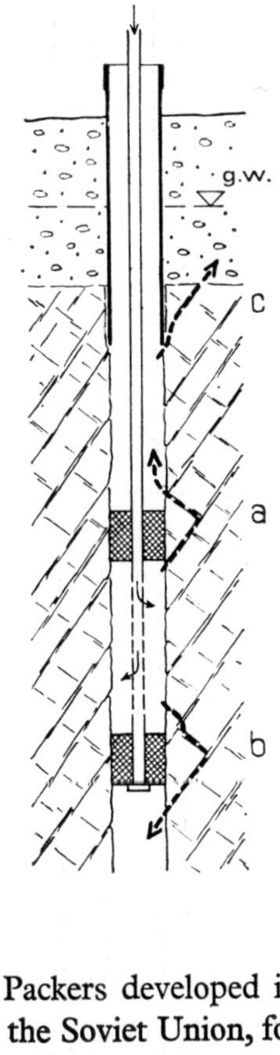

13-57. Leakage of water or grouting mixture through fissures around an obturator.

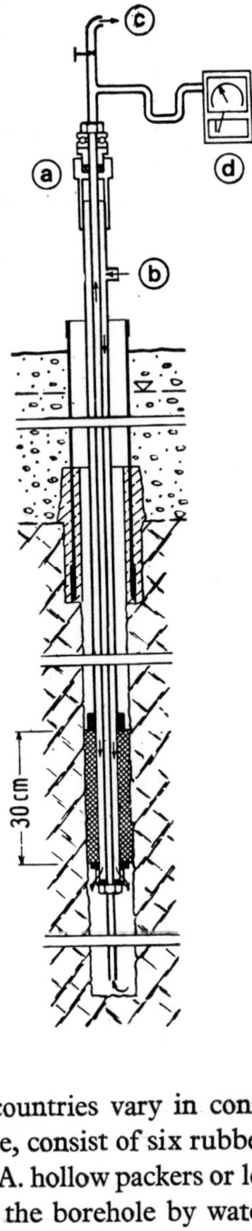

13-58. Circulation obturator (J. Verfel, 1957). a – Chucking head, b – grout feed, c – off-take, d – manometer. ←

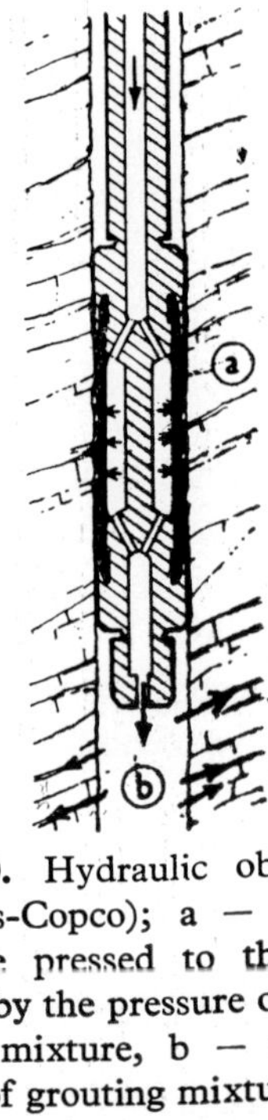

13-59. Hydraulic obturator (Atlas-Copco); a – rubber sleeve pressed to the hole wall by the pressure of grouting mixture, b – outflow of grouting mixture.

Packers developed in other countries vary in construction. The packers used in the Soviet Union, for example, consist of six rubber rings of barrel form separated by metal rings. In the U.S.A. hollow packers or leather cup packers are used, which are sealed to the wall of the borehole by water under pressure conducted by a special pipe. A hydraulic packer constructed by the Atlas-Copco firm is shown in Fig. 13-59. The rubber sleeve is sealed to the borehole wall by the pressure of water during water-pressure tests or by the pressure of the suspension during grouting.

The progress of water-pressure tests should be observed accurately on a pressure gauge and carefully recorded. The results of tests should be plotted either directly on the borehole sections or on the synoptic sections of the damsite. The results of the water-pressure test at the damsite in Krpelany on the river Váh (Slovakia) are shown in Fig. 13-60. In Palaeogene shales the water loss was negligible but in the jointed dolomites at a depth of between 36 and 40 m, the water loss was up to 6.8 l/min. per 1 m at 6 atm. Figure 13-61 shows the results of the water-pressure tests below several blocks of the Orlík Dam (Záruba, Štěpánek 1967). The permeability tests for the Sarrans Dam in France (Lugeon 1933) shown in Fig. 13-62 indicate that the left abutment was relatively watertight but the granite on the right flank was rather permeable to a depth of 50 m below the surface.

(b) *The assessment of rock permeability.* The permeability of solid rocks is due mainly to water percolation along bedding planes and other surfaces of separation such as joints originated by cooling and congealing of igneous rocks, or desiccation cracks formed during the drying out and diagenesis of sediments. Most of the fractures are of tectonic origin. Tension joints are generally open and therefore most pervious. The surface layers of solid rocks possess high permeability since

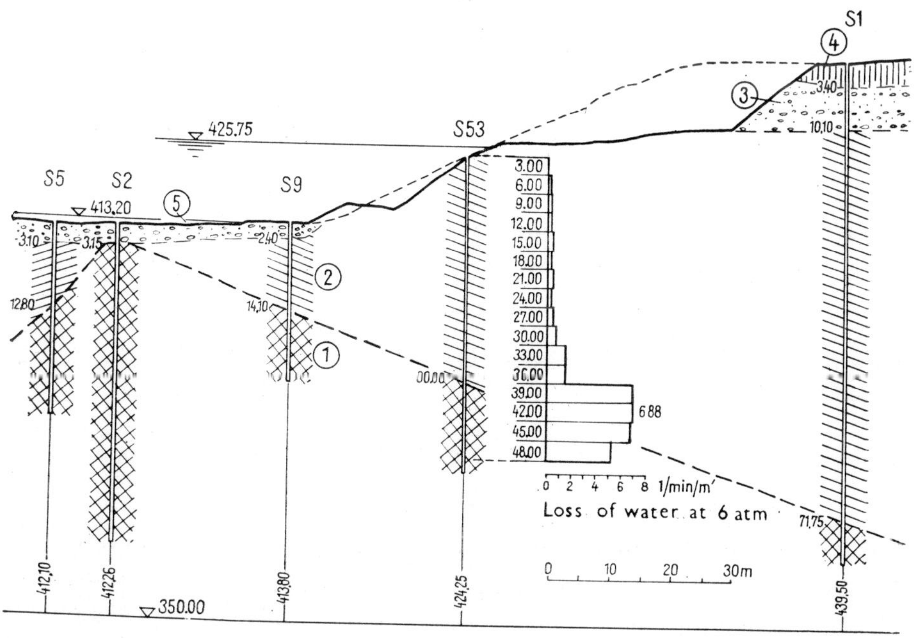

13-60. Importance of water-pressure tests for determination of rock permeability. 1 — Fractured Triassic dolomites, 2 — Palaeogene shales, virtually impermeable (see Photo 13-35), 3 — terrace sand and gravel, 4 — loess loam, 5 — sand and gravel.

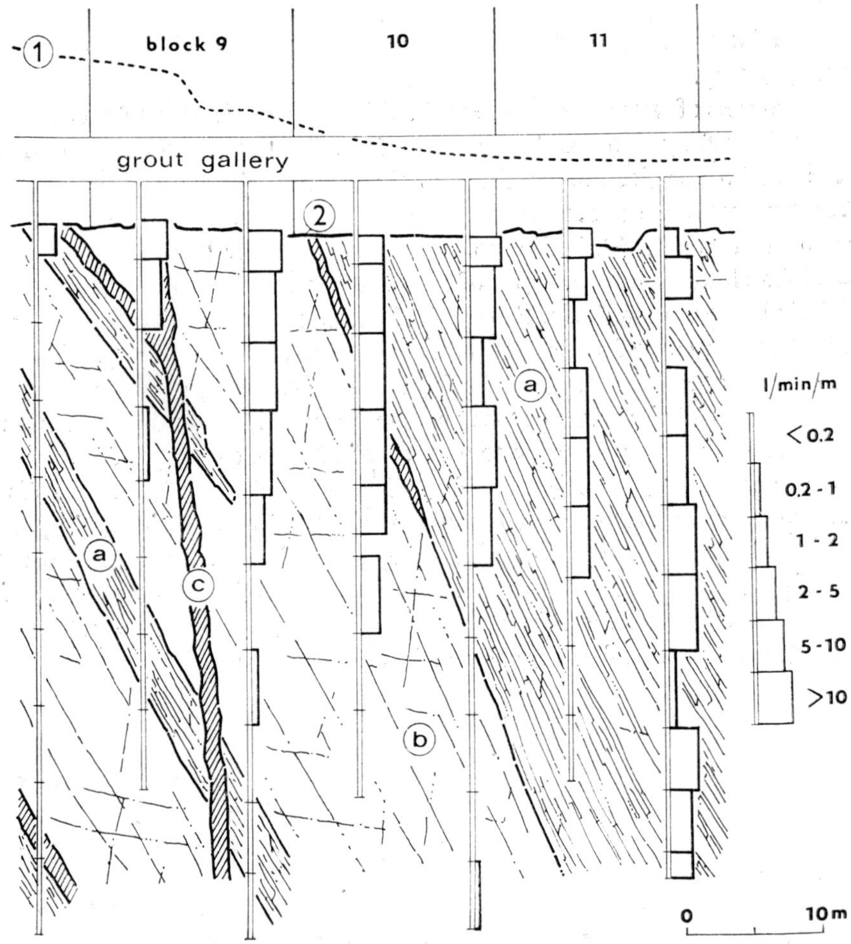

13-61. Graphical representation of water-pressure tests below several blocks of the Orlík Dam. 1 — Original land surface, 2 — depth of foundation, a — amphibolite, b — epidiorite, c — minette.

all planes of discontinuity are widened by weathering and occasionally also by slope movements.

Pore permeability is commonly negligible in solid rocks but represents the main form of permeability in unconsolidated rocks such as sand, gravel and debris.

In examining the permeability of rocks the engineering geologist should determine the principal systems of planes of separation, their strike and dip, mutual relationship, and the genesis, opening or filling of individual joints and fissures. In order to obtain test results as reliable as possible, drill holes and water-pressure tests must be distributed so as to determine the permeability of all rock types that occur in the bottom and flanks of the damsite. The sites selected for testing the permeability should be representative of the rocks tested.

The permeability of rocks is appraised on the basis of criteria established by various authors (e.g. Lugeon, Jähde, Terzaghi) with varying strictness. In Central Europe Lugeon's criterion is most commonly used as it gives reliable values for the first orientation. According to this author, in dams higher than 30 m the water loss in water-pressure tests should not exceed 1 litre in 1 min. per 1 metre of the hole at 10 atm. pressure, which should act at least 10 min. For dams below 30 m height, the water loss may rise to 3 litres under the same conditions.

These criteria of course cannot allow for all the factors that control the dimensions and arrangement of the grout curtain. In planning the grouting work, not only geological and tectonic conditions of the damsite must be taken into account but also the type of dam and the allowable uplift under the dam. Depending on the function of the reservoir, a higher seepage than normal may be permitted or, alternatively, a stricter criterion may be applied.

Another factor playing an important role is the resistance of the bedrock to the effects of flowing water, this being related to the unconsolidated rocks and to the filling of fissures in solid bedrock.

The cost of a grout curtain relative to the cost of the whole hydraulic structure is relatively high and new and more economical methods are therefore being sought. Geophysical measurements are also used for the determination of the necessary extent of the grout curtain in addition to water-pressure test.

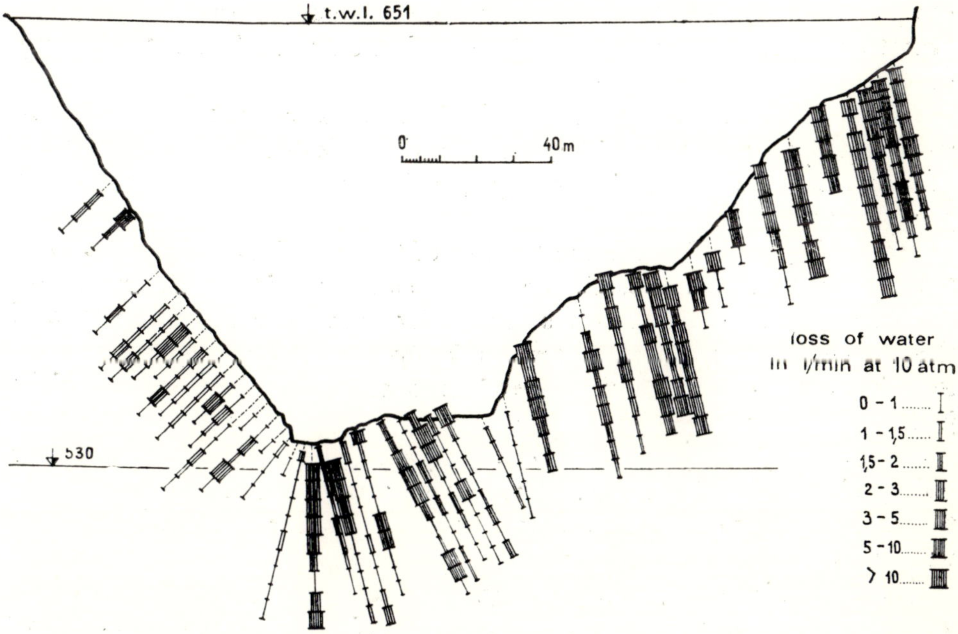

13-62. Example of permeability tests with the use of water-pressure tests, Sarrans damsite. (Lugeon, 1933.)

The criteria for the permeability of the rock mass used up to this time have recently been the subject of discussion. The functioning of the grout curtain has been examined with the use of piezometric measurements on several hydraulic structures. The results of measurement have shown that the thickness of the grout curtain is commonly far greater than has been assumed, occasionally even reaching many metres (Verfel, 1969). The hydraulic gradient in the curtain is highest below the footing base and decreases rapidly downwards with increasing length of the path of seepage. From this it follows that it is not economical to apply the criterion chosen over the whole height of the curtain. The requirements for the sealing of the bedrock must be most stringent near the dam foundation but greater permeability may be admissible further downwards.

(c) *Determination of grouting pressures.* The objective of the water-pressure tests and of the test grouting is to determine the necessary pressures for grouting at various depths below the ground surface. The choice of the grouting pressure is very critical; on the one hand it must not be so high that it might cause heaving of the foundation and rock disturbance, but, on the other hand, the highest possible pressures are desirable for maximum effect and speedy operation.

It is conventionally assumed that the grouting pressure should correspond to the hydrostatic head exerted by the impounded water. According to empirical formulas proposed by various authors the requisite pressures corespond roughly to twice the weight of the rock above the sector being tested. The pressures thus computed, however, can be either unnecessarily high or too low. The pressures cannot be determined by empirical formulas but only in relation to geological conditions.

The pressure is controlled by the depth of the grouted sector below the ground surface and by the character of the rock, mainly its bulk density. With an increase in bulk density the weight of the overlying rock mass which resists the grouting pressure naturally increases.

Another important factor is the strength of the rock. The stronger the rock, the more it resists the grouting pressure. The strength of the rock depends on its jointing. Densely jointed rock is more easily disturbed than massive rock. If the joints are open, the grout penetrates far from the grouting hole with small loss of pressure so that the disturbance of the rock mass may be greater than if the joints are closed or filled.

The grouting pressure should also be established with due regard to the strike and dip of the planes of separation. If the beds or joints are continuous and horizontal, they are liable to be disturbed by the grouting pressure, in contrast to beds with joints dipping at a steep angle.

The effects of grouting pressures vary, moreover, with the topography of the grouted site; they may be different on the bottom of the deep valley and different again on the valley slopes. On the steep slopes the horizontal component of deformation must also be considered. The maximum pressure is commonly expressed as a function of the depth of the packer and of the coefficient of rock strength:

$$p = K \times \gamma \times h$$

p — maximum pressure, γh — weight of the rock from the land surface to the packer, K — coefficient of the rock resistance to upheaval.

The resistance to upheaval can be established from the sudden increase in water consumption during a water-pressure test. The experimental pressure is gradually increased until the rock is disturbed and this is manifested by a sudden rise in water consumption and by a drop in pressure. The coefficient of the rock resistance to upheaval at a certain depth below the surface can be deduced from the highest water pressure at which consumption remains unchanged. In order to obtain useful values, several such tests must be performed in every rock. This method cannot be applied in rocks with a low modulus of deformation which undergo slow plastic deformations.

The deformation (upheaval) of the area can also be measured directly during a water-pressure test from a control point at depth. A neighbouring deep borehole can be used, for example, as a control point unless it is not farther than 4 m.

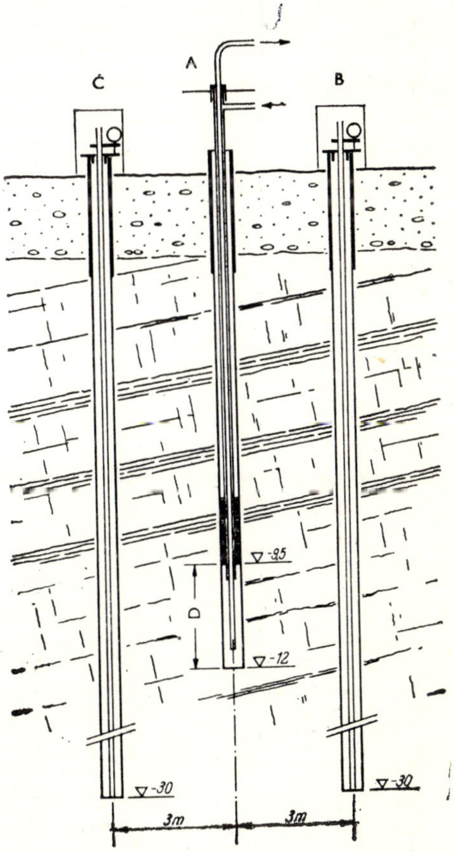

13-63. Upheaval of the ground surface in the neighbourhood of grouted borehole A is measured from fixed reference points set up in boreholes B and C.

Pipes are sunk into such a hole and are left projecting above the casing. The difference in height between the bottom of the borehole (Fig. 13-63) and the margin of casing which is concreted-in to the surface of the bedrock is accurately measured. If during the water-pressure test the rock is disturbed, the surface around the tested hole will heave (Fig. 13-64). Numerous tests carried out with

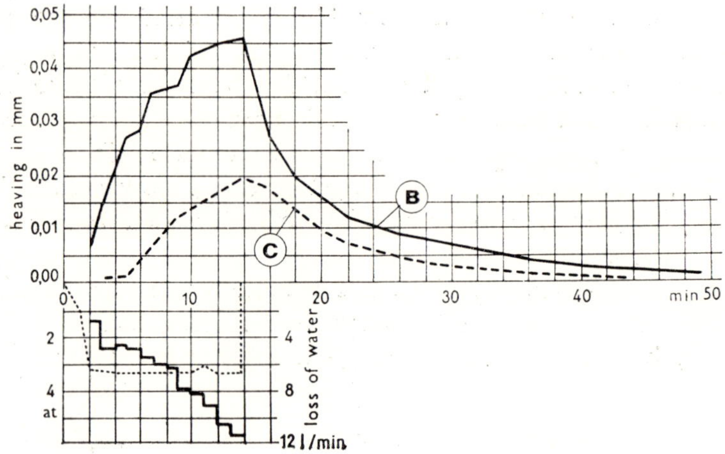

13-64. Upheaval of the bedrock surface caused by a water-pressure test in moderately inclined Cretaceous sandstones (Verfel, 1957), B – Deformation measured in borehole B (see Fig. 13-63), C – upheaval measured in borehole C.

a similar device (Verfel, 1969) have confirmed that the resistance of the rock towards uplift increases downwards, greatly depending on the dip of beds or continuous joints.

Based on our own experience, grouting pressures can be determined roughly on the basis of the diagram (Fig. 13-65) showing the relation between grouting pressures and depth for various dips of beds. The line A_1 shows pressures ($p = \gamma h$) which correspond to the weight of the overlying rock mass at $\gamma = 2.4$. Curve B is valid for solid rocks with steeply inclined joints and gives as the pressure:

5.1 atm. (ca. $2\gamma h$) at a depth 10 m,
14.2 atm. (ca. $3\gamma h$) to 20 m,
44.4 atm. (ca. $4.6\gamma h$) to 40 m

Curve C shows this relationship for bedded rocks with horizontal or moderately dipping bedding planes. The maximum pressures are in this case substantially smaller:

3 atm. (ca. $1.2\gamma h$) at a depth to 10 m,
7 atm. (ca. $1.4\gamma h$) to 20 m,
18 atm. (ca. $1.8\gamma h$) to 40 m

Similar curves for sedimentary rocks are given in a leaflet published by the firm Craelius-Atlas Copco, but the pressures recommended from the 30-m depth downwards are somewhat lower (B′, C′).

For comparison, lines A2, A3 and A4 are plotted in Fig. 13-65; they represent pressures corresponding to twice, three and four times the weight of the overlying rock mass to $\gamma = 2.4$. The comparison of these straight lines with curves B and C show that it is not correct to determine the pressures from simple linear relations only. The pressure $p = 3\gamma h$, for example, would be 7.5 atm. at a 10-m depth. This pressure is fairly high and in sedimentary rocks with moderately inclined planes of separation it would necessarily heave the overlying beds.

The tendency to heaving is particularly great in thinly bedded rocks with horizontal or subhorizontal bedding planes and joints. An example is provided by the disturbance of Cretaceous shales at a damsite near Žermanice in the Beskydy Mts. (Moravia). Grouting pressures equalling twice to three times the weight of the overlying rocks were originally planned. The measurement of deformation, however, showed that pressures exceeding $1.3\gamma h$ induced heaving of the upstream face of the dam blocks under construction.

(d) *Determination of the depth and extent of the grout curtain.* The permeability of rocks causes not only loss of reservoir water but may also impair the stability of the dam by increasing the uplift under the dam foundation. Measurements on some dams have shown that this uplift may be reduced appreciably by grouting.

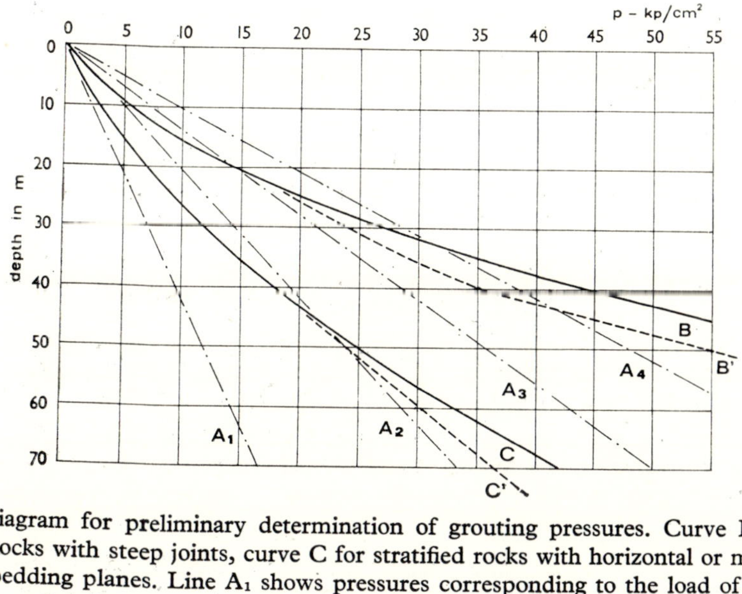

13-65. Diagram for preliminary determination of grouting pressures. Curve B is valid for firm rocks with steep joints, curve C for stratified rocks with horizontal or moderately dipping bedding planes. Line A_1 shows pressures corresponding to the load of overlying rocks at $\gamma = 2.4$. Curve B′ is given by the Atlas-Copco firm for massive rocks and curve C′ for sediments.

Three kinds of grout holes are generally drilled below a gravity dam, which are of different depth and are grouted under different pressures.

As a normal procedure in construction, contact grouting is first carried out (Fig. 13-66) with the use of boreholes 3—15 m deep and a pressure of 1.5—10 atm.

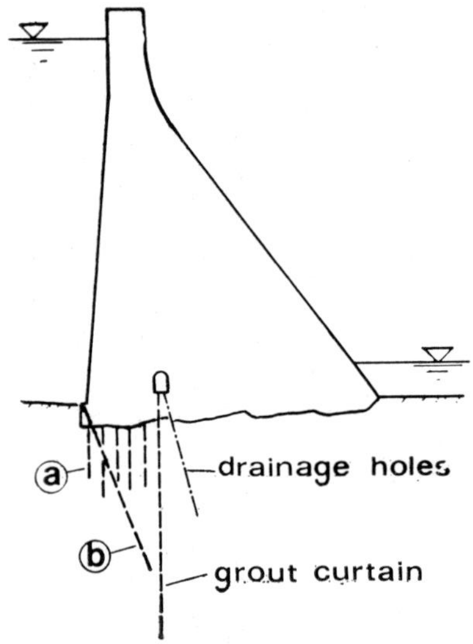

13-66. Grout pattern below a gravity dam. a — Contact grout holes, b — sealing of upstream face.

The grout holes are located in the upstream third of the foundation area; only when rock is strongly jointed are they distributed over the whole foundation area. The holes are usually drilled at right angles to the foundation. Where the beds dip steeply, the holes are inclined so as to intersect the main systems of bedding planes or joints. Contact grouting is carried out after the first concrete lift has been placed so that higher pressure may be applied.

Grouting under medium pressure is used for sealing foundation at a greater depth below the upstream face of the dam. The depth of grout holes varies between 15 and 30 m, depending on local conditions. The pressure applied depends on the weight of concrete placed at the time of grouting and on the strength of the foundation rock; a normal value is 5—20 atmospheres.

The principal grout curtain consists of one or more rows of deep boreholes which are generally drilled from a gallery near the foundation level at the upstream face. The design of this grout curtain should be made with regard to the results of water-pressure tests and experimental grouting and in relation to the geological setting of the damsite.

In constructing a grout curtain, the operation starts with grouting the holes of the first order. They are the deepest and are spaced 8—16 m apart. Grout holes of the second order are then placed between the former holes, and the third-order grout holes are drilled midway between the holes of the second order.

If necessary, fourth-order grout holes can be drilled so that in a one-row grout curtain the drill holes may be only 1.5 m apart. Grout holes of the first order are deepest and those of the successive orders are progressively shallower. The depths of the grout holes of the individual orders are determined on the basis of experimental grouting.

The depth of a grout curtain should be based on the local geological structure, and on the character of the rock and its jointing, which cannot be expressed by an empirical formula. If an impermeable bed occurs at a reasonable depth below the dam foundation the grout curtain should be keyed into it. The depth of grout holes and the choice of a grouting method should be determined from the results of in situ tests.

In Czechoslovakia the grout curtain for the Orlík gravity dam, constructed on crystalline rocks, extends to 40—50 m (the height of water impounding is 72 m), the curtain for an earth dam on the river Morávka in Mesozoic sandstones to 40—75 m (height of impounded water 40 m), and that for the earth dam near Koryčany in the flysch complex had to be extended to a depth of 70 m for an impounded height of water of 19 m. From these examples it is clear that empirical formulas derived from the height of dams would be misleading.

When grout holes in the curtain are deeper than 40 m, special attention should be paid to any departures of the drill holes from the original direction and inclination. Unforeseen deviations are hazardous to the continuity of the grout curtain, particularly at greater depths.

It is difficult in some rocks to maintain the direction and inclination of drill holes of small diameter. Even vertical holes may be deflected during drilling. The amount of departure is related to a number of factors connected with the character of rocks and structural conditions on the one hand and with technical procedure on the other. The deflection of vertical drill holes may be caused, for example, by the alternation of hard and soft rocks, chiefly where the beds or joints are steeply inclined. Technical faults can also be responsible for the deviations when the initial casing pipe is not set accurately, or when a core barrel is too short and high pressure on the bit is applied. Shot boring, in particular, leads to considerable deviations.

Our own experience has shown that the greatest departures occur when boring in a complex of steeply dipping beds of varying hardness. At one damsite in Romania, where grout holes for a 120 m deep curtain were drilled in Eocene sandstones dipping at a steep angle, the departures were as much as 9°, which would lead to a lateral deviation of 16 m at 100 m depth. Such large deviations could endanger the continuity of the grout curtain and involve the necessity of putting down additional drill holes. The question of maximum deviations must therefore be studied during engineering geological investigations, so that a suitable drilling method can be chosen. In drill holes of appreciable depth the inclinations and directions of the holes must be measured during drilling operations. Such measurements are made with inclinometers of various types or with a special television camera.

(e) *Grouting*. Successful grouting demands not only a suitable working procedure but also the use of a proper grouting mixture. Open joints are generally grouted with cement suspension or fine cement mortar of varying density. Rock with closed joints is grouted first with a thin mixture (e.g. 1:8) so as to fill minute cracks at a greater distance from the hole. To seal larger cracks near the drill hole a thicker mixture (1:4 to 1:1) is used. The procedure in rocks with wide cracks is the opposite: a thick mixture (e.g. 1:4) is used at the beginning and grouting is then completed with a thin mixture (1:8 to 1:10). Consumption of grout is given in kg per 1 m² of grout curtain or in kg per 1 m of grout hole. Cement grout is usable for cracks at least 0.2 mm wide. Finer cracks are filled either with clay or chemical substances, such as a solution of sodium silicate which coagulates into gel on the addition of a suitable coagulator. Organic resins have also been used recently.

(f) In the following examples, the importance of geological conditions and the permeability of rocks for the design of a grout curtain is clearly shown.

A layout of grout holes for slender arch dams is shown in Fig. 13-67. It represents the cross-section of the Mauvoisin Dam in Switzerland which is founded in calcareous shales of Upper Triassic to Jurassic age. These rocks provide fairly good and homogeneous foundation but their permeability is high — up to 20 l/min per 1 m drill hole at 30 atm. pressure. Three systems of grout holes were installed.

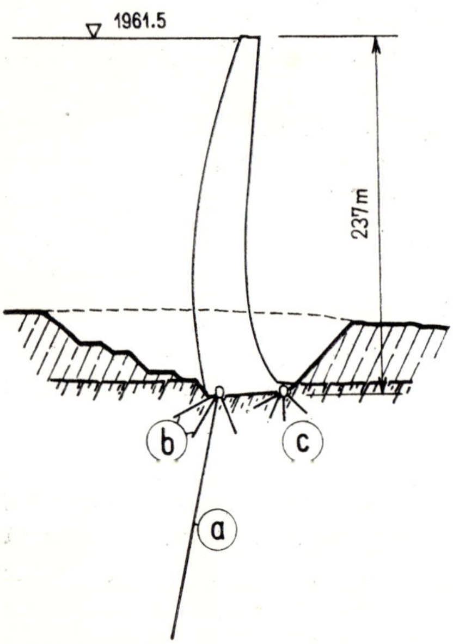

13-67. Grout holes below the Mauvoisin arch dam. a — Main curtain, b — grout holes below the upstream face, c — grout holes below the downstream face. (Gilg, 1964.)

The principal grout curtain consists of a line of holes extending as much as 220 m below the dam foundation. The drill holes were inclined 12° upstream in order to intersect more joints in the rock mass. Exploration holes were first drilled 36 m apart, and were then tested by water pressure in three-meter stages along their whole length. After they had been grouted, grout holes of the second order, 12 m apart, were put down and eventually grout holes with 6 m spacing were drilled and grouted. The average consumption of cement grout was 172 kg per 1 m borehole. The grout holes for the principal curtain were drilled from a grouting gallery so as not to disturb the work on the building site.

An additional grout curtain was constructed at the heel of the dam with the objective of preventing the access of water to the foundation. The holes of this curtain were 25 m long, 3—6 m apart and arranged fanwise around the lower gallery below the dam (Fig. 13-68). The consumption of grout mixture was 87 kg cement per 1 m grout hole.

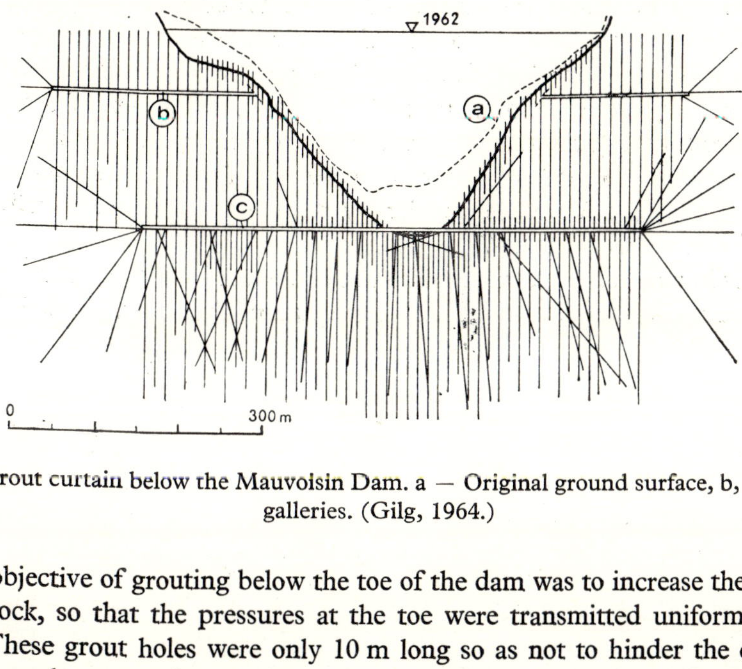

13-68. Grout curtain below the Mauvoisin Dam. a — Original ground surface, b, c — grout galleries. (Gilg, 1964.)

The objective of grouting below the toe of the dam was to increase the strength of the rock, so that the pressures at the toe were transmitted uniformly to the rocks. These grout holes were only 10 m long so as not to hinder the discharge of the seeped water and not to increase the uplift below the dam. The holes were 3—6 m apart and the consumption of cement was 60 kg per 1 m borehole.

Figure 13-69 shows the geological section of the Beauregard damsite in the Italian Alps. After a detailed geological investigation an arch dam 132 m high was constructed there during 1955—1957. Geological conditions were difficult for the construction of such a high arch dam and extensive grouting was necessary. The damsite is made up of mica-schist and schistose gneiss, which are strongly fractured on the left flank. The valley bottom is covered by glaciofluvial deposits up to 40 m thick. The valley was deepened by a glacier which had cut a deep gorge extending under the foot of the left flank. The shape of this pocket and strong compression of glaciofluvial deposits suggest that the gneiss moved partly downslope. The design of the dam had to be developed so as to reduce the specific pressure on the rock in the left valley side, the modulus of deformation of which was only 26,000 kg/cm^2. The rocks in the right slope had a modulus equal to 300,000 kg/cm^2. After grouting the left flank, the modulus of deformation of the disturbed schistose gneiss rose to 120,000 kg/cm^2. To increase the stability of the left slope glaciofluvial sediments were removed from the deep gorge and replaced by concrete. The disturbed rock was consolidated and sealed by grouting. Portland cement was used for consolidation of the rock and a mixture of bentonite and slag milled wet and activated by lime milk was used for sealing.

In the firm schistose gneiss on the right flank the grout curtain was 60 m deep (Fig. 13-69). For the left abutment the curtain was extended 80 m from the crest of the dam into the slope. It consisted of three lines of grout holes up to 150 m deep. The grout curtain was constructed from three levels: from a gallery driven at the height of the crest of the dam, and from a gallery driven at the level of the diversion tunnel, while the lower part of the curtain was grouted from a gallery running at the level of the valley floor. Many different types of measuring instruments were set up to record the statical behaviour of the dam. Results obtained after filling the reservoir were satisfactory despite the unusually unfavourable geological conditions (Fig. 13-70).

In many instances the grout curtain had to be extended into both abutments to seal the permeable rocks through which seepage might occur. A section of the

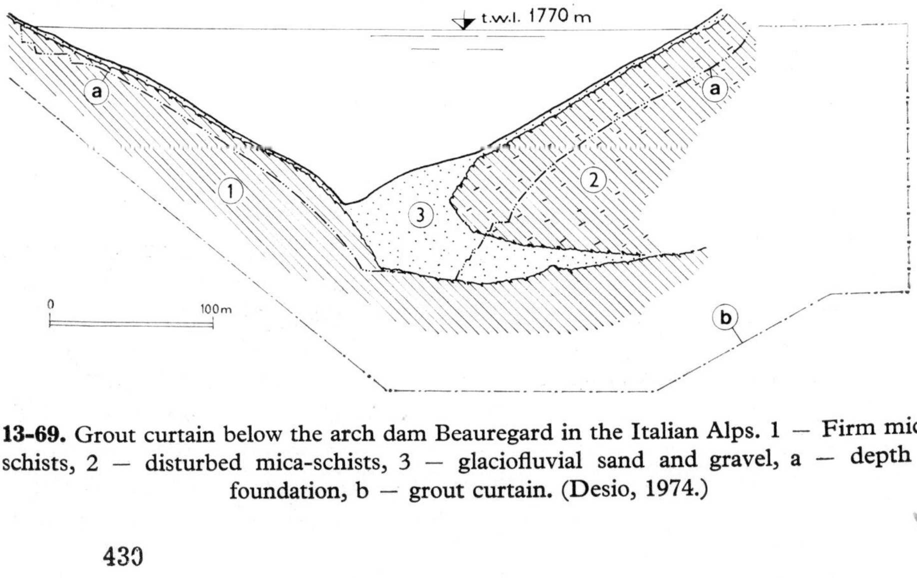

13-69. Grout curtain below the arch dam Beauregard in the Italian Alps. 1 — Firm mica-schists, 2 — disturbed mica-schists, 3 — glaciofluvial sand and gravel, a — depth of foundation, b — grout curtain. (Desio, 1974.)

13-70. The Beauregard Dam after completion. (Photograph by Záruba.)

damsite at Foum el Gherza, 15 km east of Biskra (Algeria) is shown in Fig. 13-71. A concrete arch dam 65 m high was constructed in a deep valley cut in jointed limestones of Late Cretaceous age. The bearing capacity of the limestones was adequate but their high permeability made extensive grouting necessary. The limestone complex is about 200 m thick and rests on impermeable marls. The Cretaceous sequence makes up a wide syncline, the axis of which is roughly parallel to the valley. The impermeable marls in the two limbs of the syncline reach almost to the level of the impounded water. Subsurface exploration revealed that the limestones are jointed over the whole width of the syncline and that it would be necessary to seal them on both valley sides up to the outcrops of marls.

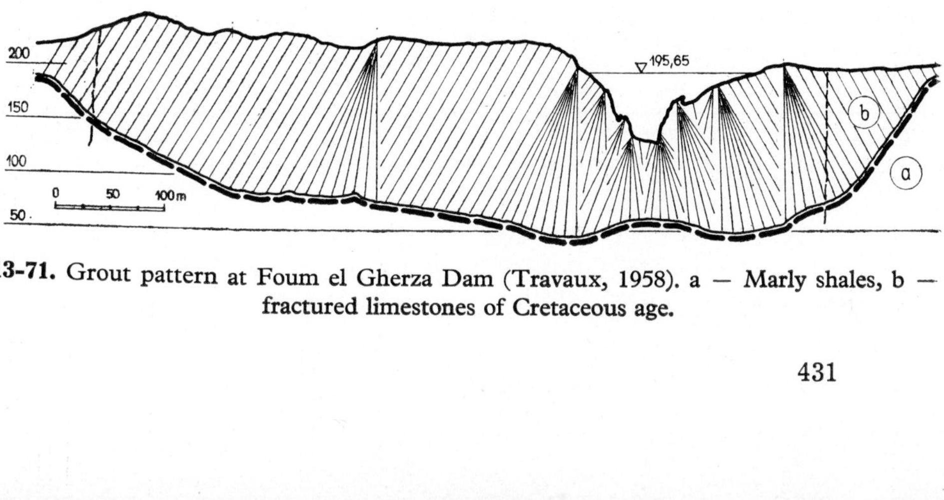

13-71. Grout pattern at Foum el Gherza Dam (Travaux, 1958). a — Marly shales, b — fractured limestones of Cretaceous age.

The grout curtain was constructed as a continuous line of grout holes under the dam and in the left and right abutments, respectively. In the abutments the holes were inclined 30° from the vertical to intersect a larger number of joints. At the points of change in the direction of the curtain the grout holes were arranged fanwise.

The total length of grout holes was 18.6 km and the area of the grout curtain was 88,800 m² so that grouting was a major economic problem. The results of the primary grouting were not sufficient, since the water loss after partial filling of the reservoir attained 370 l/sec. Water losses were reduced below the admissible limit only after the second and third grouting stages had been completed.

The success of grouting depends on the careful attention given to the prosecution of the work. Any insufficiencies result in expensive reconstructions of the grout curtain. During 1938—1947 very great reconstructions of a grout curtain had to be carried out on the Boulder Dam in the U.S.A. (Simons 1951). The dam was founded in the deep canyon of the Colorado River cut in volcanic breccias. After the reservoir had been filled, the 40-m deep grout curtain did not prove to be sufficiently watertight; at several places considerable seepage appeared and the uplift below the dam was greater than had been anticipated. A new grout curtain up to 130 m deep was therefore constructed (Fig. 13-72). The perviousness

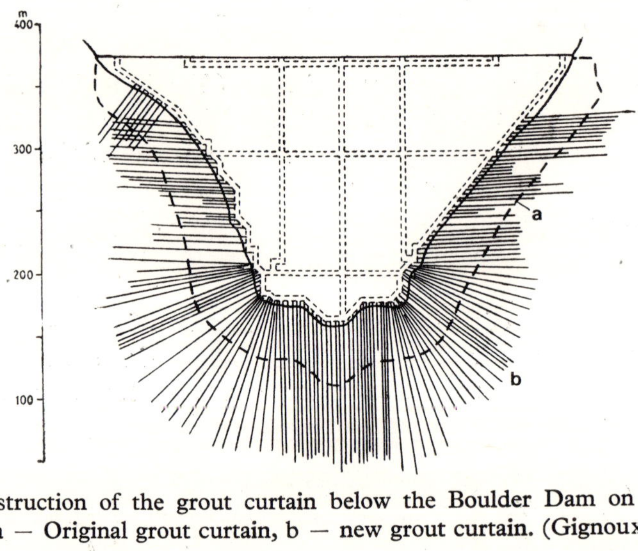

13-72. Reconstruction of the grout curtain below the Boulder Dam on the Colorado River. a — Original grout curtain, b — new grout curtain. (Gignoux, 1955.)

of the original curtain was caused by its inadequate depth and imperfect execution. During the construction of the original curtain the cement mortar leaked from some holes into the joints between the blocks of the dam. The grouting of some deep holes was abandoned for this reason, but as a result gaps originated in the grout curtain. In addition, the shatter zone on the Nevada bank was not sealed completely because of the excessive intake of cement. The cost of the reconstruction of the grout curtain amounted to 2.38 % of the expenses on the whole dam. As can be seen from Fig. 13-73 the reconstruction was successful in reducing greatly the uplift beneath the dam.

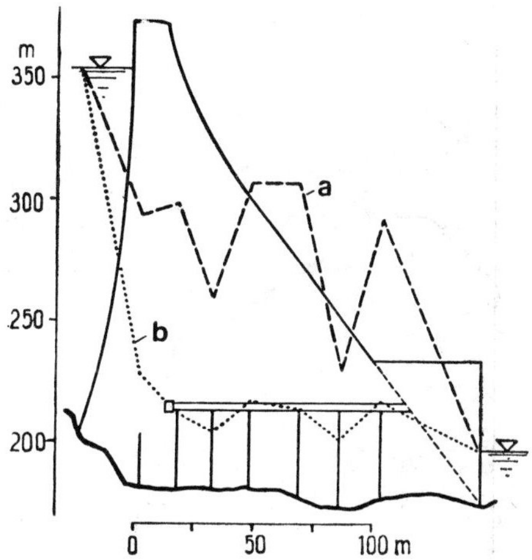

13-73. Diagram showing the course of uplift under the Boulder Dam. a — After filling the reservoir, b — after reconstruction of the grout curtain. (Simonds, 1951.)

(g) *Grouting of sand and gravel beds.* For sealing the foundation of the Serre-Ponçon Dam a new method was developed by the Solétanche firm (Cambefort 1955). The constricted part of the Durance river valley at Serre Ponçon has been studied since 1857 as it provided an exceptionally suitable damsite from the hydrological point of view. The geological conditions, however, were so unfavourable that the construction of a dam could be contemplated only in recent years when progress in dam construction engineering made it possible to overcome the extremely unsuitable natural conditions. The bedrock consits of firm marly limestones but an erosive gorge more than 100 m deep and filled with river alluvium extends below the valley bottom (Fig. 13-74). The alluvium is heterogeneous in both the grain size and permeability. There is an alternation of coarse sandy gravel with tongue-shaped layers of slope detritus consisting of limestone and marl fragments. It was established by boring that the alluvium is fed by thermal sulphate water with a temperature of 60° and having a corrosive character. An exploratory gallery and shaft driven in 1913 were flooded by an outburst of this water at a depth of 55 m.

Perforated casing provided by rubber sleeves was used for grouting. The casing with sleeves was lowered into the hole 90—110 mm in diameter, drilled by a percussion rig, and surrounded by a special semiplastic substance. Short segments (30—50 cm, sealed by double packers) were successively grouted. The grout mixture being forced out of the holes expands the rubber sleeve, breaks through the plastic substance and enters the rock. This method is advantageous in that

13-74. View of the Serre Ponçon damsite on the river Durance. The valley is overdeepened 100 m below the present river level and filled with river deposits. (Courtesy Solétanche.)

adequate pressures can be used, corresponding to the character of the rock in the respective segment. The most permeable gravel beds were injected first at a relatively small pressure so that the grout did not penetrate too far. Finer sand beds were grouted under a higher pressure (Fig. 13-75).

The vertical curtain constructed by this method below the Serre-Ponçon Dam extended from the impervious core to the bedrock (Fig. 13-76). The curtain consists of a system of grout holes arranged into the form of letter T. Under the dam it is 35 m wide and composed of 12 rows of grout holes; the width is gradually reduced to 15 m and the number of borehole rows to 4 (Fig. 13-77). The alluvial deposits were grouted with a bentonite-cement mixture. In all, 17,200 m grout holes were filled with 11,000 tons of cement and 24,000 tons of bentonite. The presence of thermal corrosive water necessitated the use of special slag cement. The method proved successful: the permeability of the alluvium in the valley bottom was reduced nearly 1000 times, i.e. from 5×10^{-2} cm/sec. to 5×10^{-5}cm/sec.

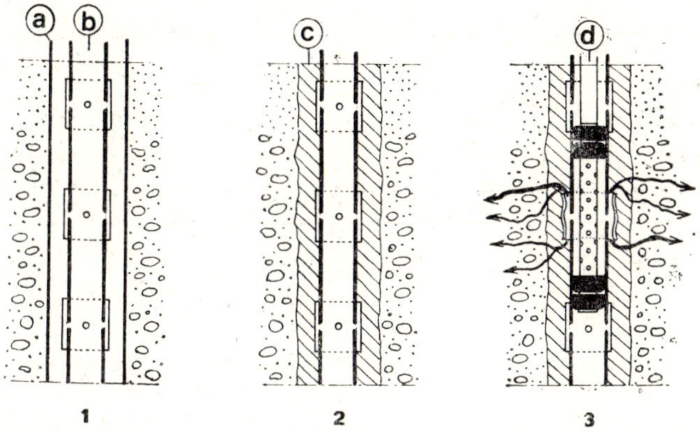

13-75. Grouting of sandy gravel by Solétanche method. 1 — After drilling and casing a bore (a) perforated tubes with rubber sleeves (b) are driven in; 2 — the space between casing and perforated tube is filled with clay-cement mixture (c) and casing is withdrawn; 3 — injection pipes with double packer are set into the perforated tube (d). The equipment allows for grouting individual gravel layers with different pressure.

The method was also used with excellent results at the Sylvenstein Dam on the river Isar (Bavaria) for grouting 100-m deep permeable alluvium and at the construction site of the Fessenheim power station on the Rhine (Fig. 13-108). The Solétanche procedure is commonly used for grouting loose soils. It has proved successful at the construction of the Metro in Paris (Sec. 12.10) and is also employed for hardening and sealing sections of the Prague Underground driven in water-bearing sands.

13.12. Investigation of the reservoir area. In geological investigations for hydraulic structures, attention is often concentrated principally on the problems

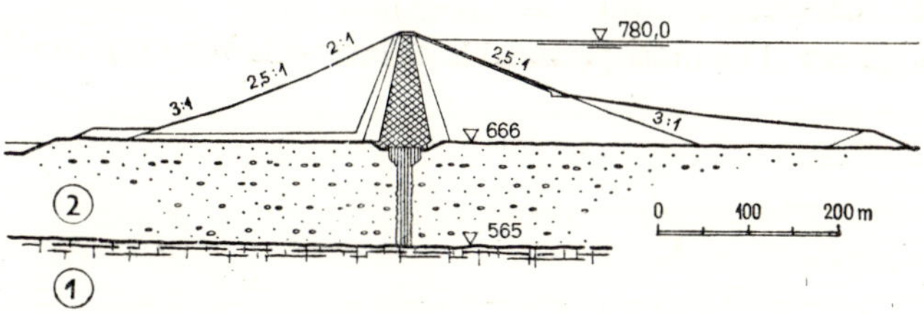

13-76. Section of the Serre-Ponçon Dam on the river Durance, France. 1 — Marly limestones, 2 — river deposits and debris material.

of the damsite. The watertightness and silting of the reservoir, however, are no less important for the economy and durability of the work. The engineering-geological investigation must therefore solve additional problems concerning

(a) watertightness of the reservoir (Sec. 13.13),
(b) stability of the reservoir banks (Sec. 13.14),
(c) siltation of the reservoir (Sec. 13.15),
(d) economic results of impounding water (Sec. 13.16).

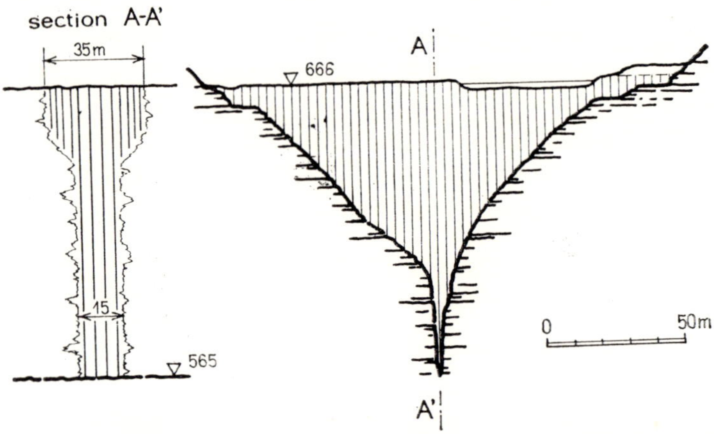

13-77. Grout pattern under the Serre-Ponçon Dam. (Cambefort, 1955.)

13.13. Watertightness of the reservoir. Water can escape from the reservoir either into a neighbouring valley or into its own channel below the dam. The latter may be caused either by the permeability of the damsite or of the flanks of the reservoir. In this paragraph, we shall deal with water seepage through reservoir banks.

Water can escape from the reservoir into a neighbouring valley when the slopes are formed of permeable rocks (jointed sandstones) which continue into the adjacent valley. The loss of water is greatest when the permeable beds are horizontal or slightly inclined, because the impounded water can infiltrate them over the whole length of the reservoir (Fig. 13-78). If the permeability is high, the base of this complex may define the elevation of the economical water level, since sealing

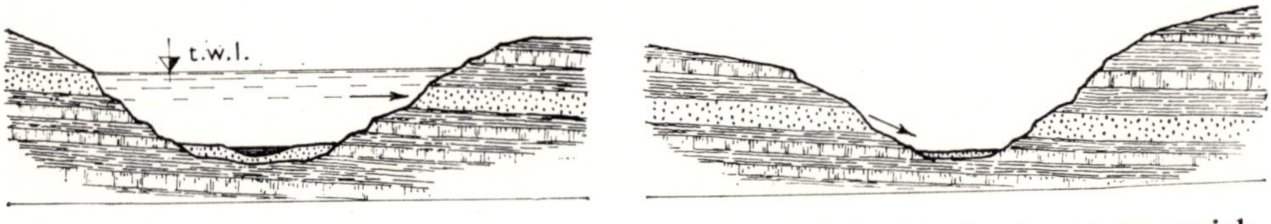

13-78. The impounded water can leak through the permeable bed in the slope into a neighbouring valley.

of banks along the whole reservoir is generally very expensive. It is the task of engineering geological investigation in this case to find out whether the valley sides are covered with slope deposits which are either of low permeability or impermeable so that they can appreciably reduce the loss of water. It should be ascertained at the same time whether the stability of slope deposits would be disturbed by raising the water level (see Secs. 13, 14).

Seepage of water from the reservoir into a neighbouring valley is also possible when rocks are cut by continuous fissures trending approximately at right angles to the valley. In the shattered rocks of broad fault zones or in limestones with joints widened by karstification the loss of water may be very great. To solve this problem, hydrogeological research of the reservoir area and the adjacent valleys must be carried out. Systematic observation of the ground-water table and yield of springs generally provides more complete and more comprehensive information on the watertightness of the reservoir than local tests of permeability in boreholes.

In some areas terrace gravels make up a continuous cover between two stream valleys. If the impounded water reaches this gravel layer, water losses from the reservoir could be excessive because gravel is usually very permeable unless it is cemented with impervious material (Fig. 13-79).

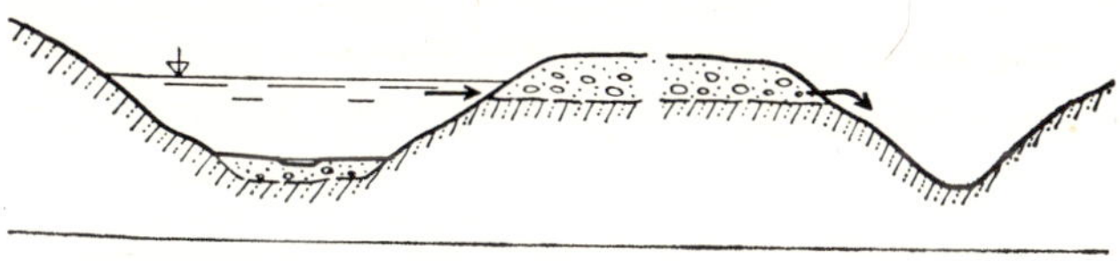

13-79. Percolation of water into a neighbouring valley when the impounded water extends into gravel terrace which connects the two valleys.

In assessing the watertightness of the reservoir area, attention should be paid to ancient *buried valleys*, which are often filled with permeable fluviatile or glaciofluvial deposits. These buried valleys are generally cut deeply into the bedrock, and when water is impounded to the level of the permeable sediments filling the overdeepened channel, the loss of water may be again considerable. The engineering geologist should examine the character and permeability of the alluvium, the depth to bedrock, and the course of the buried valley and thus also the length of the route of possible seepage. From these data he can deduce whether seepage is imminent and, if so, can propose measures for sealing the buried valley. It should be remarked that buried valleys were often eroded along fault zones so that the investigation should also cover the bedrock below the valley bottoms.

Considerable seepage through a buried channel occurred in the reservoir impounded by the Sautet Dam on the river Drac in France (Gignoux, Barbier 1955). The damsite is located in a gorge in firm Liassic limestones and marly shales but the present Drac river valley is connected above the dam with the

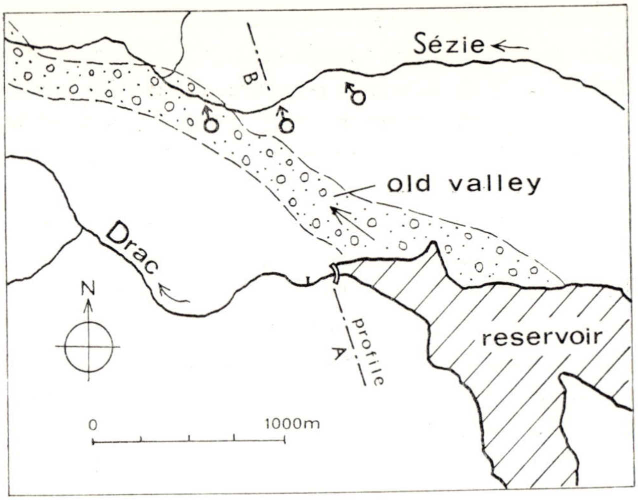

13-80. Leakage of water from the Sautet reservoir through a preglacial valley filled with gravel. (Gignoux, 1955.)

Sézie river valley by an old preglacial valley (Fig. 13-80). The bottom of the old valley is filled with permeable coarse sandy gravel, overlain with sediments of a glacial lake. The water escaped from the reservoir through alluvial sediments into the Sézie river valley where it discharged in copious springs, at a distance of 900—1000 m from the reservoir (Fig. 13-81). The coefficient of permeability of the sediments was 2 cm/sec; the sealing of the old valley was very difficult.

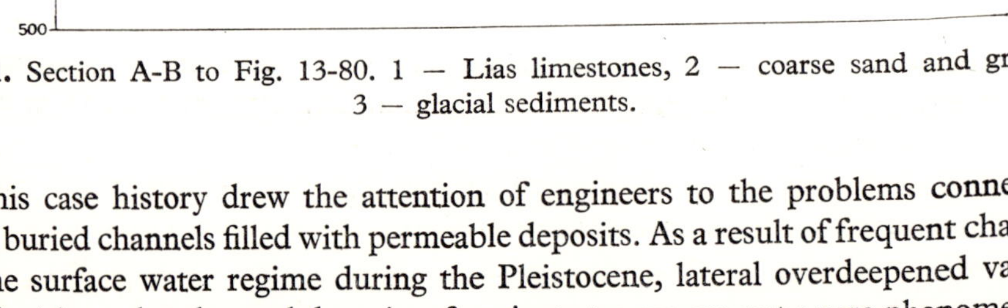

13-81. Section A-B to Fig. 13-80. 1 — Lias limestones, 2 — coarse sand and gravel, 3 — glacial sediments.

This case history drew the attention of engineers to the problems connected with buried channels filled with permeable deposits. As a result of frequent changes of the surface water regime during the Pleistocene, lateral overdeepened valleys filled with sand and gravel deposits of ancient streams are not a rare phenomenon. If, in such cases, the impounded water rises above the impermeable rock divide, seepage through the gravel into the neighbouring drainage area could occur.

The construction of the dam on the river Senaiga in the Italian Alps revealed an interesting example of an ancient valley filled by a large landslide (Ciampi 1958). An arch dam, 70 m high, is located in a gorge cut in firm Lower Cretaceous limestone. The bottom and banks of the reservoir are formed of fairly impermeable rocks, except for a short sector on the left bank close to the damsite where the limestones are strongly disturbed. It was found in boreholes and galleries that the rock was not in situ but that it had slipped down the slope, filled the valley and pushed the river to the right bank, where a new channel was eroded. The limestones were shifted a relatively short distance so that they preserved a stratigraphic continuity; only in places did they show the character of rock debris. Additional exploration defined the precise form of the ancient valley and the character of the slipped material (Fig. 13-82).

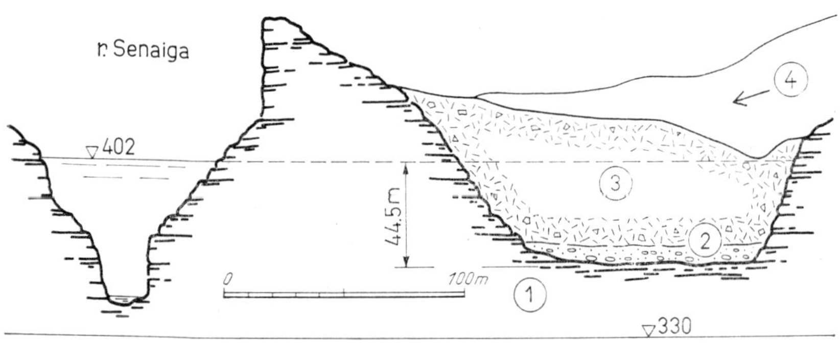

13-82. Section of the damsite on the river Senaiga and of the old valley blocked by a landslide. 1 — Lower Cretaceous limestones, 2 — glaciofluvial deposits, 3 — slipped limestones filling the old valley, 4 — direction of movement. (Ciampi, 1958.)

The valley filling is 5—6 m thick and consists of very permeable glaciofluvial deposits with boulders. The slipped limestones, particularly in the lower part, are almost impermeable, because the fissures are filled with clayey material. Figure 13-83 shows the extent of the landslide and the approximate course of the ancient valley. From Fig. 13-84 it is apparent that the valley was clogged by a natural dam which, at the level of impounded water (el. point 402), is flooded to a height of 44.50 m. At this level, the natural dam is more than 90 m wide; at the valley bottom its width is 270 m.

Permeability tests showed that apart from glaciofluvial deposits the disrupted limestones along the old surface were also highly pervious. In contrast, the water losses in the slipped material were so small that it could be considered to be virtually impermeable. It was thus not necessary to seal the ancient valley for the whole height but only within the range of glaciofluvial deposits. The grout curtain consisted of three galleries, one above the other, with interior diameters large

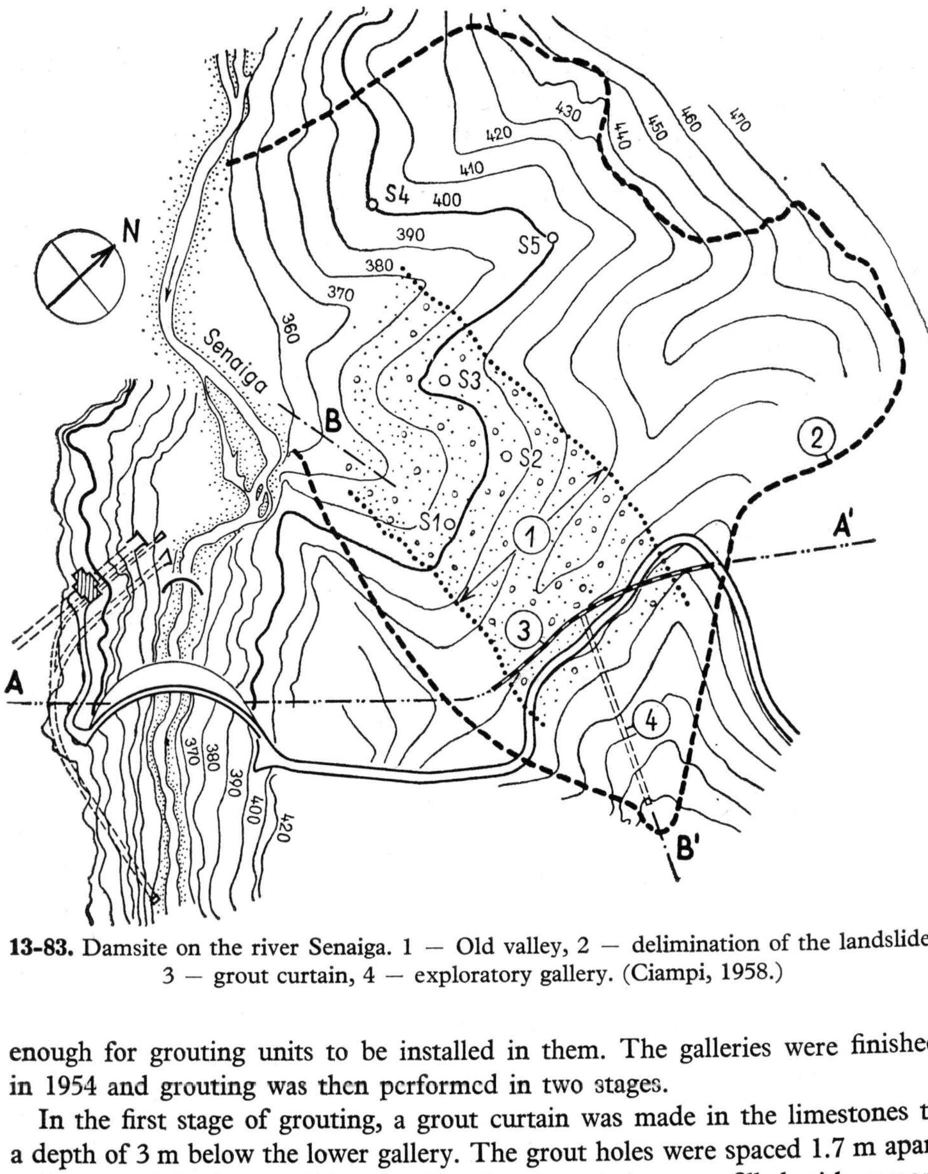

13-83. Damsite on the river Senaiga. 1 — Old valley, 2 — delimination of the landslide, 3 — grout curtain, 4 — exploratory gallery. (Ciampi, 1958.)

enough for grouting units to be installed in them. The galleries were finished in 1954 and grouting was then performed in two stages.

In the first stage of grouting, a grout curtain was made in the limestones to a depth of 3 m below the lower gallery. The grout holes were spaced 1.7 m apart and arranged in two parallel rows. The 3-m long holes were filled with cement grout. After the first stage of grouting the reservoir was filled on trial but, when the water loss attained 35 l/sec with partial impounding, it was again emptied. It was established that the water escaped mainly through the limestones below the ancient valley.

In the second stage the grout curtain was sunk 25 m below the lower gallery. The grout holes were spaced 3 m apart and the control holes were inclined at 30°. For injections, cement grout was used and a mixture of cement and fine sand

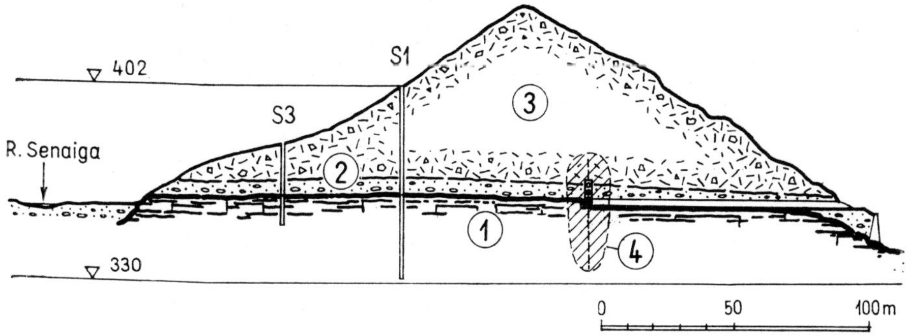

13-84. Section across the old valley filled with landslide material. 1 — Firm limestones, 2 — glaciofluvial deposits, 3 — material of the landslide, 4 — grouted area.

in sectors of very high permeability. The average consumption of dry mixture was 245 kg/1 m of hole in the curtain, and 83 kg/1 m of control hole. The control holes exhibited considerable improvement. The slipped limestones above the curtain were grouted from the upper gallery upwards to a height of 10 m above its ceiling. The holes were inclined at 45° and spaced 3.5 m apart. Drilling in downslid limestones was very difficult since the walls of the holes were not stable and collapsed. In the limestone debris the consumption of grout was 35 kg/1 m, whereas it amounted to 582 kg/1 m in the holes at the contact with the bedrock. The layout of grout pattern is shown in Fig. 13-85.

Seepage of water downstream from a reservoir may occur in a winding valley when the impounded water reaches the terrace gravels in the meander core, or

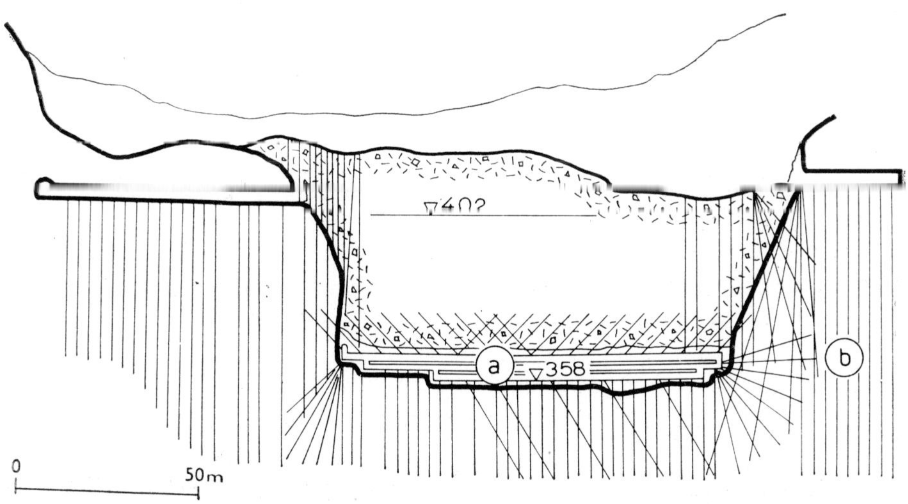

13-85. Grout curtain and sealing of the old valley. a — Grout curtain consisting of three galleries driven one above the other, b — grout holes.

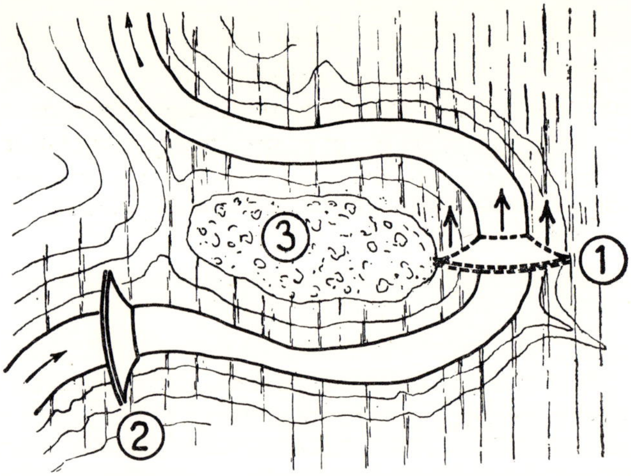

13-86. Unfavourable damsite (1) in a river meander. Water may escape from the reservoir through the joints in bedrock across the meander core, and through the terrace gravel if it were impounded to the terrace level (3). Damsite (2) is more suitable geologically.

where the river had developed encased meanders in fractured rocks and the fractures run across the valley bend (Fig. 13-86).

Figure 13-87 shows two sites studied for the Dívčí kámen Dam on the river Vltava (Bohemia). The sites at topographically suitable places are sometimes

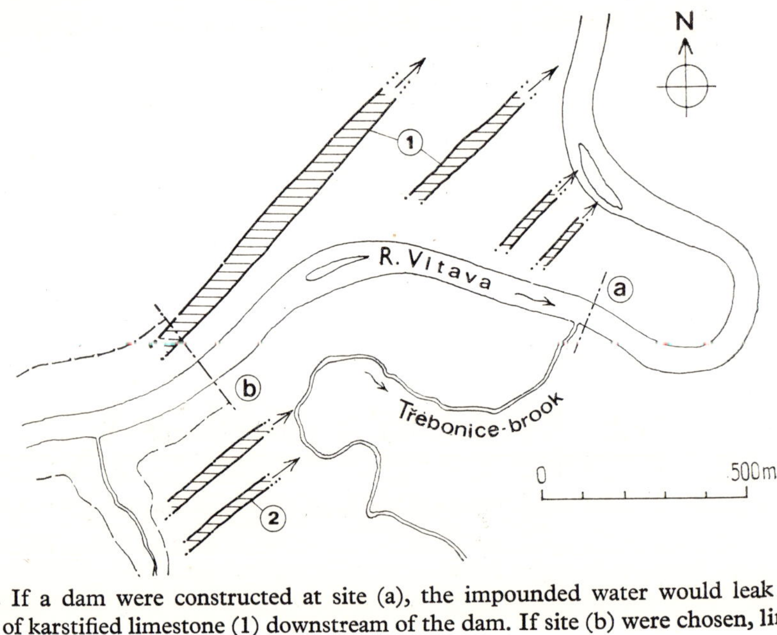

13-87. If a dam were constructed at site (a), the impounded water would leak through bands of karstified limestone (1) downstream of the dam. If site (b) were chosen, limestones should be sealed on both the left bank of the river and on the right bank (2), where leakage into the neighbouring brook valley would occur.

invalidated by karstified, permeable crystalline limestones which form stripes in a biotite gneiss complex. If a dam were erected at site *a* it would be necessary to seal several limestone strips on the left bank through which the water could escape downstream below the dam. If site *b* were chosen, it would be necessary to grout not only a wide strip on the left bank but also two limestone strips on the right bank through which the water could seep into the deep valley of the neighbouring stream.

Appreciable seepage may also take place where the banks of the reservoir are made up of jointed young effusive rocks. The Slezská Harta damsite occurs in the valley of the river Moravice (Silesia), cut in firm Lower Carboniferous shales and greywackes (Fig. 13-88). The valley is separated from the valley of a neighbouring brook by a ridge formed by a basalt flow of the Velký Roudný volcano. The flow is 5.4 km long and 50—57 m thick. The basalt makes up the upper part of the right slope and fills the Early Pleistocene valley of the river Moravice in which a bed of sandy gravel, 2—5 m thick, has been preserved. The right abutment of the dam would have to be founded partly on shales and partly on basalt. The basalt flow is disrupted into large blocks separated by fissures up to 5 m wide (Fig. 13-89). The strike of the fissures, their course and width have been established by geophysical measurements. As the loss of water would be of the order of tens of litres, and the grouting of fissures in the basalt forming the

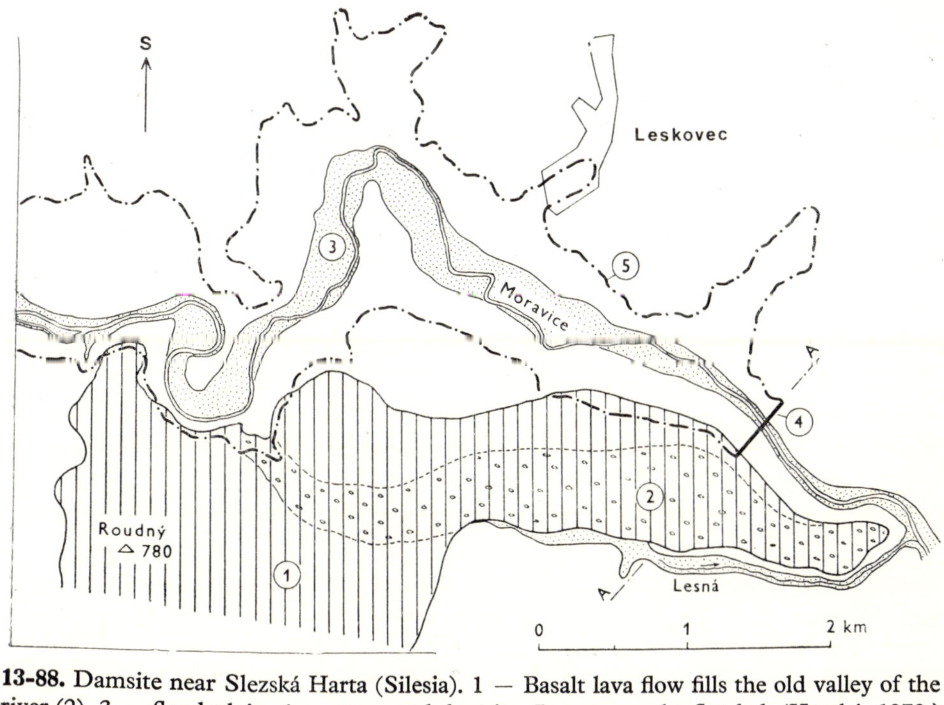

13-88. Damsite near Slezská Harta (Silesia). 1 — Basalt lava flow fills the old valley of the river (2), 3 — flood plain, 4 — presumed damsite, 5 — area to be flooded. (Horský, 1972.)

bank of the reservoir over a considerable length would be very difficult, the construction of the dam was postponed.

In mining areas there is danger of water seepage through unrecorded, abandoned, partly caved-in galleries and cross-cuts, possibly connected with a neighbouring valley or opening into the valley to be dammed, downstream of the damsite.

In exceptional cases, an absolute watertightness is not required for reservoirs in a karst area. The permeability of limestones is not generally dangerous for the stability of the dam itself so that if the reservoir is built for retaining flood waters complete watertightness is not a prerequisite. In the Dinaric Karst (Jugoslavia) the sealing of the Peruča reservoir had to be estimated from the viewpoint of economics. In essence, it had to be decided whether the high cost of grouting of the karstified limestones would be balanced by the value of the amount of electric power produced by water escaping from the reservoir through the karst cavities. Geological investigation showed that the water would escape from the reservoir only downstream, so that it was captured by the next dam. Since the power plant utilizing the main head was located below this second dam, it was not necessary to perform the sealing of the first reservoir at an excessively high cost (Jevdjević 1955).

13.14. Stability of reservoir banks. In order to assess the safety of reservoir banks, engineering geologists should study the rocks which make up the valley sides, particularly within the range of water fluctuation. The disturbance of banks is generally due to one of two causes:

(a) As a result of flooding, the stability conditions of some rocks are changed; when the rocks are saturated with water, their weight decreases due to uplift but their loading above the water level remains unchanged. Simultaneously, the cohesion of fine-grained soils may be appreciably reduced. On an abrupt lowering of the water level the rock is under the pressure of water seeping into the reservoir. All these factors may cause sliding of the reservoir banks.

(b) Wave action due to wind and motorboats disturbs the banks by erosion which, analogously to marine abrasion, creates a typical platform terminated by a scarp. "Beach" sediments are deposited on the shore. In contrast to natural lakes, reservoirs may display great fluctuations in water level depending on the

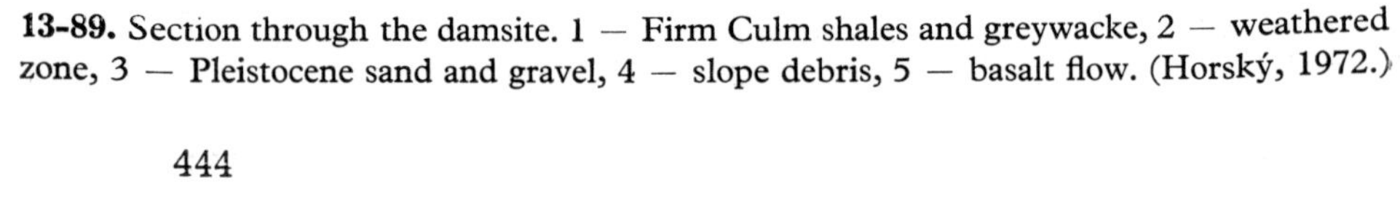

13-89. Section through the damsite. 1 — Firm Culm shales and greywacke, 2 — weathered zone, 3 — Pleistocene sand and gravel, 4 — slope debris, 5 — basalt flow. (Horský, 1972.)

operation of a power-plant. The fluctuation of the water level may be regular within a day or season. As a result, beach sediments may be exposed to erosion and their material displaced downslope towards the bottom of the reservoir. The slope can be undercut by a steep scarp and may collapse.

If the banks are made up of firm bedrock, there is no danger of their disturbance by impounded water, even with considerable fluctuation. If, however, the banks are covered with rock debris, terrace sand or clayey sediments, particular attention should be paid to the stability of the banks. If the cover deposits are of great thickness and inclined valleywards or if they have been affected by sliding before the construction of the reservoir, fluctuations of water level may provoke or renew movements disturbing the slopes even high above the impounded water.

Slipping of reservoir flanks are very frequent in the Apennines and their study is one of the important tasks of engineering geological investigations. The flooding of slopes and fluctuations in water level trigger off movements in temporarily stable sliding areas, so that slope failures are a serious cause of reservoir siltation.

The question of the stability of mountain slopes flanking the reservoir became a subject of special attention after the disaster of the Vaiont Dam in the Italian Alps (see Chapter 7, p. 189). An example of detailed study of slope movements caused by water-level fluctuation is provided by the Gepatsch Dam in the Tyrol

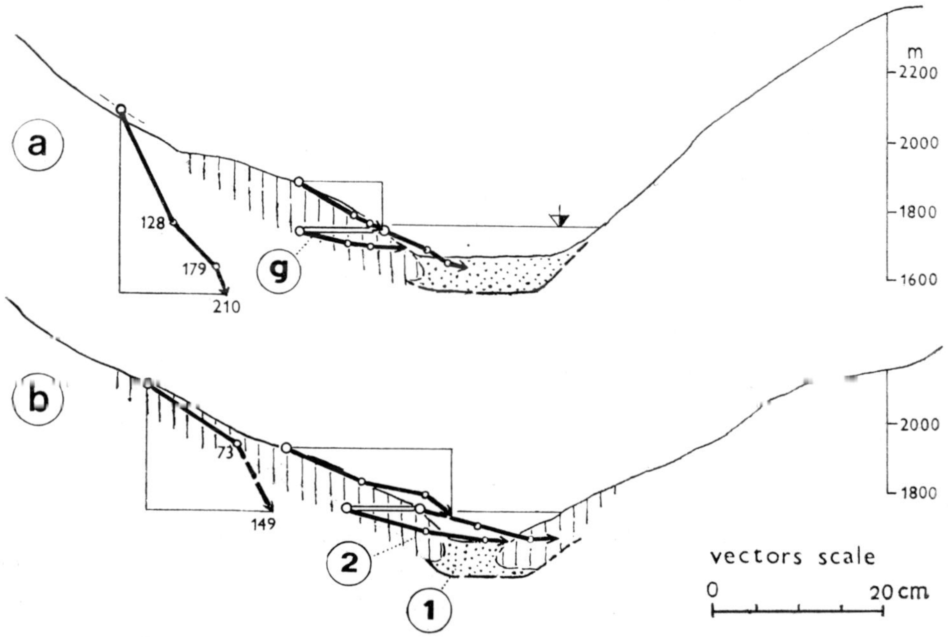

13-90. Gepatsch reservoir in the Tyrol; gravitational deformations on the northern slope monitored in 1966–1969. 1 – Glaciofluvial stressed valley sediments, 2 – gneiss and weathered rock disturbed by gravitational movement, g – galleries. (After Neuhauser and Schober, 1970.)

(Austria). After the reservoir was filled in 1965, the movement of slope debris, mica-schist and gneiss was renewed on the left flank. The mountain slope had undergone gravitational sliding in Postglacial time when about 100 million m^3 of rock moved and compressed the glaciofluvial deposit on the valley bottom. Recent movements were measured systematically during 1966—1969 with the use of a line of stakes. Measuring devices were also set up in several exploratory galleries driven into the slope. Figure 13-90 shows the results of measurement of two profiles. According to the vectors, the vertical component of movement predominated in the upper part of the disturbed slope, and the horizontal component at the foot of the slope. The vectors measured in the gallery show that the frontal part and the end of the gallery (in profile *b*) moved at roughly the same rate, which means that the disturbance of the slope reached beyond the end of the gallery at a distance of 185 m from the surface of the slope. The gallery in profile *a* is 244 m long; the rock was strongly shattered and loosened to a distance of 90 m; open or loam-filled fissures were frequent and mylonitized zones were found 94—99 m and 227—234 m from the slope surface. Both shattered zones were almost parallel with the inclination of the slope. Observations have shown that the slope movements were connected with fluctuations in water level and that they decrease gradually.

These experiences should be borne in mind in designing reservoirs in the Flysch Carpathians, for example, where there are many potential slide areas. On filling the reservoir near Rożnów on the river Dunajec in Poland, the flanks formed of weathered flysch material were strongly disturbed by slope movements, which extended high above the impounded water.

The stability of flanks of the reservoir as designed should always be examined. In considering alternative designs, differing in the top water level to be maintained in the reservoir, the final decision may depend on the cost of stabilization of the banks, especially where slope movements would damage valuable or built-on land.

A detailed investigation of reservoir flanks was conducted for the Slapy Dam on the river Vltava (Bohemia). The banks are made up mainly of bedrock, except for several sectors where there are terrace deposits and fine sand, covered with loess loam. The stability of the flanks was confirmed in 1954, when flood water filled the reservoir almost to full height. As the construction of the control weir was not yet completed, the water level dropped rapidly to the level of the crest of the dam (el. 263). During the following two months the level was gradually lowered by additional 9 m, so that construction might be completed. The drop in the water level had no unfavourable effects on the rocky flanks (only at a few places did blocks slip along joint planes of downslope dip), and on banks composed of terrace sand and gravel. The flanks were disturbed, as had been expected, where they were built up of fine sand and loam. Figure 13-91 shows such a damaged sector, where loess loam up to 7 m thick lies on fine sand covering the terrace gravel. Fine sand started to move due to the pressure of the water flowing out from the reservoir flank on thc suddcn drawdown. The loamy cover was disrupted into blocks that slid into the reservoir. At the same time, landslides occurred on the steep slopes of erosion gullies cut in loam and sand.

Extensive experience with the behaviour and treatment of flanks of large reservoirs have been gained by Soviet geologists. The changes in the configuration of a reservoir flank formed of compact Upper Jurassic clay are shown in Fig. 13-92.

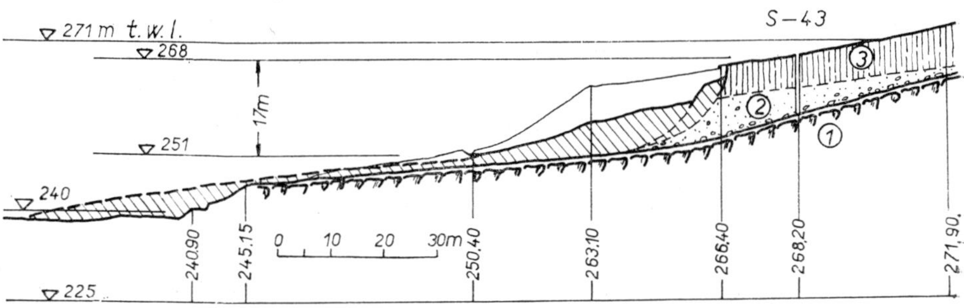

13-91. The bank of the Slapy reservoir (Bohemia) disturbed by sliding after lowering the water level by 17 m (in 1954). 1 — Amphibolite, 2 — terrace sand with gravel, 3 — loess loam.

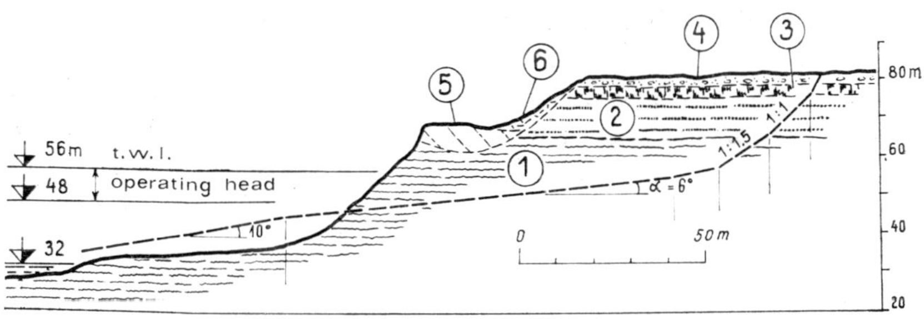

13-92. Development of a reservoir bank. 1, 2 — Jurassic clay, 3 — Jurassic sandstones, 4 — sand and gravel, 5 — slipped clay, 6 — slope debris. (After Savarenskii.)

The flat part of the bank begins below the lowest water level, at a depth to which wave erosion is still effective. The inclination of the flat part changes, depending on the properties of the rocks involved. At the bottom, where small fragments from the upper parts of the flank are deposited, the angle is about 10°; higher up it is 6°. The flat bank ends above the highest water level. The inclination of the slope is steeper upwards from this elevation depending on the character of rocks.

In Czechoslovakia, this question has been studied for the densely settled banks of the Kníničky reservoir on the river Svratka (Bayer, Mencl, Pelikán 1955).

The aim of this study was to find out to what extent structures were threatened by this phenomenon and to develop from this experience guide-lines for future dam construction. It was established that the wave-cut platform transects the line of the original slope surface at the height of the most frequent lowest water levels or some tens of centimetres above them (Fig. 13-93). Hence, it rises moderately

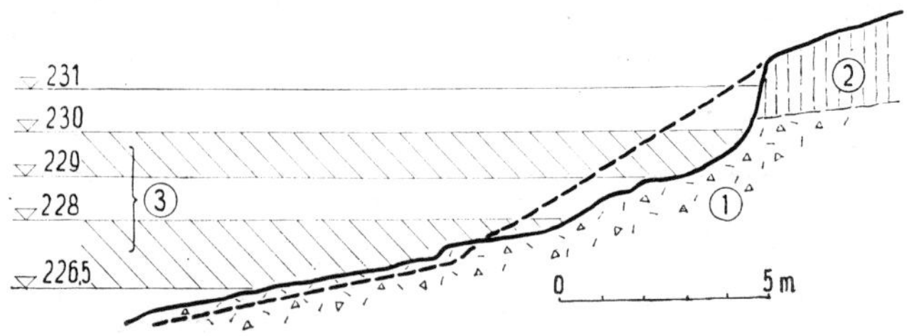

13-93. The bank of the Knínický reservoir (Moravia) disturbed by erosion. 1 — Loamy diorite detritus, 2 — loess loam, 3 — most frequent levels of impounded water.

at an angle depending on the rocks which make up the slope or which fall on the platform from the higher parts of the slope. In stony detritus the inclination is 1:2 to 1:2.5; in terrace gravel and detritus with loamy admixture it is 1:2.5; in loess loam 1:4, in Neogene clay 1:7 and in alluvial loam about 1:11. The upper end of the platform, i.e. the toe of the scarp, occurs some tens of centimetres above the most frequent highest level. If the scarp develops in solid rock, it is nearly vertical; if it is formed in superficial deposits, the material will slip down to be removed by wave action until equilibrium conditions are restored.

In the Nechranice reservoir (north-western Bohemia), the flanks of which are formed of Neogene clays and silts, extensive sliding recurs after each lowering of the water level and the newly formed cliffs cave headwards (Figs. 13-94, 13-95).

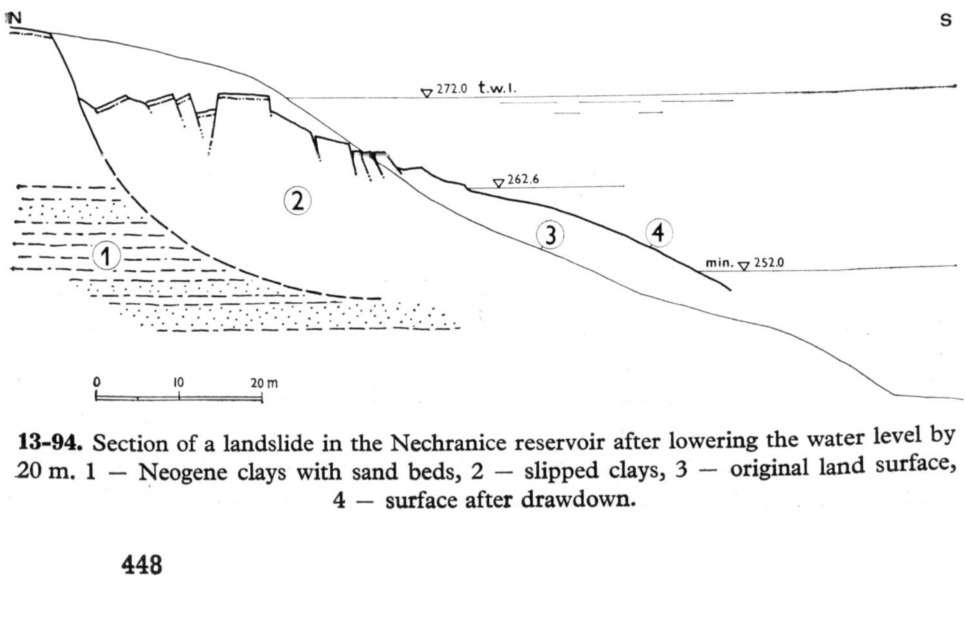

13-94. Section of a landslide in the Nechranice reservoir after lowering the water level by 20 m. 1 — Neogene clays with sand beds, 2 — slipped clays, 3 — original land surface, 4 — surface after drawdown.

13-95. Banks of the Nechranice reservoir disturbed by sliding. (Photograph by Záruba.)

The following procedure can be recommended for the investigation of bank stability:

On the basis of topographical features, flanks should be divided into sectors of roughly the same character. In these sectors typical cross-sections are chosen to be examined in detail by boring. Geological sections are constructed, and the forms and inclinations of natural river banks are plotted on them. Where the reservoir flanks will slope at a greater angle, slope disturbances are to be expected. Appropriate measures can then be proposed, depending on the economic value of the structures involved. The whole flank can then be divided in the plan and cross-sections into:

(a) Sectors liable to extensive slope movements which cannot be prevented either because of technical unfeasibility or excessive cost;

(b) sectors threatened with sliding, where controlling measures can and should be made, e. g. buttresses, decrease of inclination;

(c) sectors endangered by erosion which cannot be controlled by planting appropriate vegetation. It is necessary to estimate the height to which the slope will be disturbed and the final form of equilibrium profile;

(d) sectors which can be protected from erosion by planting appropriate vegetation;

(e) stable sectors where the flanks consist of hard rocks.

The stability of banks of a lake which should be utilized for water power development must be thoroughly studied. Slope movements have been recorded in many cases when a permanent water level has been lowered. The level of Davos Lake was lowered during the construction of the Davos-Kloster water-power project in Switzerland. Some minor landslides occurred on the banks that emerged and in 1923 when the drawdown attained 11.30 m, a large part of the alluvial cone of the Todtalbach stream (about 900,000 m^3) slid into the lake. The sudden fall of such a large mass caused a strong wave surge in the lake which broke an 80-cm thick ice layer into small floes and caused great damage to human life and property. The alluvial cone was composed of sand and pea gravel; at its toe fine silt and clay had been laid down, forming a delta-like tongue extending into the lake. These fine deposits had been gradually overlaid by coarser sand and gravel. The slide was due to the squeezing out of fine sediment from beneath the alluvial cone, the weight of which was greatly increased on emerging from the water.

Similar slide movements resulted from the drawdown of the Spullersee in the Tyrol. This lake provided a good opportunity for studying this problem, since its flanks are composed of various rocks (Fig. 13-96). The north-eastern and south-eastern shores of the lake, which are made up of fine-grained micaceous clay were seriously disturbed. Steep shores formed of stream alluvium in the eastern part were also affected by sliding. The banks in the north-western sector remained intact since they are in part rocky and in part composed of compact morainic deposits (Ampferer and Ascher 1925).

The drawdown of the level of a lake can provoke the disturbance of alluvial cones and deltas by streams discharging into the lake. The lowering of the water level results in the lowering of the lower base of erosion of streams, the erosive power of which will increase to level down the knickpoint in the length profile. Regressive erosion of streams produces deep canyon-like gorges in the alluvial cones. The slopes of these gorges can be very steep and resulting progressive lateral sliding can endanger the stability of the surrounding area.

In the water-power utilization scheme on the Aachensee in the Tyrol, a drawdown of 5—10 m was planned for winter seasons. After unfortunate experiences in other Alpine lakes, the problem was carefully studied. On the basis of the geological investigation, the following safety measures were carried out:

(a) Slopes liable to sliding were stabilized by several rows of wooden piles driven to solid bedrock;

(b) hazardous slopes were protected against wave action by wire nets filled with rip-rap, which were fixed with wooden piles;

(c) regressive erosion was prevented by draining all brooks and permanent springs into the lake by wooden troughs so that they would not flow down the steep slopes that emerged. Through these measures damage to banks was kept to a minimum. No failures of structures or transportation routes were recorded.

13.15. Siltation of reservoirs. The creation of a reservoir causes a change in the original length profile of the river and, consequently, a change in conditions

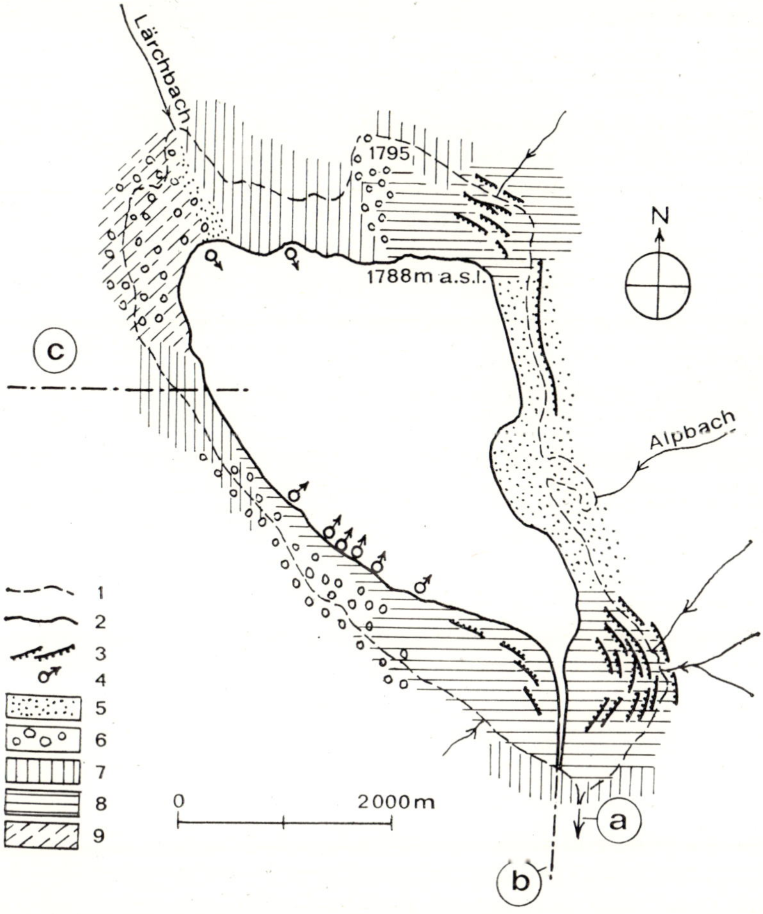

13-96. Disturbance of banks of the Spullersee reservoir after lowering the water level by 7 m (Ampferer, Ascher, 1925). 1 — Original water level, 2 — lowered level, 3 — cracks formed by slope movements, 4 — springs, 5 — sand and gravel, 6 — coarse boulders, 7 — bedrock, 8 — micaceous clay, 9 — moraine, a — original outflow, b — drainage gallery, c — tailrace tunnel.

controlling the erosion and accumulation processes in the part of the valley that is flooded. Geologically, the new reservoir forms a new lower base of erosion. In its upper part deltaic deposition will occur, the rate being governed by a number of factors. Their assessment is, for the most part, yet another of the purposes of the engineering geological investigation.

The siltation of a reservoir depends on its size and shape, on the amount and rate of inflow and on the amount of solid material supplied. This amount is variable, being controlled by the nature of the drainage basin and by climatic conditions. The main factors are the amount of precipitation and its distribution throughout the year, the gradient of the stream, the geology of the area, the inclination of slopes, their character (if they are bare or overgrown), as well as some specific factors for each drainage basin.

Material transported by streams generally consists of fine clay and silt particles carried in suspension and coarser sand and gravel. The latter are moved along the bottom but usually only during floods.

Suspended particles, representing the main part of material transported by high water flow, will not usually settle in minor reservoirs but are generally carried as far as the mouths of streams in lakes or seas. They will be, however, deposited in large reservoirs and lakes, where the velocity of the water is very small.

In Europe, the largest amount of suspended load is carried by the streams of the Toscano-Emiliani Apennines, the basins of which are built up mainly of soft, poorly consolidated rocks readily disintegrated by water. Visentiny (1937) recorded annual means calculated from 3—9 years' records as 5,280 m^3/km^2 for the river Enzo and 4,305 m^3/km^2 for the Panaro. The river Arda supplied to the Mignano reservoir 2,300 m^3/km^2 suspended material yearly in addition to 157 m^3/km^2 of bed load.

Fine material settles on the floor of lakes and large reservoirs forming a continuous cover or moving with density currents along the bottom. When the Sautet reservoir (France) was emptied in 1948, 13 years after filling, a 30-m thick mud layer was found to have accumulated to the right of the dam, probably through density currents (Gignoux, Barbier 1955). The muddy water, which is cooler and thus heavier, can flow a considerable distance along the lake bottom. Such density currents have also been observed in the Boulder reservoir on the Colorado River, U.S.A., and have been systematically studied.

In minor reservoirs the dominant part of the suspended load will usually be carried by flood water over the dam, since the water velocity will not decrease sufficiently for the suspended particles to settle. In large reservoirs, outlets should be installed near the base of the dam for intermittent sluicing out of sediments.

Siltation of reservoirs can be very rapid when there is a possibility of retaining flood waters in the reservoir since these may contain enormous amounts of muddy particles.

The amount of coarse sand and gravel moved along the bed generally at times of high water is difficult to measure directly. It is particularly difficult to establish mean values because one flood may bring more material into the basin than is brought by normal flow in the course of several years. A precise determination of the amount of coarse material would be possible only by measuring the volume of the deposit after damming the river for several years, in other words by building the projected dam itself.

A comparative method is therefore used for the estimation of the amount of bed load. The regimen of the stream to be dammed is compared with those of similar rivers which empty into lakes or on which dams have been erected.

The volume of deposited sediment is determined by a survey of cross sections of the reservoir. Surveying should be performed at intervals of at least 5—10 years, so as to record the effects of floods as well as of normal flows. The depth of the surface of the deposit below the water level can be measured by ultrasonic method. This method of measurement is suitable for coarse material accumulations forming deltas in the upper part of a lake or reservoir. The corresponding volume of fine muddy sediments forming a thin cover of large extent is difficult to measure.

Siltation of reservoirs in some regions is so extensive that the capacity of reservoirs can be greatly reduced. The question of reservoir siltation was one of the topics discussed at the International Congress for Large Dams in New Delhi, 1950. According to the reports then submitted, the supply of material into reservoirs varies greatly, depending on local conditions. Drouhin (1951) listed in his summary report the results of measurement for some drainage basins in m^3/km^2 per year. For small torrential basins (about 20 km^2) in southern California a maximum of 25,000 $m^3/1\ km^2$ during one flood was established. In Algeria, values ranging from 3,500 to 10,000 m^3/km^2 per year have been recorded. In France, the amount of downwashed material is only 30 m^3/km^2 for the Massif Central and 70 m^3/km^2 for the Pyrenees. In the Sautet reservoir in the French Alps, the measured values attained 800 m^3/km^2 per year during 1927—1930 and 166 m^3/km^2 per year during 1935—1945. The drainage basin of the Sautet river covering 987 km^2 is made up of the following rocks:

Granite, gneiss and crystalline schists	30 %
solid limestones	10 %
marls and clayey shales liable to weathering	25 %
sandstones	15 %
moraines and rock detritus	20 %

The mean annual precipitation is 1450 mm, mean annual discharge 33 m^3/sec, at low water 7—8 m^3 and 150—200 m^3/sec at normal high water.

Fig. 13—97 shows siltation of the trial Pont du Loup reservoir created on the river Drac and subsequently flooded by the Sautet reservoir. The masonry Pont du Loup Dam is 30 m high and the original capacity of the reservoir was 1,750,000 m^3. During 1927—1930 its siltation was measured systematically in preparation for the Sautet Dam project. Transverse profiles through the reservoir, 50 m apart, were surveyed regularly and the depth was measured at 20 m distances. In 1928, an extraordinary flood with 750 m^3/sec discharge supplied 536,000 m^3 material, i. e. about 550 m^3/km^2. An additional 464,000 m^3 material was supplied the same year by a flood with a discharge of about 700 m^3/sec. In 1929 in relatively dry weather the accretion was only 230,000 m^3. From Fig. 13-97 a great rise of the bottom after the 1928 floods is evident. The material deposited in 1930 at the end of the reservoir attained a height of 6 m above the normal level of impounded water and by 1932 the reservoir was virtually silted up. After several vain attempts to unplug the outlets, a tunnel was driven 23 m below the water level. For blasting of the last section of the tunnel a charge of 400 kg dynamite was used. Only five minutes after firing the water began to carry away the highly consolidated material and within four days the part of the reservoir immediately upstream of the dam was cleaned up.

The Sautet reservoir was filled in 1935 and after ten years 12 transverse profiles were surveyed and examined by boring. The boreholes showed that the deposits are on the average 14 m thick and their surface is roughly parallel with the original bottom. The lower part of the alluvial cone has a steep gradient (1 : 1). In cross-section, the cone surface is subhorizontal. The deposited material had a volume of 1,330,000 m^3, i. e. the yearly supply was 166 $m^3/1\ km^2$, much less than during 1927—1930, when two disastrous floods greatly increased the bed load.

The Gebiden reservoir on the river Massa, a tributary of the Rhone in Switzerland, was also threatened by early silting. An enormous amount of rock material was supplied by summer flood waters from the Aletsch glacier. Figure 13-98 shows the relationship between the volume of transported material and the discharge. In 1958 the total amout reached 500,000 m^3, out of which only 20 % was sand and gravel, the remainder being carried in suspension. The

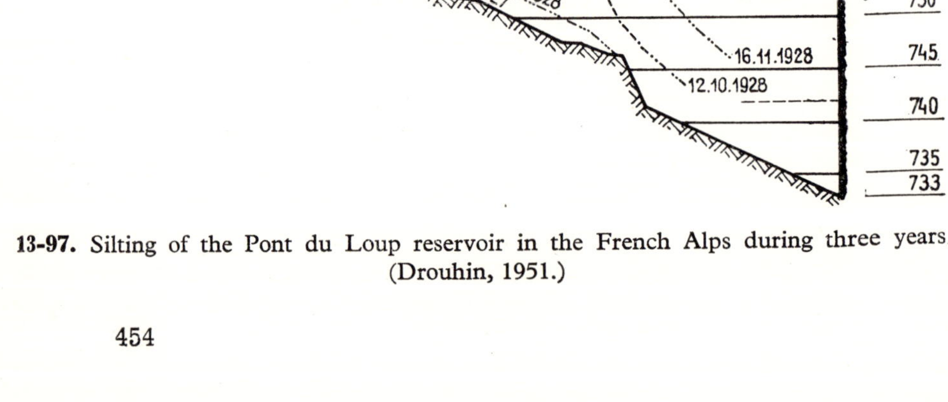

13-97. Silting of the Pont du Loup reservoir in the French Alps during three years. (Drouhin, 1951.)

drainage area covers 200 km^2 so that 2500 m^3 of material falls to 1 km^2 per year. The dam was provided by special outlets for sluicing accumulated sediment.

On La petite Rhue Dam (Dordogne river in the Massif Central area) measurements during 1927–1938 gave 60,000 m^3 of deposits from the drainage basin of 182 km^2 — ca. 30 m^3/1 km^2 per year. This far smaller amount can be accounted for by the dense vegetation cover and favourable geological conditions of the drainage basin[1]:

Andesite	30 %
basalt	35 %
gneiss	30 %
alluvium and glacial deposits	5 %

Through the history of dams cases are known when reservoirs were silted ud within a very short period. The Austin Dam on the Colorado River was erectee in 1913 and by 1926 95 % of the reservoir volume was filled with sediment. Th. Cismon reservoir in Italy was silted up to one half of its volume within ten years

In order to estimate the amount and the source of material carried by floods a synoptic geological map of the drainage basin should be made. Attention should be paid particularly to detritus fans, slope debris, glacial and glaciofluvial deposits which can supply the stream with the largest amount of material. On the basis of geological investigation, directions can be compiled as to how to reduce the amount of sediment, by afforestation of slopes, stabilization of potential sliding areas, or the damming of torrential streams.

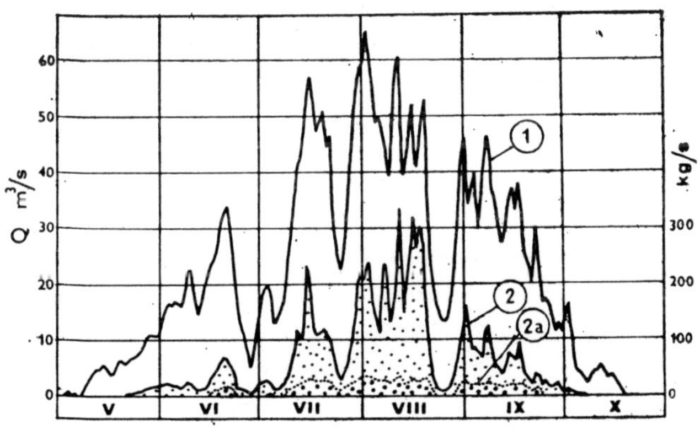

13-98. Relationship between the amount of material supplied into the Gebiden reservoir (Switzerland) and the discharge from May to October 1958. 1 — Discharge in m^3/sec, 2 — total amount of transported material in kg/sec, 2a — material of grain-size > 1 mm in kg/sec. (Ullmann, 1970.)

[1] Approximate values according to the geological map.

13.16. Economic effects of impounding water. Before the construction of a dam, it is advisable to establish whether important mineral deposits will be flooded by the reservoir. Engineering geological investigation should provide reliable data for the assessment of economic loss derived from the flooding of any such deposits in relation to the benefit derived from the construction of the dam. If only deposits of minor extent are involved (e. g. sand and gravel usable as building material, or peat), their exploitation is advisable. A partial depreciation of mineral deposits can also be caused by the rise of the ground-water table after filling the reservoir.

Flooding of peat deposits may raise another problem. When a dam is erected on the upper part of a stream, the water is generally very soft. Addition of humic acids gradually leached from inundated peat can make it harmful to concrete structures. The speed of leaching and the amount of humic acid will depend on the type of peat bog, on its size, the character of its cover, the degree of peat decomposition, and the amount of percolating water.

Before the reservoir is filled, the vegetation growth, trees and shrubs, the decay of which could increase the water acidity, should be removed. In the area to be flooded the largest possible amount of material containing calcium hydroxide or carbonate in soluble form (limestone, dolomites) should be left behind. Material from buildings that have been demolished should be strewn over a large area for the water to have access to mortar. Theoretically, it would be possible to enrich the reservoir area with quick lime etc., but this procedure on a large scale would be too expensive.

The impounded water may also inundate quarries for building stone which have been opened up in fresh rocks in the bottom parts of valley sides. In the middle reach of the Vltava valley (Bohemia) the Orlík Dam flooded many quarries of granite of high quality. These were located near the foot of the slopes because

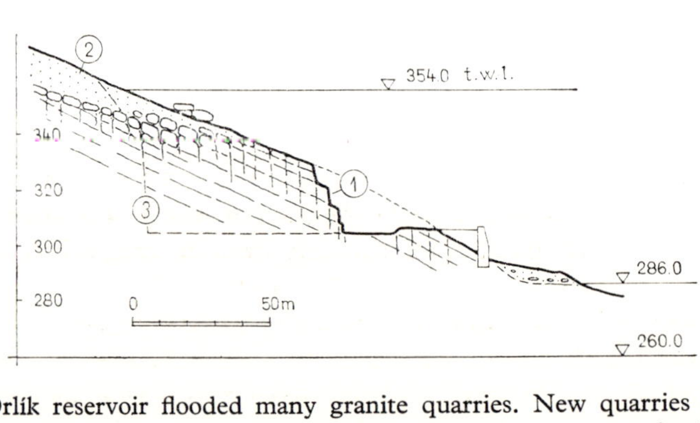

13-99. The Orlík reservoir flooded many granite quarries. New quarries could not be opened above the impounded water, since granite is weathered to a great depth and covered with residual soil. 1 — Firm granite, 2 — granite eluvium, 3 — limits of economic exploitation before flooding.

in their upper part the granite was weathered to sandy eluvium to a considerable depth (Fig. 13-99).

Serious problems may arise when the impounded water inundates the sources of ground water supplying communities and towns situated above the water level of the reservoir. Very often a new water-supply conduit must be installed when water cannot be obtained from wells sunk into dry rocky slopes.

The rise of the ground-water table along the reservoir can have unfavourable effects with respect to communities, industrial plants and the quality of agricultural land. Buildings can be depreciated by the flooding of cellars or foundations since the masonry will be permanently moist due to water rising by capillary force.

The extent of the loss of the agronomic value of lands depends on the form and composition of banks and the width of the belt of land affected. The influence of impounded water must be studied from this aspect, particularly in plains made up of permeable sand and gravel where the ground-water regimen may be changed over an extensive area.

13.17. Discharge of water from the reservoir. The necessary extent of geological investigation in relation to the discharge of water from the reservoir or for the passage of floods is governed by the specific layout of the project. In water power plants the water is discharged through a gallery or a tailrace and floodwater is passed either over a dam or through side-channel spillways.

The main objective of the engineering geological investigation is an assessment of the resistance of rocks to the erosive power of water below the overflow blocks or spillways. The effect of spilling water on the rocks immediately below the overflow must be examined thoroughly since many a dam has been threatened by undercutting owing to the inadequate resistance of foundation rocks. In the case of the Wilson Dam on the Tennessee River, the river bed at the downstream toe, made up of firm limestone, was deepened by 10 m through water flow and back-eddies so that costly remedial works had to be carried out. Failure of a dam near Austin on the Colorado River, founded on apparently firm rock (jointed limestone), is accounted for by the gradual undermining of the foundations on the downstream side.

Erosive power of overflowing water can be controlled by the shape of the spillway, by limiting the amount of water discharged and by the shape and size of the stilling pool. The design of water discharge from the reservoir should therefore be studied in co-operation with a hydraulic laboratory.

Intakes should be located with due regard to possible siltation of the reservoir so that their function cannot be endangered by rapid sedimentation of fine material at the upstream face of the dam. At damsites where slopes are not sufficiently stable, the intakes may be buried by slope movements caused by water-level fluctuation in the reservoir. This possibility should also be kept in mind in locating the intake installations.

Discharge of water from the reservoir through canals. To increase the head obtained by damming a valley, water may be conducted to more distant places where the elevation of the valley floor is lower than at the damsite. For this purpose either tunnels are driven (Chapter 12) or canals are constructed parallel to the stream. One or more power plants can be installed at suitable places to utilize the increased head. Above the power station the water is impounded high in the approach canal, whereas it will flow deep below the ground surface in the tailrace.

(1) *Intake canal.* Since the water level in the intake canal will be above the valley floor, it is always of advantage to locate the canal on the valley sides and not on its floor where high embankments would have to be constructed.

The task of the necessary engineering geological investigation is to gather reliable information for laying out the route of the headrace canal. This problem must be studied from both the geological and economical points of view. Excavation of a canal in solid bedrock is rather expensive but, on the other hand, soft rocks do not guarantee the stability of the sides of the canal. Gravel terraces and alluvial cones provide favourable conditions.

For a canal traversing moderate elevations along valley slopes and lateral valleys, a number of cuttings and fills may be necessary. Just as for transportation routes, each of these will be a separate structure presenting its own problems. Canal construction is more difficult since the slopes cannot be drained into the cutting. Ground water or water seeping through the canal lining must be drained into lateral valleys.

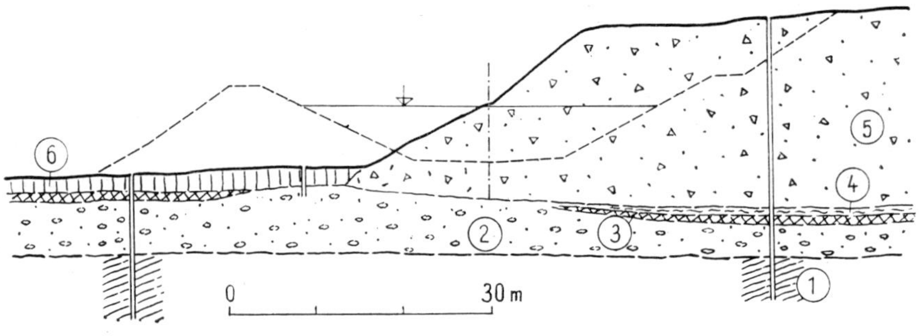

13-100. Favourable position of the intake canal near Lipovec (Slovakia) at the foot of an alluvial fan. 1 — Neogene marls, 2 — river gravel, 3 — fens, 4 — clay, 5 — granite debris of the alluvial fan, 6 — Holocene river deposits.

Figure 13—100 shows a suitable location for a diversion canal at the foot of a large alluvial cone truncated by lateral erosion by the river Váh near Lipovec (Slovakia). Geologically, there are no objections to locating the canal in a cutting since the granite debris furnishes solid foundations, stable slopes and good material for a valley dam. In contrast, near Turany (Fig. 13-101) the canal could not be located in a cutting on the valley slope because excavation would have disturbed the temporary equilibrium of fossil slides; neither could the clayey-sandy detritus obtained from the excavation be used for the valley dam.

In order to determine the permissible loss of water in the canal, the permeability of soils forming its floor and flanks and the height of the ground-water table in

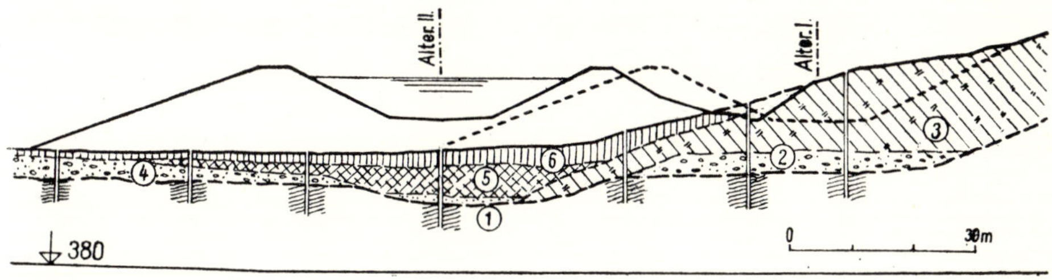

13-101. Influence of geological conditions on the location of a headrace canal. Canal in position I would disturb old landslides, for canal II compressible fen soils had to be removed. 1 — Marly Palaeogene shales, 2 — terrace sand and gravel, 3 — clayey-sandy material of an old slide, 4 — river gravel, 5 — fen, 6 — sandy loam.

the adjacent area should be established. In order to prevent the loss of water, the flanks and bottom of the intake canal will usually be sealed. The canal near Púchov on the river Váh shown in Fig. 13-102 (Smrček 1937) has been sealed by a concrete lining; this arrangement is suitable where the canal is constructed in sandy gravel which is stable at a fairly steep angle.

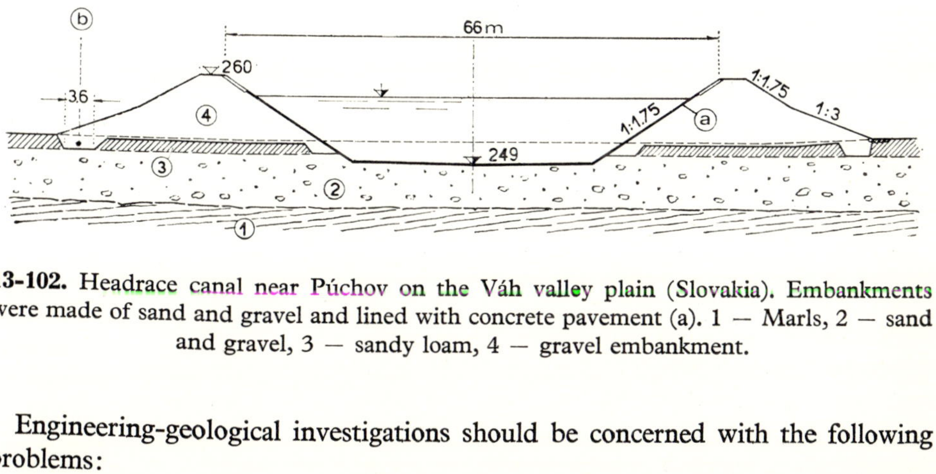

13-102. Headrace canal near Púchov on the Váh valley plain (Slovakia). Embankments were made of sand and gravel and lined with concrete pavement (a). 1 — Marls, 2 — sand and gravel, 3 — sandy loam, 4 — gravel embankment.

Engineering-geological investigations should be concerned with the following problems:

(a) Settlement of the canal after its filling; in alluvial cones and gravel-built valley plains the settlement is relatively small and because of high internal friction the differences in settlement are transmitted to a greater distance. Alluvial deposits with organic substances, muddy sands and fens are unfavourable. Similar strongly compressible beds are therefore removed before a fill is placed. Near Turany in Slovakia (Fig. 13-101, II) the intake canal was relocated to the floor of the valley, since on the slope it could have been endangered by sliding but in the flood plain an ancient stream channel was discovered. The highly compressible peat filling this channel had to be partly removed to prevent differential settlement of the bottom of the canal and disturbance of the concrete sealing.

(b) The permeability of soils below gravel dams; water percolating through fissures or contraction joints in sealing could undermine the downstream toe of the embankment. Figure 13-103 shows an arrangement to ensure that the seepage water will be led off the downstream face of the dam. The outcrop of impermeable loam on the upstream face must also be covered with gravel.

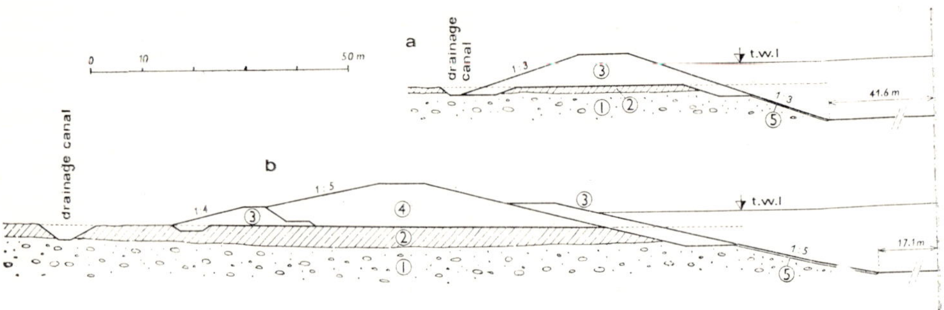

13-103. Sections of the Donzère-Mondragon headrace canal. 1 — Sand and gravel, 2 — alluvial loam, 3 — gravel embankment, 4 — embankment of mixed loam and gravel, 5 — bituminous blanket. (Bouvet, 1950.)

(c) The rise of the ground-water table in the surrounding area; if there is a possibility that uplift pressure may cause damage to concrete sealing (when the canal is empty) a way of reducing the uplift must be sought.

(d) The general geological structure of the area; the deposits of suitable material for earthworks can then be appraised.

The clay sealing of canals, if sufficiently thick, can withstand settlement of up to a matter of decimetres but assiduous observation is necessary when the canal is empty since the impermeable clay will prevent water discharge from the sides.

Unsealed intake canals can be used where the water loss is permissible and where the materials building up the canal bottom and sides are not too permeable. Silting by alluvial deposits is expected to contribute to the watertightness of the canal. Remarkable results have been achieved on the intake canal of the Donzère-Mondragon hydraulic structure (Bouvet 1950). Figure 13-103 shows two types of cross sections through this canal 17.3 km long. For economic reasons, several other designs were also used so as to adapt the canal to the kinds of material in which it was excavated and from which the embankments were made. The shape of the canal excavated in terrace gravel, where the cross-section was developed in relation to the loam cover of less favourable properties, is shown in Fig. 13-103a. The part of the canal constructed in a floodplain with loamy material is seen in Fig. 13-103b. It was expected in both these cases that the mixing of the fill material would somewhat incresae the watertightness. Tests showed that permeability would drop from the mean value $k = 2 \times 10^{-1}$ cm/sec to a quarter of this. Due to rapid siltation by stream deposits, the water loss decreased to 25 per cent in two years, i. e. to about 1 m³/1 km of the canal, which proved economically acceptable. Canal slopes are covered by a thin permeable bituminous cover which protects them from the effects of wave action and erosion. Water seeping from the intake canal is discharged by a special canal deepened to the gravel bed into the tailrace so that it does not depreciate the agricultural land.

(2) *The tailrace* is generally cut deep into the valley bottom, often well below the valley deposits. The water level in the tailrace, at least immediately downstream of the power plant, will sometimes be lower than the river level. It may therefore be necessary to examine the type and properties of the rocks in which this canal will be excavated and plan the protection of its banks from being washed out by percolating river water with a gravel layer or pavement.

An important hydrogeological problem is connected with the construction of tailraces. The ground-water table in the surrounding area may be so depressed by the excavation of the canal that the soil is not supplied by capillary water. Since the agricultural value of the land may be thus greatly impaired, the land may have to be irrigated. This was the case of the outlet canal of the Donzère-Mondragon water scheme, along which the agricultural land had to be irrigated by means of trenches issuing from the intake canal. Water from trenches infiltrated into the gravel either through special soaking pits or bored wells (Fig. 13-104).

13-104. Section of the tailrace canal of the Donzère-Mondragon scheme. 1 — Neogene marls, 2 — sand and gravel, 3 — loamy alluvium, a — dump of marly material, b — gravel dikes, c — gravel backfill, d — infiltration canal.

13.18. Engineering-geological investigation of sites for power plants. The power plant should be constructed on sound, and as far as possible, impermeable foundation beds and at a suitable site topographically, permitting the best utilization of the hydraulic head. From Fig. 13-105 the ingenious lay-out of the Donzère-Mondragon hydraulic construction will be apparent. The weir and intake as well as the power plant and navigation lock were located in solid rock, whereas the canal runs in river alluvium and marls. The location of the ridge of Cenomanian sandstones was pinpointed by an electric resistivity method.

Morphological conditions of the valley often compel the designer to locate a power plant at the foot of a slope so that the intake canal will terminate in the slope and the tailrace will be excavated in the valley floor. A power plant in such a position, however, may be threatened by sliding or its foundation may be difficult owing to faults running in the valley bottom.

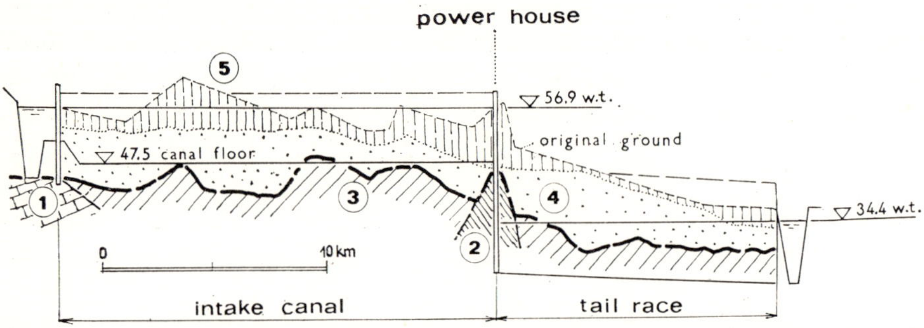

13-105. Longitudinal geological section of the Donzère-Mondragon scheme. The site of the power plant was chosen with respect to the configuration of the bedrock. 1 — Lower Cretaceous limestone, 2 — Upper Cretaceous sandstone, 3 — Pliocene marls, 4 — sand and gravel, 5 — loamy alluvium. (Bouvet, 1950.)

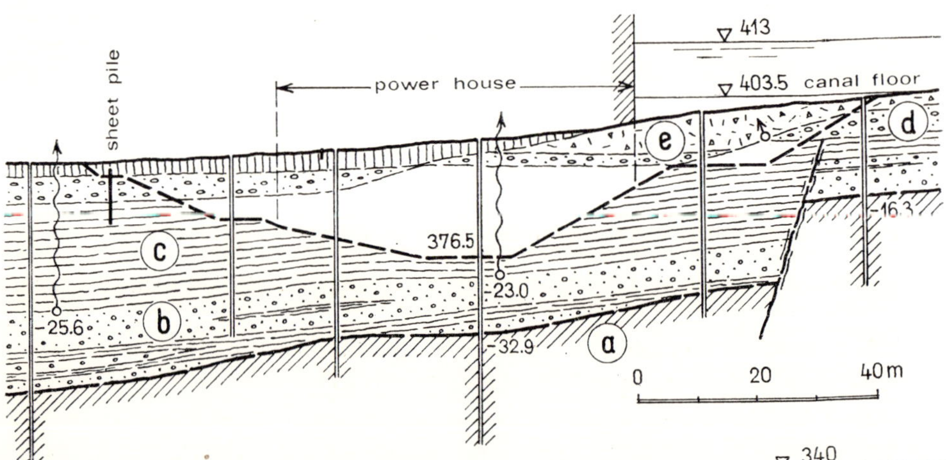

13-106. Geological section of the site of the Sučany power plant on the river Váh (Slovakia) (Záruba, Mencl, 1956). a — Palaeogene clayey shales, b — Neogene conglomerates, c — Neogene sandy marls, d — sand and gravel, e — granite debris of an alluvial cone.

The power plant near Sučany on the river Váh, for example, was located at a topographically suitable site, where the intake canal leaves an alluvial cone (Fig. 13-106). The foundation of the structure in the valley plain was rather difficult since the plain is formed of Neogene marly silts and permeable conglomerates with confined ground water. The conglomerates appeared to be fed by water rising along the faults which disturb the Neogene sediments at the foot of the slope. Since the conglomerate beds occur immediately below the bottom of the foundation excavation, the confined water had to be pumped continuously so as to prevent the upward rupture of the bottom of the excavation. It was necessary to drain the water discharging from the granite detritus and gravel covering the Neogene sediments in the slopes of the foundation excavation.

The founding of a power plant in a wide valley plain is often difficult when the bedrock is deep below the ground surface. The construction of the Fessenheim power plant on a lateral canal of the river Rhine (Lefoulon, Ischy 1955) provides an instructive example. Below Basel the Rhine valley is cut into solid limestones, which submerge below the valley deposits northwards of the Kembs power station (Fig. 13-107). In the foundation excavation for the Ottmarsheim power

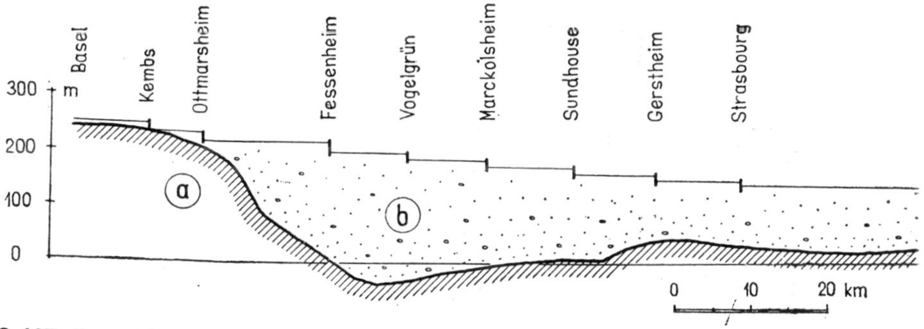

13-107. Synoptic longitudinal section of the Rhine valley between Basel and Strasbourg. a — Bedrock, b — Quaternary sediments.

plant the bedrock was found at 30 m below the surface and sheet-piling had to be sunk to this depth. The designers presumed that the same foundation method would be usable for the Fessenheim power plant. Geological investigations, however, have shown that the bedrock is there at a depth exceeding 100 m since the Rhine valley is a graben in which downfaulting persisted until the Pleistocene. Owing to this great thickness of Rhine deposits, the 27 m deep excavation for the power plant had to be sealed both in the sides and bottom (Fig. 13-108). After a detailed grain-size analysis and permeability tests, young alluvial deposits were grouted with the use of method as described in Sec. 13.11.

Geological conditions have also influenced the location of the Isola Serafini hydraulic structure on the river Po (Italy). The wide valley of the river Po is filled with permeable gravel, the thickness of which is so great that it cannot be established by boring. The power station utilizes a head of 6.5 m obtained by cutting off a large meander of the river. The power plant was built in the cut-through meander neck, in a location where an impermeable clay layer occurred

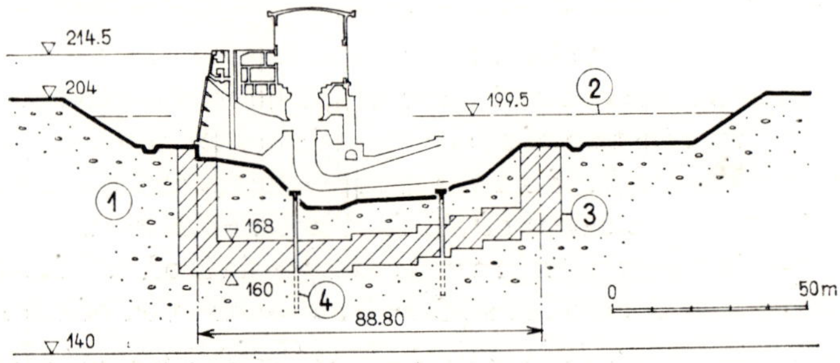

13-108. Section of the site of the Fessenheim power plant on the Rhine. 1 — Sand and gravel, 2 — original ground-water table, 3 — foundation pit sealed by grouting, 4 — relief wells. (Lefoulon and Ischy, 1955.)

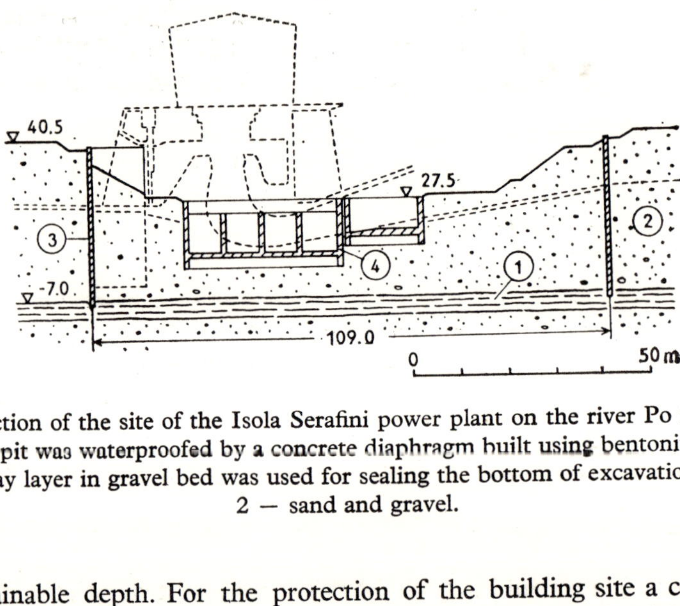

13-109. Section of the site of the Isola Serafini power plant on the river Po in Italy. The foundation pit was waterproofed by a concrete diaphragm built using bentonite suspension (3, 4). A clay layer in gravel bed was used for sealing the bottom of excavation. 1 — Clay, 2 — sand and gravel.

at an attainable depth. For the protection of the building site a concrete diaphragm was constructed with the use of bentonite suspension (ICOS method — Fig. 13-109).

The location of *pumped storage reservoirs* demands specific morphological and geological conditions which allow the construction of an adjacent upper reservoir of the necessary capacity and at a sufficient altitude. Owing to morphological conditions, the external embankment must often be erected on flanks covered with slope sediments. Their slow creep may be hazardous to the stability of the embankments and so they are usually removed. The walls and bottom of the reservoir are covered with lining easily maintained (e. g. asphalt) to protect the embankment from impairment by water seepage. Any seepage water is dis-

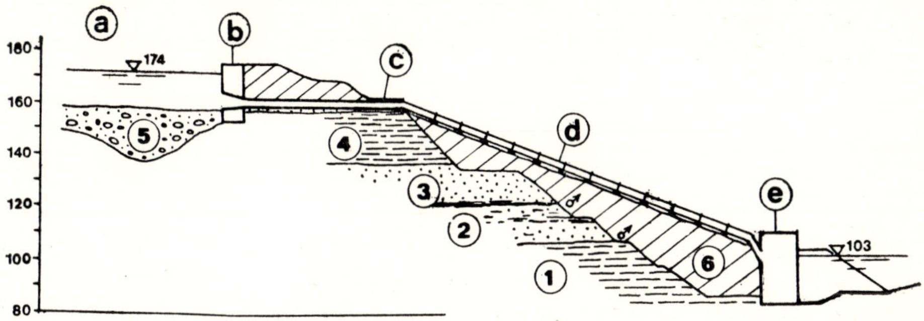

13-110. Schematic section of a storage power plant on the Dniepr near Kiev (Shmulson, 1972, exaggerated 2×). 1 — Palaeogene marlstone and siltstone, 2 — Palaeogene sands, 3 — Neogene sands with a lignite bed at the base, 4 — Neogene clays, 5 — fluvioglacial and morainic sediments, 6 — sandy-gravel fill; a — open headrace canal, 700 m long, b — intake, c — headrace gallery of reinforced concrete, d — pipe line, e — power house 220 MW.

charged through drains placed below the reservoir and embankment. If the reservoir is located above a slope formed of rocks liable to sliding, the slope must first be stabilized. The final construction of the Kiev pumped storage station (Shmulson, 1972) is shown on Fig. 13-110. The power plant is located on a 70-m high bank of the river Dniepr. The valley slope is formed of Tertiary sediments and disturbed by deep-reaching slides. The pumped storage reservoir was therefore shifted farther from the slope edge and connected by a 700-m-long canal with the intake conduit to the power plant. Drains were constructed below the reservoir and the intake canal. To avoid slope failure during construction, ground water was discharged through two drainage galleries and a strong buttress of river sand was constructed at the toe of the slope.

Chapter 14

TASKS OF GEOLOGICAL INVESTIGATION IN REGIONAL PLANNING AND ENVIRONMENTAL POLICY

The objective of regional planning is to ensure the optimum use of the area and its natural resources with regard to the natural conditions of the landscape and to the conservation of the natural environment. Regional planning is thus not only limited to urban development, building of industrial plants, transportation facilities, but it also involves the right balance of allocation of land for agricultural and forestry purposes, for the extraction of mineral resources, recreation and Natural Reserves. This complex planning can be successful only with the help of a proper understanding of the natural, and particularly the geological, conditions and of the impact of man's work on the environment.

14.1. Man as a geological agent. In long-settled countries the land surface is no longer in the original state but has been strongly altered by human activity. Deforestation of mountain slopes, for example, led in many areas to more intensive erosion and eventually to complete devastation. As a result of systematic ploughing, conditions allowing increased runoff and sheet erosion over large areas have been created (Fig. 14-1). The interference with hydrological conditions of the area may have very unpropitious effects. Hasty regulation of streams and abolition of ponds accelerate the discharge of surface waters, and the systematic capture of springs and pumping of ground water permanently lower the water table. The resulting desiccation of the landscape unfavourably affects its climate. In recent years both air and water courses are polluted by ever-increasing amounts of industrial waste and effluents.

Human activity increases progressively with technical advances. Let us realize, for example, the enormous extent of earthworks, which are carried out for transportation routes and hydraulic structures, or the volumes of material displaced in extraction of sand, gravel, coal or ores.

The work of man and his machines in some places drastically changes the character of the landscape, i. e. the vegetation cover, form of the land surface, drainage conditions of the surface and ground waters, as well as the climatic conditions. Most of these interferences have disturbed the geological, hydrogeological and

14-1. Deep erosion gullies in basalt tuffs can develop after deforestation of slopes also under temperate climatic conditions (NW Bohemia). (Photograph by Záruba.)

biological relationships, since man did not comprehend the true impact of his actions and the complex interrelations of all natural processes. With the steadily increasing human populations, all further damage must be forestalled and what has been damaged must, as far as possible, be set right. The necessary measures involve the reclamation of spoiled agricultural land, recultivation of derelict areas and nature conservation.

Every industrial country needs a large quantity of various minerals, which were and are not always worked reasonably. It is an important task of geological investigation to ensure that resources are exploited economically, are used to the best advantage and are not depreciated by being built upon. Although many instructions and regulations concerning the protection and reasonable development of natural resources have been published, they are not always observed in selecting building sites. And yet, in many cases, it is quite easy to choose an alternative site, enabling the intact deposit to be conserved for future use.

The casual and unpropitious management of the natural environment should be replaced by an integrated land-use plan which would control man's work in accordance with natural conditions. Modern economy demands a systematic and adequate urban development and usage of natural resources based on regional planning which takes into consideration preservation or improvement of the natural environment.

To accomplish this task the planner and engineer need much help and it is the geologist who is able to lend them assistance. Thanks to his training the engineering geologist can assess and fully comprehend the interrelations which affect the incorporation of the designed structure into the landscape. The collaboration of engineers and geologists should aim at making man a creative and not a destructive geological agent.

14.2. Regional planning. In recent years, in almost all highly developed countries, official instructions and codes on regional planning and conservancy of natural environment have been published.

For this purpose land-use plans are being prepared, involving the following types:

(a) Regional plans of major economic units (zones or raions),

(b) plans of settlements and their wide environment,

(c) detailed plans of cities and towns.

The regional plans (a) are prepared for districts of economic importance; they present basic instructions for the urban and industrial development of the area so as to take advantage of all possibilities provided by nature. Regional plans s. s. are prepared for a long interval, 10—15 years, and they allocate priorities for certain planned structures.

Plans of settlements (b) are compiled for the individual settlements and the adjacent areas. The objective of these plans is to delimit the area to be occupied by the settlement and to solve the relationships between the settlement and its surroundings. In these plans, the areas to be reclaimed and sites and buildings to be protected as natural or historic monuments, should be plotted. At the same time, the water supply, potential drainage, recreational and social amenities, utility services, etc. should be decided upon essentially in this stage.

Detailed plans of cities and towns (c) are prepared on the basis of principles approved in regional plans of higher order. In these plans the type, character and extent of building up are defined. Accurate situation and altitude position of all buildings, streets, communications, parks and services are also given in these plans.

The construction of regional plans requires a sound knowledge of the area to be developed. Any discussion on the optimum usage and the subdivision of the area is useless unless it has been thoroughly investigated first. The complicated geology of the ground, diversified surface relief, differences in soil, hydrogeological and climatic conditions, all cause the heterogeneity of individual regions and pose great demands on the compilation of the regional plan. All regional plans should be based on scientific analyses of the landscape which provide information concerning the inherent physical properties of the area, its present utilization and the possibilities for its development. The basic analyses necessary for the assessment of the environment are as follows:

(a) *Geographical analysis.* The geographical picture of the area and the land forms are shown in detailed topographical maps. Where these are unavailable, they should be compiled or aerial photographs should be taken. Geographical analysis thus comprises the study of maps and the inspection of the terrain. The geomorphological conditions controlling the energy of the relief are a relevant criterion for the selection of sites for residential and industrial structures and communications, and should also be considered in questions related to recreation facilities, agriculture, forestry and biology of the landscape.

(b) *Geological analysis.* The principal basis of the regional plan is a good engineering geological map constructed from the results of detailed investigation. Where the engineering geological maps are as yet unavailable, the analysis must be based on geological, pedological and hydrological maps, maps of slope movements, archive records and field investigations. Another important part of the geological analysis is the inventory and map of mineral deposits. In order to assess their economic importance it is necessary to know their present utilization, and the state of the unexploited mineral resources which are decisive for the potential development of the area (see Sec. 14.3).

(c) *Water-management analysis.* In addition to hydrological and hydrogeological data, the analysis should provide information about the permeability of the surficial beds, the size of the forest-covered area and other data which are important for the accumulation of rain and ground water (see Sec. 14.4).

(d) *Agropedological analysis.* The regional plan for rural districts takes into consideration also agropedological conditions, i. e. the appraisal of soils from the point of view of their productivity. The results of investigations are expressed in pedological maps constructed with regard to the genesis of soils, their types and depth, and to the present state of physical, chemical and biological properties of soils. The agropedological analysis should be based on the methods and working procedures of pedology.

(e) *Climatological analysis* studies the factors which control local climatic conditions. The objective of climatological studies is to define suitable and less suitable areas for settlement, recreation amenities, location of industrial plants and large civil-engineering structures.

In addition to these basic analyses of natural factors, socio-economic aspects should be considered for the compilation of regional plans (analysis of population, settlement, industry), as well as biological (vegetation, forest growth) aspects, transportation facilities, and others.

In the text below we shall discuss the individual analyses as far as they concern geological and hydrogeological conditions of the area.

14.3. Geological analysis is based on an integrated engineering geological investigation of the area. The investigation results are summarized in an engineering geological map which generally consists of several sheets and presents all data necessary for the elaboration of the regional plan (see Chapter 2).

From the basic sheets of engineering geological maps „special-purpose" maps are constructed, in which areas having similar properties with respect to a certain structural type are grouped into zones (raions). For new settlements in geologically complicated areas maps of foundation conditions are commonly prepared. In these maps there are demarcated areas of suitable and unsuitable foundation grounds and areas that should not be built upon (e. g. mineral deposits, Natural Reserves).

The geological analysis also comprises an inventory of deposits of building materials, quarries for building stone and aggregates, sand pits and loam pits.

Mineral deposits are of primary importance for the regional plan as their situation is decisive for the location of the pertinent industry which furnishes working possibility for the inhabitants and thus for the distribution of settlements, lay-out of transport routes, etc. From the above-said it follows that the presence of economic mineral deposits predetermines the future development of the area.

The areal extent of the mineral deposit must be demarcated in the regional plan so that it is not depreciated by being built up. When a settlement is located within the area of an unopened deposit, its future exploitation is more difficult and less economic. Since, for example, underground mining often results in subsidence of the ground surface, it is necessary to leave some coal pillars unextracted; this implies serious economic losses of material and labour.

Mineral resources should not be depreciated in any other way, as by flooding with impounded water of reservoirs. They should be managed reasonably and circumspectly because they are non-renewable and the demand for them rises with the increase in population and industrial development.

In the past many deposits were worked unsystematically and predatorily. The owners often worked only the most accessible part of the deposit in order to achieve maximum profit at a minimum cost. In this way many occurrences of valuable minerals were exhausted within a short time, although a responsible exploitation could have conserved sufficient material for future generations. Let us recall, for example, the non-economical method of pillar-and-stall mining, which was applied until recently in many countries. As is well known, only 40 per cent (or less) of the coal was exploited, whereas 60 per cent remained unworked and was virtually fully depreciated (Fig. 14-2). Many oil fields have also been exhausted prematurely by unsystematic and plundering exploitation.

An uneconomic procedure is often used in the exploitation of peat, the importance of which is generally underrated. Peat is today a useful raw material and, in addition, the peat bogs area a significant factor in the regulation of the surface drainage and microclimate of the landscape. It is therefore wrong to drain the bog by unskilled working and to use only a surface layer a few metres thick from a large

14-2. Ancient room-and-pillar working of brown coal exposed by an opencast. Old exploited spaces are filled with overlying clay. (Photograph by Záruba.)

deposit, as afterwards the bog partially caves in, the peat becomes loosened and muddy and is practically lost for further usage. The land becomes inaccessible and many a time even dangerous to man and animals (Fig. 14-3).

Another cause of economic loss occurs when peat deposits are left unexploited in newly built reservoirs. Moreover, this is also undesirable as the humic acids leaching from the peat will deteriorate the reservoir water. It must therefore be resolved whether the economic value produced by a new dam will compensate the loss caused by flooding or depreciation of mineral resources.

The rash and ignorant opening of quarries for building materials at a time of high demand likewise had unfavourable economic and environmental consequences. The overburden material was often stored on the reserves and the wide surroundings was severely devastated.

The planned exploitation of mineral deposits should be directed to their most advantageous usage. Economically speaking, it is quite wrong to break up high-grade limestones for building material and road aggregate even if they are suitable for this purpose. These limestones are a highly appreciated chemical material and to use them in, for example, the manufacture of cement (for which limestones with a far lower $CaCO_3$ amount would be satisfactory) is very imprudent. Similarly,

14-3. Non-economical working of peat in southern Bohemia. (Photograph by Záruba.)

marbles, especially when their deposits are relatively scarce, should not be used for the manufacture of cement, fillers or aggregates, as other limestones that are unsuitable as marble facing stone can serve this purpose. These few examples show the importance of prudential management of mineral resources.

Protection of agriculturally productive soils. In selecting a building site serious consideration should be given to the protection of good agricultural land, and, other conditions being equal, the planner should avoid building on it but should seek to conserve as much as possible for food production. In some countries severe restrictions are laid down for the protection of agricultural land, because thousands of hectares are lost every year due to urban and industrial development on the one hand and in result of extraction of minerals on the other.

This problem is of great relevance mainly to densely inhabited countries where a sufficient food supply should be considered primarily. Therefore, it is advisable to provide necessary areas for new buildings by the demolition of old unsuitable structures and re-development of old town districts rather than to occupy further areas at the expense of agricultural and forest land.

14.4. Hydrogeological analysis generally relates to a larger area than the other analyses because the whole drainage basin of the stream should be studied. Its completion requires a longer time, since reliable results demand several years' measurements of the distribution and amount of precipitation, evaporation and infiltration, discharge of springs, oscillation of the ground-water table, etc. From the gathered data hydrogeological maps are constructed. In these maps are plotted streams, lakes, ponds, artificial reservoirs, wet grounds, wells and springs. The ground-water table and chemical composition of subsurface water are expressed by symbols. If sufficient records are available, the depth of subsurface water below the ground surface is shown by isolines. Protection zones of major water sources and mineral springs are also plotted.

On the basis of long-term observations and hydrogeological maps it is possible to determine the amount of water available for the area to be developed, and what proportions may be used for the production of electric power, for water supply, navigation and irrigation.

The availability of water for towns and industrial plants is an important point of the hydrogeological analysis. Town residents who have never experienced lack of water cannot realize how important it is to ensure a sufficient supply; many poeple believe water sources to be virtually inexhaustible and very few are aware of the highest priority of water management. The adequate supply of water is a serious world problem, as the increase in population gives rise to greater demand and the improvement of hygienic conditions also increases the use of water per person[1]). The amazing development of industry and large-scale agricultural production contribute to the steadily increasing demands on water supply.

Accompanying the demands on water supply, the amount of effluents and liquid waste from settlements and industrial and agricultural establishments rises. Since filtration plants are not established everywhere, polluted waters often flow out into surface streams or soak into the ground and percolate through the permeable rocks by which they reach the aquifers. The result of this is the deterioration of the quality of surface and subsurface waters.

Systematic drainage of the land surface in the settlement areas and regulation of rivers accelerate the runoff of surface water and permanently lower the ground-water level. The capture of springs and the pumping of ground water appreciably reduces water resources. These interferences, which are detrimental to the amount and quality of water, are in direct contradiction to the increased demand. In order to ensure a satisfactory amount of potable and non-potable water for new settlements and industry, a regional plan should involve the following points:

(a) protection of purity and quality of all water sources,

(b) preservation of the quantity of water sources.

The sources of domestic and industrial water are the ground water and surface water of rivers and brooks. In many regions only ground-water sources were

---

1 After World War II the average demand was estimated at 120 l per person a day, while at present a requirement of 400 l or more per person a day in large towns and cities should be reckoned with.

used for water supply until recently. To protect the areas from which the water is or will be taken is of primary importance. They should be delimited and exempted for water supply in good time, and as far as possible, right in the regional plan.

In planning new settlements or industrial structures the potential hazards for water supply merit primary consideration. The water resources can be reduced or impaired by building up the catchment area (Fig. 14-4) or by surficial contamination of the area feeding the significant aquifers. It is therefore wrong to build establishments the operation of which pollutes the environment, such as an oil refinery or large cattle farms, on permeable and jointed bedrock of the catchment area of an important water source.

The regimen of surface and ground waters can be disturbed greatly by mining works, as the waters can escape through old galleries and pits to the deepest level of drained mine workings. This occurred, for example, in Kutná Hora, the old mining town of Bohemia. The relief of the ground surface indicates that most of the area was composed of impermeable rocks and that in the valleys between

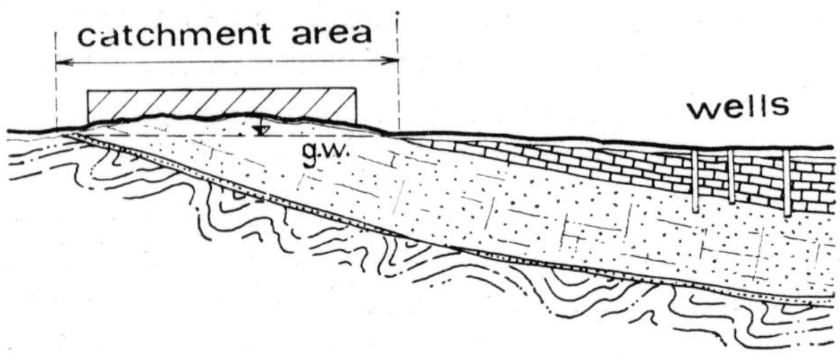

14-4. The yield of a ground-water horizon is reduced by building up of the catchment area.

the elevations there flowed small brooks, mainly in rainy periods. There was a rich aquifer at the base of Cretaceous sediments on the impermeable gneiss bedrock. Both the surface and ground waters were drained by numerous pits, galleries and open joints into old mine workings. At present, when the silver mines are abandoned and flooded, the ground water table is roughly at the altitude of the water level in the Vrchlice river. In the large part of the undermined area the hydrogeological conditions are similar to those of a karst region: most rainwater soaks rapidly into the ground and percolates through cavities and open joints from which it flows out on the bottom of deep valleys.

The hydrogeological analysis also shows which areas have an excess of ground water which can be supplied to neighbouring towns and districts having limited resources. Otherwise, surface water must be used for their water supply. With regard to the increasing demand, surface water, mainly from reservoirs, has begun to be used extensively.

In the regional plan provision should therefore be taken for the protection of the drainage basins of the streams that can be used for water supply. Along the rivers, and mainly on the banks of the planned reservoirs, urban and industrial structures should be prohibited when necessary. This relates in particular to establishments the effluents of which would contaminate surface waters.

The problem of waste disposal and pollution of water courses and its solution must also be considered in hydrological analysis. Very many of the European rivers are strongly polluted by noxious effluents, often because the industrial interest dominated over public welfare. The improvement of their drainage basins is among the foremost tasks of intelligent planning.

Hydraulic structures seriously interfere with geological and biological conditions of the landscape. Experience gathered in recent years has shown that many of these works, such as large-scale regulations of water courses, drying out of ponds and marshes and extensive drainage of land for agricultural and mining purposes unfavourably affected the regimen of subsurface water in that it speeded up the runoff of surface water and permanently lowered the ground-water table. This resulted in excessive desiccation of lands, an appreciable decrease of their fertility and of the ground-water reserves. As is seen, all hydraulic works must be carefully considered so that the water regime, which is prerequisite for the biological balance of the landscape, would not be disturbed.

The design of large dams and reservoirs should correspondingly be considered from a wider viewpoint. It would be essentially wrong to keep in view only the purpose of the work as, for example, production of electric power, and the least cost. The construction of a dam is connected with a number of problems which should be duly considered in the early stage of planning. These involve adequate protection of banks of the future reservoir, lay-out of communications, relocation of communities, recreation requirements and others. The slopes of the reservoirs often consist of loose surficial deposits, such as slope loams, sand, loess, which can be disturbed and undermined as a result of water-level oscillation and wave action. The protection of banks must be designed so as to correspond to the character of the landscape. The oscillation of the water level is necessary with regard to power-plant operation, but the lowering of water lays bare many sectors which are unattractive and occasionally even hygienically harmful. In inhabited areas the protection of banks in these sectors should be ensured, otherwise submerged dams which maintain the water level in the end parts of the reservoir at an adequate altitude should be constructed.

14.5. Geological investigation in planning and building of settlements. In planning a new settlement we are faced with a series of problems intimately connected with the natural environment. The planner must be aware that a settlement is not a superior unit to which the landscape should be subordinated, but that the construction of every object should be conceived as a constituent part of the natural environment and designed according to the regional or settlement plan.

He should aim at achieving harmony between the requirements of the planned work and the landscape.

The settlement plan should be elaborated with regard to the relief of the land surface, to the geological and foundation conditions, the agricultural value of soils and to the demands of environment protection. The location of a new settlement or industrial establishment should be determined by the regional plan, taking into consideration the proximity of mineral deposits necessary for the planned industry, the supply of domestic and industrial water and the drainage requirements. The exploration of mineral deposits and the delimitation of their areal extent should be effected so that the settlement is not situated on unexploited valuable mineral resources.

In planning urban development, the following areas should be excluded as unsuitable sites from the geological point of view:

(a) Recurrently flooded valley plains, marshes and peat bogs,
(b) areas of potential landslides and slope movements,
(c) plots on steep slopes and those with surface outcrops of hard rocks,
(d) undermined areas, areas adjacent to disused quarries and regions with advanced karst phenomena,
(e) zones along young faults in seismic areas.

(a) Valley plains. A flat widened valley bottom is apparently suitable for urban development, but it is generally filled with young Quaternary deposits and is recurrently inundated during flood periods. The shape and width of the valley plain is controlled by the morphological history of the area, by the gradient of the river and its annual discharge, the amount of transported material and other subordinate factors. The valley plain is an inherent part of the stream channel, through which the flood water is carried away. This is a serious argument against its being built up. Its esthetic value and agricultural productivity on the one hand and expensive foundation constructions and microclimatic unsuitability, on the other, are no less important reasons.

Foundations on valley plains are difficult and expensive; they require pumping of water during construction, insulation against ground water, protection from floods and hydrostatic uplift. Holocene deposits of valley plains are usually humic flood loams, in some places up to several metres thick. They are mostly water-bearing and muddy, mainly where they fill ancient abandoned meanders and stream channels. Their bearing capacity is small and non-uniform, so that buildings erected on them are often disturbed by uneven subsidence. Drainage of building sites on a flood plain is difficult due to the small gradient. All subsurface lines such as sewers, water lines and cables also suffer from being permanently below the ground-water table (Fig. 14-5).

Urban development of flood plains is likewise undesirable for agricultural reasons, since they are generally very fertile, having sufficient moisture even in dry spells, on account of a small depth of subsurface water below the land surface.

A town district or industrial establishment built on a flood plain does not only abstract the built-up area from agricultural use, but causes further indirect damage. The structures built upon the valley bottom reduce the area of passage for flood waters which then inundate the valley above the restricted sector. This leads necessarily to the regulation of the stream, which generally implies the deepening of the channel. In the regulated part of the river the water level will thus be permanently lowered, resulting in the sinking of the ground-water table and partial desiccation of the flood plain. A lovely river valley with overgrown banks and well-balanced natural conditions will then change progressively into a steppe with deep sunken water in a paved canal.

Microclimatic conditions also indicate the unfavourable use of flood plains as building sites. In these "inversion positions" frequent mists increase humidity and decrease temperature. Buildings are additionally damaged by water rising by capillarity and by frost action in winter, which reduces their durability and increases maintenance costs. Moreover, living in humid houses is undesirable and unwholesome.

In Central Europe building on a large scale in flood plains began since 1850s concurrently with the rash and uncontrolled growth of cities. Old mediaeval town districts are generally built on higher river terraces free of floods. The old town builders reasonably avoided sites threatened by inundation and providing difficult foundation ground.

Surprisingly, even at the present time there are planners who do not respect such serious reasons and devise to use flood plains for urban development. The apparent saving on excavations in a flat terrain is inevitably absorbed by the higher cost of foundation constructions and the protection against ground water.

The Vltava river valley in Prague provides an illustrative example of unreasonable management of a flood plain and river banks. Towards the end of the last century high embankments were constructed to protect the banks from flooding, and behind them rows of buildings were erected on man-made ground several metres thick, lying on water-bearing Holocene deposits (Fig. 14-6). The houses stand so close to the river that there is no free space for necessary traffic improvement, or for at least a narrow "green belt". The original character of the valley

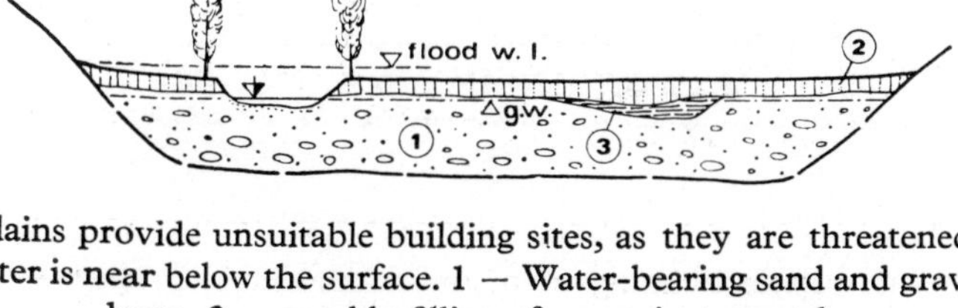

14-5. Flood plains provide unsuitable building sites, as they are threatened by floods and the ground water is near below the surface. 1 — Water-bearing sand and gravel, 2 — alluvial loam, 3 — muddy filling of an ancient meander.

was fully effaced, many historical monuments were destroyed and the excessively high embankment changed the river into an unseemly canal in some sections.

(b) Landslide and potential landslide areas. Slopes used for urban development should be stable, free of slope movements of whatever kind. It is therefore one of the main tasks of the engineering geological investigation to demarcate areas that were disturbed by slope movements in the past and those which are prone to sliding, especially when affected by construction works.

In Czechoslovakia a large landslide near Handlová in 1961 instigated a systematic investigation and inventory of all hazardous slopes for sliding. During 1962—1963 more than 9100 landslides occupying an area of about 60,000 hectares were recorded and plotted on maps at a scale of 1 : 25,000. From these data it is evident that in some regions slope movements indicate an imminent danger to settlements, highways, railways, agricultural land and forests.

In many cases movement of apparently stable slopes can be initiated by construction works or excavations which disturb equilibrium conditions. It is therefore necessary to be acquainted with hazardous slopes before the elaboration of urban plans so that they can be avoided. Landslides and slopes movements are discussed in detail in Chapter 7, to which we refer.

(c) Building on slopes. In a hilly environment some precautions should be taken in laying-out buildings and streets even when the geological conditions are favourable. The plans are often conceived only with respect to the smallest possible amount of earthwork and to the easiest ascent up the valley sides. The omission of geological relations causes economic losses during construction and maintenance of structures.

The importance of the geological structure of slopes for city planning can be demonstrated by reference to many examples in Prague. Because of the diversified relief, many parts of the city were built on fairly steep slopes made up of Ordovician shales. Their stability is controlled by the strike and dip of beds relative to the inclination of the slope and is readily disturbed by excavations parallel to the

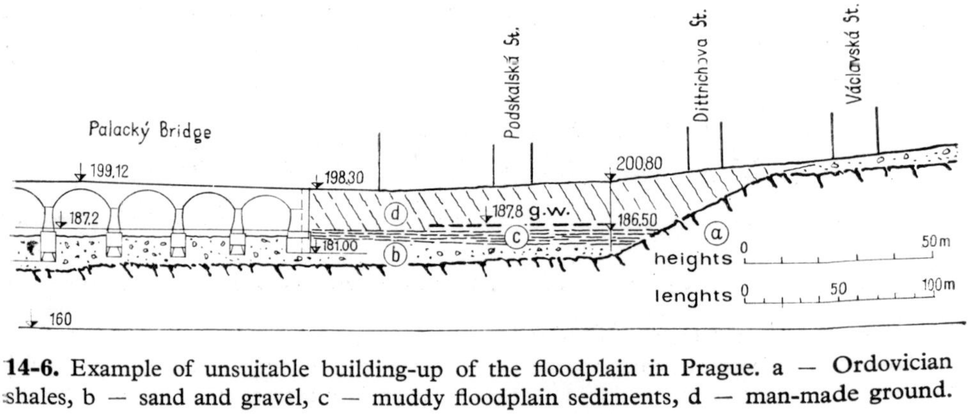

14-6. Example of unsuitable building-up of the floodplain in Prague. a — Ordovician shales, b — sand and gravel, c — muddy floodplain sediments, d — man-made ground.

strike of beds dipping downslope. At several places the cuttings of roads drawn in this direction, and even more the deep excavations for sewers, resulted in serious sliding movements along the bedding planes, and in the fracturing of retaining walls and houses (Fig. 14-7). To avoid these problems streets should be laid out so as to cross the strike of beds; this will also make excavation easier to carry out.

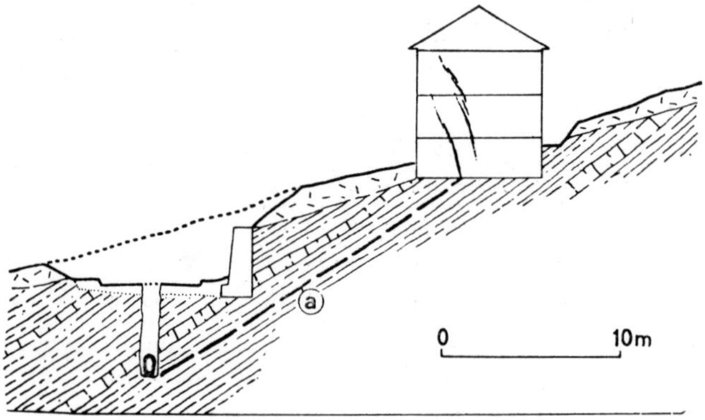

14-7. The lay-out of streets should respect the dip and strike of beds. Excavation for sewerage caused movement of shales along bedding surface a.

In some Prague districts the edges of Cretaceous sandstone elevations are unstable. They are cut by open vertical joints and disturbed by block landslides. The marginal sandstone blocks were progressively sinking into soft underlying clays and moving downslope. These phenomena are mostly of Pleistocene age but they are still active where the consistency of clays is impaired by subsurface water (Záruba, 1943, 1968). In the north-eastern part of Prague (Fig. 14-8) the situation is complicated by the fact that in the Middle Ages sandstones were quarried underground and since the entries of the quarries are caved-in, the

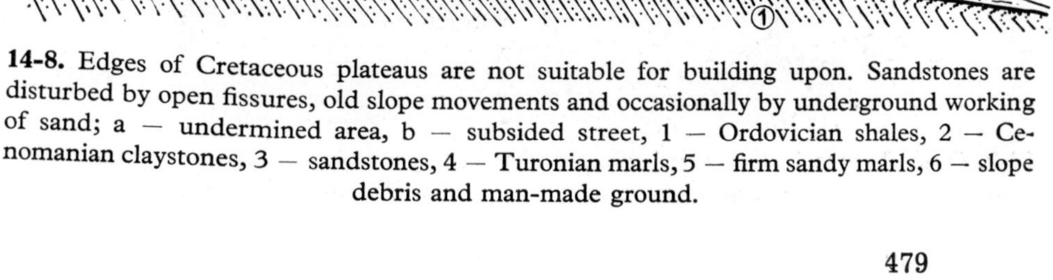

14-8. Edges of Cretaceous plateaus are not suitable for building upon. Sandstones are disturbed by open fissures, old slope movements and occasionally by underground working of sand; a — undermined area, b — subsided street, 1 — Ordovician shales, 2 — Cenomanian claystones, 3 — sandstones, 4 — Turonian marls, 5 — firm sandy marls, 6 — slope debris and man-made ground.

extent of open spaces is not known precisely (Fig. 14-9). The old cavities have led to surface subsidence endangering the safety of structures. It is advisable not to build on steep slopes and edges of elevations but rather to protect them from erosion and conserve their natural character.

On steep slopes hard rocks generally crop out at the surface making the excavation for foundations and water and other service lines very expensive. Excavations are particularly difficult to carry out when the area is partly built up, as explosives cannot be used for fear of damaging neighbouring buildings.

(d) Abandoned quarries and loam and sand pits occurring at the

14-9. Open fissures in Cenomanian sandstones at the margin of the Cretaceous plateau (see Fig. 14-8) in Prague. (Photograph by Rybář.)

14-10. Sliding of a steep slope of the Cretaceous plateau caused by brick-clay excavation in Prague suburb threatened the buildings and streets in the neighbourhood. (Photograph by Záruba.)

periphery of many towns demand a detailed geological investigation before urban development planning. In the past the disused quarries were commonly partly filled with waste and refuse and their reclamation was not thought of. It is therefore necessary to propose a suitable treatment of such derelict areas in the urban plans. Some of the quarries can be filled and levelled and used as playing grounds; others with, for example, interesting rock outcrops, can be arranged as parks and left open as natural monuments.

The walls of abandoned quarries and loam pits have often been left at such a steep inclination that the area adjacent to the slope is not permanently stable. In the urban plan, engineering geologists should suggest a suitable slope gradient and the width of a belt which should not be built up. (Fig. 14-10).

Urban development planning has to face many problems in areas where coal and ore deposits were or are intensively worked. Undermined areas suffer from irregular subsidence and caving on the one hand and are disfigured by hideous spoil heaps on the other. These areas cannot be used either for building or agricultural purposes. The reclamation of such derelict areas and the restoration of industrial regions are an urgent economic problem in all industrial countries. In northern Bohemia the brown-coal basin is most severely affected, more than 10,000 hectares of farming soil being destroyed by coal mining. The Ostrava-Karviná coal district in Moravia has been devastated to a similar extent. Some areas are extremely difficult to reclaim because of the permeation of soil by noxious chemicals.

The reclamation of land should be a self-evident obligation of present-day mining corporations. Particularly where large earth-moving machinery is applied, such working procedures should be chosen that would make reclamation feasible. In planning open cast mines and quarries the engineering geologist should determine which material of the overburden to use for land restoration.

In karst areas, limestones penetrated by open crevasses and cavities do not provide suitable foundations. These areas should not be built up if possible. Figure 14-11 demonstrates how difficult it may be to find a suitable site for a building whose function is linked to a certain place. In a health spa in Czechoslovakia a site for the central heating plant was sought; the site originally chosen (a) was found to be situated just above some caves. The borehole on the second site (b) encountered open spaces below the foundation. The last alternative (c) could be

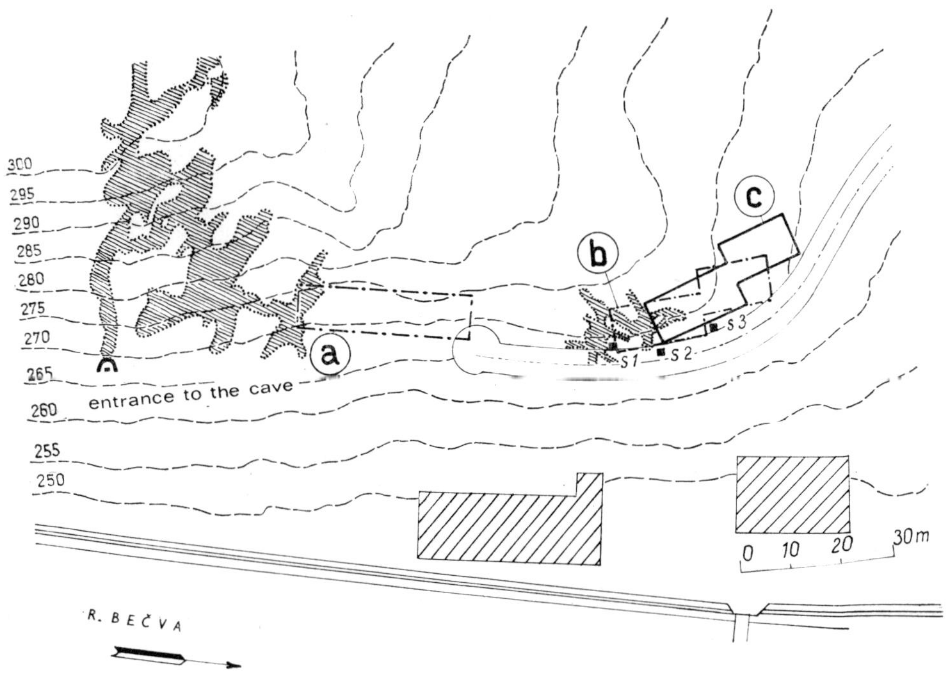

14-11. Alternative sites of a boiler house on a slope formed of karstified Devonian limestones (for a, b, c see the text).

used, but reinforced concrete slabs had to be applied to bridge over open fissures in limestones.

Careful engineering-geological investigation is necessary in areas with gypsum and salt deposits. In Germany, for example, whole communities had to be re-located because of disturbances caused by irregular surface subsidence after these soluble rocks had been leached.

(e) In areas affected by earthquakes in historical times, new settlements should not be placed near young faults. The effects of earthquakes are generally stronger along these young tectonic lines.

We have outlined some of the relevant tasks that await engineering geologists in land-use planning. If the engineering geologist is to play an important role in a country's development, he must use his knowledge of nature to assess the probable impact of the proposed action on the environment. In this way he can assist the planner in the harmonious incorporation of the work into the landscape so that it will not disturb the natural character and ecological balance. The pre-servation of a sound natural environment for future generations is a great challenge and responsibility that engineering geologists can share with planners and en-gineers.

BIBLIOGRAPHY

ALMAGIO, R., 1907. Studi geografici sulle frane in Italia. Soc. geogr. italiana, Roma.

AMBRASEYS, N. N., ZÁTOPEK, A., 1968. The Mudurnu Valley earthquake. UNESCO, Paris, No. 622 /BMS. RD/ AVS.

AMPFERER, O., ASCHER, H., 1925. Über geologisch-technische Erfahrungen beim Bau des Spullerseewerkes. Jahrb. der Geol. Bundesanstalt, Wien.

ANDREAE, C., 1926. Der Bau langer tiefliegender Gebirgstunnel. Berlin.

AVERIN, N. D., 1953. Kariernoe khozyaistvo. Moskva.

BALDOVIN, G., FATTORE, A., 1974. Example of slope stabilization in marly-sandstone Flysch. Advances in Rock Mechanics, Proc. Third CISRM, Nat. Acad. Sci., Vol. I, I, Th. 3 B, R. 1, Washington, D. C.

BALLY, R. J. L., ANTONESCU, I. P., PERLEA, V. D., 1965. Loess as foundation soil for irrigation systems. Proceedings 6th Internat. Conference S. M. F. E., University Toronto, II, 8–12.

BAYER, M., MENCL, V., PELIKÁN, V., 1955. Erosivní zjevy na březích nádrže na řece Svratce v Knínickách. Sborník VŠS, 1, Brno.

BĚLAŘ, J., ZAJÍC, J., 1958. Zajištění stěn při stavbě velkých podzemních prostor. Sborník konference geomechaniky, Brno.

BENSON, W. N., 1946. Landslides and their relation to engineering in the Dunedin District. Economic Geology, New Zealand.

BERKEY, CH. P., 1933. New York City and vicinity. 16th Int. Geol. Congress, Washington, 151 p.

BERTACCHI, P. E., CARABELLI, E., SAMPAOLO, A., 1967. A contribution to the use of geophysical methods for investigation of rock masses. Proc. Ist Int. Congress Rock Mechanics, Lisabon, I, p. 45–53.

BJERRUM, L., 1955. Stability of natural slopes in quick clay. Norwegian Geotechn. Inst. No. 10, Oslo.

— 1968. Progressive failure in slopes of overconsolidated plastic clays and clay shales. Norwegian Geotechn. Inst., No. 77, 29 p., Oslo.

BJERRUM, L., JORSTAD, F., 1966. Stability of rock slopes in Norway. Norwegian Geotechn. Inst., No. 67, p. 60–78, Oslo.

BLATTER, CH. F., 1956. Sondierung und Projektionen für Talsperren. Wasser- und Energiewirtschaft, Zürich.

BLOOM, A. L., 1969. The surface of the earth. Prentice-Hall, Inc., New Jersey, 152 p.

BONTE, A., 1969. Introduction à la lecture des cartes géologiques. Masson et Cie., Paris, 277 p.

BORECKI, M., KIDYBINSKI, A., 1966. Problems of stress measurements in rocks taken in the Polish coal industry. Proc. 1st Int. Congress Rock Mechanics, II, p. 9–16.

BOUVET, M. J., 1950. Les terrassements du Canal de Donzère-Mondragon. Ann. Inst. techn., Paris.

BRAIT, P. I., KOSAKOV, A. I., 1950. Nablyudeniya za osadkami fundamentov mnogoetazhnogo zdaniya na Smolenskom ploshchadi v Moskve. Stroit. promyshlennosť.

BURWELL, E. B., MONEYMAKER, B. C., 1950. Geology in dam construction. Geol. Soc. Am., Berkey Vol., p. 12–43, Washington.

BUXTORF, A., 1916. Prognosen und Befunde beim Hauensteinbasis- und Grenchenbergtunnel und die Bedeutung der letzteren für die Geologie des Juragebirges. Verhandl. d. Naturforsch. Gesell. in Basel, 27, p. 184–254.

CAMBEFORT, H., 1963. Etanchement et consolidation des roches. Felsmech. u. Ingenieurgeologie, Vol. 1/2.

— 1966. Forages et sondages, leur emploi dans les travaux publics. Paris, 426 p.

— — 1971. Géotechnique de l'ingénieur et reconnaissance des sols. Ed. Eyrolles, Paris, 345 p.

CAMPO DEL, A., PEIRONCELY, J. M., MAROTO, L., 1967. Foundation of the Santa Eulalia Arch Dam. Trans. 9th ICOLD,. Q 32, R 32.

CARATI, F., 1958. L'impianto idroelettrico di Premadio. L'Energia Elettrica, Milano.

CASAGRANDE, A., 1948. Classification and identification of soils. Trans. A. S. C. E., 113, New York.

CIAMPI, G., 1958. Diaframma nell alveo epigenetico del bacino del Senaiga. Geotecnica, Milano.

COATES, D. F., 1964. Classification of rocks for rock mechanics. Inter. Journal of Rock Mechanics and Mining Sciences, 1, No. 3, p. 421–429.

CORNS, C. F., NESBITT, R. H., 1967. Sliding stability of three dams on weak rock foundation. Trans. 9th ICOLD, 1, p. 463–486 Q 32, R 29.

COLLIN, A., 1846. Glissements spontanés des terrains argileux, Carilian–Goeury et Dalmont, Paris.

CYTOVICH, N. A., 1963. Mekhanika gruntov. Moskva, 528 p.

DANDUROV, M. I., 1952. Tonneli. Moskva.

DAPPLES, E. C., 1959. Basic Geology for Science and Engineering. John Wiley and Sons, London, 609 p.

DEERE, D. V., 1968. Geological considerations. In Stagg, K. G., Zienkiewicz, O. C., Rock Mechanics in Engineering Practice. J. Wiley and Sons, London.

DEERE, D. V., HENDRON, A. J., PATTON, F. D., CORDING, E. J., 1966. Design of surface and near-surface constructions in rock. Proc. 8th Symposium of Rock Mechanics, Ed. Ch. Fairhurst, p. 237-302, Univ. of Minnesota.

DESIO, A., 1961. Geological features of the reservoirs and of the dam foundation rocks, in Dams for hydroelectric power in Italy, Milano.

— 1973. Geologia applicata alla Ingegneria. Hoepli, Milano, 1194 p.

DODD, J. S., 1967. Morrow Point Underground Power plant rock mechanics investigations. U. S. Bureau of Reclamation, Denver, Col.

DOW, R. B., 1956. Some rheological properties under high pressure. Rheology 1. Acad. Press, New York.

DROUHIN, G., 1951. Sedimentation des reservoirs et problèmes connexes. 4t Congrès des Grands Barrages, New Delhi.

DROUHIN, G., GAUTIER, M., DERVIEUX, F., 1948. Slide and subsidence of the hills of St. Rafael-Télemly. Proc. of the II. Int. Conf. on Soil Mech. and Found. Eng. V, p. 104, 106.

DVOŘÁK, A., 1966. Test of anisotropic shales for foundation of large bridges. Proc. I. Int. Congress on Rock Mechanics, p. 537–541, Lisabon.

— 1970. Seismic and static modulus of rock masses. Proc. 2nd Congress Rock Mechanics, I, p. 313–317, Beograd.

— 1970. Über felsmechanische Untersuchungen für den Talsperrenbau. Bergakademie, Leipzig, 22, 4, p. 189–252.

ECKEL, E. B., (Editor), 1958. Landslides and engineering practice. Highway Research Board, Spec. Rep. 29, NAS-NRC 544. Washington, 232 p.

EDEN, W. J., FLETCHER, E. B., MITCHELL, R. I., 1971. South Nation River landslide 16.

EISENSTEIN, Z., MORRISON, N. A., 1973. Prediction of foundation deformations in Edmonton using an in situ pressure probe. Canadian Geotechnical Journal, 10, 2, p. 193—210. May 1971. Canad. Geotech. Journal, 8/3, p. 446—451, Ottawa.

EMELYANOVA, E. P., 1972. Osnovnye zakonomernosti opolznenykh processov. Izd. Nedra, Moskva, 310p.

Engineering geology in country planning, 1969. Proc. of Section 12 Int. Geol. Congress, Prague.

ENSLIN, J. F., 1951. Sinkholes in dolomite. Trans. South Afr. Inst. Civ. Eng., 1, No. 5.

EVDOKIMOV, P. D., 1970. Design and construction of earth and rockfill dams in the U.S.S.R. Trans. 10th ICOLD, 1, Q 36, R 8, p. 137—149.

FENCL, J., ZÁRUBA, Q., 1956. Geologické poměry okolí Lázní Teplic v Čechách. Sborník ÚÚG, 22, p. 427—484.

FLUHR, T., LEGGET, R. F. (ed.), 1962. Reviews in Engineering Geology. Geol. Soc. America, New York.

FOOKES, P. G., DEARMAN, W. R., FRANKLIN, J. A., 1971. Some engineering aspects of rock weathering with field examples from Dartmoor and elsewhere. Quart. Journ. Eng. Geol., 4, No. 3. p. 139—185, London.

FUJII, T., 1970. Fault treatment at Nagawado Dam. Trans. 10th ICOLD, II, Q 37, R 9.

Geology and Foundation Treatment, 1949. Tennessee Valley Auth., 22, Washington.

GIGNOUX, M., BARBIER, R., 1955. Géologie des barrages et des aménagemenst hydrauliques. Masson et Cie. Paris, 339 p.

GILG, B., 1964a. Mauvoisin arch dam, in Behaviour of large Swiss dams. Zürich, p. 199—211.

— 1964 b. Göschener Alp, in Behaviour of large Swiss dams. Zürich, p. 297—307.

GLOSSOP, R., 1968. The rise of geotechnology and its influence on engineering practice. Géotechnique, 18, p. 105—150, London.

GOGUEL, J., 1967. Application de la géologie aux travaux de l'ingénieur. Masson et Cie, ed., Paris, 373 p.

GOGUEL, J., HUMBERT, M., 1972. Carte des risques géologiques pour la commune de Peisey-Nancroix (Haute-Savoie). BRGM Orléans.

GOODMAN, R. E., TAYLOR, R. L., BREKKE, T. L., 1968. A model for the mechanics of jointed rock. Proc. ASCE, 5937, SM 3, p. 637—659.

GRAMBERG, J., 1963. The axial cleavage fracture in brittle materials at compressive loadings. Laboratory of Mining, T. Univ. Delft.

GRISHIN, M. M., 1954. Gidrotekhnicheskie sooruzheniya. Vols. I, II, Moskva.

HABIB, P., 1968. Introduction à la fissuration des roches. Revue de l'Industrie Minérale, 5, Paris.

HAEFELI, R., 1953. Creep problems in soils, snow and ice. Proc. of the 3rd Int. Conf. on Soil Mech. and Found. Eng., III, 238-251, Zürich.

HAFFEN, M., 1963. Grouting deep alluvial fill in the Durance river valley, Serre Ponçon Dam, France. Eng. Geology Case Histories, No. 4, p. 33—44, New York.

— 1967. Etanchement et consolidation des terrains alluvionnaires. Paris, 31 p.

HAMMER, W., 1927. Geologische Beobachtungen beim Bau des Wasserkraftwerkes bei Mallnitz. Jahrb. des Geol. Bundesamtes, Wien.

HANSEN, W. R., 1964. Effects of the earthquake of March 27 1964, at Anchorage, Alaska. Geol. Survey Prop. Paper 542-A, Washington.

HARRISON, J. V., FALCON, N. L., 1936. Gravity collapse structures and mountain ranges, as exemplified in south-western Persia. Quart. Jour. Geol. Soc., 92, p. 91—102, London.

HAST, N., 1964. Recent rock pressure measurements and their implications for dam building. Trans. 8. Int. Congress on Large Dams, 1, Edinburgh.

HEIM, A., 1882. Der Bergsturz von Elm. Zeitschr. der Deutsch. geol. Ges.

— 1932. Bergsturz und Menschenleben. Naturf. Gesell., Zürich, 218 p.

HOLLINGWORTH, S. E., TAYLOR, J. H., KELLAWAY, G. A., 1944. Large-scale superficial structures in the Northampton Ironstone Field. Quart. Journ. of Geol. Soc., 100, p. 1–44, London.

HORSKÝ, O., MÜLLER, K., TRÁVNÍČEK, L., 1972. Průzkum porušení čedičového příkrovu v přehradním místě Slezská Harta geologicko-geofyzikálními metodami. Sborník HIG, 10, p. 39–58.

HOUSKA, J., 1963. Zkoušky pevnosti hornin v laboratoři. Metodika inž. geol. výzkumu pevných hornin. Academia, Praha, p. 50–74.

HRUBAN, K., 1958. Přetvoření a napjatost nelineárního poloprostoru. Rozpravy ČSAV, 68, 1.

Hydroelectric development of river Po at Isola Serafini. 1961. Soc. Idroelettrica Medio Adige. Milano.

IDRISS, I. M., SEED, H. B., 1967. Response of earth banks during earthquakes. Jour. Soil Mech. Found. Eng., SM3, Proc. Am. Soc. Civ. Eng., p. 61–82.

ILIE, M., 1955. Cercetari geologice in Bazinul Transilvaniei. An. Com. Geol., 28, p. 354 to 358, Bucuresti.

ISCHY, E., 1948. Barrage de Castillon. Lutte contre les erosions souterraines. 3e Congrès des Grand Barrages, Stockholm.

ISCHY, E., HAFFEN, M., 1955. Barrage de Serre-Ponçon. Campagne de reconnaissance, 5e Congrès des Grands Barrages, Paris, 29 p.

JÄHDE, A., 1953. Injektionen zur Verbesserung von Baugrund und Bauwerk. Berlin.

JEVDJEVIČ, V. M., 1955. Les bassins d'accumulation en terrains karstiques. V. Congrès de grands barrages, Paris.

JOVANOVIĆ, D., 1970. Dams in Ethiopia, in World's dams today. Tokyo, p. 132–137.

KARL, F., 1954. Ein Beispiel für die Anwendung gefügeanalytischer Arbeitsmethoden in der Baugeologie. Geologie und Bauwesen, Wien.

KELLAWAY, G. A., TAYLOR, J. H., 1968. The influence of landslipping on the development of the City of Bath, England. 23rd Internat. Geol. Congr., Proceed. Sect. 12. Prague.

KERISEL, J., SIVIGNON, E. C., 1969. Géologie et géotechnique. Travaux p. 13–20, Paris.

KIERSCH, G. A., 1964. Vaiont reservoir disaster. Civil. Eng., 34, p. 32–39.

KIESLINGER, A., 1958. Restspannung und Entspannung im Gestein. Geologie u. Bauwesen, 24, p. 95–112.

KNILL, J. L., 1972. Engineering geology of the Cruachan Underground Power Station. Eng. Geol., 6/1, p. 289–312, Amsterdam.

KNILL, J. L., PRICE, D. G., HIGGINBOTTOM, I. E., 1968. Aspects of the engineering geology of the City of Bristol. 23rd Internat. Geol. Congr., Proceed. Sect. 12. Prague.

KO, H. Y., SCOTT, R. F., 1967. A new soil testing apparatus. Géotechnique, 17, p. 40–57.

KOJAN, E., HUTCHINSON, J. N., 1974. The rockslide – debris flow on the Mantaro River, Peru. Geol. Soc. of America.

KOLOMENSKII, N. V., 1952. Metodicheskie ukazaniya po izucheniyu procesov vyvetraniya gornikh porod dlya inzhenerno-geologicheskikh celei. Gosgeolizdat, Moskva.

KRANTZ, W., 1928. Einbrüche – Künstliche Erdfälle beim Bau des Weißenstein-Tunnels. Bauzeitung, 19, p. 187, 188.

KRYNINE, D. P., JUDD, W. R., 1957. Principles of Engineering Geology and Geotechnics. McGraw-Hill, New York, 730 p.

KUJUNDŽIĆ, B., ČOLIĆ, B., RADOSLAVLJEVIĆ, Ž., 1970. Komplexe Untersuchung des mechanischen Verhaltens der Felsmasse für die Gründung der Bauwerke des Wasserkraft- und Schiffahrtsystems Derdap. Proceedings of the IInd Congress of the ISRM, Beograd, p. 599–607.

LANE, D., TORCHINSKY, B. B., SPINKA, J. V. T., 1953. Determination of soil moisture and density by nuclear radiation. Eng. Journal, Montreal.

LAUFFER, H., SEEBER, G., 1966. Die Messung der Felsnachgiebigkeit mit der TIWAG-Radialpresse und ihre Kontrolle durch Dehnungsmessungen an der Druckschachtpanzerung des Kaunertalkraftwerkes. 1. Kongress der Internat. Gesel. für Felsmechanik, Lissabon, II, p. 347–356.

LEFOULON, M. R., ISCHY, M. E., 1955. Usine de Fessenheim. Protection des fouilles par injection des alluvions. 5e Congrès des Grands Barrages, Paris.

LEGGET, R. F., 1963. Geology and Engineering. McGraw-Hill, New York, 884 p.

– 1973. Cities and Geology. McGraw-Hill, New York, 624 p.

LISHAK, V. I., VRONSKII, A. V., 1971. Vliyanie neodnorodnosti gruntovogo osnovaniya na usiliya v konstrukciyakh beskarkasnych zdanii. Materialy III Vsesoyuznogo soveshchaniya, izd. „Budivelnik“, Kiev.

LUGEON, M., 1933. Barrages et géologie. Librairie de l'Univ. Lausanne. 138 p.

MAHÚT, P., 1971. Príspevok k otázke dlouhodobej nekonsolidovanej pevnosti ilovitej zeminy. Inženýrské stavby, 8–9, p. 361–365, Bratislava.

MARTY, M. H., MONTMARIN, A. DE, 1955. Le barrage de Ben-Metir, V. Congrès de grands barrages, Paris.

MATULA, M., 1972. Inženýrsko-geologická klasifikace hornin. Acta Geol. Univ. Comenianae, Bratislava.

MCCONNEL, D., MIELENZ, R. C., et al., 1950. Petrology of concrete affected by cement-aggregate reaction. Geol. Soc. Am., Berkey Volume, Washington p. 225–250.

MCCULLOCH, D. S., BONILLA, M. G., 1967. Railroad damage in the Alaska earthquake. Jour. Soil. Mech. Found. Eng., SM5, Proc. Am. Soc. Civ. Eng., p. 89–100.

MCGILL, J. T., 1964. Growing importance of urban geology. Geol. Surv., circ. 487 p. Washington.

MEAD, W. J., 1938. Engineering geology of dam sites. 2nd Congress of Large Dams, Washington.

MEIER, R. A., 1936. Gyumushskaya gidrostantsiya sevanozangiskogo kaskada. Gidrotekh. stroitel'stvo, Moskva.

MEJZLÍK, L., MENCL, V., 1973. Příspěvek ke statickému řešení sypaných přehrad. Vodní hosp., p. 309–317, Praha.

MENCL, V., 1955. Mechanika zemin. NČSAV, Praha, 254 p.

– 1965. Proportions of cohesion and of internal friction in the strength of rocks. Publ. No. 61, Norges Geotekniske Institutt, Oslo.

1966. Mechanika zemin a skalních hornin. Academia, Praha, 328 p.

– 1967. Plastizitätslehre und das wirkliche Verhalten von Gebirgen. Geot. Kolloquium, Salzburg, in Felsmechanik u. Ingenieurgeologie, 1968, Suppl. IV, p. 1–8.

– 1970. Klassifikation von festen Felsmassen. 10. Ländertreffen I. B. G., Leipzig; 122–125, Akademie-Verlag, Berlin.

– 1970. Vibration of structures caused by settlement of foundation rocks. Proc. 2nd ICRM, 3, p. 8–5.

MENCL, V., MENCL, J., 1956. Tunely. SNTL, Praha.

MENCL, V., TRÁVNÍČEK, I., 1964. Relaxationsversuche an Gesteinen. Bericht 6. Ländertreffen I. B. G., p. 81–84, Akademie-Verlag, Berlin.

MERMEL, T. W., 1970. Progress of dam construction in the world. World's dams today, p. 76–78, Tokyo.

MEYERHOF, G. G., 1951. Bearing capacity of foundations. Géotechnique II, 4, 301 p. London.

MINASSIAN, W. T., SABARLY, F., LONDE, P., 1967. Comment protéger les barrages-voûtes contre la pression de l'eau dans les appuis. Trans. 9th ICOLD, 1, Q 32, R 12.

MOOR, R., 1923. Der Uferabbruch am Davoser See. Schweizer. Bauzeit., 82 p.

MORGENSTERN, N. R., TAMULY PHUKAN, A. L., 1968. Stresses and displacements in a homogeneous non-linear foundation. Int. Symp. Rock Mech., III-7, Madrid.

MOOS, A., QUERVAIN, F., 1948. Technische Gesteinskunde. Basel.

MÜLLER, L., 1953a. Setzungen von Bauwerken auf Felsuntergrund. Vorträge der Baugrundtagung in Hannover, Hamburg.

— 1953b. Die Darstellung geologischer Flächen in Bauplänen. Geologie und Bauwesen, Wien.

— 1958. Geomechanische Auswertung gefügekundlicher Details. Geologie und Bauwesen, IV/5.

— 1963. Der Felsbau. Enke Ver. Stuttgart, 624 p.

— 1964. The rock slide in the Vaiont valley. Rock Mechanics a. Eng. Geol., 2, p. 148–212.

NEMČOK, A., 1972. Gravitational slope deformation in high mountains. 24. Inter Geol. Congress, Section 12 – Eng. Geology, p. 132–141, Montreal.

NEMČOK, A., PAŠEK, J., RYBÁŘ, J., 1972. Classification of landslides and other mass movements. Rock Mechanics, 4, p. 71–78.

NĚMEČEK, K., 1969. Sanace petřínské stráně v Praze. Geol. průzkum, 9, 12, p. 361–363, Praha.

NEUHAUSER, E., SCHOBER, W., 1970. Das Kriechen der Talhänge und elastische Hebungen beim Speicher Gepatsch. Proc. of the II. Congress of the ISRM, p. 447, 458, Beograd.

NOVOSAD, S., 1966. Porušení svahů v godulských vrstvách Moravsko-slezských Beskyd. Sborník geol. věd, HIG, 5, p. 71–86.

OLIVIER-MARTIN, M., KOBILINSKY, M., 1955. L' exécution d' un grand souterrain pour l'aménagement hydroélectrique d'Isère-Arc. Construction, X, 4, p. 145–156.

ONDRÁŠIK, R., 1968. Inžiniersko-geologický výskum kory zvetrávania skalných hornín na Slovensku. Acta geol. et geogr. univ. Comenianae, 11, p. 5–24, Bratislava.

PALSHIN, G. B., TRZHTSINSKII, YU. B., 1964. Opolzni v skalnykh i poluskalnykh porodakh na sklonakh yuzhnogo Priangarya. Geologiya i Geofizika, 6, p. 97–108, Novosibirsk.

PAŠEK, J., 1959. Užití filmu a televize v geologickém průzkumu. Geologický průzkum, 8, Praha.

PAŠEK, J., RYBÁŘ, J., 1961. Znázornění inženýrsko-geologických poměrů v mapě 1:25000. Geol. průzkum 2, p. 45–47, Praha.

PAŠEK, J., RYBÁŘ, J., 1969. Svahové pohyby na černomořském pobřeží Bulharska. Časopis pro miner. a geol., 14, 3–4, p. 273–284, Praha.

PERRET, W. R., 1948. Electrical resistivity exploration as a complement to boring in deep alluvial deposits. Proc. II ISMFE, 7, Rotterdam.

PESHKOVSKII, L. M., PERESKOVOVA, T. M., 1971. Inzhenernaya geologiya. Izdat. „Vysshaya shkola“, Moskva.

PETER, A., 1966. Essai de carte géotechnique. Sols-soils, 16, p. 13–29.

PETRÁNEK, J., 1963. Usazené horniny. Academia, Praha, 418 p.

PETRÁŠEK, J., ZÁRUBA, Q., 1959. Použitie geodetických metód pre sledovanie pohybu na zosuvnom území medzi Sučanmi a Klačanmi. Staveb. čas. SAV. 7, p. 33–42.

PFEFFER, F., 1955. Die Konstruktion des Hochhauses in Wien I, Ecke Schottenring - Kai, Österr. Bauzeitschrift, No. 6.

POLYANOV, P. F., 1956. Vibratory v razvedochnom burenii. Moskva.

POPOV, I. V., 1959. Inzhenernaya geologiya. Moskva, 510 p.

POUBA, Z., 1959. Geologické mapování. NČSAV, Praha, 523 p.

PRICE, D. G., 1971. Engineering geology in the urban environment. Quart. Jour. Eng. Geol., 4/3, p. 191–208, London.

PRIKLONSKII, V. A. et al., 1950. Inzhenerno-geologicheskie izyskaniya dlya planorovki i stroitelstva gorodov. Moskva.
PROKOPOVICH, N. P., 1972. A Comparison of Surface and Subsurface Structural Patterns at Spring Creek Conduit and Dam, Trinity River Division. Central Valley Project, California, Bulletin of the Assoc. of Eng. Geologists, Vol. IX/4, p. 359–375.
Přehradní výbor, 1967. Přehrady v Československu. Příl. čas. Inženýrské stavby, Bratislava.
RABCEVICZ, L., 1938. Die Nordrampe der Transiranischen Eisenbahn. Bautechnik.
– 1964, 1965. The new Austrian tunnelling method. Water power, Dec. 1964, Jan. 1965.
RADLEY SQUIER, L., VERSTEEG, J. H., 1971. The history and correction of the Omsi-Zoo lanslide, Proceed. of the 9th Annual Eng. Geol. a. Soils Eng. Symposium, Boise, Idaho, p. 237–256.
RAWLINGS, G. E., 1971. The role of the engineering geologist during construction. Quart. Jour. Eng. Geol., 4, p. 209–217, London.
REUTER, F., 1958a. Die Wasserdurchlässigkeitsprüfungen und Baugrundvergütung als ingenieurgeologische Untersuchungsmethode bei Talsperrenbauten. Zeitschr. für angew. Geologie, Berlin.
– 1958b. Hangrutschungen bei den Baustellen des Bodeswerkes. Zeitschr. für angewandte Geol., 2/3, p. 94–98.
ROCHA, M., LOPES, J. B., DA SILVA, J. N., 1967. A new method of the determination of stresses inside rock masses. Proc. 1st Conf. ISRM, II, 57–68, Lisabon.
ROSENQUIST, I. TH., 1953. Consideration on the sensitivity of Norwegian quick clays. Norwegian Geot. Inst., Oslo.
ROTH, Z., 1944. Skalní proudy, ledovcové kary a ledovce. Rozpravy České akad., Praha.
– 1945. Stabilita skalních zářezů a tlaky při ražení štol v pevných horninách. Technický obzor, 53, p. 201–288, 312–315, Praha.
ROZA, S. A., 1952. Sprotivlenie skalnykh porod osnovaniya gidrotekhnicheskogo sooruzheniya sdvigayushchemu usiliyu. Moskva.
RYBÁŘ, J., 1961. Shrnutí vrstev na okraji hnědouhelné pánve u Kadaně. Věstník ÚÚG, 36, p. 223–227, Praha.
– 1971. Tektonisch beeinflusste Hangdeformationen in Braunkohlenbecken. Rock Mechanics, 3, 139–158, Wien-N. York.
RYBÁŘ, J., DOBR, J., 1966. Vrásové deformace v severočeské hnědouhelné pánvi. Sbor. geol. věd, HIG, 5, p. 107–139, Praha.
ŘEPKA, L., 1963. Proudový sesuv u Riečnice. Geol. průzkum, 5, p. 8–10, Praha.
RŽIHA, F., 1889. Die Gewinnungsfestigkeit der Erd- und Felsmassen in Einschnitten. Zentralblatt der Bauverwaltung, Wien.
SAVARENSKII, F. P., 1937. Inzhenernaya geologiya. Moskva.
SEED, H. B., IDRISS, I. M., 1967. Analysis of soil liquefaction by Niigata earthquake. Jour. Soil. Mech. Found. Eng., SM3, Proc. Am. Soc. Civ. Eng., p. 83–108.
SELLI, R., TREVISAN, L., et al., 1964. La frana del Vaiont. Annali del Museo geologico di Bologna, 32, 68 p.
SELMER-OLSEN, R., BLINDHEIM, O. T., 1970. On the drillability of rock by percussive drilling. Proc. 2nd Int. Conf. Rock Mechanics, ref. 5–8, Beograd.
SHANNON, WILSON E. I., DEERE, D. V., et al., 1966. Design of surface and near-surface constructions in rock. Proc. 8th Symposium of Rock Mechanics, Univ. of Minnesota, p. 237–302.
SHARPE, C. F. S., 1938. Landslides and Related Phenomena. Columbia Univ. Press, New York, 137 p.
SHERARD, J. L., WOODWARD, R. J., GIZIENSKI, S. F., CLEVENGER, W. A., 1968. Earth and earth-rock dams. J. Wiley and Son, New York, 725 p.

SHMULSON, B. D., 1972. Kievskaya gidroakkumuliruyushaya elektrostantsiya. Gidrotekh. stroitel'stvo, 4, Moskva.

SCHARDT, H., 1924. Die geologischen Verhältnisse des Stau- und Kraftwerkes Wäggital. Eclogae Geol. Helvetiae.

SCHLOSSER, M., 1967. Une méthode nouvelle d'étude de l'évolution de la microfissuration des roches mises sous contrainte. Proc. Geot. Confer. of Shear Strength of Natural Soils and Rocks, II, p. 258–259, Norwegian Geot. Inst., Oslo, 1968.

SCHNEIDER, B., MENCL, V., 1970. Evaluación de los riesgos telúricos en el Callejón de Huaylas (geotechnical part), UNESCO, Paris, 2208 BMS – RD/SCE.

SCHULTZE, E., MUHS, H., 1967. Bodenuntersuchungen für Ingenieurbauten, Berlin, 592 p.

SIMONDS, A. W., 1951. Final foundation treatment at Hoover Dam. Trans. Am. Soc. of Civ. Eng., Vol. 116, New York.

SIMONDS, A. W., LIPPOLD, F. H., KEIM, R. E., 1951. Treatment of foundations for large dams by grouting methods. Transactions of Am. Soc. Civ. Eng., Vol. 116, New York.

SKEMPTON, A. W., 1953. Colloidal activity of clays. Proc. 3rd Int. Conf. S. M. F. E., I, Zürich.

SKEMPTON, A. W., HUTCHINSON, I., 1969. Stability of natural slopes and embankment foundations. Proc. 7th Conf. S. M. F. E., State of the Art Volume, p. 291–340, Mexico.

SNOW, D. T., 1964. Landslide of Cerro Condor-Sencca, Department of Ayacucho, Peru. Eng. geol. Case Histories, 5, p. 1–6.

Société Nouvelle de Construction et de Travaux, 1968, Fondations du Parc de Stationnement Souterrain „Velpeau - Boucicaut", Travaux Souterrains, 33, No. 153, p. 47–51.

SOLLE, G., 1967. Felsgleitungen in Oberägypten. Notizblatt des Hessischen Landesamtes für Bodenforschung, Wiesbaden Bd. 95, p. 117–147.

SOLONENKO, V. P., 1972. Seismogenic destruction of mountain slopes. 24. Inter. Geol. Congress, Montreal, Section 13, Engineering Geology, p. 284–290.

STINY, J., 1950. Tunnelbaugeologie. Springer-Verlag, Wien.

STUCKY, A., 1956. Ben Metir Dam. Water Power.

SZÉCHY, K., 1958. Vom Bau der neuen Untergrundbahn in Budapest. Österreichische Ingenieur-Zeitschrift, 1, p. 7.

ŠAMALÍKOVÁ, M., KAZDA, J., 1970. Weathering phenomena in some rock-cuts on the railway near Brno. I. Int. Congress IAEG, p. 561–572, Paris.

ŠKOPEK, J., RYBÁŘ, J., DOBR, J., 1972. Pore-water pressure observations in a landslide. 24th Inter. Geol. Congress, Montreal, Section 13, Engineering Geology, p. 150–159.

TAYLOR, R. K., 1971. The function of the engineering geologist in urban development. Quart. Jour. Eng. Geol., 4/3, p. 221–239, London.

TER-STEPANIAN, G. I., 1961. O dlitelnoi ustoichivosti sklonov. Izd. Akad. nauk A. S. S. R, Erevan.

– 1965. Über den Mechanismus des Hakenwerfens. Felsmechanik u. Ingenieurgeologie, III/2.

– 1969. Typy glubinoi polzuchesti sklonov v skalnykh porodakh. Problems of geomechanics, 3, p. 55–69, Erevan.

TERZAGHI, K., 1946a. Rock defects and loads on tunnel supports. Harvard Univ.

– 1946b. Introduction to tunnel geology. In R. V. Proctor and T. L. White, Rock Tunneling with Steel Supports. The Commercial Shearing and Stamping Co., Youngstown, Ohio.

– 1950. Mechanics of landslides. Geol. Soc. of Am., Berkey Volume, p. 83–124, New York.

– 1954. Fünfzig Jahre Baugrunduntersuchung. Schweizerische Bauzeitung, 72, No. 4–5, Zürich.

TERZAGHI, K., PECK, R. B., 1948. Soil Mechanics in Engineering Practice. Wiley and Sons, New York, 566 p.
THOMSON, T. F., 1966. San Jacino Tunnel. Engineering geology in Southern California. Assoc. Eng. Geologists, p. 104–107.
TIMOSHENKO, S., 1934. Theory of Elasticity. McGraw-Hill, New York.
TRÁVNÍČEK, I., MENCL, V., 1966. Č. Váh, zkoušky napjatosti a deformací ve stěnách štoly. Zpráva kat. geotechniky, VUT, Brno.
TREFZGER, R. F., 1966. Tecolote Tunnel. Engineering geology in Southern California. Assoc. Eng. Geologists, p. 108–113.
TROLLOPE, D. H., 1968. The mechanics of discontinua or clastic mechanics. Rock Mechanics in Engineering Practice. J. Wiley and Son, London, p. 275–320.
ULLMANN, F., 1970. Particular features of the Gebiden Dam of the Massa Hydroelectric scheme. World's Dams Today, Tokyo, p. 199–206.
UNDERWOOD, L. B., 1967. Classification and identification of shales. Journal Soil Mech. Found. Div., A. S. C. E., SM6, p. 97–116.
VARNES, D. J., 1950. Relation of landslides to sedimentary features. Applied sedimentation. J. Wiley and Sons, New York.
VERFEL, J., 1957. Provádění vodních tlakových a injekčních zkoušek. Brno.
— 1969. Vztlak vody na přehradní těleso a funkce injekční clony. Sborník Přehradní dny, Bratislava, p. 186–199.
WAGNER, C. J., 1884. Die Beziehungen der Geologie zu den Ingenieur-wissenschaften. Spielhagen und Schurich, Wien, 88 p.
WAHLSTROM, E. E., 1973. Tunneling in Rock. Elsevier, Amsterdam, 250 p.
WALTERS, R. C. S., 1971. Dam Geology. Butterworth, London, 470 p.
ZAJÍC, J., 1970. Geotechnical survey applied to underground hydroelectric plants in Czechoslovakia. Large Permanent Underground Openings, Oslo, p. 43–55.
ZÁRUBA, Q., 1929. Příspěvek k významu geologie pro městské stavby komunikační. Věstník st. geol. úst., 5, p. 294–304, Praha.
— 1931. O stabilitě svahů nad povltavskou silnicí u Štěchovic a Vraného. Techn. obzor, 39, p. 293–297, 330–333.
— 1932. Výzkumné práce geologické v inženýrském stavitelství (Geological investigations in civil engineering). Masaryk. Akad. práce, Praha, 144 p.
— 1934. Hráze na řece San Gabriel v Kalifornii. Tech. obzor, 42, p. 357–359, 394–396.
— 1941. Určování těžitelnosti hornin. Zprávy veř. služby technické. Praha.
— 1943. Periglaciální zjevy v okolí Prahy. Rozpravy České akad., 34 p., Praha.
— 1948. Geologický podklad a základové poměry vnitřní Prahy. Geotechnica 5, 81 p., Praha.
— 1956. Deformace hornin vzniklé vytlačováním podloží. Rozpravy ČSAV, 66/15, 35 p.
— 1958a. Bulged valleys and their importance for foundation of dams. VI. Intern. Congress on Large Dams, p. 509–515, New York.
— 1960. Plastische Verformung von Schichten in Tälern und ihre Bedeutung für die Gründung von Bauwerken. Zeitschr. f. angew. Geol., 60, p. 425–428.
— 1961. Glaukonit, ein Faktor der erhöhten Neigung zu Rutschungen. Bergakad. Jahrb., 13, p. 175–181, Freiberg.
— 1961. Determination of the height of grouting pressure. VII. Intern. Congr. on Large Dams, Transact. II, p. 41–45, Roma.
— 1968. The influence of geology on the development of the City of Prague. 23rd Int. Geol. Congress. Proc. Sec. 12, p. 133–144, Praha.
— 1970. The influence of hydrogeological conditions on the construction of the subway in Prague. In Large Permanent Underground Openings, p. 334–336, Oslo.

— 1970. Engineering geology — some experience and considerations. Bull. IAEG I, p. 15—21, Paris.

— 1974. Importance of Quaternary events for the geological conditions of building sites. Colloque Géologie de l' Ingénieur, Liège, p. 259—272.

ZÁRUBA, Q., ANDRUSOV, D., 1939. Srovnání geologických profilů Harmaneckým tunelem na dráze Banská Bystrica—Diviaky. Techn. obzor, Praha.

ZÁRUBA, Q., HROMADA, K., 1950. Technicko-geologický rozbor území města Kutné Hory. Geotechnica, Praha.

ZÁRUBA, Q., MENCL, V., 1956. Inženýrsko-geologické otázky při zakládání hydrocentrály u Sučan na Váhu. Inženýrské stavby, 7, p. 302—307.

— — 1958. Rozbor sesuvů u Klačan. Rozpravy ČSAV, vol. 68, Praha.

— — 1969. Landslides and their control. Elsevier-Academia, Amsterdam, Prague, 205 p.

ZÁRUBA, Q., VACHTL, J., POKORNÝ, M., 1972. Základy geologie a petrografie pro stavební fakulty. SNTL, Praha, 387 p.

ZÁRUBA, Q., ZAJÍC, J., et al., 1967. Geologie přehrad na Vltavě. Academia, Praha, 222 p.

ZÁRUBA, Q., ZEMAN, M., 1966. Vývoj a úkoly urbanistické geologie. Sborník geol. věd, HIG, 5, Praha.

ZOLOTAREV, G. S., 1971. Sovremennye zadachi inzhenernogeologicheskogo izucheniya processov i kor vyvetrivaniya. Voprosy inzhenernogeologicheskogo izucheniya processov i kor vyvetrivaniya. Izd. Mosk. univ., Moskva, p. 4—25.

ZOUBEK, V., 1953. Geologické podklady k projektu údolní přehrady u Orlických Zlákovic.

INDEX